Editorial Policy

§ 1. Lecture Notes aim to report new developments - quickly, informally, and at a high level. The texts should be reasonably self-contained and rounded off. Thus they may, and often will, present not only results of the author but also related work by other people. Furthermore, the manuscripts should provide sufficient motivation, examples and applications. This clearly distinguishes Lecture Notes manuscripts from journal articles which normally are very concise. Articles intended for a journal but too long to be accepted by most journals, usually do not have this "lecture notes" character. For similar reasons it is unusual for Ph. D. theses to be accepted for the Lecture Notes series.

§ 2. Manuscripts or plans for Lecture Notes volumes should be submitted (preferably in duplicate) either to one of the series editors or to Springer- Verlag, Heidelberg . These proposals are then refereed. A final decision concerning publication can only be made on the basis of the complete manuscript, but a preliminary decision can often be based on partial information: a fairly detailed outline describing the planned contents of each chapter, and an indication of the estimated length, a bibliography, and one or two sample chapters - or a first draft of the manuscript. The editors will try to make the preliminary decision as definite as they can on the basis of the available information.

§ 3. Final manuscripts should preferably be in English. They should contain at least 100 pages of scientific text and should include

- a table of contents;
- an informative introduction, perhaps with some historical remarks: it should be accessible to a reader not particularly familiar with the topic treated;
- a subject index: as a rule this is genuinely helpful for the reader.

Further remarks and relevant addresses at the back of this book.

Lecture Notes in Mathematics 1655

Editors:
A. Dold, Heidelberg
F. Takens, Groningen

Subseries: Institut de Mathématiques, Université de Strasbourg
Adviser: J.-L. Loday

Springer
Berlin
Heidelberg
New York
Barcelona
Budapest
Hong Kong
London
Milan
Paris
Santa Clara
Singapore
Tokyo

J. Azéma M. Emery M. Yor (Eds.)

Séminaire de Probabilités XXXI

Springer

Editors

Jacques Azéma
Marc Yor
Laboratoire de Probabilités
Université Pierre et Marie Curie
Tour 56, 3ème étage
4, Place Jussieu
F-75252 Paris, France

Michel Emery
Institut de Recherche Mathématique Avancée
Université Louis Pasteur
7, rue René Descartes
F-67084 Strasbourg, France

Cataloging-in-Publication Data applied for

Die Deutsche Bibliothek - CIP-Einheitsaufnahme

Séminaire de probabilités ... - Berlin ; Heidelberg ; New York ; ; Barcelona ; Budapest ; Hong Kong ; London ; Milan ; Paris ; Santa Clara ; Singapore ; Tokyo : Springer.
ISSN 0720-8766
31 (1997)
(Lecture notes in mathematics ; Vol. 1655)
ISBN 3-540-62634-4 (Berlin ...)
NE: GT

Mathematics Subject Classification (1991): 60GXX, 60HXX, 60JXX

ISSN 0075-8434
ISBN 3-540-62634-4 Springer-Verlag Berlin Heidelberg New York

Typesetting: Camera-ready LaTeX output by the editors
SPIN: 10520345   46/3142-543210 - Printed on acid-free paper

This volume is dedicated to David Williams, whose infectious enthusiasm needed no tunnel to spread across the Channel long before cow madness.

J. Azéma, M. Émery, M. Yor.

NOTE TO CONTRIBUTORS

Contributors to the Séminaire are reminded that their articles should be formatted for the Springer Lecture Notes series.

The dimensions of the printed part of a page without running heads should be:

15.3 cm × 24.2 cm if the font size is 12 pt (or 10 pt magnified 120%),

12.2 cm × 19.3 cm if the font size is 10 pt.

Page numbers and running heads are not needed. Author(s)' address(es) should be indicated, either below the title or at the end of the paper.

Packages of TEX macros are available from the Springer-Verlag site

`http://www.springer.de/author/tex/help-tex.html`

SÉMINAIRE DE PROBABILITÉS XXXI

TABLE DES MATIÈRES

Branching processes, the Ray-Knight theorem, and sticky Brownian motion

JONATHAN WARREN [1]

University of Bath, U.K.

1 Introduction

Diffusions with boundary conditions were studied by Ikeda and Watanabe [5] by means of associated stochastic differential equations. Here we are interested in a fundamental example. Let θ and x be real constants satisfying $0 < \theta < \infty$ and $0 \le x < \infty$. Suppose $(\Omega, (\mathcal{F}_t)_{t\ge 0}, \mathbb{P})$ is a filtered probability space satisfying the usual conditions, and that $(X_t; t \ge 0)$ is a continuous, adapted process taking values in $[0, \infty)$ which satisfies the stochastic differential equation

$$X_t = x + \theta \int_0^t I_{\{X_s = 0\}} ds + \int_0^t I_{\{X_s > 0\}} dW_s, \tag{1.1}$$

where $(W_t; t \ge 0)$ is a real valued $(\mathcal{F}_t)$-Brownian motion. We say that X_t is sticky Brownian motion with parameter θ, started from x. Sticky Brownian motion has a long history. Arising in the work of Feller [3] on the general strong Markov process on $[0, \infty)$ that behaves like Brownian motion away from 0, it has been considered more recently by several authors, see Yamada [12] and Harrison and Lemoine [4], as the limit of storage processes, and by Amir [1] as the limit of random walks.

Ikeda and Watanabe show that (1.1) admits a weak solution and enjoys the uniqueness-in-law property. In [2], Chitashvili shows that, indeed, the joint law of X and W is unique (modulo the initial value of W), and that X is not measurable with respect to W, so verifying a conjecture of Skorokhod that (1.1) does not have a strong solution. The filtration $(\mathcal{F}_t)$ cannot be the (augmented) natural filtration of W and the process X contains some 'extra randomness'. It is our purpose to identify this extra randomness in terms of killing in a branching process. To this end we will study the squared Bessel process, which can be thought of as a continuous-state branching process, and a simple decomposition of it induced by introducing a killing term. We will then be able to realise this decomposition in terms of the local-time processes of X and W. Finally we will prove the following result which essentially determines the conditional law of sticky Brownian motion given the driving Wiener process.

Theorem 1. *Suppose that X is sticky Brownian motion starting from zero, and that W is the driving Wiener process, also starting from zero. Letting $L_t = \sup_{s \le t}(-W_s)$, the conditional law of X given W satisfies*

$$\mathbb{P}(X_t \le x | \sigma(W)) = \exp\big(- 2\theta(W_t + L_t - x)\big) \qquad a.s.$$

for $x \in [0, W_t + L_t]$.

[1] jw1@maths.bath.ac.uk

Note in particular that $X_t \in [0, W_t + L_t]$ a.s.. The proof of this result is given in Section 4, and depends on the construction of the pair (X, W) discussed in Section 3. Section 2 is essentially independent, but helps provide us with the intuitive reason for believing Theorem 1.

We begin with a simple but illuminating lemma on sticky Brownian motion, and fix some notation we will need in the sequel.

We denote:

$$A_t^+ = \int_0^t I_{\{X_s>0\}} ds; \qquad \alpha_t^+ = \inf\{u : A_u^+ > t\}; \tag{1.2}$$

$$A_t^0 = \int_0^t I_{\{X_s=0\}} ds; \qquad \alpha_t^0 = \inf\{u : A_u^0 > t\}. \tag{1.3}$$

Then we have

Lemma 2. *If we time change both sides of (1.1) with α^+, the right-continuous inverse of A^+, we find that $(X_{\alpha_t^+}, t \geq 0)$ solves Skorokhod's reflection equation*

$$X_{\alpha_t^+} = W_t^+ + L_t^+.$$

where $W_t^+ = x + \int_0^{\alpha_t^+} I_{\{X_s>0\}} dW_s$ is a Brownian motion, and $L_t^+ = \sup_{s\leq t}\left((-W_s^+)\vee 0\right)$.

Proof. On time changing we have

$$X_{\alpha_t^+} = W_t^+ + \theta A^0_{\alpha_t^+}.$$

Observe that W^+ is a Brownian motion by Lévy's characterization. Now, A_t^0 is a continuous and increasing function of t, A_t^+ is a continuous and strictly increasing function of t, and so $L_t^+ = \theta A^0_{\alpha_t^+}$ is also a continuous and increasing function of t. Furthermore it is constant on the set $\{t : X_{\alpha_t^+} > 0\}$. The criteria of Skorokhod's lemma, see [9], are thus satisfied and $L_t^+ = \sup_{s\leq t}(-W_s^+)$ as claimed. □

This lemma shows us that sticky Brownian motion is just the time change of a reflecting Brownian motion so that the process is held momentarily each time it visits the origin. In this way it spends a real amount of time at the origin, proportional to the amount of local time the reflecting Brownian motion has spent there, in fact,

$$\theta A^0_{\alpha_t^+} = L_t^+. \tag{1.4}$$

The laws of A_t^+,A_t^0 and other quantities can be obtained directly from this, as has been accomplished by Chitashvili and Yor [13].

2 A decomposition of the squared Bessel process

We consider two processes $(R_t, t \geq 0)$ and $(Y_t, t \geq 0)$ satisfying

$$dR_t = 2\sqrt{R_t}\, dB_t - 2\theta R_t\, dt, \qquad R_0 = x, \tag{2.1a}$$

$$dY_t = 2\sqrt{Y_t}\, d\tilde{B}_t + 2\theta R_t\, dt, \qquad Y_0 = 0, \tag{2.1b}$$

where B and $\tilde{B}$ are independent Brownian motions.

Proposition 3. *$V_t = R_t + Y_t$ is a squared Bessel process of dimension 0 started from x.*

Proof. One need only make a simple application of Pythagoras's theorem, following Shiga and Watanabe [11]. We sum the two equations of (2.1) and note that

$$\int_0^t \frac{\sqrt{R_s}\,dB_s + \sqrt{Y_s}\,d\tilde{B}_s}{\sqrt{R_s + Y_s}}$$

is a Brownian motion. □

This simple decomposition can be thought of in the following manner. V_t is the total-mass process of a continuous-state critical branching process and R_t that of a subcritical process. But a subcritical process can be obtained from a critical process by introducing killing at some fixed rate into the latter. Y_t represents the mass of that part of the critical process descended from killed particles. The idea that 'R_t is V_t with killing at rate 2θ' will pervade this paper.

V_t has some finite extinction time $\tau = \inf\{t : V_t = 0\}$, see for example Revuz and Yor [9], and the same is true of R_t, its extinction time being denoted by σ. It is clear that $\tau \geq \sigma$; perhaps surprisingly τ can equal σ, and we will calculate the probability of this. This will be accomplished first via the Lévy-Khintchine formula and then extended using martingale techniques.

Lemma 4. *The laws of the extinction times τ and σ are given by*

$$\mathbb{P}(\tau \in dt) = \frac{x}{2t^2} \exp(-x/2t)\, dt,$$

and

$$\mathbb{P}(\sigma \in dt) = \tfrac{1}{2} x \left[\frac{\theta}{\sinh(t\theta)}\right]^2 \exp\left[\tfrac{1}{2} x\theta(1 - \coth(t\theta))\right] dt.$$

Proof. From Pitman and Yor [8],

$$\mathbb{P}(V_t = 0) = \exp(-x/2t),$$

and

$$\mathbb{P}(R_t = 0) = \lim_{\lambda\to\infty} \mathbb{E}\exp(-\lambda R_t) = \exp\left[\tfrac{1}{2} x\theta(1 - \coth(t\theta))\right].$$

The lemma follows on differentiating. □

We wish to prove the following.

Proposition 5. *The conditional law of the extinction time of the subcritical process given the extinction time of the critical process satisfies*

$$\mathbb{P}(\sigma = \tau | \tau) = \exp(-2\theta\tau) \qquad a.s..$$

This can be loosely interpreted as the probability that the last surviving particle of the critical process also belongs to the subcritical process, an event that depends on whether there has been any killing along its line of ancestry.

Let us denote the law of a process satisfying

$$dZ_t = 2\sqrt{Z_t}\, dB_t + 2(\beta Z_t + \delta)\, dt, \qquad Z_0 = y,$$

by ${}^{\beta}\mathbb{Q}^{\delta}_{y}$, and the law of the Z-process conditioned to be at x at time t by ${}^{\beta}\mathbb{Q}^{\delta,t}_{y\to x}$. Now the following Lévy-Khintchine formula comes from Yor [14],

$$\mathbb{E}[\exp(-\lambda Y_t)|\sigma(R)] = \exp\left\{-\int n^+(d\epsilon)\int_0^t ds\, 2\theta R_s\Big(1-\exp\big(-\lambda l_{t-s}(\epsilon)\big)\Big)\right\},$$

where n^+ is the restriction of Itô excursion measure for Brownian motion to positive excursions and $l_t(\epsilon)$ the local time at height t of the excursion ϵ. Letting $\lambda \uparrow \infty$, we have

$$\exp\big(-\lambda l_{t-s}(\epsilon)\big) \to \begin{cases} 0 & \text{if } \sup\epsilon > t-s, \\ 1 & \text{otherwise.}\end{cases}$$

Hence, since $n^+(\sup\epsilon > t-s) = 1/2(t-s)$ we obtain

$$\mathbb{P}\big(Y_t = 0|\sigma(R)\big) = \exp\left\{-\int_0^t ds\,\theta R_s/(t-s)\right\}. \tag{2.2}$$

From this it follows that

$$\mathbb{P}(Y_t = 0|\sigma = t) = {}^{-\theta}\mathbb{Q}^{4,t}_{x\to 0}\exp\left\{-\theta\int_0^t Z_s/(t-s)\, ds\right\}. \tag{2.3}$$

Note that, because we are conditioning to hit 0 at time t and not before, we obtain ${}^{-\theta}\mathbb{Q}^{4,t}_{x\to 0}$, and not ${}^{-\theta}\mathbb{Q}^{0,t}_{x\to 0}$ as one might expect, see [8] for a full discussion. To evaluate this we begin by observing that by the change of measure given in Pitman and Yor [8],

$$(2.4)\quad {}^{-\theta}\mathbb{Q}^{4,t}_{0\to 0}\exp\left\{-\theta\int_0^t Z_s/(t-s)\, ds\right\} = \frac{{}^{0}\mathbb{Q}^{4,t}_{0\to 0}\exp\left\{-\theta\int_0^t Z_s/(t-s)\,ds - \frac{1}{2}\theta^2\int_0^t Z_s\,ds\right\}}{{}^{0}\mathbb{Q}^{4,t}_{0\to 0}\exp\left\{-\frac{1}{2}\theta^2\int_0^t Z_s\,ds\right\}}.$$

Now from [9], under ${}^{0}\mathbb{Q}^{4,t}_{0\to 0}$, Z_t solves, for $u \le t$,

$$Z_u = 2\int_0^u \sqrt{Z_s}\, dB_s + 2\int_0^u [2 - Z_s/(t-s)]\, ds,$$

where B is a Brownian motion. Hence,

$$\theta\int_0^t Z_s/(t-s)\, ds = 2t\theta + \theta\int_0^t \sqrt{Z_s}\, dB_s,$$

but, of course, $\int_0^u \sqrt{Z_s}\,dB_s$ is a martingale with quadratic variation $\int_0^u Z_s\,ds$, so

$$\exp\left\{-\theta\int_0^u \sqrt{Z_s}\,dB_s - \tfrac{1}{2}\theta^2\int_0^u Z_s\,ds\right\}$$

is a martingale too (it's bounded above by $\exp(2\theta t)$!!). We take expectations and have succeeded in evaluating the numerator of (2.4),

$$\text{(2.5)} \qquad {}^{0}\mathbb{Q}^{4,t}_{0\to 0}\exp\left\{-\theta\int_0^t Z_s/(t-s)\,ds - \tfrac{1}{2}\theta^2\int_0^t Z_s\,ds\right\} = \exp(-2t\theta).$$

We find directly from Pitman and Yor [8] that the denominator satisfies

$$\text{(2.6)} \qquad {}^{0}\mathbb{Q}^{4,t}_{0\to 0}\exp\left\{-\tfrac{1}{2}\theta^2\int_0^t Z_s\,ds\right\} = \left[\frac{t\theta}{\sinh(t\theta)}\right]^2.$$

Next we observe, recalling (2.2),

$$\text{(2.7)} \qquad \begin{aligned} {}^{-\theta}\mathbb{Q}^{0,t}_{x\to 0}\exp\left\{-\theta\int_0^t Z_s/(t-s)\,ds\right\} &= \frac{{}^{-\theta}\mathbb{Q}^{0}_{x}\exp\left\{-\theta\int_0^t Z_s/(t-s)\,ds\right\}}{{}^{-\theta}\mathbb{Q}^{0}_{x} I_{\{Z_t=0\}}} \\ &= \frac{{}^{0}\mathbb{Q}^{0}_{x} I_{\{Z_t=0\}}}{{}^{-\theta}\mathbb{Q}^{0}_{x} I_{\{Z_t=0\}}} = \frac{\mathbb{P}(\tau\geq t)}{\mathbb{P}(\sigma\geq t)}. \end{aligned}$$

We can now proceed to

Proof of proposition 5. The Pitman-Yor decomposition, [8],

$${}^{-\theta}\mathbb{Q}^{4,t}_{x\to 0} = {}^{-\theta}\mathbb{Q}^{4,t}_{0\to 0} \oplus {}^{-\theta}\mathbb{Q}^{0,t}_{x\to 0},$$

allows us, combining (2.5),(2.6) and (2.7), to compute $\mathbb{P}(\tau = t|\sigma = t)$. Then we have

$$\mathbb{P}(\sigma = t|\tau = t) = \mathbb{P}(\tau = t|\sigma = t)\frac{\mathbb{P}(\sigma\in dt)}{\mathbb{P}(\tau\in dt)},$$

and substituting from the lemma we are done. □

We will now extend this result by conditioning on the whole of V, instead of just its extinction time. We will need the following lemma, which is perhaps of some independent interest.

Lemma 6. *Suppose M and N are continuous, orthogonal martingales with respect to a filtration $(\mathcal{F}_t; t\geq 0)$, and suppose that M has the following representation property. Any bounded, $\sigma(M)$-measurable variable Φ is of the form*

$$\Phi = c + \int_0^\infty H_t\,dM_t,$$

where H_t is $\mathcal{F}_t$-previsible, and $c\in\sigma(M_0)$. Let $\mathcal{G}_t = \mathcal{F}_t\vee\sigma(M)$, then N is a $\mathcal{G}_t$-martingale.

Proof. By an application of the monotone-class lemma, it suffices to show that for bounded $\sigma(M)$-measurable variables Φ,

$$\mathbb{E}[\Phi(N_t - N_s)|\mathcal{F}_s] = 0.$$

But, by the representation property,

$$\begin{aligned}\mathbb{E}[\Phi(N_t - N_s)|\mathcal{F}_s] &= \mathbb{E}\left[\left\{\int_0^t H_u\,dM_u\right\}(N_t - N_s)\middle|\mathcal{F}_s\right] \\ &= \mathbb{E}[(H\cdot M)_t N_t - (H\cdot M)_s N_s|\mathcal{F}_s] \\ &= 0,\end{aligned}$$

since $(H\cdot M)$ and N are orthogonal. □

Now on the stochastic interval $[0,\tau)$ we define

$$\Theta_t = \frac{R_t}{V_t}\exp(2\theta t).$$

Applying Itô's formula gives

$$d\Theta_t = \left\{2\frac{\sqrt{R_t}}{V_t}\,dB_t - \frac{R_t}{V_t^2}\,dV_t\right\}\exp(2\theta t),$$

which shows Θ_t to be a local martingale on $[0,\tau)$. Moreover, since $\Theta_t < \exp(2\theta t)$, Θ_t tends to a finite limit as $t\uparrow\tau$, and if we define $\Theta_t = \Theta_{\tau-}$ for $t \geq \tau$, then Θ_t is a martingale for $0 \leq t < \infty$.

If we continue to calculate with Itô's formula, we find that, for $t < \tau$,

(2.8a) $$d\Theta_t dV_t = 0$$

(2.8b) $$d\Theta_t d\Theta_t = 4\frac{\Theta_t}{V_t}(\exp(2\theta t) - \Theta_t).$$

Thus we have proved

Lemma 7. *Θ_t is a $\mathcal{F}_t$-martingale with quadratic variation*

$$[\Theta]_t = \int_0^{t\wedge\tau} ds\,4\frac{\Theta_s}{V_s}(\exp(2\theta s) - \Theta_s),$$

and furthermore Θ is orthogonal to V.

So if we put $\mathcal{G}_t = \mathcal{F}_t \vee \sigma(V)$, we can apply Lemma 6 to deduce that Θ_t is a $\mathcal{G}_t$-martingale. Moreover, τ is $\mathcal{G}_0$- measurable, and so for any positive constant K,

$$\mathbb{E}\Theta_\infty I_{\{\tau<K\}} = \mathbb{E}\Theta_0 I_{\{\tau<K\}},$$

since $\Theta_t I_{\{\tau<K\}}$ is a bounded $\mathcal{G}_t$-martingale. But as $K\uparrow\infty$ we obtain

$$\mathbb{E}\Theta_\infty = \mathbb{E}\Theta_0 = 1,$$

whence Θ is uniformly integrable. Now we are able to prove

Proposition 8. *The conditional law of the extinction time of the subcritical process given $\sigma(V_t; 0 \leq t < \infty)$ satisfies*

$$\mathbb{P}\big(\sigma = \tau | \sigma(V)\big) = \exp(-2\theta\tau) \qquad a.s..$$

Proof. We have already remarked that $\Theta_{\tau-}$ exists and hence $[\Theta]_\tau$ is finite almost surely. It is easy to confirm, for example by time inversion, that $\int_0^\tau V_s^{-1}\, ds = \infty$, and thus we deduce from the formula for the quadratic variation process of Θ, given in Lemma 7, that $\Theta_s(\exp(2\theta s) - \Theta_s) \to 0$ as $s \uparrow \tau$. Hence Θ_τ is either 0 or $\exp(2\theta\tau)$. Furthermore,

$$\mathbb{E}[\Theta_\tau | \mathcal{G}_0] = \mathbb{E}[\Theta_0 | \mathcal{G}_0] = 1,$$

and so,

$$\mathbb{P}\big(\Theta_\tau = \exp(2\theta\tau) | \sigma(V)\big) = \exp(-2\theta\tau).$$

Now observe that $\tau > \sigma$ implies that $\Theta_\tau = 0$ (but the converse isn't so evident!), whence

$$\mathbb{P}\big(\tau = \sigma | \sigma(V)\big) \geq \exp(-2\theta\tau).$$

But

$$\mathbb{P}(\tau = \sigma | \tau) = \mathbb{E}\big[\mathbb{P}\big(\tau = \sigma | \sigma(V)\big) | \tau\big] = \exp(-2\theta\tau),$$

implying the desired equality. □

3 A decomposition of Brownian motion

It is now well known, as excellently described by Le Gall [7], that if we interpret the squared Bessel process of dimension zero as a continuous-state branching process then the associated genealogical structure is carried by Brownian excursions. In this section we will give a decomposition of Brownian motion that corresponds to the decomposition of the squared Bessel process induced by the killing considered previously. By looking at local times we will be able to recover Proposition 3.

To begin we recall:

Theorem 9 (Ray-Knight). *If $\bar{W}_t$ is reflecting Brownian motion, starting from zero, with l_t^y its local time at level y, then, letting $\tau_x = \inf\{t : l_t^0 \geq x\}$, we have $(l_{\tau_x}^y, y \geq 0)$ is a squared Bessel process of dimension 0 started from x.*

If we introduce drift we can obtain the subcritical process of the previous section in a similar manner.

Theorem 10. *If $\bar{S}_t$ is reflecting Brownian motion with drift θ towards the origin, starting from zero, and if l_t^y is its local time at level y, then letting $\tau_x = \inf\{t : l_t^0 \geq x\}$, we have the law of the process $(l_{\tau_x}^y, y \geq 0)$ is ${}^{-\theta}\mathbb{Q}_x^0$.*

Proof. We follow Yor [14]. Let ${}^{\theta}\mathbf{W}$ denote the law of reflecting Brownian motion with drift θ towards the origin, with similar notation for the corresponding expectation. Then the Girsanov theorem gives us

$$\left.\frac{d\,{}^{\theta}\mathbf{W}}{d\,{}^{0}\mathbf{W}}\right|_{\mathcal{F}_t} = \exp\bigl(\theta(X_t - \tfrac{1}{2}l_t^0) - \tfrac{1}{2}\theta^2 t\bigr).$$

Hence for a positive measurable functional F, we have using the Ray-Knight theorem,

$$\begin{aligned} {}^{\theta}\mathbf{W}[F(l^y_{\tau_x}; y \geq 0)] &= {}^{0}\mathbf{W}[F(l^y_{\tau_x}; y \geq 0)\exp(-\tfrac{1}{2}\theta x - \tfrac{1}{2}\theta^2 \tau_x)] \\ &= {}^{0}\mathbb{Q}^0_x\left[F(Z_y; y \geq 0)\exp\left(-\tfrac{1}{2}\theta x - \tfrac{1}{2}\theta^2 \int_0^\infty Z_y\,dy\right)\right] \\ &= {}^{-\theta}\mathbb{Q}^0_x[F(Z_y; y \geq 0)], \end{aligned}$$

the last line following from the change of measure given in [8]. □

Now we give the fundamental results of this section, recalling the notation of Section 1.

Theorem 11. *Suppose that X is a sticky Brownian motion starting from x, and W is a Wiener process, with $W_0 \geq x$, so that equation (1.1) is satisfied. Define, for $t \geq 0$,*

$$\bar{W}_t = W_t + L_t,$$

where $L_t = \sup_{s \leq t}\bigl((-W_s) \vee 0\bigr)$, so $\bar{W}$ is a reflecting Brownian motion. Then

$$\bar{W}_t = \bar{S}_{A_t^0} + X_t,$$

where $\bar{S}_t$ is a reflecting Brownian motion, with drift θ towards the origin, independent of X.

Proof. Take (X, W) solving (1.1) with $W_0 \geq x$. Then

$$S_t = W_0 - x + \int_0^{\alpha_t^0} I_{\{X_s = 0\}}\,dW_s - \theta t$$

defines a Brownian motion with drift $(-\theta)$, independent by Knight's theorem from X_t. It is easy to check that $W_t = S_{A_t^0} + X_t$. Let

$$L_t = \sup_{s \leq t}\bigl((-W_s) \vee 0\bigr) \qquad \text{and} \qquad K_t = \sup_{s \leq t}\bigl((-S_s) \vee 0\bigr).$$

Now $\bar{W}_t = W_t + L_t$ is reflecting Brownian motion, and $\bar{S}_t = S_t + K_t$ is a reflecting Brownian motion with drift θ towards the origin, independent of X. Moreover, if we can show $K_{A_t^0} = L_t$ then we will have $\bar{W}_t = \bar{S}_{A_t^0} + X_t$ as required. But $W_t \geq S_{A_t^0}$, whence

$$\sup_{s \leq t}(-W_s) \leq \sup_{s \leq t}(-S_{A_s^0}),$$

and so $L_t \le K_{A_t^0}$. If there exists an $s \le t$ so that $X_s = 0$ then, putting

$$t^0 = \alpha^0_{A_t^0} = \sup\{s \le t : X_s = 0\},$$

so $X_{t^0} = 0$ and $S_{A_t^0} = S_{A_{t^0}^0}$, we have $W_{t^0} = S_{A_t^0}$, and hence

$$\sup_{s\le t}(-W_s) \ge \sup_{s \le t}(-S_{A_s^0}).$$

If no such s exists then $A_t^0 = 0$, and $W_s \ge 0$, for all $s \le t$. In either case $L_t \ge K_{A_t^0}$. □

For the rest of the section we assume that X_0 and $\bar{W}_0$ are both 0, and we are able to interpret the above result in terms of branching processes. A point $(t, \bar{W}_t)$ represents part of the subcritical process if $X_t = 0$; otherwise it is part of an excursion of the X process away from 0, and such an excursion represents mass descended from a single killed ancestor. Letting l_t^y be the local time of $\bar{W}$ and τ_x be as before, we have,

$$l^y_{\tau_x} = \int_0^{\tau_x} I_{\{X_t=0\}} dl_t^y + \int_0^{\tau_x} I_{\{X_t>0\}} dl_t^y. \tag{3.1}$$

The Ray-Knight theorem applies to the left-hand side and the following applies to the right-hand side.

Proposition 12. *For $y \ge 0$ define*

$$R_y = \int_0^{\tau_x} I_{\{X_t=0\}} dl_t^y,$$

and

$$Y_y = \int_0^{\tau_x} I_{\{X_t>0\}} dl_t^y.$$

Then R and Y satisfy the stochastic differential equations (2.1).

Proof. From the occupation-time formula, we have, for any positive Borel measurable function f,

$$\begin{aligned}\int_0^\infty f(y) R_y\, dy &= \int_0^\infty f(y) \int_0^{\tau_x} I_{\{X_t=0\}} dl_t^y dy \\ &= \int_0^{\tau_x} f(\bar{W}_t) I_{\{X_t=0\}} dt \\ &= \int_0^{\tau_x} f(\bar{W}_t)\, dA_t^0 \\ &= \int_0^{A^0_{\tau_x}} f(\bar{S}_t)\, dt.\end{aligned}$$

Next observe, since we demonstrated in the proof of the previous theorem $K_{A_t^0} = L_t$, that $A^0_{\tau_x}$ is the first time that the local time of $\bar{S}$ at 0 reaches x. Thus $(R_y; y \ge 0)$ is the family of local times of $\bar{S}$ stopped after it has spent local time x at the origin, and, appealing to Theorem 10, the first part of the result follows.

Similarly, for any positive Borel measurable function f,

$$\begin{aligned}\int_0^\infty f(y)Y_y dy &= \int_0^\infty f(y)\int_0^{\tau_x} I_{\{X_t>0\}} dl_t^y\, dy \\ &= \int_0^{\tau_x} f(\bar{W}_t) I_{\{X_t>0\}} dt \\ &= \int_0^{\tau_x} f(\bar{S}_{A_t^0} + X_t)\, dA_t^+ \\ &= \int_0^{A^+_{\tau_x}} f(\bar{S}_{\theta^{-1}L_t^+} + X_{\alpha_t^+})\, dt.\end{aligned}$$

Recall that $(X_{\alpha_t^+}; t \geq 0)$ is a reflecting Brownian motion and that L^+ half its local time at the origin. Note that $A^+_{\tau_x}$ is the first time that the local time of X_{α^+} at the origin reaches $2\theta A^0_{\tau_x}$. Now put

$$M_t^f = \exp\left\{-\int_0^t f(\bar{S}_{\theta^{-1}L_t^+} + X_{\alpha_t^+})\, dt\right\}.$$

Since, conditional on $\bar{S}$, M_t^f is a skew multiplicative functional of $(X_{\alpha_t^+}; t \geq 0)$, it is a consequence of excursion theory, see [9] and [14], that

$$\mathbb{E}\left[M^f_{A^+_{\tau_x}} \middle| \sigma(\bar{S})\right] = \\ \exp\left\{-\int_0^{2\theta A^0_{\tau_x}} ds \int n^+(d\epsilon)\left[1 - \exp\left(-\int_0^{T(\epsilon)} du\, f\left(\epsilon(u) + \bar{S}_{(2\theta)^{-1}s}\right)\right)\right]\right\},$$

where $T(\epsilon)$ denotes the lifetime of the excursion ϵ. This, by the occupation-time formula, remembering $A^0_{\tau_x}$ is the first time that the local time of $\bar{S}$ at 0 reaches x, equals

$$\exp\left\{-\int n^+(d\epsilon)\int_0^\infty 2\theta dy\, R_y \left[1 - \exp\left(-\int_0^{T(\epsilon)} du\, f\left(\epsilon(u) + y\right)\right)\right]\right\}.$$

Thus we have

$$\mathbb{E}\left[\exp\left\{-\int_0^\infty f(y)Y_y\, dy\right\} \middle| \sigma(R)\right] = \\ \exp\left\{-2\theta \int_0^\infty dy\, R_y \int n^+(d\epsilon)\left[1 - \exp\left(-\int_0^{T(\epsilon)} du\, f\left(\epsilon(u) + y\right)\right)\right]\right\},$$

and this characterises the solution to (2.1), see [14]. □

4 The conditional law of sticky Brownian motion

In this section we will prove Theorem 1; however we do not work directly with the pair of processes (X, W). Instead, motivated by the previous section, we consider a Markov process $(X, \bar{W})$ on the state space $E = \{(x, a) \in \mathbb{R}^2 : x \geq 0, a \geq x\}$, defined by taking X to be a sticky Brownian motion and $\bar{W}_t = \bar{S}_{A_t^0} + X_t$ where $\bar{S}$ is a reflecting Brownian motion, independent of X, with drift θ towards the origin. We denote by $\mathbb{P}^{(x,a)}$ the law of $(X, \bar{W})$ started from $X_0 = x$ and $\bar{W}_0 = a$, with similar notation for expectations. We will prove that

$$\mathbb{P}^{(0,0)}\big(X_t \le x | \sigma(\bar{W})\big) = \exp\big(- 2\theta(\bar{W}_t - x)\big) \qquad \text{a.s.} \tag{4.1}$$

for $x \in [0, \bar{W}_t]$. This has a clear interpretation in terms of our branching process with killing. We can think of the value of X_t as depending on whether, and if so where, killing occurs along a line of ancestry of length $\bar{W}_t$. Theorem 1 follows from (4.1) and Theorem 11, noting that $\sigma(\bar{W}) = \sigma(W)$.

We proceed by computing some resolvents. We need, first, to convince ourselves that $(X, \bar{W})$ has the strong Markov property; but this follows, conditioning on X, from the simple Markov property of S and the strong Markov property of X. The first part of the following result, the calculation of the resolvent of sticky Brownian motion, has been obtained previously by several authors, see for example Knight [6].

Proposition 13. *The resolvent operators $(\mathcal{U}_\lambda, \lambda > 0)$ and $(\mathcal{V}_\lambda, \lambda > 0)$ of X and $(X, \bar{W})$ respectively are given, letting $\gamma^2 = 2\lambda$, by*

$$\mathcal{U}_\lambda f(x) = (2\theta)^{-1} u_\lambda(x,0) f(0) + \int_0^\infty u_\lambda(x,y) f(y)\, dy,$$

where

$$u_\lambda(x,y) = \gamma^{-1}\left[e^{-\gamma|y-x|} + \frac{\theta\gamma - \lambda}{\theta\gamma + \lambda} e^{-\gamma|y+x|}\right],$$

and

$$\begin{aligned}\mathcal{V}_\lambda f(x,a) = &\int_0^\infty \int_y^\infty f(y,b) v_\lambda(x,a,y,b)\, db dy \\ &+ (2\theta)^{-1}\int_0^\infty f(0,b) v_\lambda(x,a,0,b)\, db \\ &+ \int_{\{a+y=b+x\}} f(y,b) r_\lambda^-(x,y)\, dy,\end{aligned}$$

where

$$v_\lambda(x,a,y,b) = \frac{2\theta}{\gamma+\theta} e^{\theta(a-b+y-x)-\gamma(y+x)}\left[e^{-(\theta+\gamma)|b-a+x-y|} + \frac{\gamma+2\theta}{\gamma} e^{-(\theta+\gamma)|a+b-y-x|}\right],$$

and $\quad r_\lambda^-(x,y) = \gamma^{-1}\left[e^{-\gamma|y-x|} - e^{-\gamma|y+x|}\right].$

Proof. We are guided (as always!) by Rogers and Williams [10].

We begin by supposing that $X_0 = 0$ and $\bar{W}_0 = a$, where $a \ge 0$. Take two independent exponential random variables, T_1 and T_2, both independent of X and $\bar{W}$, and both with mean λ^{-1}. Let

$$T = \alpha_{T_1}^0 \wedge \alpha_{T_2}^+,$$

this also being exponentially distributed with mean λ^{-1}. Now $X_T = 0$ precisely if $\alpha^0_{T_1} < \alpha^+_{T_2}$, or equivalently if $T_1 < A^0_{\alpha^+_{T_2}}$. But recall $\theta A^0_{\alpha^+_t}$ equals L^+_t, which is exponentially distributed with mean γ^{-1}, where $\gamma^2 = 2\lambda$, and so,

$$\mathbb{P}^{(0,a)}(X_T = 0) = \frac{\lambda}{\lambda + \theta\gamma}.$$

For $y > 0$, since $X_{\alpha^+_t}$ is reflecting Brownian motion, and hence $X_{\alpha^+_{T_2}}$ is independent of $L^+_{T_2}$,

$$\begin{aligned}\mathbb{P}^{(0,a)}(X_T \in dy) &= \mathbb{P}^{(0,a)}(X_{\alpha^+_{T_2}} \in dy)\mathbb{P}^{(0,a)}(\alpha^0_{T_1} > \alpha^+_{T_2})\\ &= \frac{2\theta\lambda}{\lambda + \theta\gamma}\exp(-\gamma y)\,dy.\end{aligned}$$

Now let us note that the resolvent of reflecting Brownian motion with drift θ towards the origin has density

$${}^{\theta}r_\lambda(x,y) = \alpha^{-1}e^{\theta(x-y)}\left[e^{-\alpha|y-x|} + \frac{\alpha+\theta}{\alpha-\theta}e^{-\alpha|x+y|}\right],$$

with respect to Lebesgue measure, where $\alpha^2 = 2\lambda + \theta^2$.

We have that A^0_T equals $T_1 \wedge \theta^{-1}L^+_{T_2}$, and hence is exponentially distributed at rate $\lambda + \theta\gamma$. Thus,

$$\begin{aligned}\mathbb{P}^{(0,a)}(X_T = 0, \bar{W}_T \in db) &= \mathbb{P}^{(0,a)}(\alpha^0_{T_1} < \alpha^+_{T_2} \text{ and } \bar{S}_{A^0_T} \in db)\\ &= \lambda\,{}^{\theta}r_{\lambda+\theta\gamma}(a,b)db.\end{aligned}$$

Similarly, and again crucially using the independence of $X_{\alpha^+_{T_2}}$ and $L^+_{T_2}$,

$$\begin{aligned}\mathbb{P}^{(0,a)}(X_T \in dy, \bar{W}_T \in db) &= \mathbb{P}^{(0,a)}(X_T \in dy)\mathbb{P}^{(0,a)}(\bar{S}_{A^0_T} \in d(b-y))\\ &= 2\theta\lambda\exp(-\gamma y)\,{}^{\theta}r_{\lambda+\theta\gamma}(a, b-y)\,dbdy.\end{aligned}$$

The above arguments have determined $\mathcal{U}_\lambda f(0)$ and $\mathcal{V}_\lambda f(0,a)$. If we now consider the process $(X, \bar{W})$ started from an arbitrary point $(x,a) \in E$, we may apply the strong Markov property at the time H_0, the first time that X_t is zero. We obtain

$$\mathcal{U}_\lambda f(x) = \mathcal{R}^-_\lambda f(x) + \psi_\lambda(x)\mathcal{U}_\lambda f(0),$$

and, defining the the function $f_{x,a}$ by $f_{x,a}(y) = f(y, a+y-x)$ for $y \ge 0$,

$$\mathcal{V}_\lambda f(x,a) = \mathcal{R}^-_\lambda f_{x,a}(x) + \psi_\lambda(x)\mathcal{V}_\lambda f(0, a-x),$$

where $\mathcal{R}^-_\lambda$ is the resolvent of Brownian motion killed at 0, which has density $r^-_\lambda(x,y)$ with respect to Lebesgue measure, and

$$\psi_\lambda(x) = \mathbb{E}^{(x,a)}[\exp(-\lambda H_0)] = \exp(-\gamma x).$$

This completes the proof. □

Let $T_1, T_2, \dots, T_n$ be independent exponential times with means $\lambda_1^{-1}, \lambda_2^{-1}, \dots, \lambda_n^{-1}$. We will show, for arbitrary bounded, measurable functions $f_1, \dots, f_n$ on E, that

$$
\begin{aligned}
(4.2)\quad \mathbb{E}^{(0,0)}\big[\mathcal{I}_{\{X_{T_1+\cdots+T_n}\le x\}} f_1(\bar{W}_{T_1})\dots f_n(\bar{W}_{T_1+\cdots+T_n})\big] &= \\
\mathbb{E}^{(0,0)}\big[\exp\big(-2\theta(\bar{W}_{T_1+\dots T_n} - x)\wedge 0\big) f_1(\bar{W}_{T_1})\dots f_n(\bar{W}_{T_1+\cdots+T_n})\big].
\end{aligned}
$$

Our argument will essentially depend on time reversal and the fact that $\bar{W}_t = 0$ implies that $X_t = 0$. We begin by making some remarks concerning the resolvent of $(X, \bar{W})$ that follow from the preceding proposition.

Define the measure m on Borel subsets of the state space E by

$$
(4.3)\qquad mA = \int_A e^{2\theta(x-a)}\,dxda + \int_{\{a:(0,a)\in A\}} e^{-2\theta a}/2\theta\,da,
$$

and then $\mathcal{V}_\lambda$ is self-adjoint with respect to m in the sense that for any bounded, measurable functions f and g on the state space

$$
(4.4)\qquad \int_E dm(y,a)\, f(y,a)[\mathcal{V}_\lambda g](y,a) = \int_E dm(y,a)\,[\mathcal{V}_\lambda f](y,a)g(y,a).
$$

We also have

$$
(4.5)\qquad
\begin{aligned}
\mathcal{V}_\lambda f(0,0) &= 2\theta\int_E dm(y,a)\, f(y,a)v_\lambda(y,a,0,0) \\
&= 2\theta\int_E dm(y,a)\, f(y,a)r_\lambda(0,a).
\end{aligned}
$$

where $r_\lambda(\cdot,\cdot)$ is the density, with respect to Lebesgue measure, of the resolvent $\mathcal{R}_\lambda$ of reflecting, driftless Brownian motion. Slightly abusing notation we will write r_λ for the function $r_\lambda(0,\cdot)$. If f is a bounded measurable function on $[0,\infty)$ let us define f^* to be the function on the state space satisfying $f^*(x,a) = f(a)$ for all $x \in [0,a]$. For such f observe that

$$
(4.6)\qquad \mathcal{V}_\lambda f^* = (\mathcal{R}_\lambda f)^*;
$$

this being nothing more than the statement that $\bar{W}$ is a reflecting Brownian motion.

We define the functions $e_x : \mathbb{R}^+ \to [0,1]$ and $i_x : E \to \{0,1\}$ by

$$
e_x(a) = \begin{cases} e^{-2\theta(a-x)} & \text{if } a > x, \\ 1 & \text{otherwise,} \end{cases}
$$

and

$$
i_x(y,a) = \begin{cases} 1 & \text{if } y \le x, \\ 0 & \text{otherwise.} \end{cases}
$$

Now, using the above observations, and that $\mathcal{R}_\lambda$ is self-adjoint with respect to Lebesgue measure, we have

$$
\begin{aligned}
[\mathcal{V}_{\lambda_1} f_1^\star \mathcal{V}_{\lambda_2} f_2^\star \dots \mathcal{V}_{\lambda_n} i_x f_n^\star](0,0) &= 2\theta \int_E dm(y,a) r_{\lambda_1}(0,a)[f_1^\star \mathcal{V}_{\lambda_2} f_2^\star \dots \mathcal{V}_{\lambda_n} i_x f_n^\star](y,a) \\
&= 2\theta \int_E dm(y,a) i_x(y,a)[f_n^\star \mathcal{V}_{\lambda_n} f_{n-1}^\star \dots \mathcal{V}_{\lambda_2} f_1^\star r_{\lambda_1}^\star](y,a) \\
&= 2\theta \int_{\{(y,a)\in E: y\le x\}} dm(y,a)\, [f_n \mathcal{R}_{\lambda_n} f_{n-1} \dots \mathcal{R}_{\lambda_2} f_1 r_{\lambda_1}](a) \\
&= \int_0^\infty da\, e_x(a)[f_n \mathcal{R}_{\lambda_n} f_{n-1} \dots \mathcal{R}_{\lambda_2} f_1 r_{\lambda_1}](a) \\
&= \int_0^\infty da\, r_{\lambda_1}(0,a)[f_1 \mathcal{R}_{\lambda_n} f_2 \dots \mathcal{R}_{\lambda_n} e_x f_n](a) \\
&= [\mathcal{V}_{\lambda_1} f_1^\star \mathcal{V}_{\lambda_2} f_2^\star \dots \mathcal{V}_{\lambda_n} e_x^\star f_n^\star](0,0).
\end{aligned}
$$

This proves equation (4.2). Moreover it follows simply from (4.6), that given further, independent, exponential times $T_{n+1}, \dots, T_{n+m}$, and bounded, measurable functions $f_{n+1}, \dots, f_{n+m}$, the stronger statement

$$
\begin{aligned}
(4.7)\quad \mathbb{E}^{(0,0)}\big[\mathcal{I}_{\{X_{T_1+\dots+T_n}\le x\}} f_1(\bar{W}_{T_1}) \dots f_{n+m}(\bar{W}_{T_1+\dots+T_{n+m}})\big] = \\
\mathbb{E}^{(0,0)}\big[\exp\big(-2\theta(\bar{W}_{T_1+\dots+T_n} - x)\wedge 0\big) f_1(\bar{W}_{T_1}) \dots f_{n+m}(\bar{W}_{T_1+\dots+T_{n+m}})\big],
\end{aligned}
$$

holds. Now, by the uniqueness of Laplace transforms,

$$
\begin{aligned}
(4.8)\quad \mathbb{E}^{(0,0)}\big[\mathcal{I}_{\{X_{t_1+\dots+t_n}\le x\}} f_1(\bar{W}_{t_1}) \dots f_{n+m}(\bar{W}_{t_1+\dots t_{n+m}})\big] = \\
\mathbb{E}^{(0,0)}\big[\exp\big(-2\theta(\bar{W}_{t_1+\dots+t_n} - x)\wedge 0\big) f_1(\bar{W}_{t_1}) \dots f_{n+m}(\bar{W}_{t_1+\dots+t_{n+m}})\big],
\end{aligned}
$$

for almost all $t_1, t_2, \dots, t_{n+m} \ge 0$. In order to extend this equality, so that it holds for all $t_1, t_2, \dots, t_{n+m} \ge 0$, we first assume that $f_1, \dots, f_{n+m}$ are continuous, and subsequently apply the monotone-class lemma. Now observe that

$$
t_1, t_2, \dots, t_{n+m} \longmapsto \exp\big(-2\theta(\bar{W}_{t_1+\dots+t_n} - x)\wedge 0\big) f_1(\bar{W}_{t_1}) \dots f_{n+m}(\bar{W}_{t_1+\dots+t_{n+m}})
$$

is continuous and bounded, so the bounded convergence theorem implies that the right-hand side of (4.8) is a continuous function of $t_1, t_2, \dots, t_{n+m}$ too. Note that, as we can check from Lemma 2,

$$
t_1 + t_2 + \dots + t_n \longmapsto \mathbb{E}^{(0,0)}\big[\mathcal{I}_{\{X_{t_1+\dots+t_n}\le x\}}\big]
$$

is continuous, and so by adding large constant multiples of $\mathcal{I}_{\{X_{t_1+\dots+t_n}\le x\}}$ we can assume that $f_1, f_2, \dots, f_{n+m}$ are all positive functions. Now

$$
t_1, t_2, \dots, t_{n+m} \longmapsto \mathcal{I}_{\{X_{t_1+\dots+t_n}\le x\}} f_1(\bar{W}_{t_1}) \dots f_{n+m}(\bar{W}_{t_1+\dots+t_{n+m}})
$$

is upper semi-continuous, and Fatou's lemma thus implies that the left-hand side of (4.8) is an upper semi-continuous function of $t_1, t_2, \dots, t_{n+m}$. But we can argue the same way with $f_1, f_2, \dots, f_{n+m}$ replaced by $-f_1, -f_2, \dots, -f_{n+m}$, and so the left-hand side of (4.8) must in fact be continuous, and hence equality holds for all $t_1, t_2, \dots, t_{n+m} \ge 0$. All that remains is, on applying the monotone-class lemma, to deduce (4.1), and the proof of Theorem 1 is finally complete.

Acknowledgements. The author was supported by an SERC Research Studentship. I would like to express my thanks for the generous and inspiring help provided by Marc Yor and David Williams.

References

[1] M. Amir. Sticky Brownian motion as the strong limit of a sequence of random walks. *Stochastic Processes and their Applications*, 39:221–237, 1991.

[2] R.J. Chitashvili. On the nonexistence of a strong solution in the boundary problem for a sticky Brownian motion. Technical Report BS-R8901, Centre for Mathematics and Computer Science, Amsterdam, 1989.

[3] W. Feller. On boundaries and lateral conditions for the Kolmogorov equations. *Annals of Mathematics, Series 2*, 65:527–570, 1957.

[4] J.M. Harrison and A.J. Lemoine. Sticky Brownian motion as the limit of storage processes. *Journal of Applied Probability*, 18:216–226, 1981.

[5] N. Ikeda and S. Watanabe. *Stochastic Differential Equations and Diffusion Processes.* North Holand-Kodansha, Amsterdam and Tokyo, 1981.

[6] F.B. Knight. *Essentials of Brownian motion and Diffusion*, volume 18 of *Mathematical Surveys.* American Mathematical Society, Providence, Rhode-Island, 1981.

[7] J.F. Le Gall. Cours de troisième cycle, Laboratoire de Probabilités, Paris 6. 1994.

[8] J.W. Pitman and M. Yor. A decomposition of Bessel bridges. *Zeitschrift für Wahrscheinlichkeitstheorie*, 59:425–457, 1982.

[9] D. Revuz and M. Yor. *Continuous martingales and Brownian motion.* Springer, Berlin, 1991.

[10] L.C.G. Rogers and D. Williams. *Diffusions, Markov processes and Martingales, vol 2: Itô calculus.* Wiley, New York, 1987.

[11] T. Shiga and S. Watanabe. Bessel diffusions as a one-parameter family of diffusion processes. *Zeitschrift für Wahrscheinlichkeitstheorie*, 27:37–46, 1973.

[12] K. Yamada. Reflecting or sticky Markov processes with Lévy generators as the limit of storage processes. *Stochastic Processes and their Applications*, 52:135–164, 1994.

[13] M. Yor. Some remarks concerning sticky Brownian motion. Unpublished, 1989.

[14] M. Yor. *Some aspects of Brownian motion, part 1: Some special functionals.* Birkhäuser, 1992.

INTEGRATION BY PARTS AND CAMERON–MARTIN FORMULAS FOR THE FREE PATH SPACE OF A COMPACT RIEMANNIAN MANIFOLD

R. LÉANDRE AND J. R. NORRIS

Faculté des Sciences, Université de Nancy I
54000 Nancy, France
Statistical Laboratory, University of Cambridge
16 Mill Lane, Cambridge CB2 1SB, UK

1. Introduction

Let $(h_s : s \geq 0)$ denote an absolutely continuous function with values in $\mathbb{R}^m$ whose derivative is square-integrable:

$$\int_0^\infty |\dot{h}_s|^2 ds < \infty.$$

The Cameron–Martin formula states that if $(x_s : s \geq 0)$ is a Brownian motion in $\mathbb{R}^m$, starting from 0, then, *provided also* $h_0 = 0$, the law of $(x_s + h_s : s \geq 0)$ is absolutely continuous with respect to that of $(x_s : s \geq 0)$ with density

$$\rho_s = \exp\left\{\int_0^\infty \langle \dot{h}_s, dx_s\rangle - \tfrac{1}{2}\int_0^\infty |\dot{h}_s|^2 ds\right\}.$$

In fact if one randomizes the starting point x_0 according to Lebesgue measure, then the formula remains valid without the assumption $h_0 = 0$. Thus we obtain a Cameron–Martin formula for the free path space of $\mathbb{R}^m$. For suitable functions F on the path space, the expectation

$$\mathbb{E}\left[F(x+th)\exp\left\{-\int_0^\infty \langle t\dot{h}_s, dx_s\rangle - \tfrac{1}{2}\int_0^\infty |t\dot{h}_s|^2 ds\right\}\right]$$

does not depend on t. So on differentiating in t at 0 we obtain an integration by parts formula:

$$\mathbb{E}[D_h F(x)] = \mathbb{E}\left[F(x)\int_0^\infty \langle \dot{h}_s, dx_s\rangle\right].$$

This may be regarded as the infinitesimal form of the Cameron–Martin formula.

In this note we shall discuss Cameron–Martin and integration-by-parts formulas for the free path space of a compact Riemannian manifold. The case of paths with a fixed starting point has already been thoroughly discussed: see [D],[H]. The results we obtain are at a technical level simple corollaries of results in [L2] or [N2]. The emphasis here is rather on the efficient calculation of densities and divergences for flows and vector fields. The integration by parts formula is proved first in §2, by a direct argument based on the methods of [L2]. Then in §3 we use the main result of [N2] to establish independently a corresponding Cameron–Martin formula. From here we can recover the integration by parts formula by differentiating.

Integration by parts formulas of a similar type are proved in [L1],[L2],[LR] by using a mixture of small time asymptotics and developments of Bismut's formula [B],[EL],[N1]. They rely deeply on the identity between the tangent spaces to path space used by Bismut [B] and Jones and Léandre [JL]. Such integration by parts formulas are also known for free twisted loops: see [LR]. In this case we do not yet have a corresponding Cameron–Martin formula.

We would like to thank David Elworthy for his warm hospitality during the Warwick Symposium on Stochastic Analysis and Related Topics 1994/5, where this work was done.

2. An integration by parts formula on the free path space

Let Ω denote the set of continuous paths $(x_s : s \geq 0)$ with values in a compact Riemannian manifold M. Let X denote a vector field on Ω, thus $X(x) = \big(X_s(x) : s \geq 0\big)$ where $X_s(x)$ belongs to the tangent space to M at x_s. We shall investigate the relationship between X and the equilibrium Wiener measure on Ω:

$$\mathbb{P}(dx) = \int_M \mathbb{P}^{x_0}(dx)dx_0$$

where $\mathbb{P}^{x_0}$ denotes the law of Brownian motion in M starting from x_0 and dx_0 denotes the normalized Riemannian volume. Let us consider the vector field X given by

$$X_s(x) = \tau_s \sum_{i=1}^{m} h_s^i X_i(x_0)$$

where, for $i = 1, \ldots, m$, X_i is a C^2 vector field over M and τ_s is the parallel transport from x_0 to x_s.

Conditional on x_0, we define a Brownian motion b_s in $T_{x_0}M$ by $b_0 = 0$ and

$$\partial b_s = \tau_s^{-1}\partial x_s$$

where ∂ denotes the Stratonovich differential.

For each $s \geq 0$ denote by $e_s : \Omega \to M$ the evaluation map $e_s(x) = x_s$. We consider the pullback T_s by e_s of the tangent bundle TM equipped with the pullback of the Levi–Civita connection of M. Then formally ∂x_s is a section of T_s, ∂b_s is a section of T_0, τ_s is a section of $T_s \otimes (T_0)^*$, and we have

$$\nabla_X \partial x_s = (\nabla_X \tau_s)\partial b_s + \tau_s \nabla_X \partial b_s.$$

This formula is justified in [L2] at (4.65). We know also by [L2] (see (4.64), (3.87)) that

$$\nabla_X \tau_s = \tau_s \int_0^s \tau_r^{-1} R(\partial x_r, X_r)\tau_r,$$

$$\nabla_X \partial x_s = \tau_s \sum_{i=1}^m h_s^i X_i(x_0)\partial s$$

so

$$\nabla_X \partial b_s = \sum_{i=1}^m h_s^i X_i(x_0)\partial s - \Big(\int_0^s \tau_r^{-1} R(\partial x_r, X_r)\tau_r\Big)\partial b_s.$$

Hence we obtain for the Itô differential

$$\nabla_X db_s = \sum_{i=1}^m h_s^i X_i(x_0) ds - \tfrac{1}{2}\tau_s^{-1}\mathrm{Ricci}(X_s) ds - \Big(\int_0^s \tau_r^{-1} R(\partial x_r, X_r)\tau_r\Big) db_s.$$

We compute now the action of X on a test functional

$$F = \sum_n \int_{0<s_1<\cdots<s_n} H(s_1, \ldots, s_n; x_0) db_{s_1} \ldots db_{s_n}$$

where the sum in n is finite. Here H is a cotensor in $T_{x_0}M$. We have

$$\langle dF, X\rangle = \sum_n \int_{0<s_1<\cdots<s_n} \sum_j H(s_1, \ldots, s_n; x_0) db_{s_1} \ldots \nabla_X db_{s_j} \ldots db_{s_n}$$

$$+ \sum_n \int_{0<s_1<\cdots<s_n} \nabla_{X_0} H(s_1, \ldots, s_n; x_0) db_{s_1} \ldots db_{s_n},$$

so

$$\mathbb{E}\langle dF, X\rangle = \int_M dx_0 \mathbb{E}^{x_0}\Big(\sum_n \int_{0<s_1<\cdots<s_n} H(s_1, \ldots, s_n; x_0) db_{s_1} \ldots db_{s_{n-1}}(\theta_{s_n} ds_n)\Big)$$

$$+ \int_M X_0 f(x_0) dx_0$$

where $f(x_0) = \mathbb{E}^{x_0}(F)$ and

$$\theta_s = \tau_s^{-1}(D/\partial s - \tfrac{1}{2}\mathrm{Ricci})X_s.$$

Here $D/\partial s$ denotes covariant differentiation along x_s. In the first term we used the fact that integrals in db_s vanish under the expectation. In the second we used the fact that the Fock space structure, being derived from the metric, is preserved by the Levi–Civita connection. Let us define

$$\operatorname{div} X(x) = \operatorname{div} X_0(x_0) + \int_0^\infty \langle \Big(\frac{D}{\partial s} - \tfrac{1}{2}\mathrm{Ricci}\Big) X_s(x), dx_s\rangle.$$

We have shown:

Theorem 2.1. *We have*

$$\mathbb{E}\langle dF, X\rangle = \mathbb{E}(F \operatorname{div} X).$$

3. A Cameron–Martin formula on the free path space

Recall that Ω denotes the set of continuous paths $(x_s : s \geq 0)$ with values in M. Let X denote a vector field on Ω, thus $X(x) = (X_s(x) : s \geq 0)$ with $X_s(x) \in T_{x_s}M$. Our object now is to compute the image of the equilibrium Wiener measure $\mathbb{P}$ under the flow determined by X.

We begin with a rough argument from which some technical points are missing. Later, in order to fill these gaps we shall specialize our choice of vector field X, which may obscure the simplicity of the basic argument. Let us assume that X_s is previsible, and that $DX_s/\partial s$ exists for almost all s, and is square-integrable. Let us assume also that for $\mathbb{P}$-almost all $x_0 \in \Omega$, we can integrate X to a flow in Ω

$$\dot{x}_t = X(x_t). \tag{1}$$

Here we use t to parametrize a family of paths $x_t = (x_{st} : s \geq 0)$. Let us suppose that x_{st} is a two-parameter semimartingale in the sense of [N2], then the two-parameter stochastic calculus provides a means to compute the law of x_t when x_0 has law $\mathbb{P}$.

We may rewrite (1) in differential form

$$\partial_t x_{st} = X_{st}\partial t$$

where $X_{st} = X_s(x_t)$. Recall that we write d_s and ∂_s for the Itô and Stratonovich differentials in s; we also write D_s for the covariant Stratonovich differential corresponding to the Levi–Civita connection. Then

$$D_s\partial_t x_{st} = \left(\frac{D}{\partial s}X_{st}\right)\partial s\partial t.$$

Let us introduce a lift v_{st} of x_{st} to the bundle OM of orthonormal frames in TM, choosing v_{00} arbitrarily and imposing the following horizontality conditions:

$$D_s v_{s0} = 0, \quad D_t v_{st} = 0,$$

which determine v_{st} uniquely, given v_{00}. In addition we introduce two further processes, q_{st} in TM over x_{st}, and b_{st} in $\mathbb{R}^n$, by the equations

$$\begin{aligned} D_t q_{st} &= \left(\frac{D}{\partial s} - \tfrac{1}{2}\text{Ricci}\right) X_{st}\partial t, \quad q_{s0} = 0, \\ d_s b_{st} &= v_{st}^{-1}(d_s x_{st} - q_{st}ds), \quad b_{0t} = 0. \end{aligned}$$

Since x_{s0} is a Brownian motion in M, it follows that b_{s0} is a Brownian motion in $\mathbb{R}^n$. Since our connection is torsion-free,

$$D_s\partial_t x_{st} = D_t\partial_s x_{st},$$

hence

$$\partial_t(\partial_s b_{st} \otimes \partial_s b_{st}) = v_{st}^{-1} D_t(\partial_s x_{st} \otimes \partial_s x_{st}) = 0$$

and so

$$\partial_s b_{st} \otimes \partial b_{st} = \partial_s b_{s0} \otimes \partial_s b_{s0} = \sum_{i=1}^{n} e_i \otimes e_i \partial s,$$

where e_i runs over the standard basis in $\mathbb{R}^n$. We recall the basic identity ([N2], (2.38))

$$D_t \partial_s x_{st} = D_t d_s x_{st} + \tfrac{1}{2} R(\partial_t x_{st}, \partial_s x_{st}) \partial_s x_{st},$$

where R denotes the curvature. But we have identified the quadratic variation in s of x_{st} as the trace, so

$$R(\partial_t x_{st}, \partial_s x_{st}) \partial_s x_{st} = \mathrm{Ricci}(\partial_t x_{st}) \partial s.$$

Hence

$$\begin{aligned} D_t d_s x_{st} &= D_t \partial_s x_{st} - \tfrac{1}{2} \mathrm{Ricci}(\partial_t x_{st}) \partial s \\ &= \left(\frac{D}{\partial s} - \tfrac{1}{2} \mathrm{Ricci} \right) X_{st} \partial s \partial t \\ &= D_t q_{st} \partial s, \end{aligned}$$

and

$$\partial_t d_s b_{st} = v_{st}^{-1} (D_t d_s x_{st} - D_t q_{st} ds) = 0.$$

Therefore $b_{st} = b_{s0}$ for all t, and $(x_{st} : s \geq 0)$ is a Brownian motion in M with drift q_{st}.

So far we have ignored what is happening to the starting point, but that is very simple. Previsibility forces $X_0(x)$ to be a function of the starting point x_0 alone, giving us a vector field on M, which we again denote X_0. Then x_{0t} obeys the autonomous equation

$$\partial_t x_{0t} = X_0(x_{0t}) \partial t.$$

If we assume that X_0 is C^1 say, then the law of x_{0t} is given by

$$0 = \frac{\partial}{\partial t} \mathbb{E}\left[f(x_{0t}) \exp\{ - \int_0^t \operatorname{div} X_0(x_{0\tau}) d\tau \} \right].$$

On the other hand, conditional on x_{0t}, the law of $(x_{st} : s \geq 0)$ is absolutely continuous with respect to $\mathbb{P}^{x_{0t}}$, at least on compact s-intervals, with density given by the Cameron–Martin formula. We have

$$\begin{aligned} &\frac{\partial}{\partial t} \left\{ \int_0^\infty \langle q_{st}, d_s x_{st} \rangle - \tfrac{1}{2} \int_0^\infty |q_{st}|^2 ds \right\} \\ &= \int_0^\infty \langle \frac{D}{\partial t} q_{st}, d_s x_{st} \rangle \\ &= \int_0^\infty \langle \left(\frac{D}{\partial s} - \tfrac{1}{2} \mathrm{Ricci} \right) X_{st}, d_s x_{st} \rangle. \end{aligned}$$

Hence the law of $x_t = (x_{st} : s \geq 0)$ is given by

$$0 = \frac{\partial}{\partial t}\mathbb{E}[F(x_t)\exp\{-\int_0^t \operatorname{div} X(x_\tau)d\tau\}]$$

for all bounded Borel functions F on Ω, where

$$\operatorname{div} X(x) = \operatorname{div} X_0(x_0) + \int_0^\infty \langle \left(\frac{D}{\partial s} - \tfrac{1}{2}\mathrm{Ricci}\right) X_s(x), dx_s\rangle.$$

This is our Cameron–Martin formula for the free path space. For suitably smooth functions F we can evaluate the derivative at $t = 0$ to recover the integration by parts formula of §2

$$\mathbb{E}\langle dF, X\rangle = \mathbb{E}[F \operatorname{div} X].$$

That concludes our rough argument.

The only serious gap in the above argument is the need to establish the existence of a flow for our vector field X, within the class of two-parameter semimartingales. Obviously, something in the nature of a Lipschitz condition looks desirable. But truly Lipschitz functions on Ω form an overly restricted class. We shall not attempt to find natural conditions on the vector field X, but restrict attention to the case already considered in §2. Let there be given C^2 vector fields $X_1, \ldots, X_m$ on M, together with an absolutely continuous function $h_s = (h_s^1, \ldots, h_s^m)$ in $\mathbb{R}^m$ satisfying

$$\int_0^\infty |\dot{h}_s|^2 ds < \infty. \tag{2}$$

Then for $\mathbb{P}$-almost all $x \in \Omega$ and all $s \geq 0$ we can define $X_s(x) \in T_{x_s}M$ by

$$X_s(x) = \tau_s \sum_{i=1}^m h_s^i X_i(x_0) \tag{3}$$

where τ_s denotes parallel translation $T_{x_0}M \to T_{x_s}M$ along x.

We state a special case of ([N2], Theorem 3.2.6) suited to our present needs.

Theorem 3.1. *Let M be a C^4 compact Riemannian manifold with Levi–Civita connection. Let*

$$\beta : OM \to TM \otimes T^*M$$

be a C^2 map of the fibres. Suppose we are given regular semimartingale boundary values $(x_{s0} : s \geq 0)$ and $(x_{0t} : t \geq 0)$ in M together with $u_{00} = v_{00} \in O_{x_{00}}M$. Then there exist unique two-parameter semimartingales x_{st} in M and u_{st}, v_{st} in OM over x_{st} such that $u_{s0} = v_{s0}$ and $u_{0t} = v_{0t}$, and satisfying

$$D_s\partial_t x_{st} = \beta(u_{st})\partial_s x_{st}\partial t,$$
$$D_s u_{st} = 0,$$
$$D_t v_{st} = 0.$$

We make some explanatory remarks. In this context regularity of the boundary values means uniformly Lipschitz quadratic variation and finite variation part. The auxilliary processes u_{st} and v_{st} are lifts of x_{st} in OM, which agree and are horizontal on the s and t-axes; then u_{st} is made horizontal along $(x_{st} : s \geq 0)$ for each $t \geq 0$, whereas v_{st} is horizontal along $(x_{st} : t \geq 0)$ for each $s \geq 0$. Parallel translation along $(x_{st} : s \geq 0)$ is then given by

$$\tau_{st} = u_{st} u_{0t}^{-1}.$$

The process v_{st} already appeared above in analysing the law of x_{st}.

In order to apply Theorem 3.1 to our present problem, we first integrate the C^2 vector field

$$X_0(x_0) = \sum_{i=1}^{m} h_0^i X_i(x_0)$$

which governs the autonomous motion of the base point

$$\dot{x}_{0t} = X_0(x_{0t}).$$

We denote by u_{0t} the horizontal lift along x_{0t} starting from u_{00}, and set

$$(k_t)_i = u_{0t}^{-1} X_i(x_{0t}),$$

k_t taking values in $(\mathbb{R}^m)^*$. The flow equation

$$\dot{x}_t = X(x_t) \tag{3}$$

is then equivalent to the system of two-parameter hyperbolic equations

$$D_s \partial_t x_{st} = u_{st} k_t (\partial h_s) \partial t,$$
$$D_s u_{st} = 0,$$
$$D_t v_{st} = 0.$$

In the case where h_s has bounded derivative and so is regular, we can now appeal to Theorem 3.1, applied to the augmented process $\tilde{x}_{st} = (x_{st}, h_{st}, k_{st})$ in $M \times \mathbb{R}^m \times (\mathbb{R}^m)^*$, with $h_{st} = h_s$ and $k_{st} = k_t$, satisfying

$$D_s \partial_t h_{st} = 0, \quad D_s \partial_t k_{st} = 0.$$

Hence (3) has a unique solution, whch is a two-parameter semimartingale. One can then pass to the case of general h_s by a time-change argument in s, as in ([N2], §4.2). Thus we obtain

Theorem 3.2. *Let $x_0 = (x_{s0} : s \geq 0)$ be a Brownian motion in M. Then there exists a unique two-parameter semimartingale $(x_{st} : s \geq 0, t \geq 0)$ satisfying*

$$\partial_t x_{st} = \tau_{st} \sum_{i=1}^{m} h_s^i X_i(x_0) \partial t.$$

The calculation of the law of x_{st} made above is now justified. The presence of the Ricci term in the drift means that (2) is not sufficient to make the law of $x_t = (x_{st} : s \geq 0)$ absolutely continuous with respect to $\mathbb{P}$, unless one restricts to compact s-intervals. The combination of (2) and

$$\int_0^\infty |h_s|^2 ds < \infty \tag{4}$$

is of course sufficient. We summarize our conclusions.

Theorem 3.3. *Let X be defined $\mathbb{P}$-almost everywhere on Ω by*

$$X_s(x) = \tau_s \sum_{i=1}^{m} h_s^i X_i(x_0)$$

where h_s satisfies (2) and (4) and $X_1, \dots, X_m$ are C^2 vector fields on M. There exists a unique two-parameter semimartingale $(x_{st} : s \geq 0, t \geq 0)$ such that the path-valued process $x_t = (x_{st} : s \geq 0)$ satisfies

(i) $x_0 = x$;
(ii) x_t *has law absolutely continuous with respect to* $\mathbb{P}$ *for all* $t \geq 0$;
(iii) $\dot{x}_t = X(x_t)$.

Moreover for every bounded Borel function F on Ω we have

$$0 = \frac{\partial}{\partial t}\mathbb{E}[F(x_t)\exp\{-\int_0^t \operatorname{div} X(x_\tau)d\tau\}]$$

where

$$\operatorname{div} X(x) = \operatorname{div} X_0(x_0) + \int_0^\infty \langle \left(\frac{D}{\partial s} - \tfrac{1}{2}\mathrm{Ricci}\right) X_s(x), dx_s \rangle.$$

Finally for every smooth cylinder function $F(x) = f(x_{s_1}, \dots, x_{s_k})$ we have the integration by parts formula

$$\mathbb{E}\langle dF, X\rangle = \mathbb{E}[F \operatorname{div} X]$$

where

$$\langle dF, X\rangle(x) = \sum_{j=1}^{k} \langle d_j f(x_{s_1}, \dots, x_{s_k}), X_{s_j}(x)\rangle.$$

References

[D1] B. K. Driver, *A Cameron–Martin type quasi-invariance theorem for Brownian motion on a compact Riemannian manifold*, J. Funct. Anal. **110** (1992), 272–377.

[D2] B. K. Driver, *A Cameron–Martin type quasi-invariance formula for pinned Brownian motion on a compact Riemannian manifold*, Preprint.

[EL] K. D. Elworthy and X. M. Li, *Formulae for the derivatives of heat semi-groups*, J. Funct. Anal. **125** (1994), 252–286.

[H] E. P. Hsu, *Quasi-invariance of the Wiener measure and integration by parts in the path space over a compact Riemannian manifold*, to appear, J. Funct. Anal..

[JL] J. D. S. Jones and R. Léandre, *L^p-Chen forms on loop spaces*, Stochastic Analysis, Eds. M. T. Barlow and N. H. Bingham, Cambridge University Press, 1991, pp. 103–163.

[L1] R. Léandre, *Integration by parts formulas and rotationally invariant Sobolev Calculus on free loop spaces*, J. Geometry and Physics II (1993), 517–528.

[L2] R. Léandre, *Invariant Sobolev Calculus on the free loop space*, Preprint.

[LR] R. Léandre and S. S. Roan, *A stochastic approach to the Euler–Poincaré number of the loop space of a developable orbifold*, to appear, J. Geometry and Physics.

[N1] J. R. Norris, *Path integral formulae for heat kernels and their derivatives*, Probab. Th. Rel. Fields **94** (1993), 525–541.

[N2] J. R. Norris, *Twisted sheets*, to appear, J. Funct. Anal. **132**.

The change of variables formula on Wiener space

A.S. Üstünel and M. Zakai

Abstract

The transformations of measure induced by a not-necessarily adapted perturbation of the identity is considered. Previous results are reviewed and recent results on absolute continuity and related Radon-Nikodym densities are derived under conditions which are 'as near as possible' to the conditions of Federer's area theorem in the finite dimensional case.

I. Introduction

Let $x \in \mathbb{R}^n$ and T a C^1 map from $\mathbb{R}^n$ to $\mathbb{R}^n$. The classical Jacobi formula yields

$$\int_{\mathbb{R}^n} \rho(x) g(Tx) |J(x)| dx = \int_{\mathbb{R}^n} g(x) \sum_{\theta \in T^{-1}\{x\}} \rho(\theta) dx \tag{1.1}$$

where J is the Jacobian determinant of T and ρ and g are bounded, positive and of compact support. Consider now the formulation of the same result with the Lebesgue measure replaced by the standard Gaussian measure on $\mathbb{R}^n$. Replacing, in (1.1), $g(x)$ with $(2\pi)^{-n/2} e^{-|x|^2/2} g(x)$ and setting

$$T(x) = x + f(x)$$

yields

$$E\left[\rho(x) g(Tx) \mid \Lambda(x)|\right] = E\left[g(x) \sum_{\theta \in T^{-1}\{x\}} \rho(\theta)\right] \tag{1.2}$$

where $E\psi(x) = \int_{\mathbb{R}^n} \psi(x)(2\pi)^{-n/2} e^{-|x|^2/2} dx$ and

$$\Lambda(x) = J(x) \cdot \exp - \sum_{i=1}^{n} f_i(x) \cdot x_i - \frac{1}{2} \sum_{1}^{n} f_i^2(x). \tag{1.3}$$

Equation (1.1) or (1.2) under the C^1 condition may be considered as "the first year calculus change of variables formula". The conditions under which (1.1) is valid have been considerably extended by Federer [9] by replacing the C^1 requirement on T with the requirement that T be Lipschitz and more generally by the condition

(A) $\mathbb{R}^n$ is the countable union of measurable sets such that the restriction of T to each set is Lipschitz and then (1.2) holds in the sense that if one of the sides of (1.2) is finite so is the other side and equality holds. Equation (1.1) or (1.2) under (A) may be referred to as the "Federer change of variables formula".

The extension of equation (1.2) to the infinite dimensional case was considered, first, by Cameron and Martin in 1949 [4] and, since then, by many authors. The purpose of this paper is twofold: to survey the results on this topic and to present an extension of (1.2) to the abstract Wiener space under conditions which are "as near as currently possible" to the Federer condition (A).

In the finite dimensional case the proof of Federer's formula is by starting with (1.1) under the C^1 condition and applying the following result of Federer which is based on Whitney's extension theorem:

Theorem 1.1. **([9])** *If* $\psi : \mathbb{R}^n \to \mathbb{R}^n$ *is Lipschitz then for any* $\epsilon > 0$ *there exists a function* $\psi_\epsilon : \mathbb{R}^n \to \mathbb{R}^n$ *which is* C^1 *on* $\mathbb{R}^n$ *and satisfies* $\mathrm{Leb}_n A_\epsilon \leq \epsilon$ *where* $A_\epsilon \subset \mathbb{R}^n$ *and*

$$A_\epsilon = \{x : \psi(x) \neq \psi_\epsilon(x)\} \bigcup \{x : \nabla\psi(x) \neq \nabla\psi_\epsilon(x)\}\,.$$

Recall the Rademacher theorem which states that if ψ is Lipschitz on $\mathbb{R}^n$ then $\nabla\psi$ exists for Leb_n a.a. x in $\mathbb{R}^n$ and $|\nabla\psi|$ is bounded by the Lipschitz constant of ψ. Note that each of these two theorems characterizes the Lipschitz property of ψ and could be used as a (very non-elegant) definition for this property.

Let (W, H, μ) be an abstract Wiener space. The notions of continuity, Lipschitz continuity and C^1 which turn out to be relevant in the problem of transformation of measure are as follows.

Definition 1.1. *Let* (W, H, μ) *be an abstract Wiener space,* $\mathcal{X}$ *a separable Hilbert space and* $F(w)$ *an* $\mathcal{X}$ *valued random variable, then*

(a) $F(w)$ *is* H*-continuous if for almost all* w*, the map* $h \mapsto F(w+h)$ *is continuous on* H*.*

(b) $F(w)$ *is* H*-Lipschitz continuous with Lipschitz constant* c *if, for a.a.* w

$$|F(w+h) - F(w)|_{\mathcal{X}} \leq c|h|_H$$

for all $h \in H$*.*

(c) $F(w)$ *is* $H - C^1$ *if for almost all* w*, the map* $h \mapsto F(w+h)$ *is continuously a Fréchet differentiable function of* $h \in H$*. A related notion of locally* $(H - C^1)$ *will be defined later.*

As will be pointed out in section III, the results obtained till now for the change of variables formula on abstract Wiener space can be considered as the infinite dimensional extension of the "first-year calculus change of variables formula". The problem of extending the Federer change of variables formula to the abstract Wiener case is delicate because an extension of theorem 1.1 to infinite dimensions (with the Lebesgue measure replaced by the Wiener measure) is not available. Several properties of Lipschitz functions have been extended to the abstract Wiener space (e.g., [8, 19]) but not theorem 1.1. In order to overcome (or bypass) this difficulty we will follow, here, the following path. Let us replace (A) by ($\tilde{\mathrm{A}}$):

($\tilde{\mathrm{A}}$) There exists a countable sequence of measurable sets B_k and C^1 functions $\psi_k(x)$ such that $\mu(\cup B_k) = 1$ and $T(x) = \psi_k(x)$ whenever $x \in B_k$.

Note that by theorem 1.1, condition ($\tilde{\mathrm{A}}$) is equivalent to (A). Instead of extending equation (1.2) under (A) to the abstract Wiener space we will extend (1.2) under ($\tilde{\mathrm{A}}$) to this space.

In the next section we will summarize some results of stochastic analysis which are needed in later sections. Previous results on the change of variables formula will be reviewed in section III. This will represent the path from 1945 till recent years and can be considered as the extension of the "first year calculus formula" to the infinite dimensional case. The extension of Federer's change of variables formula (under ($\tilde{\mathrm{A}}$)) to the infinite dimensional case will be formulated and proved in section IV and an example of a case where this result is applicable while previous results are not, will be given.

II. Preliminaries

Let (W, H, μ) be an abstract Wiener space. We start with a short summary of the notations of the Malliavin calculus. For $h \in H^* = H$, the Wiener integral $w(h)$ will also be denoted $\langle h, w\rangle$, $w \in W$. Let $\mathcal{X}$ be a real separable Hilbert space; smooth, $\mathcal{X}$-valued functionals on (W, H, μ) are functionals of the form

$$a(w) = \sum_1^N \eta_i(\langle h_1, w\rangle, \cdots, \langle h_m, w\rangle)\, x_i$$

with $x_i \in \mathcal{X}$ and $\eta_i \in C_b^\infty(\mathbb{R}^m)$, $h_i \in W^* \subset H$. For smooth $\mathcal{X}$-valued functionals, define

$$\nabla a(w) = \sum_{i=1}^N \sum_{j=1}^m \partial_j \eta_i\Big(\langle h_1, w\rangle, \cdots, \langle h_j, w\rangle\Big) \cdot x_i \otimes h_j\,,$$

and $\nabla^k, k = 2, 3, \cdots$ are defined recursively. For $p > 1$, $k \in \mathrm{N}$ the Sobolev space $\mathbb{D}_{p,k}(\mathcal{X})$ is the completion of $\mathcal{X}$-valued smooth functionals with respect to the norm

$$\| a \|_{p,k} = \sum_{i=0}^{k} \| \nabla^i a \|_{L^p(\mu, \mathcal{X} \otimes H^{\otimes i})} . \tag{2.1}$$

The gradient $\nabla : \mathbb{D}_{p,k}(\mathcal{X}) \to \mathbb{D}_{p,k-1}(\mathcal{X} \otimes H)$ denotes the closure of ∇ as defined for smooth functionals under the norm of (2.1). The gradient ∇a is considered as a mapping from H to $\mathcal{X}$ and $(\nabla a)^*$ will denote the adjoint of ∇a and is a mapping from $\mathcal{X}^*$ to H. The adjoint of ∇ under the Wiener measure μ is denoted by δ and called the divergence or the Skorohod integral or the Ito-Ramer integral (recall that it is defined by the "integration by parts formula" $E(G\delta u) = E\langle \nabla G, u\rangle_H$ for smooth real valued G and H-valued u). Also recall that if F is in $\mathbb{D}_{p,1}(H)$, for some $p > 1$, then for a.e. w, $\nabla F(w)$ is a Hilbert-Schmidt operator from H to H and for any smooth H-valued F and any complete orthonormal basis of H, say $\{e_i, i = 1, 2, \cdots\}$ we have

$$\delta F = \sum_{i=0}^{\infty} \langle F, e_i\rangle_H \langle e_i, w\rangle - \Big\langle \nabla(\langle F, e_i\rangle_H), e_i \Big\rangle_H , \tag{2.2}$$

and the Ogawa integral, if it exists, is given by

$$\delta \circ F = \sum_{i=1}^{\infty} \langle F, e_i\rangle_H \langle e_i, w\rangle . \tag{2.3}$$

An $\mathcal{X}$-valued random variable F is said to be in $\mathbb{D}^{\text{loc}}_{p,k}(\mathcal{X})$ if there exists a sequence (A_n, F_n) where A_n are measurable subsets of $W, \cup_n A_n = W$ almost surely, $F_n \in \mathbb{D}_{p,k}(\mathcal{X})$ and for every n, $F_n = F$ almost surely on A_n. It was shown in [10] that if $F(w)$ is H valued and $H - C^1$, then $F \in \mathbb{D}^{\text{loc}}_{\infty,1}(H)$.

Let K be a linear operator from H to H with discrete spectrum and let $\lambda_i, i = 1, 2, \cdots$ be the sequence of eigen-values of K repeated according to their multiplicity. The Carleman-Fredholm determinant of K is defined as:

$$\det{}_2(I + K) = \prod_{i=1}^{\infty} (1 + \lambda_i) e^{-\lambda_i} \tag{2.4}$$

and the product is known to converge for Hilbert-Schmidt operators. For $F \in \mathbb{D}^{\text{loc}}_{p,1}(H)$, ∇F is Hilbert-Schmidt and define

$$\Lambda_F(w) = \det{}_2(I + \nabla F) \exp(-\delta F - \frac{1}{2} \| F \|_H^2) . \tag{2.5}$$

The following lemma will be needed in section IV:

Lemma 2.1. *Let F_1, F_2, F_3 belong to $\mathbb{D}_{p,1}^{\text{loc}}(H)$ and let $T_i w = w + F_i(w)$, $i = 1, 2, 3$. Assume that: (i) $\mu \circ T_2^{-1} \ll \mu$ and (ii) $T_3 = T_1 \circ T_2$ (i.e. $F_3 = F_2 + F_1 \circ T_2$). Then*

(a) $I + \nabla F_3 = [I + (\nabla F_1)(T_2)](I + \nabla F_2)$

(b) $\Lambda_{F_3} = (\Lambda_{F_1} \circ T_2) \cdot \Lambda_{F_2}$.

The proof is straightforward (cf. lemma 6.1 of [10] or lemma 1.5 of [11]) and uses the fact that for $T(w) = w + u(w)$

$$(\delta F) \circ T = \delta(F \circ T) + \langle F \circ T, u \rangle_H + \text{Trace}\,((\nabla F) \circ T \cdot \nabla u\,.$$

Remark: Recall that for any measurable set A on W there exists a *σ-compact modification* of A, i.e. there exists a σ-compact set G such that $G \subset A$ and $\mu(G) = \mu(A)$.

Following Kusuoka [10] we associate with every measurable subset A of W the following random variable $\rho_A(w)$ which plays an important role in the construction of a class of mollifiers:

Definition 2.1. *Let A be a measurable subset of W, set*

$$\rho_A(w) = \inf_{h \in H} \{\| h \|_H :\ w + h \in A\} \tag{2.6}$$

and $\rho_A(w) = \infty$ if $w \notin A + H$.

Clearly, $\rho_A(w) = 0$ if $w \in A$, moreover [10], $\rho_A(w)$ is a measurable random variable and:

(i) If $A \subset B$ then $\rho_A(w) \geq \rho_B(w)$.

(ii) $|\rho_A(w) - \rho_A(w + h)| \leq \| h \|_H$.

(iii) $A_n \nearrow A$ implies $\rho_{A_n}(w) \searrow \rho_A(w)$.

(iv) If G is σ-compact and $\varphi \in C_0^\infty(\mathbb{R})$ (compact support) then $\varphi(\rho_G(w)) \in \mathbb{D}_{p,1}$ for all p and

$$\begin{aligned} \| \nabla \varphi(\rho_G(w)) \|_H &\leq \| \varphi' \|_\infty \cdot \mathbf{1}_{\{\varphi'(\rho_G) \neq 0\}} \\ &\leq \| \varphi' \|_\infty\,. \end{aligned} \tag{2.7}$$

(v) Let $Z = \{w :\ \rho_A(w) < \infty\}$. It is straightforward to see that, $A \subset Z$, and that, if $w \in Z$, then so does $w + h$, for any $h \in H$. Consequently, the distributional derivative (cf. e.g. [1] or [16]) $\nabla \mathbf{1}_Z = 0$, hence $\mathbf{1}_Z$ is almost surely a constant. Consequently $\mu(Z) = 1$ if $\mu(A) > 0$.

III. Review of previous results

In this section we present a short guided tour in the research on the change of variables formula along the "main" or "central" research path in the last 45 years, other directions will be mentioned very briefly later. Let μ be the classical Wiener measure on $C_0([0,1])$, and $f_{\cdot} = \int_0^{\cdot} f'_s ds$, with $\int_0^1 (f'_s)^2 ds < \infty$ denote the elements of the Cameron Martin space. For any $w \in C_0([0,1])$ set

$$\begin{aligned}(Tw)_{\cdot} &= w_{\cdot} + \int_0^{\cdot} f'_s(w)ds \\ &= w + f(w),\end{aligned} \tag{3.1}$$

where $f(w)$ is an H-valued measurable random variable and T is said to be the shift induced by f on the Wiener space . Let $T^*\mu$ denote the measure defined on $C_0([0,1])$

$$T^*\mu(A) = \mu(T^{-1}A). \tag{3.2}$$

Otherwise stated, for any bounded measurable function on $C_0[0,1]$, $E_\mu[g \circ T] = E_{T^*\mu}[g]$.

The first problem associated with the 'change of variables formula' is whether $T^*\mu$ is absolutely continuous with respect to μ (or equivalent to μ) and to calculate the associated Radon Nikodym derivative $L(w)$

$$\frac{dT^*\mu}{d\mu}(w) = L(w).$$

A related, but not equivalent, problem is the following: A measure ν is said to be a Girsanov measure associated with T if $T^*\nu = \mu$, i.e., $\nu(T^{-1}A) = \mu(A)$ or $E_\nu[g \circ T] = E_\mu[g]$; namely, Tw is Wiener under ν. If such ν exists and $\nu \ll \mu$, we will denote $(d\nu/d\mu)(w)$ by $|\Lambda(w)|$, where we denote the density as the absolute value of some random variable Λ since the random variable Λ itself plays an important role in the degree theory on the Wiener space (cf. [21]). Note that if T is (left) invertible

$$E_\mu[g(T^{-1}Tw)] = E_{(T^{-1})^*\mu}[g \circ T]$$

and

$$\nu = (T^{-1})^*\mu.$$

The case of T as defined by equation 3.1 where f is non-random was first considered by Cameron and Martin in 1944. This was followed in 1945 with a treatment of the case where f is linearly dependent on w and in 1949 with the case where f may depend non-linearly on w [4].

A short outline of the case where $f(w)$ is finite dimensional is as follows. Let $e_i, i = 1, \cdots$ be a complete orthonormal base on H ($e_i = \int_0^{\cdot} e_i'(s)ds, \int_0^1 e_i'(s)e_j'(s)ds = \delta_{ij}$) and $w(e_i) = \int_0^1 e_i'(s)dw_s$. Assume that

$$(Tw)_t = w_t + \sum_1^n \psi_i(w(e_1), \cdots, w(e_n))e_i(t)$$

As is well known the Wiener process $w(t)$ has the representation

$$w(t) = \sum_1^\infty \eta_i e_i(t) .$$

where $\eta_i = w(e_i)$. Therefore we have in this case

$$(Tw)_t = \sum_1^\infty w(e_i)e_i(t) + \sum_1^n \psi_i(w(e_1), \cdots w(e_n)) \cdot e_i(t) \tag{3.3}$$

and only the first n-coordinates participate in the transformation. Consequently, from equations (1.2) and (1.3) and assuming that T is bijective, it follows that

$$E\left[|\Lambda| \cdot G \circ T\right] = E[G] \tag{3.4}$$

where

$$\begin{aligned}\Lambda(w) &= \det\left(I_{\mathbb{R}^n} + \left(\frac{\partial \psi_i(w(e_1), \cdots, w(e_n))}{\partial x_j}\right)\right) \cdot \\ &\quad \cdot \exp\left(-\sum_1^n \psi_i(w(e_1), \cdots, w(e_n)) \cdot w(e_i) - \frac{1}{2}\sum_1^n \psi_i^2\right).\end{aligned} \tag{3.5}$$

Hence

$$E\left[1_A \cdot |\Lambda|\right] = \nu(A)$$

is the Girsanov measure. Note that the first sum in the exponent is an Ogawa integral (*cf.* equation (2.3)). Denoting

$$f(\cdot) = \int_0^{\cdot} f_s'(w)ds = \sum_i \psi_i(w(e_1), \cdots, w(e_n)) \cdot e_i(\cdot)$$

and denoting by λ_i the eigenvalues of the $(n \times n)$ matrix $\partial\psi_i/\partial x_j$, equation (3.5) can be rewritten as

$$\Lambda(w) = \prod_1^n (1 + \lambda_i) \exp -\delta \circ f - \frac{1}{2}\int_0^1 f_s^2(w)ds \tag{3.6}$$

$$= \det{}_1(I_H + \nabla f) \exp -\delta \circ f - \frac{1}{2}\int_0^1 f_s^2(w)ds \tag{3.7}$$

where $\det_1(I_H + \nabla f)$ is the Fredholm determinant of $(I_H + \nabla(\sum_1^n \psi_i(w(e_1), \cdots, w(e_n))e_i)$, i.e., the Fredholm determinant of $\left(I_H + \sum_{i,j=1}^n \frac{\partial \psi_i}{\partial x_j} e_j \otimes e_i\right)$.

Many papers were written in the period from 1949 till 1974 devoted to proving the validity of equation (3.4) with Λ as given by (3.7) for the infinite dimensional classical and abstract Wiener spaces. Two difficulties stood in the way of such an extension. The first being the fact that the Fredholm determinant of $(I_H + K)$ where K is a Hilbert-Schmidt operator on H, may not exist since $\det_1(1 + K) = \prod_1^\infty (1 + \lambda_i)$ where λ_i are the eigenvalues of K and the product may not converge or the convergence may depend on the order of λ_i. In order to assure the existence of the Fredholm determinant of $(I_H + K)$, K has to be of trace class and this is a strong restriction. The second serious difficulty is the Ogawa integral appearing in the exponent since this integral is not a closable operation and strong conditions are needed in order to assume its existence.

It was Ramer who pointed out in his 1974 paper [13] that equation (3.7) is the 'wrong' prototype. Following the 1965 paper of L. Shepp [14], dealing with the absolute continuity of Gaussian measures with respect to the Wiener measure, Ramer noticed that the right prototype for the change of variables formula induced by a bijective transformation is obtained by first rewriting (3.5), (3.6) as

$$\Lambda(w) = \left(\prod_1^n (1 + \lambda_i) e^{-\lambda_i}\right) \exp\left(-\sum_1^n \psi_i(w(e_1), \cdots, w(e_n)) w(e_i) + \sum_1^n \lambda_i - \frac{1}{2} \sum_1^n \psi_i^2\right). \tag{3.8}$$

Note that what is achieved by multiplying and dividing by $\exp - \sum \lambda_i$ is, (a) the Fredholm determinant becomes a Carleman-Fredholm determinant which exists for all Hilbert-Schmidt operators and (b) since $\sum \lambda_i$ is the trace of the $n \times n$ matrix $\partial \psi_i / \partial x_j$, the first two sums in the exponent of (3.8) can be written as a Skorohod or Ito-Ramer integral since by (2.2):

$$\Lambda(w) = \det_2(1 + \nabla f) \cdot \exp\left(-\sum \psi_i \cdot w(e_i) + \operatorname{trace} \partial \psi_i / \partial x_j - \frac{1}{2} \sum \psi_i^2\right) \tag{3.9}$$

hence it can be written in short as

$$\Lambda = \det_2(1 + \nabla f) \exp[-\delta f - \frac{1}{2} |f|_H^2] \tag{3.10}$$

where the elements $\det_1$ and $\delta \circ f$ are replaced by $\det_2$ and δf which exists under considerably weaker assumptions. Note that Ramer's paper appeared in 1974, Skorohod's paper introducing the Skorohod integral appeared in 1975, the Malliavin calculus made its appearance in 1975 but the fact that the Skorohod integral is the dual to the gradient was shown by Gaveau and Trauber in 1982. Ramer's paper showed very convincingly that the right the end of what may be called "the romantic period" and starts "the modern period" of research on this subject. The results of Ramer required some strong continuity assumptions, his work was considerably improved by Kusuoka [10]. The main result of Kusuoka is the following:

Theorem 3.1. **([10])** : *Let $F(w) \in H - C^1$. Further assume that $Tw = w + F(w)$ is bijective and $(I_H + \nabla F)$ is a.s. invertible, then*

$$E[g \circ T \cdot |\Lambda|] = E[g].$$

The results of [10] were generalized by Üstünel and Zakai [18] in two directions; first, the shifts $T = w + F(w)$ were not required to be invertible and the condition that $F(w)$ be $H - C^1$ was replaced by the following weaker condition.

Definition 3.1. *An H-valued random variable is said to be $(H - C^1)_{\text{loc}}$ if there exists a random variable $q(w) > 0$ a.s. such that the map $h \mapsto F(w + h)$ is continuously Fréchet differentiable for all $h \in H$ satisfying $|h| < q(w)$.*

It was shown in [18] that if F is $(H - C^1)_{\text{loc}}$ then $F \in \mathbb{D}^{\text{loc}}_{\infty,1}(H)$. The result of [18] is the following

Theorem 3.2. *Let $F : W \to H$ be an $H - C^1_{\text{loc}}$ map, $Tw = w + F(w)$. Let M denote the set*

$$M = \{w : det_2(I_H + \nabla F(w)) \neq 0\}$$

or, what is the same, M is the set on which $I_H + \nabla F$ is invertible. Then there exists a measurable partition of $(M_n;\ n = 1, 2, \cdots)$ of M and a sequence of shifts $(T_n; n = 1, 2, \cdots)$ with $T_n w = w + F_n(w)$, $F_n \in \mathbb{D}^{\text{loc}}_{p,1}$ for some $p > 1$ such that, for each n, $T_n = T$ almost surely on M_n and $T_n : W \to W$ is bijective. Moreover

$$E[g \circ T_n |\Lambda_n|] = E[g],$$

for any $g \in C_b(W)$. Consequently

(i) For almost all w, the cardinal of the set $T^{-1}\{w\} \cap M$, denoted by $N(w, M)$ is at most countably infinite.

(ii) For any $g, \rho \in C_b^+(W)$, we have

$$E[g \circ T|\Lambda|] = E[g\,.N(w, M)],$$

and

$$E[g(Tw)\rho(w)|\Lambda|] = E\left[g(w) \cdot \sum_{\theta \in T^{-1}\{w\} \cap M} \rho(\theta)\right].$$

(iii) $(\mu|_M)\circ T^{-1} = T^*(\mu|_M) \ll \mu$ *with*

$$\frac{dT^*(\mu|_M)}{d\mu}(w) = \sum_{\theta\in T^{-1}\{w\}\cap M} \frac{1}{|\Lambda_F(\theta)|}\,.$$

Theorem 3.2 was proved in [18] using the decomposition technique developed in [10] and the following result

Theorem 3.3. **([18])** *Let* $F: W \to H$ *be a measurable map belonging to* $\mathbb{D}_{p,1}(H)$ *for some* $p > 1$. *Assume that there exist constants* c, d *(with* $c > 1$*) such that for almost every* $w \in W$

$$\| \nabla F(w) \| \leq c < 1$$

and

$$\| \nabla F(w) \|_2 \leq d < \infty$$

where $\|\cdot\|$ *denotes the operator norm and* $\|\cdot\|_2 = \|\cdot\|_{H\otimes H}$ *denotes the Hilbert-Schmidt (or* $H\otimes H$*) norm (in other words, for almost all* $w \in W$, $\| F(w+h) - F(w) \|_H \leq c \| h \|_H$ *for all* $h \in H$ *where* c *is a constant,* $c < 1$ *and* $\nabla F \in L^\infty(\mu, H\otimes H)$*). Then:*

(a) Almost surely $w \mapsto T(w) = w + F(w)$ *is bijective.*

(b) the measures μ *and* $T^*\mu$ *are mutually absolutely continuous.*

(c) $E[f] = E[f\circ T \cdot |\Lambda_F|]$
for all bounded and measurable f *on* W *and in particular* $E[|\Lambda_F|] = 1$.

Theorem 3.3 extends previous results ([10, 2, 3]) and its proof is based on the result of [15] (cf. also [16]) that $\| \nabla F \|_2 \leq d$ implies that $E\exp\lambda|F|^2 < \infty$ for all $\lambda < 1/2d^2$.

A result similar to that of theorem 3.3 for the case where Tw is a monotone shift $((T(w+h) - T(w)), h)_H > 0$ a.s. has recently been derived in [20].

At the beginning of this section we referred to the problem of the change of variables formula for $Tw = w + F(w)$ as the 'central model'. Another direction of active research, considered here, is the following: Assume that $F(w)$ is parameterized by some parameter $\alpha \in [0,1]$, $T_\alpha w = w + F_\alpha(w)$, where $F_{\alpha=0}(w) = 0$ and T_α is considered as a flow ([5, 3, 17, 7]).

The reader must have noticed that we have not mentioned the celebrated Girsanov, or Cameron-Martin-Maruyama-Girsanov, theorem. This is not because of lack of respect to this result, it would be difficult to overestimate the importance of this theorem. The Girsanov theorem and its extension to martingale setup play a most important role in both the theory (e.g., weak solutions to stochastic differential equations and extending to semimartingales the results known for quasimartingales) and to applications (e.g. non linear filtering and stochastic control theory) of stochastic processes. The success of the Girsanov theorem was perhaps an important spur for the derivation of analogous results for the non-adapted case. The reason for not mentioning it here was that the techniques are quite different. For some considerations related to or motivated by Girsanov theorem and, in particular, for an explanation why the multiplicity $N(w,H)$ and $\det_2$ do not appear in the Girsanov theorem, *cf.* [12] and [22].

IV. The change of variables formula

Theorem 4.1. *Let $F \in \mathbb{D}^{\text{loc}}_{p,1}(H)$ for some $p > 1$. Suppose that there exists a sequence of measurable sets B_m such that $\mu(\cup B_m) = 1$, and a sequence of $(H - C^1)_{\text{loc}}$ random H valued functions F_m such that*

$$\mathbf{1}_{B_m}(w) \cdot (F(w) - F_m(w)) = 0 \quad a.s.$$

Let $M = \{w : \det_2(I_H + \nabla F) \neq 0\}$. Then:

(i) The cardinal of the set $T^{-1}\{w\} \cap M$, denoted $N(w,M)$ is, at most countably infinite.

(ii) For any positive measurable bounded real random variables ρ and g

$$E[\rho(Tw)g(w) \cdot |\Lambda_F|] = E\left[\rho(w) \sum_{\theta \in T^{-1}\{w\} \cap M} g(\theta)\right]$$

in the sense that if one side is finite, so is the other side and equality holds.

(iii) $T^*(\mu|_M) \ll \mu$ and

$$\frac{dT^*(\mu|_M)}{d\mu}(w) = \sum_{\theta \in T^{-1}\{w\} \cap M} \frac{1}{|\Lambda_F(\theta)|}$$

The proof of this theorem follows along the same lines as the proof of theorem 3.2 i.e. the decomposition technique developed in [10] and theorem 3.3, it will be given after the following example which presents a case which is covered by theorem 4.1 but is not covered by previous results.

Let r_n denote the rationals in $(0,1)$ arranged in some order and $\eta(x) = \exp -|x|$. Set

$$\theta(x) = \sum_{n=1}^{\infty} 2^{-n}\eta(x - r_n).$$

The function $\theta(x)$ is Lipschitz with Lipschitz constant 1 and is non differentiable on all the rationals x in $[0,1]$. Let $e_i, i = 1, 2, \cdots$ be a complete orthonormal base on H and set

$$F(w) = \sum_{i=1}^{\infty} 2^{-i}\theta(\delta e_i) \cdot e_i \tag{4.1}$$

then $F(w+h)$ is Lipschitz in h. Note first that even the case where $F(w) = \theta(\delta e_1) \cdot e_1$ is not covered by the results of the previous section, however it can be deduced from the finite dimensional Federer formula (1.1) or (1.2). Returning to $F(w)$ as defined by (4.1), let ϵ_n, $\epsilon > 0$ and $\Sigma \epsilon_n \leq \epsilon$. Let a_n denote a subset of $(-\infty, \infty)$ in which $\theta(\cdot)$ is C^1 and Lebesgue measure of a_n^c is bounded by ϵ_n. Therefore $F(w)$ is $H - C^1$ on $\{w : \delta e_i \in a_n, i \in \mathrm{N}\}$ and the conditions of theorem (4.1) are satisfied.

In order to prove theorem 4.1 we prepare the following:

Proposition 4.1. *Under the assumptions of the theorem, there exists a measurable partition $M_{m,n}$ $m,n = 1,2,\cdots$, of M and shifts $T_{m,n}w = w + F_{m,n}$, $F_n \in \mathbb{D}_{p,1}^{\mathrm{loc}}(H)$ for some $p > 0$ and such that for each m and n, $T_{m,n} = T$ on $M_{m,n}$ a.s. and the $T_{m,n} : W \to W$ are bijective. Moreover*

$$E\left[\rho(w) g(T_{m,n}w) \cdot |\Lambda_{m,n}(w)|\right] = E\left[g(w)\rho(T_{m,n}^{-1}w)\right] . \tag{4.2}$$

for all bounded and measurable ρ and g.

Proof of the proposition: (*cf.* p. 495 of [18]] for a heuristic outline of the proof):

Let $e_i, i = 1,2,\cdots$ be a complete orthonormal basis of H. Let $\lambda = \{\lambda_{i,j}, 1 \leq i,j \leq n\}$ be a real valued $n \times n$ matrix such that $I_{\mathbb{R}^n} + \lambda$ is invertible. Set $T_\lambda w = w + F_\lambda(w)$ where

$$F_\lambda(w) = \sum_{i,j=1}^{n} \lambda_{i,j} \cdot \delta e_j \cdot e_i$$

and note that $\nabla F_\lambda = \sum_{i,j} \lambda_{i,j} e_j \otimes e_i$ is deterministic, $T_\lambda^* \mu \sim \mu$ and T_λ is bijective.

Let $\gamma(\lambda)$ be the inverse of the operator norm of $(I_{\mathbb{R}^n} + \lambda)^{-1}$ and define

$$A(m,n,v,\lambda) = \left\{ w : w \in B_m, q_m(w) > \frac{4}{n}, \sup_{\|h\| < \lambda} \| F_m(w+h) - F_\lambda(w+h) - v \| \leq a\frac{\gamma(\lambda)}{n} \right.$$

$$\left. \text{and } \sup_{\|h\| < \frac{1}{n}} \| \nabla F_m(w+h) - \nabla F_\lambda \|_{H-S} \leq a\gamma(\lambda) \right\},$$

where a is a constant to be chosen below and q_m is a random variable which is the radius of the set of $H - C^1$-property of F_m. Let $G(m,n,v,\lambda)$ be a σ-compact modification of $A(m,n,v,\lambda) \cap M$. Let $\rho_A(w)$ be as defined by equation (2.6) and let $\varphi \in C_o^\infty(\mathbb{R})$, $|\varphi(x)| \leq 1$, $|\varphi(x)| = 1$ for $|x| < \frac{1}{3}$, $\varphi(x) = 0$, for $|x| > 2/3$ and $\| \varphi' \|_\infty \leq 4$. Assume, now, that v and the elements of λ are rational and λ is non-singular. The collection of such four-tuples (m,n,v,λ) is countable and $G(m,n,v,\lambda)$ will be denoted by $G_\nu, \nu = 1,2,\cdots$. Set $F_\nu(w) = v + \varphi(n\rho_{G_\nu}(w))[F(w) - F_\lambda(w) - v] + F_\lambda(w)$ and note that for $w \in G_\nu$, $\rho_{G_\nu}(w) = 0$

and $F_\nu(w) = F(w)$. On the other hand, setting

$$\begin{aligned} T_\nu(w) &= w + F_\nu(w), \\ T_\lambda(w) &= w + F_\lambda(w) \\ T_c(w) &= w + \varphi(n\rho_{G_\nu}(T_\lambda^{-1}w))[F(T_\lambda^{-1}w) - F_\lambda(T_\lambda^{-1}w) - v] \\ T_v(w) &= w + v \end{aligned}$$

it is easily verified that $T_v \circ T_c \circ T_\lambda = T_\nu$. Now, T_λ and T_v are bijective and from the definition of G_ν for $a < 1/3 \parallel \nabla F_c(w) \parallel_2 < 1$. Therefore by theorem 3.3, T_c is also bijective and consequently T_ν is bijective. Moreover, $T_\nu^*\mu \sim \mu$ since T_λ, T_v and T_c induce equivalent transformations of measure.

Now, let $T_i, i = 1, 2, 3$ be measurable transformation of W to W, $T_i^*\mu \ll \mu$ and

$$E\left[\eta_i(w) f(T_i(w))\right] = E[f]$$

then

$$\begin{aligned} E\left[g \circ T_3 \circ T_2 \circ T_1 \cdot \eta_3 \circ T_2 \circ T_1 \cdot \eta_2 \circ T_1 \cdot \eta_1\right] &= E\left[g(T_3 \circ T_2 w)\eta_3(T_2 w)\eta_2(w)\right] \\ &= E\left[g(T_3 w)\eta_3(w)\right] \\ &= E[g]. \end{aligned}$$

Therefore

$$Eg(w) = Eg(T_\nu w) \cdot \Lambda_{F_v}(T_c \circ T_\lambda w) \cdot \Lambda_{F_c}(T_\lambda(w))\Lambda_{F_\lambda}(w).$$

By a direct calculation using

$$\det{}_2(1 + a)(1 + b) = \det{}_2(I + a) \cdot \det{}_2(I + b) \cdot \exp - \text{trace } (ab)$$

and for $Tw = w + f(w)$

$$(\delta G(w)) \circ Tw = \delta G(Tw) - (G(Tw), f)_H - \text{trace } ((\nabla G \circ Tw) \cdot \nabla f),$$

we get that

$$\Lambda_{F_v}(T_c \circ T_\lambda w) \circ \Lambda_{F_c}(T_\lambda w) \cdot \Lambda_{F_\lambda}(w) = \Lambda_{F_\nu}(w)$$

(for details *cf.* lemma 6.1 of [10] or the lemma of [11]) and (4.2) follows. □

Proof of theorem 4.1: Since $\cup_{m,n} M_{m,n} = M$ we may assume without loss of generality that $M_{m,n}$ are disjoint and

$$\begin{aligned} T^{-1}\{w\} \cap M &= \{\theta \in M : T(\theta) = w\} \\ &= \bigcup_{m,n} \{\theta \in M_{m,n} : T_{m,n}\theta = w\}. \end{aligned}$$

Since the shifts are bijective, the cardinality of the above set is at most countably infinite. Now

$$
\begin{aligned}
E\left[\rho(w)g(Tw)\cdot|\Lambda_F|\right] &= \sum_{m,n} E\mathbf{1}_{M_{m,n}}(w)\rho(w)g(T_{m,n}w)\cdot|\Lambda_{m,n}(w)| \\
&= \sum_{m,n} E\mathbf{1}_{T_{m,n}M_{m,n}}(T_{m,n}w)\cdot\rho(w)g(T_{m,n}w)|\Lambda_{m,n}|\,.
\end{aligned}
$$

Applying equation (4.2) yields

$$
\begin{aligned}
E\left[\rho(w)g(Tw)|\Lambda_F|\right] &= \sum_{m,n} E\mathbf{1}_{T_{m,n}M_{m,n}}(w)\rho(T_{m,n}^{-1}w)g(w) \\
&= E\left[g(w)\sum_{\theta\in T^{-1}\{w\}\cap M}\rho(\theta)\right]
\end{aligned}
$$

which proves (ii) and (iii) follows by a similar argument (cf. [18]). □

References

[1] Bouleau, N., Hirsch, F.: Dirichlet Forms and Analysis on Wiener Space. De Gruyter Studies in Math., Vol. 14, Berlin, New York; De Gruyter 1991.

[2] Buckdahn, R.: Anticipative Girsanov transforms. Probab. Theory Relat. Fields **89**, 211–238 (1991).

[3] Buckdahn, R.: Anticipative Girsanov transformations and Skorohod stochastic differential equations. Seminarbericht No. 92–2 (1992).

[4] Cameron, R.H., Martin, W.T.: The transformation of Wiener integrals by nonlinear transformations. Trans. Amer. Math. Soc. **66**, 253–283 (1949).

[5] Cruzeiro, A.B.: Equations differentielles sur l'espace de Wiener et formules de Cameron-Martin nonlinéaires. J. Functional Anal., Vol. 52, 335–347, (1984).

[6] Enchev, O.: Nonlinear transformations on the Wiener space. The Annals of Probability, Vol. 21, 2169–2188 (1993).

[7] Enchev, O., D.W. Stroock: Anticipative diffusions and related change of measure. J. Functional Anal. **116**, 449–477 (1993).

[8] Enchev, O., D.W. Stroock: Rademacher's theorem for Wiener functionals. Annals of Probability **21** 25–33 (1993).

[9] Federer, H.: Geometric Measure Theory, Springer-Verlag, 1969.

[10] Kusuoka, S.: The nonlinear transformation of Gaussian measures on Banach space and its absolute continuity, I.J. Fac. Sci. Univ. Tokyo, Sect. IA, math. **29**, 567–598 (1982).

[11] Nualart, D.: Markov fields and transformations of the Wiener measure. The Proceedings of the Oslo-Silivri Conference on stochastic analysis and related topics, p. 45–88, Gordon and Breach Stochastic Monograph, Vol. 8, T. Lindstrom,B. Oksendal and A. S. Üstünel (editors).

[12] Nualart, D., Zakai, M.: Generalized stochastic integrals and the Malliavin calculus. Probab. Theory Relat. Fields **73**, 255–280 (1986).

[13] Ramer, R.: On nonlinear transformations of Gaussian measures. J. Funct. Anal. **15**, 166–187 (1974).

[14] Shepp, L.A.: Radon Nikodym derivatives of Gaussian measures. The Annals of Math. Stat. **37**, 321–354 (1966).

[15] Üstünel, A.S.: Integrabilité exponentielle de fonctionnelles de Wiener. CRAS, Paris, Série I **305**, 279-282 (1992).

[16] Üstünel, A.S.: An Introduction to Analysis on Wiener Space. Lecture Notes in Math. Vol. 1610, Springer, 1995.

[17] Üstünel, A.S., Zakai, M.: Transformations of Wiener measure under anticipative flows. Probab. Theory Relat. Fields **93**, 91–136 (1992).

[18] Üstünel, A.S., Zakai, M.: Transformation of the Wiener measure under non-invertible shifts, Prob. Theory Related Fields, **99**, 485–500, 1994.

[19] Üstünel, A.S., Zakai, M.: Extensions of Lipschitz functions on Wiener space. To appear.

[20] Üstünel, A.S., Zakai, M.: Measures induced on Wiener space by monotone shifts. To appear.

[21] Üstünel, A.S., Zakai, M.: Applications of the degree theorem to absolute continuity on Wiener space. Probab. Theory Relat. Fields, 95, 509-520, 1993.

[22] Zakai, M., Zeitouni, O.: When does the Ramer formula look like the Girsanov formula? The Annals of Probability, Vol., 20, pp. 1436–1440, 1992.

A.S. Üstünel, ENST, Dépt. Réseaux, 46 rue Barrault, 75013, Paris, France.

M. Zakai, Department of Electrical Engineering, Technion—Israel Institute of Technology, Haifa 32000, Israel.

Classification des Semi-Groupes de diffusion sur $I\!R$ associés à une famille de polynômes orthogonaux

Olivier Mazet
Laboratoire de Statistique et Probabilités
Université Paul Sabatier, Toulouse III
118 Route de Narbonne 31062 Toulouse Cedex

1 Introduction.

Un semi-groupe de Markov est le noyau de transition d'un processus de Markov ; un semi-groupe de diffusion est un semi-groupe de Markov particulier, dont il constitue dans certaines situations le cas limite.

Nous considérons ici les diffusions au sens de Bakry-Emery ([5]), dont la définition, un peu plus restrictive que celle usuellement utilisée dans les ouvrages de références (voir par exemple [11]), sera précisée au paragraphe suivant.

Le but de ce travail est d'étudier les semi-groupes opérant sur un intervalle de $I\!R$, qui ont la propriété de diffusion, dont la mesure réversible admette un moment exponentiel, et qui laissent stable pour tout n l'espace des polynômes de degré inférieur ou égal à n.

Les exemples connus de cette situation sont les semi-groupes d'Ornstein Uhlenbeck, Ultra-sphériques, de Laguerre, de Jacobi. Il existe une littérature abondante les étudiant en détail. On pourra se référer par exemple à [8], [6], [12], [22].

On montre, dans un premier temps, que essentiellement, ces semi-groupes sont les seuls. Dans une deuxième partie, on met en évidence, au moyen de transformations géométriques, les liens qui unissent ces semi-groupes. Fondamentalement, ces derniers peuvent se retrouver, pour certaines valeurs entières des paramètres, à partir du mouvement brownien sur les sphères, et passages à la limite sur la dimension des sphères.

Afin de se familiariser avec le cadre d'étude général, essentiellement algébrique, que l'on va se fixer, voici un exemple simple :

Exemple : Rappelons brièvement l'exemple bien connu du semi-groupe d'Ornstein Uhlenbeck (pour une présentation plus détaillée, voir par exemple [15]), défini de la façon suivante :

Pour une fonction f borélienne bornée

$$P_t f(x) = \int_{I\!R} f(e^{-t}x + \sqrt{1 - e^{-2t}}y)\,\mu(dy)$$

avec $\mu(dy) = \frac{e^{-\frac{x^2}{2}}}{\sqrt{2\pi}}\,dy$ mesure gaussienne.

Le générateur infinitésimal de $(P_t)_{t\geq 0}$ s'écrit

$$Lf(x) = \lim_{t\to 0} \frac{1}{t}(P_t f - f)(x) = f''(x) - xf'(x), \quad pour f \in C_b^2.$$

(Remarquons que l'on a évité le coefficient multiplicatif $\frac{1}{2}$ dans l'expres-sion de L, qui apparaît dans la définition classique du semi-groupe d'Orn-stein-Uhlenbeck, par une simple homothétie de rapport $\frac{1}{2}$ sur t.)

Si on considère la suite $(H_n)_{n\in \mathbb{N}}$ des polynômes d'Hermite, on voit que

$$\{H_n, n \in \mathbb{N}\} \text{ est totale dans } L^2(\mathbb{R}, d\mu) \tag{1.1}$$

$$\forall n \in \mathbb{N}, \quad LH_n = -nH_n \tag{1.2}$$

Comme on va le voir, L, (1.1) et (1.2) suffisent à caractériser le semi-groupe d'Ornstein-Uhlenbeck, (ou plus précisément la famille des semi-grou-pes d'Ornstein-Uhlenbeck, de paramètres quelconques, moyennant une hypothèse selon laquelle on travaille à affinité près).

En fait, la donnée du générateur infinitésimal et de son image sur une partie dense de son domaine permet de caractériser le semi-groupe associé. On va maintenant dresser, ci-dessous, une liste exhaustive des semi-groupes de diffusion sur un intervalle de $\mathbb{R}$, associés à des familles de polynômes orthogonaux, avec les hypothèses naturelles que l'on précise dans le paragraphe suivant.

2 Notations - Hypothèses.

Soit I intervalle de $\mathbb{R}$, $\mathcal{B}(I)$ tribu des boréliens sur I. On s'intéresse à l'espace mesuré $E = (I, \mathcal{B}(I), \mu)$, où μ est une mesure absolument continue par rapport à la mesure de Lebesgue, et vérifiant la propriété suivante :

μ admet un moment exponentiel, i.e. :

$$\exists \lambda > 0 \int_I e^{\lambda|x|}\mu(dx) < +\infty \tag{Pme}$$

μ est alors une mesure finie sur I, que l'on peut supposer ramenée à une mesure de probabilité par normalisation. On montre, au moyen de la transformée de Laplace, que (Pme) est une condition suffisante pour que l'ensemble des polynômes soit dense dans $L^2(E)$.

Il existe donc une suite $(Q_n)_{n\in \mathbb{N}}$ de polynômes orthonormés sur I, dense dans $L^2(E)$, telle que degré $(Q_n) = n$, unique au signe près : on suppose que le coefficient du plus haut degré est positif, ce qui caractérise entièrement (Q_n). En fait, cette suite (Q_n) est obtenue à partir de $(x^n)_{n\in \mathbb{N}}$ par le procédé d'orthogonalisation de Schmidt.

On s'intéresse aux semi-groupes de Markov, symétriques dans $L^2(E)$, qui admettent la suite (Q_n) comme décomposition spectrale. On définit un semi-groupe de Markov admettant μ comme mesure stationnaire par une famille $(P_t)_{t\geq 0}$ d'opérateurs agissant sur les fonctions boréliennes bornées, et vérifiant :

- $P_t P_s = P_{t+s}$; $\quad P_0 = Id$,
- $f \geq 0 \Rightarrow P_t f \geq 0$; $\quad P_t 1 = 1$,

- $\forall f \in L^1(E)$ $\int_I P_t f \, d\mu = \int_I f \, d\mu$ (hypothèse d'invariance).

On voit alors que (P_t) est une contraction de $L^p(E)$ dans lui-même, et ce $\forall p \in [1, +\infty[$. On exige alors de plus, que

- $\forall f \in L^2(E)$ $\lim_{t\to 0} P_t f = f$ dans $L^2(E)$,

- $\forall f, g \in L^2(E)$ $\int_I (P_t f) g \, d\mu = \int_I f(P_t g) \, d\mu$ (hypothèse de symétrie).

(En ce qui concerne les semi-groupes de diffusion sur I, sur lesquels on va porter notre intérêt, l'hypothèse de symétrie est en fait équivalente à l'hypothèse d'invariance, et ce grâce à une propriété spécifique de la dimension 1, où tout champ de vecteurs est un champ de gradient).

L'hypothèse de symétrie montre alors que le semi-groupe admet une décomposition spectrale :

$$P_t = \int_0^{+\infty} e^{-\lambda t} \, dE_\lambda,$$

où (E_λ) est une famille croissante de projecteurs orthogonaux. Ce que l'on demande ici, c'est que cette décomposition spectrale soit en fait une décomposition en vecteurs propres, et que ces vecteurs propres soient les polynômes Q_n :

$$\forall n \in \mathbb{N}, \quad P_t Q_n = e^{-\lambda_n t} Q_n,$$

d'où

$$f \in L^2(E) \Rightarrow f = \sum_{n=0}^{+\infty} f_n Q_n \Rightarrow P_t f = \sum_{n=0}^{+\infty} f_n e^{-\lambda_n t} Q_n.$$

On introduit alors le générateur infinitésimal de P_t dans $L^2(E)$:

Définition 1

$$Lf(x) = \lim_{t\to 0} \frac{P_t f(x) - f(x)}{t}$$

$\forall f \in \mathcal{D}_2(L)$ ensemble des fonctions de $L^2(E)$ telles que la limite existe.

On a alors

$$\forall n \quad LQ_n = -\lambda_n Q_n$$

et

$$\mathcal{D}_2(L) = \{f = \sum_{n=0}^{+\infty} f_n Q_n \in L^2(E), \quad \sum_{n=0}^{+\infty} \lambda_n^2 f_n^2 < +\infty\}.$$

L'étude des familles (Q_n) et des semi-groupes de Markov associés est un sujet assez difficile qu'il est impossible d'aborder ici. On renvoie à [16] et à [10], pour des caractérisations dans les cas simples des polynômes d'Hermite et de Jacobi, des suites (λ_n) pour lesquelles le semi-groupe associé est un semi-groupe de Markov.

On se restreint donc à l'étude des semi-groupes de diffusion :

Définition 2 *Pour tout couple de polynômes* (f, g) *on définit la quantité*

$$\Gamma(f,g) = \frac{1}{2}[L(fg) - fLg - gLf].$$

Cet opérateur bilinéaire est appelé opérateur carré du champ.

Il reste à définir la propriété de diffusion du semi-groupe, et enfin à poser le cadre d'étude de son générateur infinitésimal.

Dans ce cadre, on se donne une définition de la propriété de diffusion adaptée à la famille des polynômes ; c'est une définition "algébrique".

Définition 3 *On dit que* $(P_t)_{t\geq 0}$ *est un semi-groupe de diffusion si, pour* f *et* ϕ *polynômes, on a :*

$$L[\phi(f)] = \phi'(f)Lf + \phi''(f)\Gamma(f,f) \qquad (Pd).$$

Remarque 1 On n'aura besoin en fait de cette hypothèse que lorsque $f = x$.

En effet, si on écrit (Pd) en prenant en particulier $\phi = Q_n$ et $f = x$, on obtient

$$L(Q_n(x)) = Q_n'(x)Lx + Q_n''(x)\Gamma(x,x).$$

Par linéarité, on voit donc que, sur les polynômes, L s'écrit:

$$L = \Gamma(x,x)\frac{d^2}{dx^2} + L(x)\frac{d}{dx}.$$

◇

Si on traduit maintenant le fait que $\forall n \quad LQ_n = -\lambda_n Q_n$ sur Q_1 et Q_2, on obtient que

$$\Gamma(x,x) = Ax^2 + Bx + C, \quad L(x) = ax + b.$$

Finalement, L se met sous la forme

$$L = (Ax^2 + Bx + C)\frac{d^2}{dx^2} + (ax+b)\frac{d}{dx},$$

avec A, B, C, a, b, réels.

Tout le travail de classification va être basé sur l'étude de ce générateur infinitésimal, par la discussion des valeurs des paramètres A, B, C, a et b, et de la forme de l'intervalle I.

Le cadre de cette étude est posé par l'énoncé des hypothèses suivantes :

On travaille à affinité près pour la "variable d'espace", et à homothétie près pour la "variable de temps", c'est-à-dire que l'on se permet, pour simplifier l'étude, et l'on peut vérifier facilement qu'il n'y a aucune perte de généralité, de travailler "modulo" les opérations suivantes :

$$\begin{array}{lll} x \mapsto mx & x \in I,\ m \in \mathbb{R}^* & (H.1) \\ x \mapsto x + p & x \in I,\ p \in \mathbb{R} & (H.2) \\ L \mapsto \lambda L & \lambda \in \mathbb{R}^* & (H.3) \end{array}$$

L étant le générateur infinitésimal.

Remarque 2 D'une part, nous avons :

$$\forall x \in I \quad \Gamma(x,x) \geq 0 \Rightarrow Ax^2 + Bx + C \geq 0.$$

D'autre part, nous avons, en écrivant $\mu(dx) = a(x)\,dx$:

$$\begin{aligned}\int_I fLg\,d\mu &= \int_I (\Gamma(x,x)g''(x) + L(x)g'(x))f(x)a(x)\,dx \\ &= [\Gamma(x,x)g'(x)f(x)a(x)]_I - \int_I \Gamma(x,x)a(x)f'(x)g'(x)\,dx \\ &\quad + \int_I L(x)g'(x)f(x)a(x)\,dx - \int_I (\Gamma(x,x)a(x))'f(x)g'(x)\,dx.\end{aligned}$$

Le fait que L soit symétrique dans $L^2(E)$, montre que si l'on prend f et g à support compact dans I, on obtient que

$$(\Gamma(x,x)a(x))' = L(x)a(x),$$

d'où

$$a(x) = \exp \int_{x_0}^{x} \frac{L(u) - \Gamma'(u,u)}{\Gamma(u,u)}\,du.$$

Donc la symétrie de L sur les polynômes implique que le terme entre crochets de l'intégration par parties doit s'annuler aux bords de I, pour tout f et g. $a(x)$ ne s'annulant aux bords que si $\Gamma(x,x)$ fait de même, on obtient l'équivalence suivante :

$$x \in I \Leftrightarrow Ax^2 + Bx + C > 0 \qquad (Pp).$$

◇

3 Résultat Principal.

Nous allons maintenant démontrer la

Proposition : Les seuls semi-groupes de diffusion[1] satisfaisant toutes les hypothèses énoncées – donc en particulier à une affinité près, $((H.1)$ et $(H.2))$ – sont ceux associés aux générateurs suivants :

a) $L = \dfrac{d^2}{dx^2} - x\dfrac{d}{dx}$.

$I = \mathbb{R}$ muni de $\mu(dx) = \frac{1}{\sqrt{2\pi}} e^{-\frac{x^2}{2}}\,dx$ mesure gaussienne.

$(Q_n)_{n\in\mathbb{N}}$ est alors la suite des polynômes d'Hermite, définis par leur série génératrice :

$$\exp(tx - \frac{t^2}{2}) = \sum_n \frac{t^n}{n\,!} Q_n(x),$$

et on a $\forall n \in \mathbb{N}, \quad LQ_n = -nQ_n$.

Le semi-groupe associé est ici celui d'Ornstein-Uhlenbeck.

[1]Pour une étude détaillée de ces semi-groupes, et en particulier du cas b), on peut se référer, outre les ouvrages cités dans l'introduction, à [17] et [14].

b) $L = x\frac{d^2}{dx^2} + (\gamma + 1 - x)\frac{d}{dx}, \quad \gamma > -1.$
$I =]0, +\infty[$ muni de $\mu(dx) = K_\gamma e^{-x} x^\gamma \, dx$.
$(Q_n^\gamma)_{n\in \mathbb{N}}$ est alors la suite des polynômes de Laguerre, définis par leur série génératrice :

$$(1-t)^{-\gamma-1} \exp(-\frac{xt}{1-t}) = \sum_n t^n Q_n^\gamma(x),$$

et on a $\forall n \in \mathbb{N}, \quad LQ_n^\gamma = -nQ_n^\gamma$.

c) $L = (1-x^2)\frac{d^2}{dx^2} + (\beta - \gamma - (\beta + \gamma + 2)x)\frac{d}{dx}, \quad \beta > -1, \gamma > -1.$
$I =]-1, 1[$ muni de $\mu(dx) = K_{\beta,\gamma}(1-x)^\gamma(1+x)^\beta \, dx$.
$(Q_n^{\gamma,\beta})$ est alors la suite des polynômes de Jacobi, définis par leur série génératrice :

$$2^{\gamma+\beta}(1-2xt+t^2)^{-\frac{1}{2}}(1-t+(1+2xt+t^2)^{\frac{1}{2}})]{-\gamma}(1+t1(1-2xt+t^2)^{\frac{1}{2}})^{-\beta} =$$

$$\sum_n t^n Q_n^{\gamma,\beta}(x),$$

et on a $\forall n \in \mathbb{N}, \quad LQ_n^{\gamma,\beta} = -n(n+\gamma+\beta+1)Q_n^{\gamma,\beta}$.

Preuve.

1ère étape :

L étant le générateur infinitésimal d'un semi-groupe de diffusion sur $E = (I, \mathcal{B}(I), \mu)$, il possède la propriété suivante : les valeurs propres (λ_n) associées à la suite orthogonale de vecteurs propres $(Q_n)_{n\in \mathbb{N}}$ sont négatives. Or si l'on identifie les coefficients des termes du plus haut degré de l'équation :

$$LQ_n = \lambda_n Q_n \quad \forall n \in \mathbb{N},$$

on obtient que

$$\forall n \quad An(n-1) + an = \lambda_n,$$

d'où

$$\forall n \geq 1, \quad A(n-1) + a \leq 0 \Rightarrow A \leq 0,$$

donc

$$\text{si } A \neq 0, \text{ alors } \forall n \geq 1 \quad \frac{a}{A} \geq 1 - n \Rightarrow a \leq 0,$$

$$\text{si } A = 0, \text{ alors } a \leq 0.$$

On obtient donc

$$A \leq 0 \text{ et } a \leq 0. \tag{3.1}$$

2ème étape :

On distingue trois formes fondamentales que peut revêtir l'intervalle I :

- $I = \mathbb{R}$,
- $I =]x_0, +\infty[$ ou $I =]-\infty, x_0[$ avec $x_0 \in \mathbb{R}$,
- $I =]x_0, y_0[$, x_0, y_0 réels distincts.

On applique dans chaque cas, une partie ou la totalité des hypothèses $(H.1)$, $(H.2)$, $(H.3)$ et la propriété (Pp) :

a) $I = \mathbb{R}, \quad \Gamma(x,x) = Ax^2 + Bx + C.$

La propriété (Pp) entraîne que l'on a nécessairement $A = 0$ et $B = 0$. D'autre part, on applique $(H.3)$ pour se ramener à la valeur $C = 1$, d'où le cas se réduit à l'étude de

$$I = \mathbb{R}, \quad \Gamma(x,x) = 1.$$

b) $I =]x_0, +\infty[$ ou $I =]-\infty, x_0[, \quad \Gamma(x,x) = Ax^2 + Bx + C.$

On utilise $(H.1)$ et $(H.2)$ pour ramener I à la forme $I =]0, +\infty[$.
On remarque que l'on n'a utilisé $(H.1)$ qu'en partie, et que l'on peut encore transformer x en mx avec $m > 0$).
La propriété (Pp) nous montre alors que $A = 0$, et $C = 0$; enfin $(H.3)$ nous ramène à $B = 1$, d'où la cas se réduit à l'étude de

$$I =]0, +\infty[, \quad \Gamma(x,x) = x.$$

c) $I =]x_0, y_0[, \quad \Gamma(x,x) = Ax^2 + Bx + C.$

De la même manière, on obtient :
$(H.1)$ et $(H.2) \Rightarrow$ on se ramène à $I =]-1, 1[$,
$(Pp) \Rightarrow B = 0$ et $C = -A$,
$(H.3) \Rightarrow C = -A = 1$, d'où le cas se réduit à l'étude de

$$I =]-1, 1[, \quad \Gamma(x,x) = 1 - x^2.$$

3ème étape :

On introduit l'outil principal de calcul, qui consiste à effectuer sur I un changement de variables bijectif $y = \phi(x)$, de façon à ce que L se mette sous la forme classique

$$L = \frac{d^2}{dy^2} + \alpha(y)\frac{d}{dy}$$

sur $J = \phi(I)$. On doit donc poser

$$\phi'(x) = \frac{1}{\sqrt{\Gamma(x,x)}}.$$

On vérifie alors facilement que la mesure ν de densité $\exp(\int_K^y \alpha(t)\,dt)$ par rapport à la mesure de Lebesgue, satisfait à la propriété de symétrie pour l'opérateur L, c'est-à-dire que

$$\langle Lf, g\rangle_{L^2(\nu)} = \langle Lg, f\rangle_{L^2(\nu)}, \quad \forall f, g \in \mathcal{D}_2(L).$$

Le changement de variables permet de trouver la mesure réversible associée au semi-groupe. Il suffit donc de revenir à la forme première de L, et de retenir les valeurs des paramètres telles que la mesure image $\mu = \phi^{-1}(\nu)$ vérifie la propriété (Pme). Ce changement de variables a essentiellement été utilisé pour ramener l'opérateur L à sa forme fondamentale, et il est à remarquer que les mêmes calculs auraient pu être faits en gardant tel quel le coefficient de la dérivée d'ordre 2.

Enfin, il ne reste plus qu'à appliquer celles des hypothèses $(H.1), (H.2)$ ou $(H.3)$ qui n'auraient pas encore été utilisées, dans chaque cas, pour clore la classification :

a) $I = \mathbb{R}$

L se trouve déjà sous la forme désirée

$$L = \frac{d^2}{dx^2} + (ax+b)\frac{d}{dx}, \quad a \leq 0.$$

On obtient alors

$$\mu(dx) = K_1 e^{\frac{a}{2}x^2+bx}\,dx,$$

le fait que μ vérifie (Pme) entraîne que $a \neq 0$, et on utilise $(H.1)$ et $(H.2)$ pour poser $y = -(ax+b)$, et pour obtenir enfin

$$\mu(dy) = Ke^{-\frac{y^2}{2}}\,dy,$$

avec

$$L = \frac{d^2}{dy^2} - y\frac{d}{dy}.$$

b) $I =]0, +\infty[$.

$$L = x\frac{d^2}{dx^2} + (ax+b)\frac{d}{dx}, \quad a \leq 0 \;;$$

on pose $y = \sqrt{x}$, d'où

$$L = \frac{d^2}{dy^2} + (\frac{a}{2}y + \frac{2b-1}{y})\frac{d}{dy},$$

donc

$$\nu(dy) = K_1 e^{\frac{a}{4}y^2} y^{2b-1}\,dy,$$

d'où

$$\mu(dx) = Ke^{ax}x^{b-1}\,dx,$$

et $(Pme) \Rightarrow a \neq 0$ et $b > 0$.

On utilise enfin complètement $(H.1)$ pour se ramener à

$$\mu(dx) = Ke^{-x}x^{b-1},$$

ce qui est le résultat annoncé avec $\gamma = b - 1$.

c) $I =]-1, 1[$.

$$L = (1 - x^2)\frac{d^2}{dx^2} + (ax + b)\frac{d}{dx}, \quad a \leq 0.$$

On pose $y = \phi(x) = \arcsin(x)$, d'où[2]

$$L = \frac{d^2}{dy^2} + [(a+1)\tan(y) + \frac{b}{\cos(y)}]\frac{d}{dy},$$

d'où on obtient

$$\nu(dy) = K_1(1 + \tan^2(y))^{\frac{a+1}{2}}(\frac{\sin y + 1}{\cos y})^b\, dy,$$

donc

$$\mu(dx) = K(1 - x)^{-\frac{a}{2} - \frac{b}{2} - 1}(1 + x)^{-\frac{a}{2} + \frac{b}{2} - 1}\, dx,$$

ce qui est le résultat annoncé avec $\gamma = -1 - \frac{a}{2} - \frac{b}{2}$, et $\beta = -1 - \frac{a}{2} + \frac{b}{2}$, enfin $(Pme) \Rightarrow \gamma > -1$ et $\beta > -1$.

Remarque 3 Plus généralement, considérons E un espace de probabilité quelconque, L un opérateur de diffusion admettant comme décomposition spectrale une suite $(f_n)_{n \in \mathbb{N}}$ de fonctions satisfaisant les propriétés suivantes :

- $\forall n, p \in \mathbb{N}, \quad \exists \lambda_i$ réels, $\quad f_n f_p = \sum_{i=0}^{n+p} \lambda_i f_i,$
- f_1 bijection de E sur $f_1(E)$.

Par récurrence, on montre facilement que

$$\forall k \in \mathbb{N}, \quad \exists \mu_i \text{ réels}, \quad f_k = \sum_{i=0}^{k} \lambda_i f_1^i,$$

et après le changement de variables $y = f_1(x)$, on est ramené à notre cadre d'étude basé sur les polynômes de la variable réelle. ◇

4 Remarques sur l'interprétation géométrique de ces processus de diffusion particuliers.

Il est intéressant de remarquer que tous ces processus de diffusions peuvent se retrouver, pour certaines valeurs entières des paramètres, au moyen de diverses transformations géométriques, à partir du mouvement brownien sur les sphères de $\mathbb{R}^n$. On peut trouver certaines de ces remarques dans [12], que l'on généralise notamment aux semi-groupes de Jacobi dissymétriques.

[2] Il est intéressant de rapprocher la famille des processus relative à ce générateur infinitésimal, avec celle dégagée dans [1], voir la remarque 5 page 11.

On note :

- $S_r^{n-1} = \{x \in I\!R^n, \|x\| = r\}$
- $D_r^n = \{x \in I\!R^n, \|x\| \leq r\}$
- $\Phi_{n,p}^r$ projection de S_r^{n-1} sur un sous-espace vectoriel de dimension $p \leq n-1$ passant par l'origine :

$$\Phi_{n,p}^r(S_r^{n-1}) = D_r^p.$$

Le mouvement brownien sur S_1^{n-1} se projette par $\Phi_{n,p}^1$ sur un processus de Markov, sur D_1^p, de générateur infinitésimal qui s'écrit dans la base canonique :

$$L_{n,p} = \sum_{i=1}^{p}(\delta^{ij} - x^i x^j)\frac{\partial^2}{\partial x_i \partial x_j} - n\sum_{i=1}^{p} x_i \frac{\partial}{\partial x_i}.$$

On pose $\rho = \sqrt{\sum_{i=1}^p x_i^2}$ norme euclidienne de $I\!R^p$.

On considère les fonctions sur D_1^p qui ne dépendent que de ρ, et on obtient alors que

$$L_{n,p}^{(\rho)} f(\rho) = (1-\rho^2)f''(\rho) + \frac{p-1}{\rho}f'(\rho) - n\rho f'(\rho) \quad \text{avec} \quad \rho \in [0,1].$$

On effectue alors le changement de variables

$$\begin{array}{ccc} [0,1] & \to & [-1,1] \\ \rho & \mapsto & x = 2\rho^2 - 1, \end{array}$$

pour obtenir

$$L_{n,p}^{(x)} f(x) = 4[(1-x^2)f''(x) + (\frac{n+1-2p}{2} - \frac{n+1}{2}x)f'(x)].$$

L'opérateur

$$(1-x^2)\frac{d^2}{dx^2} + (\frac{n+1-2p}{2} - \frac{n+1}{2}x)\frac{d}{dx}$$

est l'opérateur de Jacobi de paramètres $\beta = \frac{n-p-1}{2}$ et $\gamma = \frac{p-2}{2}$.

On vient d'obtenir l'ensemble des opérateurs de Jacobi dissymétriques dont les paramètres sont des demi-entiers ; de plus on remarque que les conditions $\beta > -1$, $\gamma > -1$ se traduisent par $n > p-1$, $p > 0$ sur les dimensions des espaces de départ.

Par ailleurs, avant d'opérer le changement de variables $x = 2\rho^2 - 1$, le cas $p = 1$ nous fournit directement $L_{n,1}f(\rho) = (1-\rho^2)f''(\rho) - n\rho f'(\rho)$ qui est l'opérateur de Jacobi symétrique (ou opérateur ultrasphérique) de paramètre $n \in I\!N^*$. Puis, si l'on effectue alors le changement de variables $x = 2\rho^2 - 1$ sur cet opérateur $L_{n,1}$, on obtient une classe particulière des opérateurs de Jacobi dissymétriques qui sont ceux de paramètres $\beta = \frac{n}{2}, \gamma = -\frac{1}{2}$.

Remarque 4 Cet opérateur $L_{n,1}$, qui est la partie radiale du Laplacien sur la sphère, est un cas particulier d'un phénomène plus général : la partie radiale du Laplacien sur un espace symétrique compact de rang 1 est un opérateur de Jacobi. Ces espaces ont été classifiés dans [21] (voir aussi [9]), et outre les sphères, ils comprennent aussi les espaces projectifs réels, complexes et quaternioniques, ainsi que le plan elliptique de Cayley. ◇

Remarque 5 En faisant le changement de variables $y = \arcsin x$ décrit plus haut, l'opérateur $\frac{1}{4}L^{(x)}_{n,p}$ s'écrit

$$\frac{d^2}{dy^2} + \frac{a}{\cos y}\frac{d}{dy} + b\tan y\frac{d}{dy},$$

avec $a = \frac{n+1-2p}{2}$, $b = \frac{1-n}{2}$.
Dans [1], les auteurs font apparaître un opérateur similaire qui semble correspondre pour les cas où a et b sont des demi-entiers, à la même opération sur les parties radiales, où l'espace hyperbolique remplace la sphère. ◇

D'autre part, il est bien connu que le semi-groupe d'Ornstein-Uhlenbeck en dimension p est la limite quand n tend vers l'infini de l'image du semi-groupe du mouvement brownien sur $S^{n-1}_{\sqrt{n}}$ par $\Phi^{\sqrt{n}}_{n,p}$, avec $p < n$. On obtient le semi-groupe d'Ornstein-Uhlenbeck classique (i.e. unidimensionnel) pour $p = 1$. (On utilise ici l'outil communément appelé "lemme de Poincaré", qui serait plutôt dû à Mehler, voir [19], page 77).

De plus, si l'on considère les fonctions radiales sur $\mathbb{R}$, c'est-à-dire les fonctions $f : \mathbb{R} \to \mathbb{R}$ vérifiant

$$\exists g : \mathbb{R}^* \to \mathbb{R} \quad \forall x \in \mathbb{R} \quad g(|x|) = f(x),$$

et si l'on définit le semi-groupe (Q_t) sur $\mathbb{R}^*$ de la façon suivante :

$$Q_t g(|x|) = P_t f(x),$$

où (P_t) est le semi-groupe d'Ornstein-Uhlenbeck, alors (Q_t) est le semi-groupe de diffusion associé aux polynômes de Laguerre.

Les semi-groupes d'Ornstein-Uhlenbeck et de Laguerre peuvent aussi être retrouvés par leurs générateurs infinitésimaux à partir des opérateurs de Jacobi. En effet, en effectuant une homothétie de rapport $\sqrt{n}$ sur l'opérateur de Jacobi symétrique $L_{n,p}$, (ce qui équivaut à considérer l'image du Laplacien sur $S^{n-1}_{\sqrt{n}}$ par $\Phi^{\sqrt{n}}_{n,p}$), et en faisant tendre n vers l'infini, on obtient le générateur infinitésimal du semi-groupe d'Ornstein-Uhlenbeck.

De la même façon, en dimension 1, en considérant l'opérateur de Jacobi dissymétrique

$$(1-x^2)\frac{d^2}{dx^2} + \left(\frac{n+1-2p}{2} - \frac{n+1}{2}x\right)\frac{d}{dx},$$

en utilisant le changement de variable

$$x \mapsto n\frac{x-1}{2},$$

et en faisant tendre n vers l'infini, on obtient le générateur infinitésimal du semi-groupe de Laguerre.

Enfin, en ce qui concerne les valeurs propres des générateurs infinitésimaux sur la base des polynômes dans L^2, on remarque que la suite $\lambda_k = -k(k+n-1)$, énoncée au paragraphe 3, des valeurs propres de l'opérateur de Jacobi symétrique $L_{n,1}$ correspond exactement à la suite des valeurs propres de l'opérateur de Laplace-Beltrami sur S^n ; alors que l'opérateur de Jacobi dissymétrique $L^{(x)}_{n,1}$ admet la suite

$$\lambda'_k = 4\left(-k\left(k + \frac{n+1}{2} - 1\right)\right) = -2k(2k+n-1) = \lambda_{2k}$$

de valeurs propres, donc ne reprend qu'une valeur propre sur deux du même Laplacien sur S^n.

Cette remarque, vraie pour toutes les valeurs des paramètres, s'explique aisément pour les valeurs entières en remarquant que l'opérateur de Jacobi dissymétrique $L_{n,p}^{(\rho)}$ peut être considéré comme la restriction de l'opérateur de Jacobi symétrique $L_{n,p}$ aux fonctions radiales sur D_1^p, donc en l'occurrence, pour $p=1$, aux fonctions paires sur $[-1,1]$, donc la suite des vecteurs propres de $L_{n,1}^{(\rho)}$ est la suite des polynômes de Jacobi de degré pair, d'où le résultat sur les valeurs propres.

La remarque peut se généraliser en dimension supérieure : Soit $P_k(x)$ le k-ième polynôme de Jacobi sur $[-1,1]$; alors P_k est vecteur propre de l'opérateur $L_{n,1}$. On pose maintenant

$$\begin{array}{rccc} f_k : & D^p & \to & \mathbb{R} \\ & (x_1,\dots,x_p) & \mapsto & P_k(x_1). \end{array}$$

Alors clairement f_k est vecteur propre de l'opérateur $L_{n,p}$. De plus, le Laplacien sur S^n étant invariant par les rotations de $\mathbb{R}^n$, il en est de même pour sa projection $L_{n,p}$ sur D^p, les rotations de $\mathbb{R}^p$ pour $p \le n$ étant des rotations de $\mathbb{R}^n$ particulières. Donc si $R \in SO_p$ groupe des rotations de $\mathbb{R}^p$, $f_k \circ R$ reste vecteur propre de $L_{n,p}$.

On munit maintenant SO_p de la mesure de Haar dR, et on pose

$$\begin{array}{rccc} h_k : & D^p & \to & \mathbb{R} \\ & x & \mapsto & \int_{SO_p} f_k(Rx)\,dR\,; \end{array}$$

pour $x_0 \in D^p$, $h_k(x_0)$ est en fait la valeur moyenne de f_k sur $S^{p-1}_{\|x_0\|_p}$. h_k est donc une fonction radiale, i.e.

$$\exists h'_k : [0,1] \to \mathbb{R}, \quad h_k(x) = h'_k(\|x\|_p).$$

De plus, il est facile de vérifier que $h_k(x)$ reste vecteur propre de $L_{n,p}$, donc que $h'_k(\rho)$ et vecteur propre de $L_{n,p}^{(\rho)}$. On effectue enfin le changement de variables

$$\begin{array}{rccc} \phi : & [0,1] & \to & [-1,1] \\ & \rho & \mapsto & x = 2\rho^2 - 1, \end{array}$$

et on retrouve que $h'_k(x)$ est vecteur propre de $L_{n,p}^{(x)}$, donc est polynôme de Jacobi dissymétrique ; on peut constater en effet que les diverses modifications subies par $P_k(x)$ n'ont pas altéré sa forme polynômiale. Enfin, on retrouve que $L_{n,p}^{(x)}$ n'admet comme valeurs propres qu'une valeur propre sur deux de l'opérateur $L_{n,p}$ par le fait que dans un cas sur deux, $h_k(x) \equiv 0$.

Vérifions le sur un exemple simple : $p=2$

$$SO_2 = \left\{ \begin{pmatrix} \cos\theta & -\sin\theta \\ \sin\theta & \cos\theta \end{pmatrix}, \theta \in [0, 2\pi[\right\},$$

d'où

$$h_k(x,y) = \int_0^{2\pi} P_k(x\cos\theta + y\sin\theta)\,\frac{d\theta}{2\pi},$$

on pose alors $x = r\cos\varphi$, $y = r\sin\varphi$:

$$\begin{aligned} h_k(x,y) = h'_k(r) &= \int_0^{2\pi} P_k(r\cos\theta\cos\varphi - r\sin\theta\sin\varphi)\,\frac{d\theta}{2\pi} \\ &= \int_0^{2\pi} P_k(r\cos(\theta+\varphi))\,\frac{d\theta}{2\pi} \\ &= \int_0^{2\pi} P_k(r\cos\theta)\,\frac{d\theta}{2\pi}. \end{aligned}$$

Or k impair $\Rightarrow h'_k(r) = 0$ (car P_k est un polynôme impair).

En conclusion, tous les opérateurs de Jacobi dont les paramètres sont des demi-entiers, s'obtiennent par projection du mouvement brownien sur les sphères, (et même, en ce que concerne les opérateurs symétriques, de deux façons différentes). Les semi-groupes d'Ornstein-Uhlenbeck et de Laguerre s'obtiennent par des limites quand n et p tendent vers l'infini, des semi-groupes de Jacobi, moyennant un changement d'échelle et de temps.

References

[1] L. Alili, D. Dufresne, and M. Yor. Sur l'identité de Bougerol pour les fonctionnelles exponentielles du mouvement Brownien avec drift. A paraître, 1996.

[2] D. Bakry. La propriété de sous-harmonicité des diffusions dans les variétés. In *Séminaire de probabilité XXII, Lectures Notes in Mathematics*, volume 1321, pages 1–50. Springer-Verlag, 1988.

[3] D. Bakry. L'hypercontractivité et son utilisation en théorie des semi-groupes. In *Lectures on Probability Theory*, volume 1581. Springer-Verlag, 1994.

[4] D. Bakry. Remarques sur les semi-groupes de Jacobi. In *Hommage à P.A. Meyer et J. Neveu*, volume 236, pages 23–40. Astérisque, 1996.

[5] D. Bakry and M. Emery. Hypercontractivité de semi-groupes de diffusion. *C.R.Acad. Paris*, 299, Série I(15):775–778, 1984.

[6] S. Bochner. Sturm-Liouville and heat equations whose eigenfunctions are ultraspherical polynomials or associated Bessel functions. *Proc. Conf. Differential Equations*, pages 23–48, 1955.

[7] W. Feller. The parabolic differential equations and the associated semi-groups of transformations. *Ann. of Math.*, 55:468–519. 1952.

[8] W. Feller. Diffusion processes in one dimension. *Trans. Amer. Math. Soc.*, 77:1–31, 1954.

[9] R. Gangolli. Positive definite kernels on homogeneous spaces and certain stochastic processes related to Lévy's Brownian motion of several parameters. *Ann. Inst. Henri Poincaré*, III(2):9–226, 1967.

[10] G. Gasper. Banach algebras for Jacobi series and positivity of a kernel. *Ann. of Math.*, 2(95):261–280, 1972.

[11] K. Ito and H. P. McKean. *Diffusion processes and their sample paths*, volume 125. Springer-Verlag, 1965.

[12] S. Karlin and J. McGregor. Classical diffusion processes and total positivity. *Journal of mathematical analysis and applications*, 1:163–183, 1960.

[13] H. Koornwinder. Jacobi functions and analysis on nomcompact semisimple Lie groups. In R.A. Askey et al. (eds.), editor, *Special functions: group theoretical aspects and applications*, pages 1–85. 1984.

[14] A. Korzeniowski and D. Stroock. An example in the theory of hypercontractive semigroups. *Proc. A.M.S.*, 94:87–90, 1985.

[15] PA. Meyer. Note sur le processus d'Ornstein-Uhlenbeck. In *Séminaire de probabilités XVI*, volume 920, pages 95–133. Springer-Verlag, 1982.

[16] O.V. Sarmanov and Z.N. Bratoeva. Probabilistic properties of bilinear expansions of Hermite polynomials. *Teor. Verujatnost. i Primenen*, 12:470–481, 1967.

[17] T. Shiga and S. Watanabe. Bessel diffusions as a one-parameter family of diffusion processes. *Z. Wahrscheinlichkeitstheorie verw. Geb.*, 27:37–46, 1973.

[18] E.M. Stein and G. Weiss. *Introduction to Fourier Analysis on Euclidean Spaces.* Princeton University Press, 1971.

[19] D. Stroock. *Probability Theory: an analytic view.* Cambridge University Press, 1993.

[20] G. Szegö. *Orthogonal Polynomials.* American Mathematical Society, 4th edition, 1975.

[21] H.C. Wang. Two-point homogeneous spaces. *Annals of Mathematics*, 55:177–191, 1952.

[22] E. Wong. The construction of a class of stationary Markov processes. *Amer. Math. Soc., Proc. of the XVIth Symp. of App. Math.*, pages 264–276, 1964.

[23] K. Yosida. *Functional Analysis.* Springer-Verlag, 1968.

A DIFFERENTIABLE ISOMORPHISM BETWEEN WIENER SPACE AND PATH GROUP

Shizan FANG and Jacques FRANCHI

Abstract: Given a compact Lie group G endowed with its left invariant Cartan connection, we consider the path space $\mathcal{P}$ over G and its Wiener measure $\mathbb{P}$. It is known that there exists a differentiable measurable isomorphism I between the classical Wiener space (W,μ) and $(\mathcal{P},\mathbb{P})$. See [A], [D], [S2], [PU], [G].

In this article, using the pull-back by I we establish the De Rham-Hodge-Kodaira decomposition theorem on $(\Lambda(\mathcal{P}),\mathbb{P})$.

I. Introduction and Main Result

Ten years ago Shigekawa [S1] proved on an abstract Wiener space an infinite dimensional analog of the de Rham-Hodge-Kodaira theorem. The key point for that is to get an expression of the de Rham-Hodge-Kodaira operator dd^*+d^*d acting on n-forms in terms of the Ornstein-Uhlenbeck operator $\nabla^*\nabla$, expression that we may call Shigekawa identity. This expression in particular supplies spectral gap and de Rham-Hodge-Kodaira decomposition.

Our first aim in the present article was to extend the Shigekawa identity (and then the de Rham-Hodge-Kodaira theorem) to the path group over a compact Lie group.

To reach this end, we use the pull back I^* by the Itô map I. It is well known that I realizes a measurable isomorphism between Wiener space (W,μ) and path group $(\mathcal{P},\mathbb{P})$; now there is something more: having noticed the flatness of $\mathcal{P}$, we show that I^* indeed supplies a diffeomorphism between the differentiable structures of the exterior algebras $\Lambda(W)$ and $\Lambda(\mathcal{P})$.

Take the group $\mathcal{P}$ of continuous paths over a compact (or compact x $\mathbb{R}^N$) Lie group G, endow it with its Wiener measure $\mathbb{P}$ (induced by the Brownian motion on G), and consider its Cameron-Martin space $\mathbb{H}$ as its universal tangent space; the exterior algebra $\Lambda(\mathcal{P})$ is then the space of step functions from $\mathcal{P}$ into $\Lambda(\mathbb{H})$.

Following [A], [D], [S2], we introduce on $\Lambda(\mathcal{P})$ the Levi-Civita connection ∇, that we show to be flat.

We define in a classical way Hilbert-Schmidt norm $|\ |$, covariant derivative ∇, and coboundary d on $\Lambda(\mathcal{P})$.

Let I denote the Itô map from the classical Wiener space (W,μ) onto $(\mathcal{P},\mathbb{P})$. We consider the pull back by I : I^* pulls $\Lambda(\mathcal{P})$ towards $\Lambda(W)$, and we show that this I^* is in fact an isomorphism between these two differentiable (in the sense of Malliavin) structures.
More precisely, we get:

Theorem *We have for any* $\omega \in \Lambda(\mathcal{P})$ *and any* $z \in \mathbb{H}$, μ-a.s. :

a) $I^*(\nabla^{\mathcal{P}}_z \omega) = \nabla^{W}_{I^*z}(I^*\omega)$;

b) $|\nabla^{\mathcal{P}}\omega|\circ I = |\nabla^{W} I^*\omega|$;

c) $I^*(d\omega) = d(I^*\omega)$ *and* $I^*(d^*\omega) = d^*(I^*\omega)$.

This allows for example to transport the Shigekawa identity ([S1]) on $\Lambda(\mathcal{P})$:

Corollary *We have on* $\Lambda_n(\mathcal{P})$: $dd^*+d^*d = \nabla^*\nabla + n\ \mathrm{Id}$.

[FF2] gives a direct complete proof of Shigekawa's identity on $\Lambda(\mathcal{P})$, different from Shigekawa's proof (that is valid only on $\mathbb{R}^N$), and not using I*.

In the loop group case, the Levi-Civita connection is no longer flat, so there exists no differentiable isomorphism with the Wiener space. A direct approach is worked out in [FF3], in the same vein as in [FF2].

[L] and [LR] also deal with connections, de Rham-Hodge-Kodaira operator and Ornstein-Uhlenbeck operator, on path space and on free loop space, but over a compact manifold and with different preoccupations.

II. Notations, and Flatness of the Path Group

Let G be a compact Lie group, with unit e and Lie algebra $\mathcal{G} = T_eG$, endowed with an Ad-invariant inner product < , > and its Lie bracket [,] .

Let $\mathcal{P}$ be the group of continuous paths with values in G, defined on [0,1] and started from e ; let $\mathbb{H}$ be the corresponding Cameron-Martin space, that is to say:

$$\mathbb{H} = \{ h : [0,1] \to \mathcal{G} \mid \int_0^1 \langle\dot h(s),\dot h(s)\rangle\, ds < \infty \text{ and } h(0)=0 \} ;$$

we denote (,) the inner product of $\mathbb{H}$: $(h_1,h_2) = \int_0^1 \langle\dot h_1(s),\dot h_2(s)\rangle\, ds$, and we identify $h \in \mathbb{H}$ with $(h,.) \in \mathbb{H}^*$.

Let $W = \mathcal{C}_0([0,1],\mathcal{G})$ be the classical Wiener space, endowed with its Wiener measure μ. Denote I the one-to-one Itô application from W onto $\mathcal{P}$, defined by the following Stratonovitch stochastic differential equation :

$dI(w)(s) = \partial w(s)I(w)(s)$, for $w\in W$ and $s\in[0,1]$.

The Wiener measure $\mathbb{P}$ on $\mathcal{P}$ is the law of I under μ.

A functional $F \in L^{\infty-}(\mathcal{P},K)$, taking its values in some Hilbert space K, is said to be strongly differentiable when there exists DF belonging to $L^{\infty-}(\mathcal{P},K\otimes\mathbb{H})$ such that for all $h\in\mathbb{H}$ and $\gamma\in\mathcal{P}$:
the derivative $D_hF(\gamma)$ at 0 with respect to ε of $F(\gamma e^{\varepsilon h})$ exists in $L^{\infty-}(\mathcal{P},K)$ and equals $(DF(\gamma),h)$.

We denote $\mathcal{C}(\mathcal{P})$ the space of cylindrical functions on $\mathcal{P}$, that is to say of functions of the form $\gamma\mapsto f(\gamma(s_1),..,\gamma(s_m))$, m being a variable integer, f being C^∞ from G^m into $\mathbb{R}$, the s_j's being in [0,1] .
Note that a cylindrical function is strongly differentiable .

We extend the Lie bracket from $\mathcal{G}$ to $\mathbb{H}$ in setting for h and k in $\mathbb{H}$ and $s\in[0,1]$: $[h,k](s) = [h(s),k(s)] = h(s)k(s)-k(s)h(s)$.

Viewing $\mathbb{H}$ as the universal tangent space of $\mathcal{P}$, we define an affine connection ∇ on $\mathbb{H}$, following [A], [D], [S2] :

Definition 1 For y and z in $\mathbb{H}$, let $\nabla_z y$ be the unique element in $\mathbb{H}$ whose derivative $(\nabla_z y)^\cdot$ is $[z,\dot y]$.

The following proposition of [FF1] will not be used in the sequel, but explains why our theorem could be true.

Proposition 1 *∇ is the Levi-Civita connection on $\mathcal{P}$, and moreover it is flat; that is to say : for* h,k,y,z *in* $\mathbb{H}$ *, we have :*

a) $(\nabla_h y,z) = -(y,\nabla_h z)$ *ie ∇ preserves the metric ;*

b) $\nabla_h k - \nabla_k h = [h,k]$ *ie the torsion is null ;*

c) $[\nabla_h,\nabla_k] = \nabla_{[h,k]}$ *ie the curvature is null .*

Proof a) is due to the skew-symmetry of ad( ) in $\mathcal{G}$ with respect to < , > ;

$(\nabla_h k)^{\cdot} - (\nabla_k h)^{\cdot} = [h,\dot{k}]-[k,\dot{h}] = ([h,k])^{\cdot}$ shows b) ;

finally c) is due to the Jacobi identity:

$(\nabla_h\nabla_k z)^{\cdot} - (\nabla_k\nabla_h z)^{\cdot} - (\nabla_{[h,k]}z)^{\cdot} = [h,[k,\dot{z}]] + [k,[\dot{z},h]] + [\dot{z},[h,k]] = 0$. ■

III. Exterior Algebra $\Lambda(\mathcal{P})$

X will denote either W or $\mathcal{P}$, and for each $n\in\mathbb{N}$ $\Lambda_n = \Lambda_n(X)$ will denote the space of step n-forms on X, that is to say the vector space spanned by the elementary n-forms : $F\, h_1\wedge..\wedge h_n$, where $F\in\mathcal{C}(X)$ is cylindrical and $h_1,..,h_n$ are in $\mathbb{H}$.

The Malliavin derivative D_h defined in II above is indeed $D_h^{\mathcal{P}}$, whereas $D_h^W F(w)$ will be the derivative at $\varepsilon=0$ of $F(w+\varepsilon h)$.

We now extend $\nabla=\nabla^X$ to $\Lambda = \Lambda(X) := \sum_{n\in\mathbb{N}} \Lambda_n$, following Aida ([A]) :

Definition 2 For $\omega \in \Lambda_n$ and $z,h_1,..,h_n$ in $\mathbb{H}$, set :

a) $\partial_z\omega(h_1,..,h_n) = - \sum_{j=1}^{n} \omega(h_1,..,\nabla_z h_j,..,h_n)$;

b) $\nabla_z\omega = D_z\omega + \partial_z\omega$, where $D_z(F\, h_1\wedge..\wedge h_n) = (D_zF)\, h_1\wedge..\wedge h_n$.

Remarks 1 For $\omega \in \Lambda_n$, $\omega' \in \Lambda_m$, and $z\in\mathbb{H}$, we have :

a) $\nabla_z\omega \in \Lambda_n$;

b) $\nabla_z(\omega\wedge\omega') = (\nabla_z\omega)\wedge\omega' + \omega\wedge(\nabla_z\omega')$;

c) $\nabla_z(F\, h_1\wedge..\wedge h_n) = (D_zF)\, h_1\wedge..\wedge h_n + \sum_{j=1}^{n} F\, h_1\wedge..\wedge\nabla_z h_j\wedge..\wedge h_n$;

d) For X=W , we have of course $\nabla_z h_j = [z,\dot{h}_j]=0$, and hence $\nabla_z^W=D_z^W$.

Indeed, the verifications are straightforward from the definition; so ∇_z is determined by definition 1, remark (1,b) , and : $\nabla_z=D_z$ on Λ_0 .

We now introduce the gradient on Λ and the normalized Hilbert-Schmidt norms :

Definition 3 For $\omega \in \Lambda_n$, $\nabla\omega$ is the one element of $\Lambda_n\otimes\mathbb{H}$ defined by :

$(\nabla\omega(z_1,..,z_n) , h) = \nabla_h\omega(z_1,..,z_n)$, for all $h,z_1,..,z_n$ in $\mathbb{H}$.

<u>Definition 4</u> $\mathcal{B}$ being any Hilbertian basis of $\mathbb{H}$ and ω being in Λ_n :

$$|\omega|^2 = (n!)^{-1} \times \sum_{z_1,\dots,z_n \in \mathcal{B}} \omega(z_1,\dots,z_n)^2 \quad \text{and} \quad |\nabla\omega|^2 = \sum_{h\in\mathcal{B}} |\nabla_h \omega|^2 .$$

<u>Remark 2</u> This norm on Λ_n extends the norm of $\mathbb{H}$, and we have:

$$|F\, h_1\wedge..\wedge h_n|^2 = F^2 \sum_{\sigma\in\mathcal{S}_n} \varepsilon(\sigma) \prod_{j=1}^{n} (h_j, h_{\sigma_j}) = F^2\, h_1\wedge..\wedge h_n(h_1,..,h_n) .$$

We now classically skew-symmetrize the gradient to get the coboundary :

<u>Definition 5</u> For $\omega \in \Lambda_n$ and $z_0,..,z_n$ in $\mathbb{H}$, set :

$$d\omega(z_0,..,z_n) = \sum_{j=0}^{n} (-1)^j\, \nabla_{z_j} \omega(z_0,..,\hat{z}_j,..,z_n) ,$$

where $\hat{z}_j$ means that z_j is absent .

<u>Remark 3</u> Using proposition (1,b), we easily get :

$$d\omega(z_0,..,z_n) = \sum_{j=0}^{n} (-1)^j\, D_{z_j} \omega(z_0,..,\hat{z}_j,..,z_n) + \sum_{0\le i<j\le n} (-1)^{i+j}\, \omega([z_i,z_j],z_0,..,\hat{z}_i,..,\hat{z}_j,..,z_n) .$$

<u>Lemma 2</u> *For any $\omega \in \Lambda$ and any Hilbertian basis $\mathcal{B}$ of $\mathbb{H}$:* $d\omega = \sum_{h\in\mathcal{B}} h\wedge\nabla_h\omega$.

<u>Proof</u> Remarking that for $h,h_1,..,h_n,z_0,..,z_n$ in $\mathbb{H}$:

$$h\wedge h_1\wedge..\wedge h_n(z_0,..,z_n) = \sum_{j=0}^{n} (-1)^j\, h(z_j)\, h_1\wedge..\wedge h_n(z_0,..,\hat{z}_j,..,z_n) \text{ , we get :}$$

$$d\omega(z_0,..,z_n) = \sum_{h\in\mathcal{B}} \sum_{j=0}^{n} (-1)^j (h,z_j)\, \nabla_h\omega(z_0,..,\hat{z}_j,..,z_n) = \sum_{h\in\mathcal{B}} h\wedge\nabla_h\omega(z_0,..,z_n). \blacksquare$$

Let $\bar{\Lambda}^r_n$ be the completion of Λ_n with respect to the norm

$$\|\omega\|_r^2 = \mathbb{E}\left(\sum_{k=0}^{r} |\nabla^k\omega|^2 \right) \text{ , for } r\in\mathbb{N} \text{ , and set } \quad \bar{\Lambda}^r = \sum_{n\in\mathbb{N}} \bar{\Lambda}^r_n .$$

<u>Remark 4</u> ∇_h , ∇ , d clearly extend continuously to $\bar{\Lambda}^r$ for $r\in\mathbb{N}^*$.
D_h and ∇_h still make sense for h depending on w , for example $h \in \bar{\Lambda}^0_1$.

<u>Corollary 1</u> *For $\omega \in \bar{\Lambda}^r_n$ and $\omega' \in \bar{\Lambda}^r_m$, we have $d\omega \in \bar{\Lambda}^{r-1}_{n+1}$ and*
$d(\omega\wedge\omega') = (d\omega)\wedge\omega' + (-1)^n\, \omega\wedge(d\omega')$, *whence*

$$d(F\, h_1\wedge..\wedge h_n) = DF\wedge h_1\wedge..\wedge h_n - F \sum_{j=1}^{n} (-1)^j\, h_1\wedge..\wedge(dh_j)\wedge..\wedge h_n .$$

<u>Proof</u> $d(\omega\wedge\omega') = \sum_{h\in\mathcal{B}} h\wedge\nabla_h(\omega\wedge\omega') = \sum_{h\in\mathcal{B}} h\wedge(\nabla_h\omega)\wedge\omega' + \sum_{h\in\mathcal{B}} h\wedge\omega\wedge\nabla_h\omega'$

(by remark (1,b)) $\qquad = (d\omega)\wedge\omega' + (-1)^n\, \omega\wedge \sum_{h\in\mathcal{B}} h\wedge\nabla_h\omega'$. $\blacksquare$

IV. The isomorphism I^* between $\Lambda(\mathcal{P})$ and $\Lambda(W)$

Lemma 3 (Malliavin [M],[MM]) *For* $h\in\mathbb{H}$, *we have :*

$$D_h^W I(w)(t) = I(w)(t)\int_0^t \mathrm{Ad}\left(I(w)(s)^{-1}\right)\dot h(s)ds \quad \text{in } L^{\infty-}(W) .$$

Proof Set $I_\varepsilon(w)=I(w+\varepsilon h)$; we have :

$$d(I^{-1}I_\varepsilon) = -I^{-1}\partial I\, I^{-1}I_\varepsilon + I^{-1}\partial I_\varepsilon = -I^{-1}\partial w\, I_\varepsilon + I^{-1}(\partial w + \varepsilon dh)I_\varepsilon$$

$$= \varepsilon I^{-1}dh\, I_\varepsilon \;,\quad \text{whence } (I^{-1}D_h^W I)^\cdot = \mathrm{Ad}(I^{-1})\dot h \quad \text{by derivation at } \varepsilon=0.\blacksquare$$

We now introduce our pull back by I :

Definition 6

a) For $h\in\mathbb{H}$ and $w\in W$: $\tilde I h(w) = I(w)^{-1}D_h^W I(w)$, or : $(\tilde I h(w))^\cdot = \mathrm{Ad}(I(w)^{-1})\dot h$;

b) For $\omega\in\Lambda_n(\mathcal{P})$ and $h_1,..,h_n$ in $\mathbb{H}$: $(I^*\omega)(h_1,..,h_n) = (\omega\circ I)(\tilde I h_1,..,\tilde I h_n)$.

Remarks 5

a) $\tilde I h$ maps W into $\mathbb{H}$, and I^* maps $\Lambda_n(\mathcal{P})$ into $\bar\Lambda_n(W)$;
definition (6,b) agrees with the usual one in finite dimensions.

b) $(\tilde I h_1,\tilde I h_2) = (h_1,h_2)$ μ-a.s. for all h_1,h_2 in $\mathbb{H}$: $\tilde I$ is an isometry.

c) I^* is invertible from $\bar\Lambda_n(\mathcal{P})$ onto $\bar\Lambda_n(W)$.

d) $I^*h(k) = (h,\tilde I k) = (\tilde I^{-1}h,k)$ μ-a.s. for all h,k in $\mathbb{H}$, whence
$I^*h = \tilde I^{-1}h = \int_0^\cdot \mathrm{Ad}\left(I(.)\right)\dot h$, μ-a.s. for all h in $\mathbb{H}$.

e) $I^*(F\, h_1\wedge..\wedge h_n) = F\circ I\,(I^*h_1)\wedge..\wedge(I^*h_n)$, whence $I^*(\omega\wedge\omega') = (I^*\omega)\wedge(I^*\omega')$.

Lemma 4 a) $I^*(\nabla_z^{\mathcal{P}}h) = \nabla_{I^*z}^W(I^*h)$ μ-a.s. , *for all* z,h *in* $\mathbb{H}$;
b) $|I^*\omega| = |\omega|\circ I$ μ-a.s. , *for each* ω *in* $\Lambda(\mathcal{P})$.

Proof a) We use remarks (1,d),(5,d), definition 6 and lemma 3 to get :

$$(\nabla_z^W I^*h)^\cdot = D_z^W(\mathrm{Ad}(I)\dot h) = (D_z^W I)\dot h I^{-1} - I\dot h I^{-1}(D_z^W I)I^{-1}$$

$$= I(\tilde I z)\dot h I^{-1} - I\dot h(\tilde I z)I^{-1} = \mathrm{Ad}(I)[\tilde I z,\dot h] = \mathrm{Ad}(I)(\nabla_{\tilde I z}^{\mathcal{P}}h)^\cdot \;\mu\text{-a.s.}$$

whence $\nabla_{I^*z}^W I^*h = \int_0^\cdot \mathrm{Ad}(I)(\nabla_z^{\mathcal{P}}h)^\cdot = I^*(\nabla_z^{\mathcal{P}}h)$.

b) For $\omega=F\,h_1\wedge..\wedge h_n$, we have after remark 2 and remark (5,b,d) :

$$|I^*\omega|^2 = F^2\circ I \sum_{\sigma\in\mathcal{S}_n}\varepsilon(\sigma)\prod_{j=1}^n(I^*h_j,I^*h_{\sigma_j}) = F^2\circ I\sum_{\sigma\in\mathcal{S}_n}\varepsilon(\sigma)\prod_{j=1}^n(h_j,h_{\sigma_j})$$

$$= |\omega|^2\circ I \quad \mu\text{-a.s.}.\blacksquare$$

In finite dimensions the pull back of Levi-Civita connection by an isometry classically is still Levi-Civita connection. Lemma 4 in fact shows that we have the same situation in our infinite dimensional setting. The following proposition proves that this invariance property extends to n-forms.

Proposition 2 $I^*(\nabla^{\mathcal{P}}_z\omega) = \nabla^W_{I^*z}(I^*\omega)$ μ-a.s. , *for all* z *in* $\mathbb{H}$ *and* ω *in* $\Lambda(\mathcal{P})$.

Proof For $F(\gamma)=f(\gamma(s_1),..,\gamma(s_m))$ in $\mathcal{C}(\mathcal{P})$ and w in W, we have :

$$D_{I^*z}(F\circ I)(w) = \sum_{j=1}^{m}\partial_j f(I(w)(s_1),..,I(w)(s_m))(D_{I^*z}I(w)(s_j))$$
$$= \sum_{j=1}^{m}\partial_j f(I(w)(s_1),..,I(w)(s_m))(I(w)(s_j)z(s_j)) \quad \text{by remark (5,d) and lemma 3}$$
$$= (D_zF)\circ I(w) \ ;$$

then for $\omega = F\ h_1\wedge..\wedge h_n$ we have by remarks (1,c) and (5,e) and lemma 4 :

$$I^*(\nabla^{\mathcal{P}}_z\omega) = (D_zF)\circ I\ (I^*h_1)\wedge..\wedge(I^*h_n) + F\circ I\ \sum_{j=1}^{n}(I^*h_1)\wedge..\wedge I^*(\nabla^{\mathcal{P}}_z h_j)\wedge..\wedge(I^*h_n)$$
$$= D_{I^*z}(F\circ I)\ (I^*h_1)\wedge..\wedge(I^*h_n) + F\circ I\ \sum_{j=1}^{n}(I^*h_1)\wedge..\wedge\nabla^W_{I^*z}(I^*h_j)\wedge..\wedge(I^*h_n)$$
$$= \nabla^W_{I^*z}(F\circ I\ (I^*h_1)\wedge..\wedge(I^*h_n)) = \nabla^W_{I^*z}(I^*\omega) \ .\blacksquare$$

We can now precise in which sense I^* really is a differentiable isomorphism from $\Lambda(\mathcal{P})$ onto $\Lambda(W)$:

THEOREM *For each* ω *in* $\Lambda(\mathcal{P})$ *, we have* μ*-a.s. :*

a) $|\nabla^{\mathcal{P}}\omega|\circ I = |\nabla^W I^*\omega|$; b) $I^*d\omega = dI^*\omega$; c) $I^*d^*\omega = d^*I^*\omega$.

Proof We fix an Hilbertian basis $\mathcal{B}$ of $\mathbb{H}$, and use the fact that, after remark (5,b,d), $I^*\mathcal{B}$ is μ-a.s. an Hilbertian basis of $\mathbb{H}$ also .

a) $$|\nabla^{\mathcal{P}}\omega|^2\circ I = \sum_{z\in\mathcal{B}}|\nabla^{\mathcal{P}}_z\omega|^2\circ I = \sum_{z\in\mathcal{B}}|I^*\nabla^{\mathcal{P}}_z\omega|^2 \quad \text{by definition 4 and lemma (4,b)}$$
$$= \sum_{z\in\mathcal{B}}|\nabla^W_{I^*z}I^*\omega|^2 = |\nabla^W I^*\omega|^2 \quad \text{by proposition 2 and definition 4 ;}$$

b) $$I^*d\omega = I^*(\sum_{z\in\mathcal{B}} z\wedge\nabla^{\mathcal{P}}_z\omega) = \sum_{z\in\mathcal{B}}(I^*z)\wedge(I^*\nabla^{\mathcal{P}}_z\omega) \quad \text{by lemma 2 and remark (5,e)}$$
$$= \sum_{z\in\mathcal{B}}(I^*z)\wedge(\nabla^W_{I^*z}I^*\omega) = dI^*\omega \quad \text{by proposition 2 and lemma 2 ;}$$

c) $$\mathbb{E}((d\omega',\omega)) = \int_W(I^*d\omega',I^*\omega)\,d\mu = \int_W(dI^*\omega',I^*\omega)\,d\mu = \int_W(I^*\omega',d^*I^*\omega)\,d\mu$$
$$= \mathbb{E}((\omega',I^{*-1}d^*I^*\omega)) \quad \text{for any } \omega' \text{ in } \Lambda(\mathcal{P}) \ .\blacksquare$$

Corollary 2 $d^2 = 0 = d^{*2}$ on $\Lambda(\mathcal{P})$.

Remark that this is not immediate, since d and d^* are not local on $\Lambda(\mathcal{P})$.

Corollary 3 *The (Shigekawa) identity of* [S1] *:* $dd^* + d^*d = \nabla^*\nabla + n\,\mathrm{Id}$ *is valid on* $\Lambda_n(\mathcal{P})$ *, for any* n *in* $\mathbb{N}$.

Proof For any ω in $\Lambda_n(\mathcal{P})$, we have by lemma (4,b) and by the above theorem :

$$\mathbb{E}(|d^*\omega|^2) + \mathbb{E}(|d\omega|^2) - \mathbb{E}(|\nabla\omega|^2) - n\,\mathbb{E}(|\omega|^2) =$$
$$= \int_W\Big(|I^*d^*\omega|^2 + |I^*d\omega|^2 - |\nabla I^*\omega|^2 - n\,|I^*\omega|^2\Big)\,d\mu$$

$$= \int_W \left(|d^*I^*\omega|^2 + |dI^*\omega|^2 - |\nabla I^*\omega|^2 - n\,|I^*\omega|^2 \right) d\mu$$

$$= \int_W \left((dd^*+d^*d-\nabla^*\nabla-n\mathrm{Id})I^*\omega\,,\ I^*\omega \right) d\mu$$

$= 0$ by [S1] , whence the result by polarization.■

See [FF2] for another proof of this, not using [S1] nor I^* , very different from Shigekawa's proof and valid directly on $\Lambda(\mathcal{P})$.

Corollary 4 *The De Rham-Hodge-Kodaira operator on* $\Lambda(\mathcal{P})$: $\square = dd^* + d^*d$ *is hypoelliptic and selfadjoint on* $\bar{\Lambda}^2(\mathcal{P})$ *, with eigenvalues* $\geq n$ *on* $\bar{\Lambda}^2_n(\mathcal{P})$ *; moreover for any* $\omega\in\bar{\Lambda}^2(\mathcal{P})$: $\square\omega = 0 \Leftrightarrow d\omega = d^*\omega = 0 \Leftrightarrow \omega \in \Lambda_0(\mathcal{P})$ *is constant , and for any* n *in* $\mathbb{N}^*$ *we have on* $\bar{\Lambda}^0_n(\mathcal{P})$ *equivalence between closedness and exactness, and the De Rham decomposition :* $\bar{\Lambda}^0_n(\mathcal{P}) = \mathrm{Im}(d) \oplus \mathrm{Im}(d^*)$.

Remark 6 It is also possible to consider an other Itô application, defined by: $dJ(w) = J(w)\partial w$; the results are the sames, once the definitions of $D^{\mathcal{P}}$ and $\tilde{J}$ are modified as follows: $D^{\mathcal{P}}_h F(\gamma) = \frac{d}{d\varepsilon} F(e^{-\varepsilon h}\gamma)\Big|_{\varepsilon=0}$ and $(\tilde{J}h)^{\boldsymbol{\cdot}} = -\mathrm{Ad}(J)\dot{h}$.

REFERENCES

[A] AIDA S. *Sobolev spaces over loop groups.* J.F.A. 127, p. 155-172, 1995.

[AM] ARAI A. and MITOMA I. *De Rham-Hodge-Kodaira decomposition in infinite dimension.* Math. Ann. 291, p. 51-73, 1991.

[D] DRIVER B. *The non-equivalence of Dirichlet forms on path spaces.* Stochastic Analysis on Infinite Dimensional Spaces, p. 75-87, Proceedings Baton Rouge 1994, H. Kunita and H.H. Kuo ed.

[FF1] FANG S. and FRANCHI J. *Platitude de la structure riemannienne sur les groupes de chemins et identité d'énergie pour les intégrales stochastiques.* C.R.A.S. Paris, t. 321, S.1, p. 1371-1376, 1995.

[FF2] FANG S. and FRANCHI J. *Flatness of the path group over a compact Lie group and Shigekawa identity.* Prepublication n° 310 du laboratoire de probabilités de Paris VI, 1995.

[FF3] FANG S. and FRANCHI J. *De Rham-Hodge-Kodaira operator on loop groups.* Prepublication n° 341 du laboratoire de probabilités de Paris VI, 1996.

[G] GROSS L. *Uniqueness of ground states for Schrödinger operators over loop groups.* J.F.A. 112, p. 373-441, 1993.

[L] LEANDRE R. *Integration by parts formulas and rotationally invariant Sobolev calculus on free loop spaces.* J. Geom. Phys. 11, p. 517-528, 1993.

[LR] LEANDRE R. and ROAN S.S. *A stochastic approach to the Euler-Poincaré number of the loop space over a developpable orbifold.* J. Geom. Phys. 16, p. 71-98, 1995.

[M] MALLIAVIN P. *Hypoellipticity in infinite dimension.* Diffusion processes and related problems in Analysis, vol. 1, Progress in Probability 32, p. 17-33, M. Pinsky ed., Birkhäuser 1991.

[MM] MALLIAVIN P. and M.P. *Integration on loop groups I : Quasi invariant measures.* J.F.A. 93, p. 207-237, 1990.

[PU] PONTIER M. and USTUNEL A.S. *Analyse stochastique sur l'espace de Lie-Wiener.* C.R.A.S. 313, p. 313-316, 1991.

[S1] SHIGEKAWA I. *De Rham-Hodge-Kodaira's decomposition on an abstract Wiener space.* J. Math. Kyoto Univ. 26-2, p. 191-202, 1986.

[S2] SHIGEKAWA I. *A quasi homeomorphism on the Wiener space.* Proceedings of symposia in pure mathematics 57, Stochastic Analysis, p. 473-486, M. Cranston and M. Pinsky ed., 1995.

S. FANG Institut de Mathématiques, boîte 172, tour 46-0
Université Paris VI
4, place Jussieu, 75232 Paris cedex 05, France.

J. FRANCHI Laboratoire de probabilités, tour 56, 3° étage
Université Paris VI
4, place Jussieu, 75232 Paris cedex 05, France;
et Université Paris XII, département de mathématiques,
61, Avenue du général De Gaulle, 94010 Créteil cedex.

On martingales which are finite sums of independent random variables with time dependent coefficients

Jean Jacod
and
Víctor Pérez-Abreu

1 Introduction

We consider the following problem: for a positive integer $n \geq 1$, let $U_1, ..., U_n$ be n independent, integrable, centered, non-degenerate random variables. We are looking for conditions on a family of n càdlàg functions $f_1, ..., f_n$ on $\mathbb{R}_+$ with $f_i(0) = 0$, under which the following process:

$$X_t = \sum_{i=1}^{n} f_i(t) U_i \tag{1}$$

is a martingale, with respect to its own filtration $(\mathcal{F}_t)_{t \geq 0}$.

This (apparently) simple problem has a general solution given in Section 1. However, the answer is not quite satisfactory, since for example it does not allow to recognize whether there is a unique (up to the obvious multiplication by constants and time-changes) set (f_i) meeting our condition.

To get more insight, we specialize in Section 3 to the case where $n = 2$ and (for the most interesting results) with U_1 and U_2 having the same law. In this very particular situation we are able to give a complete description of all martingales of the form (1). This description emphasizes the particular role played by the stable distributions.

For the case $n \geq 3$, we have been unable to provide any interesting result of the same kind as for $n = 2$.

2 A general result

Here is a general theorem solving (in principle) our problem.

Theorem 1. *The process X is a martingale if and only if it satisfies the following:*

Condition [M]: *There are an integer p, $0 \leq p \leq n$, and deterministic times $0 = T_0 < T_1 < ... < T_p < T_{p+1} = \infty$, and p linearly independent vectors $a_j = (a_j^i)_{1 \leq i \leq n}$ in $\mathbb{R}^n$ (when $p \geq 1$), such that, with $V_0 = 0$ and $V_j = \sum_{1 \leq i \leq n} a_j^i U_i$ for $j \geq 1$,*

(M1) $(V_j)_{0\le j\le p}$ *is a discrete-time martingale;*

(M2) $X_t = \sum_{1\le j\le p} V_j 1_{[T_j,T_{j+1})}(t)$.

Before proving this theorem, we state some remarks on the conditions. First, Condition (M2) implies that $f_i(t) = \sum_{1\le j\le p} a^i_j 1_{[T_j,T_{j+1})}(t)$, because of the following property:

$$\alpha_i, \beta_i \in I\!R, \quad \sum_{i=1}^n \alpha_i U_i = \sum_{i=1}^n \beta_i U_i \quad a.s. \quad \Rightarrow \quad \alpha_i = \beta_i \quad \forall i. \tag{2}$$

Second, Condition (M1) is obviously difficult to verify, except when $p = 0$ (it is void) and $p = 1$ (it is obvious because V_1 is centered). Below we give an equivalent condition based on the characteristic functions φ_i of U_i. We recall that each function φ_i is C^1 with $\varphi_i'(0) = 0$. Then, when $p \ge 2$, (M1) is equivalent to the following:

Condition (M'1). For all $1 \le l \le p-1$ and all v_j in $I\!R$,

$$\sum_{i=1}^n (a^i_{l+1} - a^i_l)\varphi_i'(\sum_{j=1}^l a^i_j v_j) \prod_{k\ne i} \varphi_k(\sum_{j=1}^l a^k_j v_j) = 0. \tag{3}$$

We observe that (3) is the same as $E((V_{l+1} - V_l)\exp i \sum_{j=1}^l v_j V_j) = 0$. When the φ_i's do not vanish (so $\varphi_i = \exp \psi_i$ with ψ_i of class C^1 and $\psi_i'(0) = 0$) this condition is also equivalent to:

Condition (M"1). For all $1 \le l \le p-1$ and all v_j in $I\!R$,

$$\sum_{i=1}^n (a^i_{l+1} - a^i_l)\psi_i'(\sum_{j=1}^l a^i_j v_j) = 0. \tag{4}$$

Proof. The sufficient condition is obvious. For the necessary condition, we suppose that X is a martingale and let $F(t)$ be the vector with components $(f_i(t))_{1\le i\le n}$. Denote by E_t the linear space spanned by $(F(s) : s \le t)$, let $d_t = \dim(E_t)$, $T_{-1} = -1$, $T_j = \inf(t : d_t \ge j)$ for $0 \le j \le n$, and $T_{n+1} = \infty$. Thus $T_{-1} < 0 = T_0 \le T_1 \le ... \le T_p < T_{p+1} = \infty$ for some $0 \le p \le n$, and $d_0 = 0$.

Let $0 \le i \le p$ with $T_i < T_{i+1}$ and consider s, t such that $T_i < s < t < T_{i+1}$. Then $E_t = E_s$ is spanned by the linearly independent vectors $F(s_1), ..., F(s_i)$ with $s_j \le s$ (if $i = 0$, then $E_t = E_s = \{0\}$). Therefore, X_s and X_t are $\sigma(X_{s_1}, ..., X_{s_i})$-measurable and thus $\mathcal{F}_s = \mathcal{F}_t = \sigma(X_{s_1}, ..., X_{s_i})$ (which is the trivial σ-field when $i = 0$). The martingale property $E(X_t|\mathcal{F}_s) = X_s$ yields $X_t = X_s$ a.s., and (2) gives $F(s) = F(t)$. It follows that $F(.)$ is constant on (T_i, T_{i+1}) as well as on $[T_i, T_{i+1})$ by right-continuity. Thus

$$T_i < T_{i+1} \quad \Rightarrow \quad d_r = i \quad \forall r \in [T_i, T_{i+1}). \tag{5}$$

In fact $0 < T_1 < ... < T_p$; otherwise we would be in one of the following two situations:

a) $0 = T_j < T_{j+1}$ for some $1 \leq j \leq p$, and therefore $d_{T_j} = d_0 = 0$, which contradicts (5);

b) $T_{i-1} < T_i = T_j < T_{j+1}$ for i, j with $1 \leq i < j \leq p$, in which case $d_r = i - 1$ on $[T_{i-1}, T_j)$ by (3). This implies that $d_{T_j} \leq i$; being also impossible since $d_{T_j} \geq j$.

Since $0 < T_1 < ... < T_p$ holds, we trivially have (M2) with $a_j = F(T_j)$. Finally, (M2) and the martingale property of X yield (M1). □

3 The case n=2

Let φ_i be the characteristic function of U_i, and when φ_i never vanishes we use the notation $\varphi_i = \exp \psi_i$ without further comment. In this section we always assume that $n = 2$.

Theorem 2. *The process X is a martingale if and only if it has one of the following two (mutually exclusive) representations:*

a) For some $\alpha, \beta \in I\!R$, $S_1, S_2 \in (0, \infty]$

$$X_t = \alpha U_1 1_{[S_1,\infty)}(t) + \beta U_2 1_{[S_2,\infty)}(t). \tag{6}$$

b) For some $0 < T_1 < T_2 < \infty$, $\alpha, \alpha', \gamma, \gamma' \in I\!R^*$ *with* $\gamma \neq \gamma'$ *and*

$$\varphi_1'(v)\varphi_2(\gamma v) + \gamma'\varphi_1(v)\varphi_2'(\gamma v) = 0 \quad \forall v \in I\!R, \tag{7}$$

$$X_t = \alpha(U_1 + \gamma U_2)1_{[T_1,\infty)}(t) + \alpha'(U_1 + \gamma' U_2)1_{[T_2,\infty)}(t). \tag{8}$$

Remark. Since the coefficients in (8) do not vanish, the form (8) is indeed symmetric in (U_1, U_2). When φ_1 and φ_2 do not vanish, (7) is equivalent to $\psi_1'(v) + \gamma'\psi_2'(\gamma v) = 0$, which is the same as $\psi_1(v) + \frac{\gamma'}{\gamma}\psi_2(\gamma v) = 0$, which in turn is equivalent to

$$\varphi_1(v) = \varphi_2(\gamma v)^{-\gamma'/\gamma} \quad \forall v \in I\!R. \tag{9}$$

Proof. Sufficient condition: That (a) gives a martingale is obvious. Condition (b) implies (M2) with $a_1^1 = \alpha$, $a_1^2 = \alpha\gamma$, $a_2^1 = \alpha' + a_1^1$, $a_2^2 = \alpha'\gamma' + a_1^2$ and then (7) gives (M'1).

Necessary condition: We assume (M'1) and (M2). If $T_1 = \infty$, then (a) holds with $\alpha = \beta = 0$ and S_i arbitrary. If $T_1 < T_2 = \infty$, then (a) holds with $\alpha = a_1^1$, $\beta = a_1^2$ and $S_1 = S_2 = T_1$.

Suppose now that $T_1 < T_2 < \infty$. We have $a_1 \neq 0$, and since both (a) and (b) are symmetric in (U_1, U_2), without lost of generality we assume that $a_1^1 \neq 0$. Let $\alpha = a_1^1$ and $\gamma = a_1^2/\alpha$ and write $a_2^i = a_1^i + \beta^i$. Then the linear independence between a_1 and a_2 gives

$$\beta^2 \neq \gamma\beta^1, \tag{10}$$

while (M'1) is

$$\beta^1\varphi_1'(v)\varphi_2(\gamma v) + \beta^2\varphi_1(v)\varphi_2'(\gamma v) = 0 \quad \forall v \in I\!R. \tag{11}$$

We assume first that $\gamma = 0$. Recalling that $\varphi_i(0) = 1$, $\varphi_i'(0) = 0$ and φ_i' is not identically 0 in any neighborhood of 0 (because $P(U_i = 0) < 1$), (11) yields $\beta^1 = 0$, that is, we have (a) with $S_1 = T_1$, $S_2 = T_2$, $\beta = \beta^2$.

Next, assume that $\gamma \neq 0$. Then there exists $\theta \in \mathbb{R}^*$ with $\varphi_1'(\theta) \neq 0$, $\varphi_1(\theta) \neq 0$ and $\varphi_2(\gamma\theta) \neq 0$. Suppose for the time being that $\varphi_2'(\gamma\theta) = 0$. Then (11) yields $\beta^1 = 0$ and since there is another $\theta' \in \mathbb{R}^*$ with $\varphi_1(\theta') \neq 0$ and $\varphi_2(\gamma\theta') \neq 0$, we also have $\beta^2 = 0$, which contradicts (10). Thus $\varphi_2'(\gamma\theta) \neq 0$ and (10) and (11) yield $\beta^1 \neq 0$ and $\beta^2 \neq 0$. Hence we have (b) with $\gamma' = \beta^2/\beta^1$ and $\alpha' = \beta^1$ (note that $\gamma \neq \gamma'$ follows from (10), and (7) is the same as (11)). □

When U_1 and U_2 are arbitrary, it seems there is not much more to say. From now on we concentrate on the case where $U_1 =^d U_2$, i.e. $\varphi_1 = \varphi_2 = \varphi$. In this situation, the existence of a martingale X of the form (b) above depends on the existence of constants $\gamma, \gamma' \in \mathbb{R}^*$ with $\gamma \neq \gamma'$ and

$$\varphi'(v)\varphi(\gamma v) + \gamma'\varphi(v)\varphi'(\gamma v) = 0 \quad \forall v \in \mathbb{R}. \tag{12}$$

Let D denote the set of all $\gamma \in \mathbb{R}^*$ for which (12) holds for some $\gamma' \in \mathbb{R}^*$ with $\gamma' \neq \gamma$. If $\gamma \in D$ there is a unique $\gamma' = \delta(\gamma)$ satisfying (12), because we have seen before that for each $\gamma \neq 0$ there is $v \in \mathbb{R}$ with $\varphi(v) \neq 0$ and $\varphi'(\gamma v) \neq 0$.

Theorem 3. **a)** *If U_1 is symmetric about 0, then one of the following three cases is satisfied:*

(Cs-1) $D = \{-1, 1\}$.

(Cs-2) $D = \{r^n, -r^n : n \in \mathbb{Z}\}$ *for some* $r > 1$ *and* φ *never vanishes.*

(Cs-3) $D = \mathbb{R}^*$. *This is the case if and only if U_1 is stable with index $\rho \in (1, 2]$, i.e. $\varphi(u) = e^{-a|u|^\rho}$ for some $a > 0$.*

b) *If U_1 is not symmetric about 0, we are in one of the following five situations:*

(Ca-1) $D = \{1\}$.

(Ca-2) $D = \{-1, 1\}$. *This is the case if and only if $\varphi = \rho e^\eta$, where ρ and η are real-valued, $\eta(0) = 0$, and η is constant on each open interval on which ρ (or φ) does not vanish (necessarily φ vanishes somewhere, and η is not identically 0, otherwise we would be in the symmetric case).*

(Ca-3) $D = \{r^n : n \in \mathbb{Z}\}$ *for some $r > 1$ and φ never vanishes.*

(Ca-4) $D = \{r^n, -r^{n+1/2} : n \in \mathbb{Z}\}$ *for some $r > 1$ and φ never vanishes.*

(Ca-5) $D = (0, \infty)$. *This is the case if and only if U_1 is asymmetric strictly stable with index $\rho \in (1, 2)$, i.e., $\varphi(u) = e^{-a|u|^\rho(1+ib\,\mathrm{sign}(u))}$ for some $a > 0$, $b \neq 0$, $|b| \leq \tan(\frac{\pi}{2(2-\rho)})$.*

c) *There is a constant $\theta \in (1, 2]$ such that $\delta(\gamma) = -\gamma/|\gamma|^\theta$ (so $\delta(1) = -1$, and $\delta(-1) = 1$ if $-1 \in D$), and $\theta = \rho$ in cases (Cs-3) and (Ca-5).*

Therefore the martingales X of the form (8) are indeed represented as

$$X_t = \alpha(U_1 + \gamma U_2)1_{[T_1,\infty)}(t) + \alpha'(U_1 - \gamma U_2/|\gamma|^\theta)1_{[T_2,\infty)}(t), \tag{13}$$

where $\alpha, \alpha' \in \mathbb{R}^*$, $0 < T_1 < T_2 < \infty$, *and* $\gamma \in D$.

Remark. There are of course examples of variables satisfying (Cs-1) or (Cs-3) in the symmetrical case, (Ca-1) in the asymmetrical case. We presume that (Cs-2) and (Ca-3) are not empty, and believe that (Ca-2) is empty (but we have been unable to prove these facts).

Before giving the proof of Theorem 3 we present some useful lemmas. First we note that $\gamma = 1$ and $\gamma' = -1$ always satisfy (12), so $1 \in D$ and $\delta(1) = -1$.

Lemma 4. *We have* $-1 \in D$ *if and only if* $\varphi = \rho e^{\eta}$, *where* ρ *and* η *are real-valued and* $\eta(0) = 0$ *and* η *is constant on each open interval on which* ρ *(or* φ*) does not vanish. Moreover,* $\delta(-1) = 1$.

Proof. Let (x, y) be a maximal interval on which φ does not vanish, so φ does not vanish either on $(-y, -x)$ (we may have $(x, y) = \mathbb{R}$, of course). We can write $\varphi = e^{\psi}$ with ψ of class C^1 on (x, y) and $(-y, -x)$, and since $\psi(-v) = \overline{\psi(v)}$ the property $-1 \in D$ and (12) yield

$$\psi'(v) = \gamma' \overline{\psi'(v)} \quad \forall v \in (x, y).$$

Since $\gamma' \in \mathbb{R}^*$, we deduce that $\psi'(v) \in \mathbb{R}$ and thus $\gamma' = 1$ (because ψ' cannot be identically 0). Therefore, if $v_0 \in (x, y)$, we have $\psi(v) - \psi(v_0) \in \mathbb{R}$ for all $v \in (x, y)$ and hence $\varphi = \rho e^{\eta}$ with $\eta(v) = \eta(v_0) \in \mathbb{R}$ for all $v \in (x, y)$. The converse is obvious.□

Lemma 5. *Let* $\gamma \in \mathbb{R}^*$ *with* $|\gamma| \neq 1$. *Then* $\gamma \in D$ *if and only if* φ *does not vanish, and satisfies for some* $C(\gamma) > 0$

$$\varphi(v) = \varphi(\gamma v)^{C(\gamma)} \quad \forall v \in \mathbb{R}. \tag{14}$$

Moreover,

- **a)** $\mathcal{R}e\psi(v) < 0$ *for all* $v \in \mathbb{R}^*$.
- **b)** $\delta(\gamma) = -\gamma C(\gamma)$.
- **c)** *For all* $n \in \mathbb{Z}$ *we have* $\gamma^n \in D$ *and* $C(\gamma^n) = C(\gamma)^n$.
- **d)** $-\gamma \in D$ *if and only if* φ *is real-valued, and then* $C(-\gamma) = C(\gamma)$.

Proof. The sufficient condition is obvious, as well as (b).

Conversely, assume that $\gamma \in D$. Let $(-x, x)$ be the maximal interval on which φ does not vanish. We have $\varphi = e^{\psi}$ with ψ of class C^1 on $(-x, x)$. For simplicity we set $\psi_r = \mathcal{R}e\psi$, and we have $\psi_r(u) \to -\infty$ as $|u| \uparrow x$ if $x < \infty$. On $(-x, x)$, (12) yields $\psi'(v) + \gamma'\psi'(\gamma v) = 0$, so $\psi(v) + \frac{\gamma'}{\gamma}\psi(\gamma v) = 0$, since $\psi(0) = 0$.

If $|\gamma| > 1$ and $x < \infty$, then $|\psi_r(v)| = |\frac{\gamma'}{\gamma}||\psi_r(\gamma v)| \to \infty$ as $|v| \uparrow x/|\gamma|$, contradicting the fact that ψ is continuous on $(-x, x)$. Similarly, if $|\gamma| > 1$ and $x < \infty$, $|\psi_r(\gamma v)| = |\frac{\gamma}{\gamma'}||\psi_r(v)| \to \infty$ as $|v| \uparrow x$, bringing up the same contradiction; therefore $x = \infty$, and φ does not vanish. It follows that $\varphi = e^{\psi}$ everywhere and, with $C(\gamma) = -\gamma'/\gamma$,

$$\psi(v) = C(\gamma)\psi(\gamma v) \quad \forall v \in \mathbb{R}, \tag{15}$$

that is, we have (14). Since U_1 is non-degenerate, ψ is not identically 0 and thus $C(\gamma) \neq 0$. Note also that (c) is obvious from (14).

We always have that $\psi_r \leq 0$ and that ψ_r is even. Assume that $\psi_r(v) = 0$ for some $v > 0$. Then (15) and (c) imply $\psi_r(v|\gamma|^n) = 0$ for all $n \in \mathbb{Z}$. It follows that the characteristic fonction of the symmetrized random variable $U = U_1 - U_2$ equals 1 for all $v|\gamma|^n$, $n \in \mathbb{Z}$, so U is supported by $\{2k\pi/v|\gamma|^n : k \in \mathbb{Z}\}$, for all $n \in \mathbb{Z}$, which implies that $U = 0$ a.s., contradicting again the non-degeneracy assumption. Thus (a) holds and (15) yields $C(\gamma) > 0$.

Finally, it only remains to prove (d). If φ is real-valued, it is even and (14) is satisfied with $-\gamma$ and $C(-\gamma) = C(\gamma)$. Suppose conversely that $-\gamma \in D$, then (15) gives $\overline{\psi(v)} = C(\gamma)\psi(-\gamma v)$, while $-\gamma \in D$ yields $\psi(v) = C(-\gamma)\psi(-\gamma v)$. Comparing the real parts of these two equalities and using (a) we obtain $C(-\gamma) = C(\gamma)$. Then $\overline{\psi} = \psi$ and φ is real-valued. □

Lemma 6. *With $D_+ = D \cap \mathbb{R}_+$, one of the following three cases is satisfied:*

(C_+1) $D_+ = \{1\}$.

(C_+2) $D_+ = \{r^n : n \in \mathbb{Z}\}$ *for some* $r > 1$.

(C_+3) $D_+ = \mathbb{R}_+^*$.

Moreover, we are in case (C_+3) if and only if either $\varphi(u) = e^{-a|u|^2}$ *for some* $a > 0$ *or* $\varphi(u) = e^{-a|u|^\rho(1+ib\,sign(u))}$ *for some* $a > 0$, $\rho \in (1,2)$, $|b| \leq \tan(\frac{\pi}{2(2-\rho)})$.

Proof. Due to the fact that $1 \in D$ and to Lemma 5, if we are not in case (C_+1), D_+ contains at least a $\gamma > 0$, $\gamma \neq 1$, and then $\varphi = e^\psi$ satisfies (14). Indeed, D_+ is the set of all $\gamma > 0$ such that (15) holds for some $C(\gamma) > 0$. Then D_+ is clearly a multiplicative group, therefore it is closed since ψ is continuous and thus it is of the form (C_+2) or (C_+3).

Assuming (C_+3), for each $\gamma > 0$ there is $C(\gamma) > 0$ such that, if f denotes either the real or the imaginary part of ψ, we have $f(0) = 0$ and

$$f(v) = C(\gamma)f(\gamma v) \quad \forall v \geq 0.$$

Then f is either identically 0, or everywhere positive, or everywhere negative, on $(0,\infty)$. In the last two cases, $g(u) = \log|f(e^u)/f(1)|$ satisfies $g(u + \log\gamma) = g(u) + g(\log\gamma)$ for all $u \in \mathbb{R}$, $\gamma > 0$, i.e., $g(u + u') = g(u) + g(u')$ for all $u, u' \in \mathbb{R}$. Since g is continuous, we obtain $g(u) = Ku$. Thus, in all cases we have $f(v) = \eta v^\rho$ for some $\eta, \rho \in \mathbb{R}$, and furthermore $\gamma^\rho C(\gamma) = 1$ for all $\gamma > 0$ (hence ρ is the same for both the real and imaginary parts of ψ). We then deduce that $\psi(v) = (\alpha + i\beta)v^\rho$ for some $\alpha, \beta, \rho \in \mathbb{R}$, if $v > 0$. By (a) of Lemma 5 we have $\alpha < 0$ and since $\psi(-v) = \overline{\psi(v)}$, we also have $\psi(v) = (\alpha - i\beta)|v|^\rho$ for $v < 0$. Then $\psi(v) = -a|v|^\rho(1 + ib\,sign(v))$ for $a > 0$, $b \in \mathbb{R}$, $\rho \in \mathbb{R}$. Conversely, each such ψ satisfies (15) for all $\gamma > 0$, with $C(\gamma 1) = \gamma^{-\rho}$, implying $D_+ = \mathbb{R}_+^*$.

It remains to examine under which conditions on (a, b, ρ) the function $\varphi = e^\psi$ with ψ as above is a characteristic function. Observe that for all $\alpha, \alpha' > 0$ we have $\psi(\alpha v) + \psi(\alpha' v) = \psi(\alpha'' v)$ with $\alpha''^\rho = \alpha^\rho + \alpha'^\rho$. Then, if it is the case, the corresponding distribution will be strictly stable, with a first moment equal to 0. As is well known, this will be the case if and only if either $\rho = 2$ and $b = 0$ (normal case), or $\rho \in (1,2)$ and $|b| \leq \tan(\frac{\pi}{2(2-\rho)})$. □

Proof of Theorem 3. a) When U_1 is symmetric, so is D, and (Cs-i) = (C$_+$i). Therefore Lemma 5 yields that one of (Cs-1), (Cs-2) or (Cs-3) is satisfied. Moreover, (Cs-2) implies that φ never vanishes (by Lemma 5), and (Cs-3) holds if and only if $\varphi(v) = e^{-a|v|^\rho}$ (because here φ is real-valued).

b) Now we suppose that U_1 is not symmetric. It suffices to prove that if $D \neq \{1\}$, then we are in one of the cases (Ca-i) for i=2,3,4,5.

First, by Lemma 4, $-1 \in D$ if and only if the necessary and sufficient condition in (Ca-2) is satisfied. Then φ vanishes somewhere, and D contains no γ with $|\gamma| \neq 1$ by Lemma 5. Thus $-1 \in D$ if and only if (Ca-2) holds.

Next, suppose that we are not in any of the cases (Ca-1) and (Ca-2). If $D = D_+$, we are then in cases (Ca-3) or (Ca-5) by Lemma 5. Otherwise there exists $\gamma > 0$ with $\gamma \neq 1$ and $-\gamma \in D$. Then $\gamma^2 \in D$ and $\gamma^2 \neq 1$ and by Lemma 5 either (C$_+$2) or (C$_+$3) holds. However, under (C$_+$3) we also have $\gamma \in D$, hence Lemma 5(d) contradicts the assumption that U_1 is non-symmetric and indeed we have (C$_+$2) with some $r > 1$. It then follows that $\gamma^2 = r^k$ for some $k \in I\!N^*$, while Lemma 5(c) gives $C(r^n) = C(r)^n$ and $C(\gamma) = C(r)^{k/2}$. Furthermore if k were even we would have $r^{k/2} \in D$ and $-r^{k/2} = -\gamma \in D$, again a contradiction by Lemma 5(d), so $k = 2p+1$ with $p \in \mathbb{Z}$ and $\gamma = r^{p+1/2}$. In order to obtain (Ca-4), it thus remains to prove that $-r^{n+1/2} \in D$ for all $n \in \mathbb{Z}$. For this, a repeated use of (15) yields

$$\psi(v) = C(r^{n-p})\psi(r^{n-p}v) = C(r^{n-p})C(\gamma)\psi(-\gamma r^{n-p}v) = C(r)^{n+1/2}\psi(-r^{n+1/2}),$$

and the result follows.

c) Since $\delta(1) = -1$ and $\delta(-1) = 1$ (Lemma 4), (c) is obvious in cases (Cs-1), (Ca-1) and (Ca-2). Also, (c) with $\theta = \rho$ follows from Lemma 5(b) and from a comparison between (14) and the explicit form of φ in cases (Cs-3) and (Ca-5).

Under (Ca-4) we have seen that $C(r^n) = C^n$ and $C(-r^{n+1/2}) = C^{n+1/2}$, where $C = C(r)$. Thus, for all $\gamma \in D$, $C(\gamma) = C^{\log(|\gamma|)/\log(r)} = |\gamma|^{-\theta}$ with $\theta = -\log(C)/\log(r)$ (so $\delta(\gamma) = -\gamma/|\gamma|^\theta$). The same holds for (Cs-2) and (Ca-3). Now (15) yields $\psi'(v) = Cr\psi'(rv)$, hence $\psi'(r^{-n}v) = (Cr)^n\psi'(v)$ and since $\psi'(0) = 0$ and $\psi'(v) \neq 0$ for some $v \neq 0$ we must have $Cr < 1$ and thus $\theta > 1$.

Suppose that $\theta > 2$, i.e. $A := Cr^2 < 1$. Then $\psi'(r^{-n}v)/(r^{-n}v) = A^n\psi'(v)/v$ and if $|w| \leq r^{-n}$ there is $m \geq n$ and $v \in (1/r, 1]$ with $w = vr^{-m}$ or $w = -vr^{-m}$. Therefore $\sup_{|w|\leq r^{-n}} |\psi'(w)/w| \leq A^n \sup_{1/r<|v|\leq 1} |\psi'(v)|$. It follows that ψ' is differentiable at 0, with $\psi''(0) = 0$. Hence U_1 is square-integrable, with variance 0, which contradicts once more the non-degeneracy assumption and therefore $\theta \leq 2$. □

J. Jacod: Laboratoire de Probabilités (CNRS, URA 224), Université Paris VI, Tour 56, 4, Place Jussieu, 75252 Paris Cedex 05, France.

V. Pérez-Abreu: Department of Probability and Statistics, Centro de Investigación en Matemáticas A. C., Apdo. Postal 402, Guanajuato, Gto. 36000, México

Oscillation presque sûre de martingales continues

Jean-Marc Azaïs.
Laboratoire de Statistique et Probabilités
UMR-CNRS C55830, Université Paul Sabatier
Toulouse France.
Mario Wschebor.
Centro de Matemática, Facultad de Ciencias
Universidad de la Republica. Montevideo - Uruguay.

Nous établissons une limite presque sûre en loi pour les variations de martingales continues. Ce résultat généralise un résultat précédent de Azaïs et Wschebor qui demandait des conditions techniques sur les martingales. On en déduit une approximation presque sûre faible de la mesure d'occupation à partir du nombre de franchissements.

Mathematics Subject Classification (1991): 60F05, 60G44.

1 Introduction

Soit $X = \{X_t : t \in \mathbb{R}\}$ un processus stochastique à valeurs réelles sur un espace de probabilités $(\Omega, \mathcal{F}, \mathcal{P})$. L'article [3] étudie le comportement asymptotique des variations normalisées du processus X_t. Plus précisément nous définissons :

$$Z_h(t) = \frac{X_{t+h} - X_t}{a(h)} \tag{1}$$

où $a(.)$ est une certaine fonction de normalisation. On montre dans [3] que pour plusieurs classes de processus : processus gaussiens, P.A.I. stables (voir également [6]), intégrales stochastiques par rapport au mouvement brownien, si l'on définit

$$\mu_h(B) = \frac{1}{\lambda(I)} \lambda(\{t \in I, Z_h(t) \in B\}),$$

où I est un intervalle borné dans $\mathbb{R}$, B un borélien et λ la mesure de Lebesgue, alors il existe une normalisation $a(h)$ telle que μ_h converge faiblement vers $\mu^* \neq \delta_0$, lorsque h tends vers zéro. Ces résultats sont généralisés au cas où $Z_h(t)$ est défini par

$$Z_h(t) = \frac{h\dot{X}_h(t)}{a(h)}, \tag{2}$$

où $\dot{X}_h$ est la dérivée de X_h, $X_h = \psi_h * X$ est la régularisation de X par convolution avec une approximation de l'unité ψ_h, $\psi_h(t) = \frac{1}{h}\psi(\frac{t}{h})$, $h > 0$, ψ étant une fonction fixée. La formule (1) correspond au cas particulier $\psi = \mathbb{1}_{[-1,0]}$.

L'objet de cet article est d'étendre les résultats de [3] à toutes les martingales continues. Dans [3] on n'étudie que des martingales de la forme

$$X_t = \int_0^t b(s)dW_s,$$

où W_s est un brownien standard et l'intégrand $b(s)$ est adapté et satisfait certaines conditions de régularité. La démonstration consiste à se ramener au résultat correspondant pour le brownien. Dans le présent article on utilise des techniques différentes.

Dans le paragraphe 3 nous utilisons la convergence de μ_h pour construire une approximation de la mesure d'occupation basée sur les nombres de franchissements. Ceci permet d'étendre en un certain sens les résultats de [1] [2] [4] [5].

Dans tout ce qui suit, $M = \{M_t : t \geq 0\}$ est une martingale locale à valeurs réelles et à trajectoires continues sur un espace de probabilité filtré $(\Omega, \mathcal{F}, \{\mathcal{F}_t, t \geq 0\}, P)$. On note $\{A_t : t \geq 0\}$ le crochet de M et on définit la décomposition de Lebesgue de A_t.

$$A_t = S_t + \int_0^t \dot{A}_s ds, \tag{3}$$

où

- S_t correspond à une mesure étrangère à la mesure de Lebesgue λ
- $\dot{A} \in L^1([0,T], \lambda)$ pour chaque T positif.

ψ est une fonction à variation bornée dont le support est inclus dans $[-1,1]$, $\int_{\mathbb{R}} \psi(t)dt = 1$. $\|\psi\|_p$ est la norme de ψ dans $L^p(\mathbb{R}, \lambda)$. On note $\Psi(v)$, la variation totale de la mesure signée de distribution $\psi(-v)$ sur l'intervalle $[-1, v]$. La martingale locale M est prolongée par la valeur M_0 sur $\mathbb{R}^-$.

2 Oscillation de martingales

Sans perte de généralité, nous posons $I = [0,1]$ et nous avons le résultat suivant :

Théorème 2.1 *Soit $M_h = \psi_h * M$, alors presque sûrement, pour tout réel x non nul,*

$$\lambda(\{t \in I, h^{1/2}\dot{M}_h(t) \leq x\}) \overset{h \to 0}{\longrightarrow} \int_0^1 P(\dot{A}_t^{1/2}\|\psi\|_2 \eta \leq x)dt,$$

η suivant une loi normale centrée réduite et indépendante de M.

Démonstration: Par un argument de localisation, on peut supposer que M et A sont bornés sur $[0, 1+\delta]$, $\delta > 0$, uniformément en ω. En remarquant que

$$\lambda(\{t \in [0,1], h^{1/2}\dot{M}_h(t) \leq x\}) - \lambda(\{t \in [0,1], h^{1/2}\dot{M}_h(t+h) \leq x\}) \longrightarrow 0, \ h \to 0,$$

on se ramène au cas où ψ est à support dans $[-1, 0]$. Soit t un instant donné et C une constante positive, on définit

$$T_{t,C} = \inf\{s \ :\ s > 0\ ;(A_{t+s} - A_t)/s \geq C\},$$

avec la convention $\inf(\emptyset) = +\infty$. $T_{t,C}$ est un temps d'arrêt pour la filtration $\{\mathcal{G}_s^t = \mathcal{F}_{t+s}\ ; s \geq 0\}$. On définit le processus

$$X_h(t,z) = Z_{h\wedge T_{t,C}}(t,z),$$

avec

$$Z_s(t,z) = exp\left(iz\int_0^1 \psi(-v)d_v(M_{t+sv} - M_t) + \frac{1}{2}z^2\int_0^1 \psi^2(-v)d_v(A_{t+sv} - A_t)\right). \quad (4)$$

Pour chaque $t, z \in \mathbb{R}$, $s \to X_s(t,z)$ est une $\{\mathcal{G}_s^t\ ;\ s \geq 0\}$ martingale, donc

$$E\{X_s(t,z)/\mathcal{F}_t\} = 1.$$

Le deuxième terme dans l'exposant de (4) s'écrit

$$-\frac{1}{2}z^2\int_0^1 (A_{t+sv} - A_t)d\psi^2(-v),$$

ce qui entraine l'inégalité

$$\begin{aligned}|X_h(t,h^{-\frac{1}{2}}z)| &= exp\left[-\frac{1}{2}\frac{z^2}{h}\int_0^1 (A_{t+(h\wedge T_{t,C})v} - A_t)d\psi^2(-v)\right] \leq \\ &\leq exp\left[\frac{z^2}{h}(A_{t+(h\wedge T_{t,C})} - A_t)\int_0^1 \Psi(v)d\Psi(v)\right],\end{aligned}$$

$$|X_h(t,h^{-\frac{1}{2}}z)| \leq exp\left[\frac{1}{2}z^2C(\Psi(1))^2\right].$$

De plus,

$$\begin{aligned}&E\left(\left|\int_0^1 (X_h(t,h^{-1/2}z) - 1)dt\right|^2\right)\\ &= \iint_{|t-s|>h} E\left[(X_h(t,h^{-1/2}z) - 1)(\overline{X_h(s,h^{-1/2}z) - 1})\right] dsdt +\\ &+ \iint_{|t-s|\leq h} E\left[(X_h(t,h^{-1/2}z) - 1)(\overline{X_h(s,h^{-1/2}z) - 1})\right] dsdt.\end{aligned}$$

Dans la première intégrale, si $t > s+h$ on conditionne par $\mathcal{F}_t$. La relation $E\{X_h(t,z)/\mathcal{F}_t\} = 1$ montre que l'intégrand est nul. Le second terme est borné par $(cte)h$.

Le lemme de Borel-Cantelli implique que si $h_n = n^{-a}, a > 1$, on a presque sûrement

$$\int_0^1 (X_{h_n}(t,h_n^{-1/2}z) - 1)dt \to 0\ ,\ (n \to +\infty).$$

On obtient le même résultat si l'on remplace $[0,1]$ par un intervalle d'intégration $J \subset [0,1]$ fixé. Par un argument de densité, $X_h(t,h^{-1/2}z)$ étant borné par $exp\left[\frac{1}{2}z^2C(\Psi(1))^2\right]$, on obtient que presque sûrement

$$\forall J \subset [0,1]\ \int_J (X_{h_n}(t,h_n^{-1/2}z) - 1)dt \to 0\ ,\ (n \to +\infty).$$

On en déduit que presque sûrement pour toute fonction $g(.) \in L^1([0,1],\lambda)$, on a

$$\int_0^1 g(t)(X_{h_n}(t,h_n^{-1/2}z)-1)dt \longrightarrow 0\ ,\ n\to+\infty. \tag{5}$$

Maintenant nous introduisons les notations

$$\phi_{n,z,C}(t) = exp\left[\frac{1}{2}z^2\int_0^1 \frac{A_{t+(h_n\wedge T_{t,C})v}-A_t}{h_n}d\psi^2(-v)\right]$$

$$\phi_{z,C}(t) = exp\left[-\frac{1}{2}z^2\dot{A}_t\|\psi\|_2^2\right]\mathbf{1}_{\{T_{t,C}>0\}}+\mathbf{1}_{\{T_{t,C}=0\}}.$$

Remarquons que la dernière fonction n'est définie que pour presque tout t. Nous utilisons la representation suivante:

$$h\dot{M}_h(t) = \int_0^1 \psi(-v)d_v(M_{t+hv}-M_t).$$

Nous allons d'abord prouver que la convergence dans l'énoncé du théorème a lieu sur la suite $\{h_n\}$. Soit

$$F_h(z) = \int_0^1 exp(izh^{1/2}\dot{M}_h(t))dt\ ,\ z\in\mathbb{R},$$

la transformée de Fourier de la fonction $t \longrightarrow h^{1/2}\dot{M}_h(t)$ définie dans l'espace mesuré (I,λ). On a:

$$\left|F_{h_n}(z)-\int_0^1\phi_{n,z,C}(t)X_{h_n}(t,h_n^{-1/2}z)dt\right| \le 2\lambda(\{t\in I,\ T_{t,C}=0\})+$$

$$+\left|\int_{\{t\in I,\ T_{t,C}>0\}}\left[exp(izh_n^{1/2}\dot{M}_{h_n}(t))-exp(izh_n^{-1/2}(h_n\wedge T_{t,C})\dot{M}_{h_n\wedge T_{t,C}}(t))\right]\right|dt.$$

Or si $T_{t,C}>0$ et n est suffisamment grand, $h_n\wedge T_{t,C}=h_n$ et l'intégrand dans le dernier terme s'annule. Une application directe du théorème de Lebesgue montre que ce dernier terme tend vers zéro, lorsque $n\longrightarrow+\infty$ et par conséquent:

$$\limsup_{n\longrightarrow\infty}\left|F_{h_n}(z)-\int_0^1\phi_{n,z,C}(t)X_{h_n}(t,h_n^{-1/2}z)dt\right| \le 2\lambda(\{t\in I,\ T_{t,C}=0\}). \tag{6}$$

D'autre part,

$$\int_0^1\phi_{n,z,C}(t)X_{h_n}(t,h_n^{-1/2}z)dt = \int_0^1(\phi_{n,z,C}(t)-\phi_{z,C}(t))\,X_{h_n}(t,h_n^{-1/2}z)dt+$$

$$+\int_0^1\phi_{z,C}(t)\left(X_{h_n}(t,h_n^{-1/2}z)-1\right)dt+\int_0^1\phi_{z,C}(t)dt.$$

Le premier terme tend vers zéro par le théorème de Lebesgue, le second à cause de (5). La relation (6) implique que

$$\limsup_{n\longrightarrow\infty}\left|F_{h_n}(z)-\int_0^1\phi_{z,C}(t)dt\right| \le 2\lambda(\{t\in I,\ T_{t,C}=0\}). \tag{7}$$

La fonction croissante A_t étant dérivable λ−presque partout on a bien que

$$p.s \ \lambda(\{t \ : \ \forall C > 0, T_{t,C} = 0\}) = 0,$$

ce qui permet de déduire de (7) et de la définition de $\phi_{z,C}$ pour tout réel z on a presque sûrement

$$F_{h_n}(z) \longrightarrow \int_0^1 exp[-\frac{1}{2}z^2 \dot{A}_t \|\psi\|_2^2]dt. \tag{8}$$

Le théorème de Fubini permet de conclure que p.s. la convergence dans (8) a lieu pour λ−presque tout $z \in \mathbb{R}$. Maintenant, une modification standard du théorème de Cramér-Lévy implique que p.s., pour tout $x \neq 0$:

$$\lambda(\{t \ : \ t \in I, h_n^{1/2}\dot{M}_{h_n}(t) \leq x\}) \longrightarrow \int_0^1 P(\dot{A}_t^{1/2}\|\psi\|_2\eta \leq x)dt, \ (n \to \infty). \tag{9}$$

La démonstration sera achevée si l'on remplace la suite $\{h_n\}$ par $h \to 0$. Soit donc $0 < h < \delta$ et $n = n(h)$ tel que $h_{n+1} \leq h < h_n$. Comme $h \simeq h_n$ quand h tend vers zéro, il suffit de montrer que presque sûrement

$$\overline{M}_n(.) = \sup_{h_{n+1} \leq h < h_n} h_n^{-1/2}(h_n\dot{M}_{h_n}(.) - h\dot{M}_h(.))$$

converge vers zéro en mesure dans l'espace $\{I, \lambda\}$. Ce qui découle à son tour de

$$p. \ s. \ \int_0^1 |\overline{M}_n(t)|^\gamma dt \to 0,$$

pour un certain $\gamma > 0$. Or nous savons qu'il existe un mouvement brownien $B = (B(t), t \geq 0)$ tel que $M_t = B(A_t)$. Soit maintenant $0 < \alpha < 1/2$; presque sûrementles trajectoires de B sont α-hölderiennes sur tout compact. Posons $\gamma = 1/\alpha$, nous avons, pour $h_{n+1} \leq h \leq h_n$:

$$\left|h_n\dot{M}_{h_n}(t) - h\dot{M}_h(t)\right| = \left|\int_0^1 [(M_{t+vh_n} - M_t) - (M_{t+vh} - M_t)](-d\psi(-v))\right| \leq$$

$$\leq C_\omega \int_0^1 (A_{t+vh_n} - A_{t+vh_{n+1}})^\alpha d\Psi(v),$$

où C_ω désigne une variable aléatoire positive presque sûrement finie. Donc,

$$\int_0^1 |\overline{M}_n(t)|^\gamma dt \leq C_\omega^\gamma h_n^{-\gamma/2} \int_0^1 dt \int_0^1 \left(A_{t+vh_n} - A_{t+vh_{n+1}}\right)^{\gamma\alpha} d\Psi(v)(\Psi(1))^{\gamma-1} =$$

$$= C'_\omega \ h_n^{-\gamma/2} \int_0^1 d\Psi(v) \int_0^1 \left(A_{t+vh_n} - A_{t+vh_{n+1}}\right) dt \leq$$

$$\leq C''_\omega \ h_n^{-\gamma/2}(h_n - h_{n+1}) \leq C'''_\omega n^{-(a+1)} n^{\frac{a}{2\alpha}}.$$

C'_ω, C''_ω et C'''_ω désignent également des variables aléatoires positives presque sûrement finies. Le terme de droite ci-dessus tend vers zéro si α a été choisi suffisamment proche de 1/2.

□

Corollaire 2.1 *Presque sûrement, pour tout réel x non nul,*

$$\lambda(\{t \in I, \frac{M_{t+h} - M_t}{h^{1/2}} \leq x\}) \stackrel{h\to 0}{\longrightarrow} \int_0^1 P(\dot{A}_t^{1/2}\eta \leq x)dt,$$

η suivant une loi normale centrée réduite indépendante de M.

Démonstration: Appliquer le théorème 2.1 avec $\psi = \mathbf{I}_{[-1,0]}$.

3 Approximation de la mesure d'occupation

Lemme 3.1 *Avec les notations précédentes, soit $1 < \beta < 2$ et $\delta > 0$, alors*

$$p.s. \quad \sup_{0<|h|<\delta} \int_0^1 \left| \frac{M_{t+h} - M_t}{\sqrt{h}} \right|^\beta dt < +\infty.$$

Démonstration: Il suffit de démontrer la relation pour $h > 0$. Comme dans la démonstration du théorème 2.1 on peut supposer M et A bornés sur $[0, 1+\delta]$ uniformément en ω. Soit $G : \mathbb{R} \longrightarrow \mathbb{R}^+$ une fonction de classe $\mathcal{C}^2$ telle que $G(x) = x^2$ pour $|x| \leq \frac{1}{2}$ et $G(x) = |x|^\beta$ pour $|x| > 1$. On vérifie les inégalités suivantes pour tout $x \in \mathbb{R}$

$$|G(x)| \leq (cte)|x|^\beta \tag{10}$$

$$|G'(x)| \leq (cte)|x|^{\beta-1} \tag{11}$$

$$|G''(x)| \leq (cte). \tag{12}$$

Il est clair que

$$\int_0^1 \left| \frac{M_{t+h} - M_t}{\sqrt{h}} \right|^\beta dt \leq 1 + \int_0^1 G\left(\frac{M_{t+h} - M_t}{\sqrt{h}} \right) dt. \tag{13}$$

Pour chaque t, en appliquant la formule d'Ito, on obtient pour $h > 0$:

$$\begin{aligned} G\left(\frac{M_{t+h} - M_t}{\sqrt{h}} \right) &= \int_0^h h^{-\frac{1}{2}} G'\left(\frac{M_{t+s} - M_t}{\sqrt{h}} \right) d_s(M_{t+s} - M_t) \\ &+ \frac{1}{2} \int_0^h h^{-1} G''\left(\frac{M_{t+s} - M_t}{\sqrt{h}} \right) d_s(A_{t+s} - A_t) = X_{t,h} + Y_{t,h}. \end{aligned}$$

D'une part, en vertu de (12)

$$\int_0^1 |Y_{t,h}| dt \leq (cte) \int_0^1 h^{-1}(A_{t+h} - A_t) dt \leq (cte) A_{1+\delta} \leq (cte), \tag{14}$$

si $0 < h < \delta$. D'autre part puisque $E(X_{t,h}/\mathcal{F}_t) = 0$, on a bien :

$$E(\int_0^1 X_{t,h} dt) = 0$$

et

$$E\left(\left[\int_0^1 X_{t,h} dt \right]^2 \right) = 2 \iint_{0 \leq t \leq s \leq t+h \leq 1} E(X_{t,h} X_{s,h}) ds.$$

Nous majorons l'intégrand en utilisant (11)

$$E[(X_{t,h})^2] = 1/h \; E\left[\int_0^h G'^2 \left(\frac{M_{t+h} - M_t}{\sqrt{h}} \right) d_s(A_{t+s} - A_t) \right] \leq (cte)\; h^{-\beta} E(A_{t+h} - A_t).$$

Ce qui entraîne

$$E\left(\left[\int_0^1 X_{t,h} dt \right]^2 \right) \leq (cte) h^{-\beta} \iint_{0 \leq t \leq s \leq t+h \leq 1} (E(A_{t+h} - A_t) E(A_{s+h} - A_s))^{\frac{1}{2}} ds$$

$$\leq (cte)h^{-\beta} \iint_{0\leq t\leq s\leq t+h\leq 1} E(A_{t+2h} - A_t)ds \leq (cte)h^{1-\beta} \int_0^1 E(A_{t+2h} - A_t)dt \leq (cte)h^{2-\beta},$$

si h est inférieur à $\delta/2$.

Le lemme de Borel-Cantelli implique que pour toute suite $h_n = n^{-a}$, $a(2-\beta) > 1$, presque sûrement

$$\int_0^1 X_{t,h_n} dt \to 0 \quad (n \to \infty). \tag{15}$$

Les relations (13) (14) et (15) impliquent que

$$p.s. \quad \sup_{n\in\mathbb{N}} \int_0^1 \left| \frac{M_{t+h_n} - M_t}{\sqrt{h_n}} \right|^\beta < +\infty.$$

Pour obtenir le résultat du lemme, il suffit d'appliquer l'argument utilisé à la fin de la démonstration du théorème 2.1.

□

Théorème 3.1 *Si I est un intervalle borné de $\mathbb{R}^+$, p.s. pour toute fonction f continue $\mathbb{R} \longrightarrow \mathbb{R}$ on a :*

$$\sqrt{\pi/2}\|\psi\|_2^{-1} \int_{-\infty}^{+\infty} h^{1/2} f(u) N_u^{M_h}(I) du \longrightarrow \int_I f(M_t) \dot{A}_t^{1/2} dt \ ; \ (h \to 0),$$

où $N_u^g(I)$ est le nombre de franchissements du niveau u par la fonction g : $\mathbb{R} \to \mathbb{R}$ durant l'intervalle de temps I :

$$N_u^g(I) = \#\{t \in I, g(t) = u\}.$$

Démonstration : Sans perte de généralité nous pouvons supposer $I \subset [0,1]$ et comme précédemment, M et A bornés uniformément en ω sur $[0, 1+\delta]$, $\delta > 0$. L'identité suivante est vraie pour f continue et g de classe C^1 ([5])

$$\int_{-\infty}^{+\infty} f(u) N_u^g(I) du = \int_I f[g(t)] |\dot{g}(t)| dt.$$

Posons

$$C_\psi = \sqrt{\pi/2}\|\psi\|_2^{-1}.$$

On a,

$$C_\psi \int_{\mathbb{R}} h^{\frac{1}{2}} f(u) N_u^{M_h}(I) du = C_\psi \int_I f(M_h(t)) |h^{\frac{1}{2}} \dot{M}_h(t)| dt$$

$$= C_\psi \int_I f(M_t) |h^{\frac{1}{2}} \dot{M}_h(t)| dt + C_\psi \int_I [f(M_h(t)) - f(M_t)] |h^{\frac{1}{2}} \dot{M}_h(t)| dt. \tag{16}$$

Soit $1 < \beta < 2$, alors,

$$p.s. \quad \sup_{0<h<\delta} \int_0^1 \left| h^{\frac{1}{2}} \dot{M}_h(t) \right|^\beta dt < +\infty. \tag{17}$$

En effet :

$$\int_0^1 \left| h^{\frac{1}{2}} \dot{M}_h(t) \right|^\beta dt = \int_0^1 \left| \int_{-1^-}^{1^+} \frac{M_{t+hu} - M_t}{\sqrt{h}} d\psi(-u) \right|^\beta dt$$

$$\leq (cte)\int_{-1^-}^{1^+} d\Psi(u)\int_0^1 \left|\frac{M_{t+hu}-M_t}{\sqrt{h}}\right|^\beta dt \leq C_\omega \int_{-1^-}^{1^+} |u|^{\beta/2} d\Psi(u),$$

en utilisant le lemme 3.1 (C_ω est une certaine constante aléatoire). Ceci prouve (17). Il s'ensuit que p.s. le second terme dans (16) tend vers zéro , puisque $\int_I |h^{1/2}\dot{M}_h(t)|dt$ est borné et que $f(M_h(t)) - f(M_t)$ tend vers zéro uniformément sur $t \in I$ lorsque $h \to 0$.

En ce qui concerne le premier terme dans (16), la relation (17) implique que p.s. l'ensemble des fonctions $\{h^{1/2}\dot{M}_h(.)\ ; 0 < h < \delta\}$ est uniformément intégrable sur $([0,1],\lambda)$. Pour chaque intervalle $J \subset [0,1]$ le théorème 2.1 appliqué à l'intervalle J au lieu de $[0,1]$ entraîne que p.s.

$$\int_J |h^{1/2}\dot{M}_h(t)|dt \longrightarrow \int_{\mathbb{R}} |x|F_J(dx) = (2/\pi)^{1/2}\|\psi\|_2 \int_J \dot{A}_t^{1/2}dt, \ (h \to 0), \qquad (18)$$

où

$$F_J(x) = \int_J P(\dot{A}_t^{1/2}\|\psi\|_2\ \eta \leq x)dt.$$

Par un argument de monotonie, p.s. la convergence dans (18) a lieu simultanément pour tout intervalle $J \subset [0,1]$. On en déduit que p.s.

$$\int_I g(t)|h^{1/2}\dot{M}_h(t)|dt \longrightarrow C_\psi^{-1}\int_I g(t)\dot{A}_t^{1/2}dt, \ (h \to 0),$$

simultanément pour toutes les fonctions g qui sont des combinaisons linéaires d'indicatrices d'intervalles, donc pour toutes les fonctions continues. En posant $g(t) = f(M_t)$ on a bien le théorème.

□

4 Bibliographie

[1]Azaïs, J-M. (1989). "Approximation des trajectoires et temps local des diffusions". Ann. Inst. Henri Poincaré, Vol.25,2,175-194.

[2]Azaïs , J-M. (1990). "Conditions for convergence of number of crossings to the local time. Application to stable processes with independent increments and to Gaussian processes". Prob. and Math. Statistics, Vol.11,1,19-36.

[3] Azaïs J-M. & Wschebor M., (1996). "Almost Sure Oscillation of Certain Random Processes". A paraître dans Bernoulli.

[4] Berzin, C. & Wschebor,M. (1993). "Approximation du temps local des surfaces gaussiennes". Probab. Theory Relat. Fields",96,1-32.

[5] Nualart, D. & Wschebor, M. (1991). "Integration par parties dans l'espace de Wiener et approximation du temps local". Probab. Th. Rel. Fields,90,83-109.

[6] Wschebor, M. (1995). "Almost sure weak convergence of the increments of Lévy processes". Stochastic Processes and their Applications,55,253-270.

A NOTE ON CRAMER'S THEOREM

GAO Fuqing*

Let X be a locally convex vectorial space, Polish with a metric ρ. Let $(\xi_n)_{n\geq 1}$ be a sequence of X-valued i.i.d.r.v., defined on a probability space $(\Omega, \mathcal{F}, \mathbb{P})$. Consider the empirical means

$$L_n := S_n/n = \sum_{k=1}^{n} \xi_k / n \ , \ n\geq 1 \ .$$

The Cramer functional of ξ_1 is given by

(1) $\quad \Lambda(y) := \log \mathbb{E} \exp \langle \xi_1, y\rangle \in (-\infty, +\infty]$, for $y\in X'$,

where X′ is the topological dual space of X with the dual relation denoted by $\langle x,y\rangle$. The Legendre transformation is defined by

(2) $\quad \Lambda^*(x) = \sup\{ \langle x,y\rangle - \Lambda(y) \mid y\in X'\}$ *for all* $x\in X$.

The purpose of this Note is to prove

Theorem 1: *As* $n\to+\infty$, $\mathbb{P}(L_n\in\bullet)$ *satisfies the large deviation principle (in abridge: LDP) on* (X,ρ) *(i.e.,*

(i) $\exists\ I : X \to [0, +\infty]$ *such that* $[I\leq L]$ *is compact for any* $0\leq L<+\infty$;

(ii) for any Borel subset A in (X,ρ),

(3) $$-\inf_{x\in A^0} I(x) \leq \lim_{n\to+\infty} \begin{pmatrix}\inf\\ \sup\end{pmatrix} \frac{1}{n} \log \mathbb{P}(L_n\in A) \leq -\inf_{x\in \bar{A}} I(x) \ ;$$

where A^0 *and* $\bar{A}$ *are respectively the interior and the closure of A), if and only if there is a compact convex balanced subset* $\mathbf{K}$ *in X such that*

(4) $$\mathbb{E} \exp q_{\mathbf{K}}(\xi_1) < +\infty \ ,$$

where $q_{\mathbf{K}}(x) = \inf\{\lambda>0 \mid x/\lambda \in \mathbf{K}\}$ *is the Minkovski functional of* $\mathbf{K}$.

In this case, $\Lambda^*(x) = I(x)$ *over X.*

Before giving its proof, let us make some remarks.

(a) If $\dim(X)<+\infty$, the condition (4) becomes

(5) $$\exists\ \delta>0 \ \text{ such that } \ \mathbb{E} \exp (\delta\|\xi_1\|) < +\infty \ .$$

The sufficiency of (5) to the LDP is the well-known (improved) Cramer theorem, contained in Azencott [Az, 1980]. The necessity of (5) is already noted in [W].

(b) If $\dim(X)=+\infty$, and X is a separable Banach space, Donsker & Varadhan [DV, 1976] proved that the condition

(6) $$\forall\ \lambda>0 \ , \ \mathbb{E} \exp (\lambda\|\xi_1\|) < +\infty \ ,$$

is sufficient to the Cramer theorem (the LDP above).

(c) de Acosta gave another proof of the Cramer theorem due to Donsker and Varadhan by showing that (6) implies (4). One of his further remark is that (5) does not imply (4) in the infinite dimensional case in the following sense: for any separable Banach space $(X,\|\cdot\|)$ with $\dim(X)=+\infty$, there is always a X-valued r.v. ξ_1 which satisfies (5), but not (4).

Hence by Theorem 1 above, (5) is not enough to the Cramer theorem, illustrating an essential difference between the finite and infinite dimensional situations.

Proof of Theorem 1. The sufficiency. That the condition (4) implies the LDP with $I=\Lambda^*$ is a direct consequence of [St, Corollary 3.27], because (4) implies the exponential tightness of $\mathbb{P}(L_n\in\bullet)$.

The necessity. If the LDP holds, by [LS, Lemma 2.6], $\mathbb{P}(L_n\in\bullet)$ is exponentially tight. In particular, there is a compact subset K' in (X,ρ) such that

$$\text{(7)}\qquad \limsup_{n\to+\infty} \frac{1}{n} \log \mathbb{P}(L_n \notin K') \le -5 .$$

Let K be the closed, convex, and balanced hull of K', which is still compact ([Sc, p50]). We have,

$$[\xi_1/n \notin 3K] \subseteq [S_n/n \notin K] \cup [(\xi_2 +\dots+ \xi_n)/n \notin 2K]$$
$$\subseteq [S_n/n \notin K] \cup [(\xi_2 +\dots+ \xi_n)/(n-1) \notin 2K] .$$

Hence by (7), $\exists$ $N\ge 0$ such that for all $n\ge N$,

$$\mathbb{P}[\xi_1/n \notin 3K] \le 2e^{-4n} , \text{ or } \mathbb{P}(q_K(\xi_1)>3n) \le 2e^{-4n} .$$

This last estimation implies

$$\mathbb{E}\exp(q_K(\xi_1)) \le \sum_{n=0}^{\infty} e^{3n+3}\, \mathbb{P}(3n\le q_K(\xi_1)<3(n+1)) < +\infty ,$$

the desired condition (4). ∎

Additional notes (due to the referee):

1) That the LDP implies the exponential tightness (due to [LS, Lemma 2.6]) holds in any Polish space.

2) Instead of the Polish property of the global space X, we assume that ξ_1 takes values in a convex **Polish** subspace Z of a locally convex quasi-complete vector space X (see Stroock [St]). Theorem 1 still holds in this situation. In fact, only the necessity requires a little more attention. By the previous note 1), we can always find a compact $K'\subseteq Z$ such that (7) holds. By the quasi-completeness of X, the convex balanced closed hull K of K' is compact ([Sc,p50]). The rest is the same.

In this situation, it is in further known that $[\Lambda^*<+\infty]\subseteq Z$ (see [W]).

Acknowledge: The author is grateful to the referee for the above notes, and to Mr. WU Liming for the useful discussions.

References

[Az]**R. Azencott:** *Grandes déviations et applications,* In "Ecole d'Eté de Probabilités de Saint-Flour (1978)", edited by P.L. Hennequin. *Lect. Notes in Math. N°774, pp.1-176, Springer, Berlin, 1980.*

[deA] **A. de Acosta:** Upper bounds for large deviations of dependent random vectors, *Z. Wahrsch. verw. Geb., 69, 1985.*

[DV] **M.D. Donsker & S.R.S. Varadhan :** Asymptotic evaluation of certain Markov process expectations for large time, III.
Comm. Pur. Appl. Math. **29**, *p.389-461 (1976).*

[LS] **J. Lynch & J. Sethuraman:** Large deviations for processes with independent increments, *Ann. Probab. 15, N°2, pp.610-627, 1987.*

[Sc] **Schaefer H.H.:** *Topological Vector Spaces. Macmillan Serie in Advanced Math. and Theoretical Phys., 1966.*

[St] **Stroock D.W.:** *An introduction to the theory of large deviations Springer, Berlin 1984*

[W] **L.M. Wu :** An introduction to large deviations, *Academic Press of China, 1996 (in press).*

* Department of Mathematics, University of Hubei, Province HUBEI. China

The Hypercontractivity of Ornstein–Uhlenbeck Semigroups with Drift, Revisited*

Sheng-Wu He and Jia-Gang Wang

1. In Qian, He[3] the hypercontractivity of Ornstein– Uhlenbeck semigroup with drift was established in the framework of white noise analysis. Let $(S) \subset (L^2) \subset (S)^*$ be the Gel'fand's triple over white noise space $(S'(\boldsymbol{R}), \mathcal{B}(S'(\boldsymbol{R})), \mu)$. Let H be a strictly positive self-adjoint operator in $L^2(\boldsymbol{R})$. Then

$$P_t^H \varphi(x) = \int_{S'(\boldsymbol{R})} \varphi(e^{-tH}x + \sqrt{1-e^{-2tH}}y)\mu(dy), \ \varphi \in (S), t \geq 0,$$

determines a diffusion semigroup in (L^p), $p \geq 1$, called the Ornstein–Uhlenbeck semigroup with drift operator H. It was shown that if

$$\alpha = \inf_{0 \neq \xi \in S(\boldsymbol{R})} \frac{(H\xi, H\xi)}{(H\xi, \xi)} > 0, \tag{1.1}$$

then (P_t^H) is hypercontractive : for any $p \geq 1$, $q(t) = 1+(p-1)e^{2\alpha t}$ and nonnegative $f \in (L^p)$,

$$\|P_t^H f\|_{q(t)} \leq \|f\|_p.$$

The proof there was based on Bakry–Emery's local criterion for hypercontractivity by computing Bakry–Emery's curvature of the semigroup $(P_t^H)_{t\geq 0}$. In this note we shall point out that Neveu's probabilistic proof ([2]) remains available for the Ornstein–Uhlenbeck semigroup with drift. After recalling Neveu's result, a simple proof for the hypercontractivity of the semigroup $(P_t^H)_{t\geq 0}$ is given.

The following theorem is indeed extracted from Neveu[2].

Theorem 1. *Let $\{X_t,\ t \in T,\ Y_s,\ s \in S\}$ be a Gaussian process, where T and S are two arbitrary index sets. Assume $\forall t_i \in T$, $s_j \in S$, $a_i, b_j \in \boldsymbol{R}$, $i = 1, \cdots, n$, $j = 1, \cdots, m$; $n, m \geq 1$*

$$|\rho(\sum_{i=1}^n a_i X_{t_i}, \sum_{j=1}^m b_j Y_{s_j})| \leq r, \tag{1.2}$$

$$p > 1, \qquad q > 1, \qquad (p-1)(q-1) \geq r^2. \tag{1.3}$$

Then for any $\sigma\{X_t, t \in T\}$-measurable random variable ξ and $\sigma\{Y_s, s \in S\}$-measurable random variable η

$$\mathsf{E}|\xi\eta| \leq \|\xi\|_p \|\eta\|_q. \tag{1.4}$$

*The project supported by National Natural Science Foundation of China.

2. Now we turn to Ornstein-Uhlenbeck semigroup.

Let $S(\mathbf{R})$ be the Schwartz space of rapidly decreasing functions on $\mathbf{R}$ and $S'(\mathbf{R})$ be its dual space. There exists a unique probability measure μ on $\mathcal{B}(S'(\mathbf{R}))$, the σ-field generated by cylinder sets, such that

$$\int_{S'(\mathbf{R})} e^{i\langle x,\xi\rangle}\mu(dx) = \exp\left\{-\frac{1}{2}|\xi|_2^2\right\}, \qquad \xi \in S(\mathbf{R}).$$

The measure μ is called the white noise measure, and the probability space $(S'(\mathbf{R}), \mathcal{B}(S'(\mathbf{R})), \mu)$ is called the white noise space, which is our basic probability space. Let $(S), (L^2)$ and (S^*) be the spaces of test functionals, square-integrable functionals and generalized functionals (or Hida distributions) over $(S'(\mathbf{R}), \mathcal{B}(S'(\mathbf{R})), \mu)$ respectively. A brief introduction to white noise analysis is given in [3]. More materials on white noise analysis may be refered to Hida et al.[1] and Yan[4].

Let H be a strictly positive self-adjoint operator in $L^2(\mathbf{R})$. Set

$$M_t = e^{-tH}, \qquad T_t = \sqrt{1-e^{-2tH}} = \sqrt{1-M_{2t}}, \quad t \geq 0. \tag{2.1}$$

The following assumptions are made:

(H_1) $S(\mathbf{R}) \subset \mathcal{D}(H)$ and H is a continuous mapping from $S(\mathbf{R})$ into itself.
(H_2) $\forall t > 0$ M_t and T_t are continuous operators from $S(\mathbf{R})$ into itself.

Then M_t and T_t, $t > 0$, can be extended onto $S'(\mathbf{R}) : \forall x \in S'(\mathbf{R})$, $\xi \in S(\mathbf{R})$,

$$\langle M_t x, \xi\rangle = \langle x, M_t\xi\rangle, \quad \langle T_t x, \xi\rangle = \langle x, T_t\xi\rangle. \tag{2.2}$$

Now for all $t \geq 0, x \in S'(\mathbf{R})$ and $\varphi \in (S)$ define

$$P_t^H\varphi(x) = \int \varphi(M_t x + T_t y)\mu(dy). \tag{2.3}$$

Let $\Gamma(e^{-tH}) = \Gamma(M_t)$ be the second quantization of M_t. Then the Ornstein - Uhlenbeck semigroup with drift H

$$P_t^H = \Gamma(e^{-tH}) = e^{-td\Gamma(H)}, \qquad t \geq 0, \tag{2.4}$$

is a semigroup with infinitesimal operator $-d\Gamma(H)$, where $d\Gamma(H)$ is a self-adjoint operator in (L^2):

$$d\Gamma(H) = \sum_{n=1}^{\infty} \oplus\{\underbrace{H\otimes I\otimes\cdots\otimes I}_{n \text{ terms}} + \underbrace{I\otimes H\otimes I\otimes\cdots\otimes I}_{n \text{ terms}} + \cdots + \underbrace{I\otimes\cdots\otimes H}_{n \text{ terms}}\}.$$

The properties of Ornstein-Uhlenbeck semigroup may be refered to [3].

Theorem 2. *Assume*

$$\beta = \inf_{0 \neq \xi \in \mathcal{D}(H)} \frac{(H\xi, \xi)}{(\xi, \xi)} > 0. \tag{2.5}$$

Then for any $p \geq 1$, $q(t) = 1 + (p-1)e^{2\beta t}$, $t \geq 0$ and $f \in (L^p)$ with $f \geq 0$ we have

$$\|P_t^H f\|_{q(t)} \leq \|f\|_p. \tag{2.6}$$

Proof. Let $\varphi, \psi \in (S)$. Then

$$\langle P_t^H \varphi, \psi \rangle = \iint \varphi(M_t x + T_t y)\psi(x)\mu(dx)\mu(dy). \tag{2.7}$$

Now take

$$(\Omega, \mathcal{F}, \mathsf{P}) = (S'(\boldsymbol{R}) \times S'(\boldsymbol{R}), \mathcal{B}(S'(\boldsymbol{R})) \times \mathcal{B}(S'(\boldsymbol{R})), \mu \times \mu).$$

We shall discuss on the probability space $(\Omega, \mathcal{F}, \mathsf{P})$. Put

$$X_\xi(x,y) = \langle \xi, x \rangle, \qquad \xi \in S(\boldsymbol{R}),$$
$$Y_\eta(x,y) = \langle \eta, M_t x + T_t y \rangle, \quad \eta \in S(\boldsymbol{R}).$$

It is not difficult to see that $\{X_\xi, \xi \in S(\boldsymbol{R})\}$ and $\{Y_\eta, \eta \in S(\boldsymbol{R})\}$ are jointly normally distributed, and $\forall \xi, \eta \in S(\boldsymbol{R})$

$$\mathsf{E}X_\xi = 0, \quad \mathsf{E}Y_\eta = 0,$$
$$\mathsf{E}X_\xi^2 = \|\xi\|_2^2, \quad \mathsf{E}Y_\eta^2 = \|M_t\eta\|_2^2 + \|T_t\eta\|_2^2 = \|\eta\|_2^2.$$

If $\|\xi\|_2 = \|\eta\|_2 = 1$, then

$$|\mathsf{E}X_\xi Y_\eta| = |\langle \xi, M_t\eta \rangle| = |\langle e^{-tH}\xi, \eta \rangle| \leq e^{-\beta t} = r.$$

Thus $\forall \xi, \eta \in S(\boldsymbol{R})$

$$|\rho(X_\xi, Y_\eta)| \leq r.$$

Noting that $S(\boldsymbol{R})$ is a linear space and X_ξ, Y_η are linear in ξ, η respectively, $\{X_\xi, \xi \in S(\boldsymbol{R})\}$ and $\{Y_\eta, \eta \in S(\boldsymbol{R})\}$ satisfy the condition (1.2). Denote by $\overline{q}(t)$ the conjugate index of $q(t)$:

$$\overline{q}(t) = 1 + \frac{1}{p-1}e^{-2\beta t}.$$

Then we have

$$(p-1)(\overline{q}(t) - 1) = e^{-2\beta t} = r^2.$$

Noting that $\psi(x)$ and $\varphi(M_t x + T_t y)$ are measurable with respect to $\{X_\xi, \xi \in S(\boldsymbol{R})\}$ and $\{Y_\eta, \eta \in S(\boldsymbol{R})\}$ respectively, from (2.7) and Theorem 1 we get

$$|\langle P_t^H \varphi, \psi \rangle| \leq \|\varphi\|_p \|\psi\|_{\overline{q}(t)}, \quad \forall \varphi, \psi \in (S).$$

Hence

$$\|P_t^H \varphi\|_{q(t)} \leq \|\varphi\|_p.$$

By the density of (S) in (L^p), (2.6) follows immediately. □

By Cauchy - Schwarz inequality for all $\xi \in S(\boldsymbol{R})$ we have

$$\frac{(H\xi, H\xi)}{(H\xi, \xi)} \geq \frac{(H\xi, \xi)}{(\xi, \xi)}.$$

Hence $\alpha \geq \beta$, and Theorem 2 is weaker than the result in [3]. In the cases when $\alpha = \beta$, we arrive at the same conclusion as in [3].

Lemma 1. *Let $\beta > 0$. Then*

$$\beta = \inf_{0 \neq \xi \in \mathcal{D}(H)} \frac{(H\xi,\, H\xi)}{(H\xi,\, \xi)}. \tag{2.8}$$

Proof. Denote by γ the right side of (2.8). Obviously, we have $\gamma \geq \beta$. Let $\{E_l, l > 0\}$ be the spectral system of H. $\forall \epsilon > 0$, take $0 \neq \xi \in \mathcal{D}(H)$ such that $\xi = (E_{\beta+\epsilon} - E_{\beta-0})\xi$. Then

$$\frac{(H\xi,\, H\xi)}{(H\xi,\, \xi)} = \frac{\int_{[\beta,\beta+\epsilon]} l^2 d(E_l\xi,\, \xi)}{\int_{[\beta,\beta+\epsilon]} l d(E_l\xi,\, \xi)} \leq \frac{(\beta+\epsilon)^2}{\beta}.$$

Hence

$$\gamma \leq \frac{(\beta+\epsilon)^2}{\beta}.$$

Letting $\epsilon \to 0$ yields $\gamma \leq \beta$, and (2.8) follows. □

Theorem 3. *Let $\beta > 0$. If $HS(\boldsymbol{R})$ is dense in $L^2(\boldsymbol{R})$, then*

$$\alpha = \beta. \tag{2.9}$$

Proof. By Lemma 1, it suffices to show

$$\inf_{0 \neq \xi \in S(\boldsymbol{R})} \frac{(H\xi,\, H\xi)}{(H\xi,\, \xi)} = \inf_{0 \neq \xi \in \mathcal{D}(H)} \frac{(H\xi,\, H\xi)}{(H\xi,\, \xi)}. \tag{2.10}$$

Under our assumption, H^{-1} is well defined on $HS(\boldsymbol{R})$, and indeed can be extended as a bounded positive self-adjoint operator on $L^2(\boldsymbol{R})$. Noting that

$$\frac{(H\xi,\, H\xi)}{(H\xi, \xi)} = \left[\frac{(H^{-1}H\xi,\, H\xi)}{(H\xi,\, H\xi)}\right]^{-1},$$

(2.10) is equivalent to

$$\sup_{0 \neq \eta \in HS(\boldsymbol{R})} \frac{(H^{-1}\eta,\, \eta)}{(\eta,\, \eta)} = \sup_{0 \neq \eta} \frac{(H^{-1}\eta,\, \eta)}{(\eta,\, \eta)}. \tag{2.11}$$

Then (2.10) follows from the density of $HS(\boldsymbol{R})$ in $L^2(\boldsymbol{R})$. □

Remark. If H is a continuous operator in $L^2(\boldsymbol{R})$, i.e. H is a bounded self-adjoint operator, then $HS(\boldsymbol{R})$ is dense in $L^2(\boldsymbol{R})$. In fact, in this case $HS(\boldsymbol{R})$ is dense in the range of H. But the range of H is dense in $L^2(\boldsymbol{R})$, since H is a strictly positive self-adjoint operator in $L^2(\boldsymbol{R})$.

It is also easy to see that (2.9) holds for $H = A$, the self-adjoint extension of the harmonic oscillator (cf. [3])

$$-\frac{d^2}{dx^2} + (x^2 + 1),$$

since the system of the eigenfunctions A is contained in $S(\boldsymbol{R})$, and forms an orthogonal base of $L^2(\boldsymbol{R})$.

REFENRENCES

[1] Hida, T., Kuo, H. H., Potthoff, J. and Streit, L., *White Noise - An Infinite Dimensional Calculus*, Kluwer Academic Publ., 1993.

[2] Neveu, J., Sur l'espérance conditionnelle par rapport à un mouvement brownien, Ann. Inst. Henri Poincare XII(1976), 105–109.

[3] Qian, Z. M. and He, S. W., On the hypercontractivity of Ornstein - Uhlenbeck semigroup with drift, Sém. Probab. XXIX Lecture Notes in Math. no.**1613**, 202–217, Springer, 1995.

[4] Yan J. A., Some recent developments in white noise analysis. In *Probability and Statistics*, A. Badrikian et al. (eds.), World Scientific, 1993.

Sheng-Wu He,
Department of Statistics,
East China Normal University,
200062 Shanghai, China.

Jia-Gang Wang,
Institute of Applied Mathematics,
East China University of Science and Technology,
200237 Shanghai, China.

UNE PREUVE STANDARD DU PRINCIPE D'INVARIANCE DE STOLL

B.Cadre
IRMAR, Université de Rennes I, Campus de Beaulieu, 35042 RENNES

Introduction

En utilisant l'analyse non-standard, Stoll [20] a prouvé que le temps local d'intersection renormalisé d'un mouvement brownien plan pouvait être construit par passage à la limite d'un équivalent discret construit avec des marches aléatoires. Le but de cet article est de retrouver ce résultat, en utilisant des techniques classiques.

Rappelons donc tout d'abord ce qu'est le temps local d'intersection renormalisé d'un mouvement brownien plan W issu de 0 sur $(\Omega, \mathcal{F}, P)$ (nous renvoyons à Bass-Khoshnevisan [1], Le Gall [10] et [11], Werner [22] et Yor [24] pour de plus amples informations). Pour tout A borélien borné de $\mathbb{R}_+^2$, définissons la mesure $\lambda_A(dz)$ sur $(\mathbb{R}^2, \mathcal{B}(\mathbb{R}^2))$ par $\lambda_A(f) = \int_A f(W_u - W_v)dudv$, pour toute fonction $f : \mathbb{R}^2 \to \mathbb{R}_+$ borélienne bornée. Rosen [16] a montré que la mesure λ_A admet une densité $(\alpha(z, A); z \in \mathbb{R}^2)$ par rapport à la mesure de Lebesgue, appelée temps local d'intersection de W. Cependant, $\alpha(0, A) = +\infty$ P-ps lorsque A rencontre la diagonale de $\mathbb{R}_+^2$. Le Gall [10], Rosen [17] et Yor [24] montrent alors, selon l'idée de renormalisation de Varadhan [21], que la fonction $z \mapsto \gamma(z, A)$ définie pour $z \neq 0$ par $\gamma(z, A) = \alpha(z, A) - E[\alpha(z, A)]$ se prolonge par continuité à $\mathbb{R}^2$ tout entier. Naturellement, $(\gamma(z, A); z \in \mathbb{R}^2)$ a été baptisé temps local d'intersection renormalisé de W.

Considérons alors une marche aléatoire de carré intégrable $(S_n)_{n\geq 0}$ issue de 0, à valeurs dans $\mathbb{Z}^2$, telle que $(S_{[nt]}/\sqrt{n})_{t\leq 1}$ converge en loi vers un mouvement brownien plan W issu de 0. Si, pour chaque borélien borné A, $(\gamma(z, A); z \in \mathbb{R}^2)$ désigne le temps local d'intersection renormalisé de W, on peut légitimement s'attendre, au moins pour certaines de ces marches aléatoires, à ce que la convergence en loi suivante soit vraie:

$$\frac{1}{n}\sum_{i,j=0}^{n}(I_{(S_i=S_j)} - P(S_i = S_j)) \xrightarrow[n\to\infty]{\mathcal{L}} \gamma(0, [0, 1]^2).$$

En dehors de Stoll [20] qui a prouvé cette convergence dans un contexte non-standard, ce résultat a été montré, en utilisant l'analyse standard, par Brydges et Slade [4] (lorsque $(S_n)_{n\geq 0}$ est la marche aléatoire simple symétrique sur $\mathbb{Z}^2$) ou bien par Le Gall [12] (dans un contexte très général concernant la marche aléatoire $(S_n)_{n\geq 0}$). Signalons enfin que Rosen [18] a lui aussi obtenu ce résultat avec l'analyse standard, mais il doit supposer que la marche aléatoire $(S_n)_{n\geq 0}$ est fortement apériodique, ce qui n'est pas le cas de la marche aléatoire simple symétrique sur $\mathbb{Z}^2$ par exemple.

Les méthodes utilisées par ces auteurs reposent essentiellement sur des estimations classiques de la transformée de Fourier d'une marche aléatoire. La méthode que nous allons exposer, quant à elle, repose aussi bien entendu sur de telles estimations : nous utilisons des inégalités prouvées par Stoll [20], que l'on peut obtenir aussi bien avec l'analyse standard (dans la preuve de Stoll, l'intérêt de l'analyse non-standard apparaît plus loin). Mais aussi, notre méthode repose sur un résultat de plongement. Plus précisément, nous montrons que certaines marches aléatoires à valeurs dans $\mathbb{Z}^2$ peuvent être plongées dans un mouvement brownien plan. Ce plongement est en fait obtenu par prolongement du résultat classique de plongement en dimension 1, que l'on trouve par exemple dans le livre de Revuz et Yor [15]. Enfin, mentionnons le fait que nous prouvons en réalité un résultat de convergence fonctionnelle sur les variables de temps et d'espace, alors que les auteurs précédents obtiennent des résultats de convergence ponctuelle.

Comme conséquence du théorème limite que nous venons d'évoquer (en réalité comme conséquence d'un résultat un peu plus fort), nous retrouvons un résultat prouvé par Stoll [20] (toujours en utilisant l'analyse non-standard), Le Gall [12] et Brydges et Slade [4]. Ce résultat affirme que la mesure de Domb-Joyce converge au sens de la topologie de la convergence étroite vers la mesure de polymère.

Nous préciserons en partie I les hypothèses et les notations de cet article, et nous définirons les mesures de Domb-Joyce et de polymère. Dans cette même partie, nous énonçerons le principal résultat (théorème 1) concernant l'approximation du temps local d'intersection renormalisé du mouvement brownien plan. Enfin, le résultat de convergence pour la mesure de Domb-Joyce sera énonçé (corollaire 1).

La partie II sera consacrée à rappeler quelques résultats plus ou moins classiques sur les temps locaux d'intersections associés au mouvement brownien ou à la marche aléatoire.

Dans la partie III, nous prouverons le théorème 1. C'est dans cette partie que le lemme de plongement (lemme 3) sera énonçé et prouvé.

Enfin, en partie IV, nous prouverons le corollaire 1.

I- Notations et présentation des principaux résultats

Dans tout ce qui suit, nous noterons $\to^{L^p}$ la convergence dans L^p pour $p \geq 1$ et $\to^{\mathcal{L}}$ la convergence en loi.

Si $z = (z^1, z^2) \in \mathbb{R}^2$, nous appelerons $[z]$ l'élément de $\mathbb{Z}^2$: $([z^1], [z^2])$, où $[.]$ désigne la partie entière.

D'autre part, si M est un processus stochastique, nous noterons $\mathcal{F}^M$ la filtration naturelle associée à M, $\mathcal{L}(M)$ la loi de M, et si M est à valeurs dans $\mathbb{R}^2$, nous appelerons (M^1, M^2) ses coordonnées.

Si $Z \in L^1$, nous noterons $\{Z\}$ la variable aléatoire recentrée $\{Z\} = Z - E[Z]$.

Enfin, pour chaque $t \in \mathbb{R}_+$, Δ_t désignera le triangle défini par $\Delta_t = \{(u,v) \in \mathbb{R}_+^2 : 0 \leq u < v \leq t\}$ et pour $k \in \mathbb{N}$, D_k sera l'équivalent discret de Δ_t : $D_k = \{(i,j) \in \mathbb{N}^2 : 0 \leq i < j \leq k\}$.

Soit Q une mesure de probabilité centrée sur $\mathbb{Z}^2$, à support compact, et telle que

$$Q(x,y) = Q(x,-y) \; \forall x, y \in \mathbb{Z}.$$

Si $(X_k)_{k\geq 1}$ est une suite de variables aléatoires indépendantes et même loi Q, on associe la marche aléatoire $(S_n)_{n\geq 0}$ issue de 0 définie pour chaque $n \geq 1$ par $S_n = \sum_{k=1}^{n} X_k$. Nous supposerons enfin que $(S_n)_{n\geq 0}$ est adaptée (apériodique au sens de Spitzer [19]), au sens où cette marche aléatoire ne vit pas sur un sous-groupe strict de $\mathbb{Z}^2$. Par exemple, la marche aléatoire simple symétrique sur $\mathbb{Z}^2$ vérifie ces hypothèses.

En dehors de l'hypothèse $Q(x, y) = Q(x, -y)\ \forall x, y \in \mathbb{Z}$, qui entraîne que la matrice $\int z.z^T dQ(z)$ est diagonale, toutes ces hypothèses sur la marche aléatoire $(S_n)_{n\geq 0}$ ont été supposées par Stoll [20]. Nous pourrons donc utiliser dans la suite certaines des estimations de Stoll qui peuvent être obtenues aussi bien au moyen de l'analyse standard.

Nous appelerons aussi, pour chaque $n \geq 1$, $z \in \mathbb{Z}^2$ et A sous-ensemble borné de $\mathbb{N}^2$:

$$\gamma_n(z, A) = \frac{1}{n} \sum_{(i,j)\in A} \{I_{(S_j = S_i + z)}\}.$$

Notons W un mouvement brownien plan sur $(\Omega, \mathcal{F}, P)$ issu de 0, avec $\text{cov}(W_1) = \int z.z^T dQ(z)$. Pour chaque borélien borné A de $\mathbb{R}_+^2$, nous assoçions à W son temps local d'intersection renormalisé $(\gamma(z, A); z \in \mathbb{R}^2)$.

Notre principal résultat est le suivant:

Théorème 1 *Nous avons le résultat de convergence fonctionnelle:*

$$\left(\left(\frac{S_{[nt]}}{\sqrt{n}}\right)_{t\leq 1}, (\gamma_n([z\sqrt{n}], D_{[nt]}))_{t\leq 1, z\in\mathbb{R}^2}\right) \xrightarrow[n\to\infty]{\mathcal{L}} \left((W_t)_{t\leq 1}, (\gamma(z, \Delta_t))_{t\leq 1, z\in\mathbb{R}^2}\right).$$

Remarque - Un des intérêts de l'utilisation d'un lemme de plongement (alors que les auteurs précédents ne s'en servent pas) est que l'on peut obtenir, moyennant de très légères modifications de la preuve du théorème 1, la version forte du théorème 1, prolongeant ainsi l'étude de Csàki et Révész [6] pour le temps local du mouvement brownien réel (voir [5], théorème II.3.2.1).

- Une méthode similaire à celle que nous allons exposer permet de retrouver le résultat équivalent en dimension 1, résultat qui a été obtenu par exemple par Borodin [3].

- Nous avons montré dans [5] (théorème II.2.4.1) le résultat correspondant en dimension 3 (en supposant donc que $z \neq 0$ car il n'existe pas de renormalisation de Varadhan en dimension 3).

Pour chaque $g \geq 0$ et $n \geq 1$, définissons maintenant la mesure de Domb-Joyce μ_g^n (cf. [7]) par:

$$\mu_g^n(d\omega) = \frac{1}{L_g^n} \exp\left(-\frac{g}{n} \sum_{i<j\leq n} I_{(S_j = S_i)}\right) P_n(d\omega),$$

où P_n est la loi de $(S_{[nt]}/\sqrt{n})_{t\leq 1}$ et $L_g^n = E[\exp(-g/n \sum_{i<j\leq n} I_{(S_j=S_i)}]$. Notons de plus, pour chaque $g \geq 0$, μ_g la mesure de polymère définie par:

$$\mu_g(d\omega) = \frac{1}{L_g}\exp(-g\gamma(0,\Delta_1))P_W(d\omega),$$

où P_W est la loi de W et $L_g = E[\exp(-g\gamma(0,\Delta_1))]$. Nous verrons plus loin que $\gamma(0,\Delta_1)$ possède des moments exponentiels, ce qui justifie la définition qui précède.

Il est intuitivement clair que μ_g est l'équivalent continu de μ_g^n. C'est ce que prouve le corollaire suivant:

Corollaire 1 *Pour tout $g \geq 0$, μ_g^n converge au sens de la topologie de la convergence étroite vers μ_g.*

II- Quelques rappels sur les temps locaux d'intersection

Tout d'abord, rappelons la formule classique du temps d'occupation:

Proposition 1 *Soit f une fonction borélienne bornée. Si A est un borélien borné de $\mathbb{R}_+^2$, le temps local d'intersection renormalisé $(\gamma(z,A); z \in \mathbb{R}^2)$ de W vérifie P-ps l'égalité:*

$$\int_{\mathbb{R}^2} f(z)\gamma(z,A)dz = \int_A \{f(W_u - W_v)\}dudv.$$

Le résultat qui suit a été obtenu, du moins lorsque $z = 0$, par Le Gall [10], Rosen [17] et Yor [24]. L'extension à $z \in \mathbb{R}^2$ est évidente à partir des démonstrations de ces auteurs.

Théorème 2 *Soit $(f_n)_{n\geq 1}$ une suite de fonctions bornées, d'intégrales égales à 1, tel que $f_n(y)dy$ converge étroitement vers la mesure de Dirac en $z \in \mathbb{R}^2$. Alors, si A est un borélien borné de $\mathbb{R}_+^2$, on a pour chaque $p \geq 1$:*

$$\int_A \{f_n(W_u - W_v)\}dudv \xrightarrow[n\to\infty]{L^p} \gamma(z,A).$$

Enfin, le dernier résultat a été établi par Stoll [20] dans un contexte d'analyse non-standard. La preuve est en réalité calquée sur celles de Geman, Horowitz et Rosen [8] et de Le Gall [10] dans le cas brownien, et elle ne nécessite pas l'utilisation de l'analyse non-standard. Dans la preuve de Stoll, l'intérêt de l'analyse non-standard apparaît plus tard.

Théorème 3 *Soit p un entier pair. Pour tout $\lambda \in]0,1[$, il existe une constante c telle que si $z_1, z_2 \in \mathbb{Z}^2$ et $n \geq 1$:*

i) $||\gamma_n(z_1, D_n) - \gamma_n(z_2, D_n)||_{L^p} \leq c(\frac{|z_1 - z_2|}{\sqrt{n}})^\lambda$;

ii) $||\gamma_n(z_1, D_n)||_{L^p} \leq c$.

Si de plus $A = \{(i,j) : i = a_1, \cdots, a_2$ et $j = b_1, \cdots, b_2\}$, *où* a_1, a_2, b_1, b_2 *sont des entiers tels que* $a_1 < a_2 \leq b_1 < b_2$*:*

$$\text{iii)} \; \|\sum_{(i,j)\in A} I_{(S_j - S_i = z_1)}\|_{L^p} \leq \frac{c}{n}\sqrt{(a_2-a_1)(b_2-b_1)}.$$

III- Preuve du Théorème 1

Le théorème 1 sera une conséquence des deux lemmes, que nous montrerons à la fin de cette partie:

Lemme 1 *Pour chaque* $n \geq 1$, *il existe une suite de variables aléatoires* $(T_k^n)_{k\leq n} = (T_k^{1,n}, T_k^{2,n})_{k\leq n}$ *telle que si le processus* $(W^1_{T^{1,n}_{[nt]}}, W^2_{T^{2,n}_{[nt]}})_{t\leq 1}$ *est noté de façon abusive* $(W_{T^n_{[nt]}})_{t\leq 1}$*:*

$$\mathcal{L}((W_{T^n_{[nt]}})_{t\leq 1}) = \mathcal{L}((\frac{S_{[nt]}}{\sqrt{n}})_{t\leq 1}) \text{ et } \sup_{t\leq 1} |W_{T^n_{[nt]}} - W_t| \xrightarrow[n\to\infty]{L^1} 0.$$

De plus, nous avons le résultat:

Lemme 2 *Soit* p *un entier pair. Il existe une constante* c *telle que pour chaque* $n \geq 1$, $z \in \mathbb{Z}^2$ *et* $s < t \leq 1$*:*

$$\|\gamma_n(z, D_{[nt]}) - \gamma_n(z, D_{[ns]})\|_{L^p} \leq c\sqrt{\frac{[nt]-[ns]}{n}}.$$

Preuve du théorème 1 Dans ce qui suit, c désignera une constante positive qui pourra varier de place en place.

Soit $t \leq 1$, $z \in \mathbb{R}^2$ et $f : \mathbb{R}^2 \to \mathbb{R}$ une fonction différentiable à support compact, de différentielle bornée et d'intégrale 1. Soit pour chaque $n \geq 1$, $(W_{T^n_{[nt]}})_{t\leq 1}$ le processus introduit dans le lemme 1, et $\theta(n)$ le nombre

$$\theta(n) = (E[\sup_{t\leq 1} |W_{T^n_{[nt]}} - W_t|] + \frac{1}{\sqrt{n}})^{-1/6}.$$

Toujours d'après le lemme 1, $\theta(n) \to \infty$ si $n \to \infty$. Enfin, désignons pour chaque $n \geq 1$ la fonction f_n définie pour tout $y \in \mathbb{R}^2$ par

$$f_n(y) = \theta(n)^2 f(\theta(n)(y - z)).$$

D'après le théorème 2, puisque $f_n(y)dy$ converge au sens de la topologie de la convergence étroite vers la mesure de Dirac en z:

$$\int_{\mathbb{R}^2} f_n(y)\gamma(y, \Delta_t)dy \xrightarrow[n\to\infty]{L^1} \gamma(z, \Delta_t).$$

D'autre part, d'après la proposition 1, on a pour chaque $n \geq 1$:

$$\int_{\mathbb{R}^2} f_n(y)\gamma(y,\Delta_t)dy = \int_{\Delta_t} \{f_n(W_v - W_u)\}dudv.$$

En utilisant le théorème des accroissements finis, on obtient d'après notre choix pour la suite $(\theta(n))_{n\geq 1}$:

$$\begin{aligned}
&E[|\int_{\Delta_t} \{f_n(W_v - W_u)\}dudv - \int_{\Delta_t} \{f_n(W_{T^n_{[nv]}} - W_{T^n_{[nu]}})\}dudv|] \\
&\leq c\,\theta(n)^3 E[\sup_{t\leq 1} |W_{T^n_{[nt]}} - W_t|] \\
&\leq c\,(E[\sup_{t\leq 1} |W_{T^n_{[nt]}} - W_t|])^{1/2}
\end{aligned}$$

qui tend vers 0 si n tend vers l'infini selon le lemme 1. Or, toujours d'après le lemme 1, pour chaque $n \geq 1$:

$$\mathcal{L}(\int_{\Delta_t} \{f_n(W_{T^n_{[nv]}} - W_{T^n_{[nu]}})\}dudv) = \mathcal{L}(\int_{\Delta_t} \{f_n(\frac{S_{[nv]} - S_{[nu]}}{\sqrt{n}})\}dudv).$$

Puis, on a pour chaque $n \geq 1$:

$$\begin{aligned}
\int_{\Delta_t} \{f_n(\frac{S_{[nv]} - S_{[nu]}}{\sqrt{n}})\}dudv &= \frac{1}{n^2} \sum_{j=0}^{[nt]-1} \sum_{l=j+1}^{[nt]} \{f_n(\frac{S_l - S_j}{\sqrt{n}})\} \\
&= \frac{1}{n} \sum_{y\in\mathbb{Z}^2} f_n(\frac{y}{\sqrt{n}})\gamma_n(y, D_{[nt]}) \\
&= \int_{\mathbb{R}^2} f_n(\frac{[y\sqrt{n}]}{\sqrt{n}})\gamma_n([y\sqrt{n}], D_{[nt]})dy
\end{aligned}$$

Utilisant une nouvelle fois le théorème des accroissements finis, nous trouvons pour chaque $n \geq 1$:

$$\begin{aligned}
&E[|\int_{\mathbb{R}^2} f_n(\frac{[y\sqrt{n}]}{\sqrt{n}})\gamma_n([y\sqrt{n}], D_{[nt]})dy - \int_{\mathbb{R}^2} f_n(y)\gamma_n([y\sqrt{n}], D_{[nt]})dy|] \\
&\leq c\,\theta(n)^3 \int_K |\frac{[y\sqrt{n}]}{\sqrt{n}} - y| E[|\gamma_n([y\sqrt{n}], D_{[nt]})|]dy \\
&\leq c \sup_{n\geq 1} \sup_{y\in\mathbb{Z}^2} E[|\gamma_n(y, D_{[nt]})|] \frac{\theta(n)^3}{\sqrt{n}} \\
&\leq c \sup_{n\geq 1} \sup_{y\in\mathbb{Z}^2} E[|\gamma_n(y, D_{[nt]})|] \frac{1}{n^{1/4}},
\end{aligned}$$

où K est un certain compact de $\mathbb{R}^2$ contenant le support de f. Or, $E[|\gamma_n(y, D_{[nt]})|]$ est borné uniformément en $n \geq 1$ et $y \in \mathbb{Z}^2$ d'après le théorème 3 ii), donc le dernier terme de l'inégalité ci-dessus tend vers 0 si n tend vers l'infini. Ensuite, par un changement de variable évident, on a pour chaque $n \geq 1$:

$$\int_{\mathbb{R}^2} f_n(y)\gamma_n([y\sqrt{n}], D_{[nt]})dy = \int_{\mathbb{R}^2} f(y)\gamma_n([\frac{y\sqrt{n}}{\theta(n)} + z\sqrt{n}], D_{[nt]})dy.$$

Alors, d'après le théorème 3 i), puisque $\int f(y)dy = 1$:

$$E[| \int_{\mathbb{R}^2} f(y)\gamma_n([\frac{y\sqrt{n}}{\theta(n)} + z\sqrt{n}], D_{[nt]})dy - \gamma_n([z\sqrt{n}], D_{[nt]}|]$$
$$\leq \int_{\mathrm{supp} f} E[|\gamma_n([\frac{y\sqrt{n}}{\theta(n)} + z\sqrt{n}], D_{[nt]})dy - \gamma_n([z\sqrt{n}], D_{[nt]})|]dy \leq \frac{c}{\theta(n)^{1/2}}.$$

Puisque $\theta(n) \to \infty$ si $n \to \infty$, ceci montre finalement que nous avons obtenu, à $t \leq 1$ et $z \in \mathbb{R}^2$ fixés, la convergence en loi:

$$\gamma_n([z\sqrt{n}], D_{[nt]}) \xrightarrow[n\to\infty]{\mathcal{L}} \gamma(z, \Delta_t).$$

Il est clair que la méthode que nous venons d'exposer s'applique aussi bien pour établir la convergence des répartitions fini-dimensionnelles de la suite de processus

$$((S_{[nt]}/\sqrt{n})_{t\leq 1}, (\gamma_n([z\sqrt{n}], D_{[nt]}))_{t\leq 1, z\in \mathbb{R}^2})_{n\geq 1}$$

vers celles de $((W_t)_{t\leq 1}, (\gamma(z, \Delta_t))_{t\leq 1, z\in\mathbb{R}^2})$. Il reste à obtenir un résultat de tension pour cette suite de processus. La tension de la suite de processus $((S_{[nt]}/\sqrt{n})_{t\leq 1})_{n\geq 1}$ se montre de la façon classique (voir par exemple le théorème 16.1 de Billingsley [2]). Pour simplifier, nous nous contenterons de montrer que la suite de processus $((\gamma_n([z\sqrt{n}], D_{[nt]}))_{t\leq 1})_{n\geq 1}$ à $z \in \mathbb{R}^2$ fixé. Pour obtenir en plus la tension de la suite $((\gamma_n([z\sqrt{n}], D_{[nt]}))_{z\in\mathbb{R}^2, t\leq 1})_{n\geq 1}$, on procède de la même manière en utilisant le théorème 3 i) et le lemme 2.

Soient $z \in \mathbb{R}^2$, $t_1 \leq t \leq t_2 \leq 1$ et $n \geq 1$. Selon le lemme 2, on a

$$E[(\gamma_n([z\sqrt{n}], D_{[nt]}) - \gamma_n([z\sqrt{n}], D_{[nt_1]}))^2(\gamma_n([z\sqrt{n}], D_{[nt]}) - \gamma_n([z\sqrt{n}], D_{[nt_2]}))^2]$$
$$\leq c\,\frac{[nt]-[nt_1]}{n}\,\frac{[nt_2]-[nt]}{n} \leq c\,(\frac{[nt_2]-[nt_1]}{n})^2.$$

Deux cas se présentent:
- Si $t_2 - t_1 \geq 1/n$ alors $[nt_2] - [nt_1] \leq 2(nt_2 - nt_1)$
- Si $t_2 - t_1 < 1/n$ alors, t_1 et t ou t_2 et t sont dans le même intervalle $[(i-1)/n, i/n[$ et donc $\gamma_n([z\sqrt{n}], D_{[nt]})$ est soit égal à $\gamma_n([z\sqrt{n}], D_{[nt_1]})$, soit à $\gamma_n([z\sqrt{n}], D_{[nt_2]})$.

Globalement, nous avons donc

$$E[(\gamma_n([z\sqrt{n}], D_{[nt]}) - \gamma_n([z\sqrt{n}], D_{[nt_1]}))^2(\gamma_n([z\sqrt{n}], D_{[nt]}) - \gamma_n([z\sqrt{n}], D_{[nt_2]}))^2]$$
$$\leq c(t_2 - t_1)^2,$$

ce qui, d'après le théorème 15.6 de Billingsley [2], montre la tension de la suite des processus $((\gamma_n([z\sqrt{n}], D_{[nt]}))_{t\leq 1})_{n\geq 1}$, et termine la preuve du théorème 1 ∎

Nous allons tout d'abord prouver le résultat de plongement du lemme 1, et nous prouverons que la marche aléatoire résultant de ce plongement vérifie la deuxième assertion du lemme 1. Afin d'établir le résultat de plongement en dimension 2, nous allons utiliser un résultat de plongement uni-dimensionnel montré par exemple dans Revuz et Yor [15] (théorème 5.4, chapitre VI). Rappelons ce résultat dans le contexte qui nous intéresse:

Théorème 4 *Soit B un mouvement brownien réel centré sur $(\Omega, \mathcal{F}, P)$ et ν une mesure de probabilité centrée telle que $\int |x| d\nu(x) < \infty$. Il existe un temps d'arrêt T_ν (dépendant mesurablement de ν) P-ps fini, relativement à $\mathcal{F}^B$ et vérifiant:*

i) $(B_{t\wedge T_\nu})_{t\geq 0}$ est uniformément integrable et $\mathcal{L}(B_{T_\nu}) = \nu$
ii) $E[B_1^2]\, E[T_\nu] = \int x^2 d\nu(x)$
iii) Pour tout $p \geq 1$, il existe une constant universelle c_p telle que $E[B_1^{2p}]\, E[T_\nu^p] \leq c_p \int x^{2p} d\nu(x)$.

Remarque Le point iii) ne figure pas dans Revuz et Yor [15], mais est donné par Haeusler [9].

Avant d'énoncer notre résultat de plongement, nous définissons quelques notations. Rappelons que $(X_k)_{k\geq 1}$ est la suite de variables aléatoires indépendantes et de même loi Q telles que $S_n = \sum_{k=1}^n X_k$ pour chaque $n \geq 1$. Soit C le support de la mesure $P_{X_1^1}$:

$$C = \{x \in \mathbb{Z} : P_{X_1^1}(x) > 0\},$$

et pour chaque $n \geq 1$, $C_n = (1/\sqrt{n})C$.

Lemme 3 *Soit $n \geq 2$ et $(x_k)_{k\geq 1}$ une suite d'éléments de C_n. Il existe des temps d'arrêts $\tau_1^{1,n}, \cdots, \tau_n^{1,n}, \tau_1^{2,n}(x_1), \tau_2^{2,n}(x_1, x_2) \cdots, \tau_n^{2,n}(x_1, \cdots, x_n)$ adaptés respectivement aux filtrations engendrées par des mouvements browniens réels indépendants $W^{1,1}, \cdots, W^{1,n}, W^{2,1}, W^{2,2}(x_1), \cdots, W^{2,n}(x_1, \cdots, x_{n-1})$ tels que, si $T_0^{1,n} = T_0^{2,n} = 0$ et pour chaque $k \geq 1$, $T_k^{1,n} = \sum_{i=1}^k \tau_i^{1,n}$ et $T_k^{2,n} = \sum_{i=1}^k \tau_i^{2,n}(W^{1,1}_{\tau_1^{1,n}}, \cdots, W^{1,n}_{\tau_i^{1,n}})$, on a*

$$\mathcal{L}((W^1_{T_k^{1,n}}, W^2_{T_k^{2,n}})_{k\leq n}) = \mathcal{L}((\frac{S_k}{\sqrt{n}})_{k\leq n}).$$

De plus, ces temps d'arrêts sont tels que pour chaque $p \geq 1$, il existe une constante c_p telle que $\forall n \geq 2$:

$$\sup_{k=1,\cdots,n} \Big(\sup_{x_1,\cdots,x_k \in C_n} E[\tau_k^{2,n}(x_1, \cdots, x_k)^p] + \sup_{k=1,\cdots,n} E[(\tau_k^{1,n})^p]\Big) \leq \frac{c_p}{n^p}$$

$$\text{et } \forall n \geq 2, k = 1, \cdots, n \ : \ E[\tau_k^{1,n}] = E[\tau_k^{2,n}(W^{1,1}_{\tau_1^{1,n}}, \cdots, W^{1,k}_{\tau_k^{1,n}})] = \frac{1}{n}.$$

Remarque Dans l'énonçé de ce lemme, les mouvements browniens réels $W^{1,1}, \cdots,$

$W^{1,n}, W^{2,1}, W^{2,2}(x_1), \cdots, W^{2,n}(x_1, \cdots, x_{n-1})$ dépendent en réalité de n, comme nous le verrons dans la preuve. Nous n'écrivons pas cette dépendance en n de façon à ne pas alourdir les notations.

Preuve Soit $n \geq 2$ et $(x_k)_{k\geq 1}$ une suite d'éléments de C_n. Dans un souci de simplification, notons pour chaque $k \geq 1$:

$$(U_k^1, U_k^2) = (\frac{X_k^1}{\sqrt{n}}, \frac{X_k^2}{\sqrt{n}}).$$

Appelons $W^{1,1} := W^1$ et $W^{2,1} := W^2$. D'après le théorème 4, puisque la mesure $P_{U_1^1}$ est par hypothèse centrée, il existe un temps d'arrêt $\tau_1^{1,n}$ adapté à $\mathcal{F}^{W^{1,1}}$ tel que $\mathcal{L}(U_1^1) = \mathcal{L}(W^{1,1}_{\tau_1^{1,n}})$. De même, la mesure $P_{U_1^2 | U_1^1 = x_1}$ est centrée car $Q(x,y) = Q(x,-y)$ $\forall x, y \in \mathbb{Z}$ par hypothèse et donc, d'après le théorème 4, il existe un temps d'arrêt $\tau_1^{2,n}(x_1)$ adapté à $\mathcal{F}^{W^{2,1}}$ tel que $\mathcal{L}(U_1^2 | U_1^1 = x_1) = \mathcal{L}(W^{2,1}_{\tau_1^{2,n}(x_1)})$. Appelons alors $W^{1,2}$ et $W^{2,2}(x_1)$ les mouvements browniens réels définis pour chaque $t \geq 0$ par

$$W_t^{1,2} = W^{1,1}_{t+\tau_1^{1,n}} - W^{1,1}_{\tau_1^{1,n}} \text{ et } W_t^{2,2}(x_1) = W^{2,1}_{t+\tau_1^{2,n}(x_1)} - W^{2,1}_{\tau_1^{2,n}(x_1)}.$$

Pour la même raison, il existe d'après le théorème 4 un temps d'arrêt $\tau_2^{1,n}$ adapté à $\mathcal{F}^{W^{1,2}}$ tel que $\mathcal{L}(U_2^1) = \mathcal{L}(W^{1,2}_{\tau_2^{1,n}})$. De même, la mesure $P_{U_2^2 | U_2^1 = x_2}$ étant centrée, il existe un temps d'arrêt $\tau_2^{2,n}(x_1, x_2)$ adapté à $\mathcal{F}^{W^{2,2}(x_1)}$ tel que $\mathcal{L}(U_2^2 | U_2^1 = x_2) = \mathcal{L}(W^{2,2}_{\tau_2^{2,n}(x_1,x_2)}(x_1))$. Les processus $W^{1,1}, W^{1,2}$, ainsi que $W^{2,1}, W^{2,2}(x_1)$ sont indépendants, et de même, les variables aléatoires U_1^1, U_2^1, ainsi que U_1^2, U_2^2 sont indépendantes, donc

$$\begin{aligned}
&\mathcal{L}(U_1^1, U_2^1) = \mathcal{L}(W^{1,1}_{\tau_1^{1,n}}, W^{1,2}_{\tau_2^{1,n}}), \\
\text{et } &\mathcal{L}(U_1^2, U_2^2 | U_1^1 = x_1, U_2^1 = x_2) = \mathcal{L}(W^{2,1}_{\tau_1^{2,n}(x_1)}, W^{2,2}_{\tau_2^{2,n}(x_1,x_2)}(x_1)).
\end{aligned}$$

On itère ensuite de façon évidente le procédé, ce qui nous donne l'existence de temps d'arrêts $\tau_1^{1,n}, \cdots, \tau_n^{1,n}, \tau_1^{2,n}(x_1), \tau_2^{2,n}(x_1, x_2) \cdots, \tau_n^{2,n}(x_1, \cdots, x_n)$ adaptés aux filtrations engendrées par des mouvements browniens linéaires indépendants : $W^{1,1}, \cdots,$ $W^{1,n}, W^{2,1}, W^{2,2}(x_1), \cdots, W^{2,n}(x_1, \cdots, x_{n-1})$ où pour chaque $k = 3, \cdots, n$ et $t \geq 0$:

$$\begin{aligned}
W_t^{1,k} &= W^{1,k-1}_{t+\tau_{k-1}^{1,n}} - W^{1,k-1}_{\tau_{k-1}^{1,n}}, \\
W_t^{2,k}(x_1, \cdots, x_{k-1}) &= W^{2,k-1}_{t+\tau_{k-1}^{2,n}(x_1,\cdots,x_{k-1})}(x_1, \cdots, x_{k-2}) \\
&\quad - W^{2,k-1}_{\tau_{k-1}^{2,n}(x_1,\cdots,x_{k-1})}(x_1, \cdots, x_{k-2}).
\end{aligned}$$

Ces temps d'arrêt vérifient donc les relations:

$$\begin{aligned}
&\mathcal{L}(U_1^1, \cdots, U_n^1) = \mathcal{L}(W^{1,1}_{\tau_1^{1,n}}, \cdots, W^{1,n}_{\tau_n^{1,n}}) \\
\text{et } &\mathcal{L}(U_1^2, \cdots, U_n^2 | U_1^1 = x_1, \cdots, U_n^1 = x_n) \\
&= \mathcal{L}(W^{2,1}_{\tau_1^{2,n}(x_1)}, (W^{2,i}_{\tau_i^{2,n}(x_1,\cdots,x_i)}(x_1, \cdots, x_{i-1}))_{i=2,\cdots,n}).
\end{aligned}$$

Remarquons que d'après le théorème 4, pour chaque $k = 1, \cdots, n$, l'application $(x_1, \cdots, x_k) \mapsto \tau_k^{2,n}(x_1, \cdots, x_k)$ est mesurable. D'autre part, puisque la matrice de covariance de W est diagonale, les mouvements browniens linéaires $W^{2,1}, W^{2,2}(x_1), \cdots, W^{2,n}(x_1, \cdots, x_{n-1})$ sont indépendants de $W^{1,1}_{\tau_1^{1,n}}, \cdots, W^{1,n}_{\tau_n^{1,n}}$, et donc:

$$\mathcal{L}(W^{2,1}_{\tau_1^{2,n}(x_1)}, (W^{2,i}_{\tau_i^{2,n}(x_1,\cdots,x_i)}(x_1, \cdots, x_{i-1}))_{i=2,\cdots,n})$$
$$= \mathcal{L}(W^{1,1}_{\tau_1^{2,n}(W_{\tau_1^{1,n}})}, (W^{2,i}_{\tau_i^{2,n}(W^{1,1}_{\tau_1^{1,n}},\cdots,W^{1,i}_{\tau_i^{1,n}})}(W^{1,1}_{\tau_1^{1,n}}, \cdots, W^{1,i-1}_{\tau_{i-1}^{1,n}}))_{i=2,\cdots,n} | \bigcap_{i=1}^{n} [W^{1,i}_{\tau_i^{1,n}} = x_i]).$$

Ceci étant valable pour toute suite $(x_k)_{k\geq 1}$ d'éléments de C_n, et puisque les variables aléatoires $U_1^1, \cdots, U_n^1$ sont à valeurs dans C_n, nous déduisons des trois dernières relations que:

$$\mathcal{L}((U_i^1, U_i^2)_{i=1,\cdots,n})$$
$$= \mathcal{L}((W^{1,1}_{\tau_1^{1,n}}, W^{1,2}_{\tau_1^{2,n}(W^{1,1}_{\tau_1^{1,n}})}), (W^{1,i}_{\tau_i^{1,n}}, W^{2,i}_{\tau_i^{2,n}(W^{1,1}_{\tau_1^{1,n}},\cdots,W^{1,i}_{\tau_i^{1,n}})}(W^{1,1}_{\tau_1^{1,n}}, \cdots, W^{1,i-1}_{\tau_{i-1}^{1,n}}))_{i=2,\cdots,n})$$
$$= \mathcal{L}((W^1_{T_i^{1,n}} - W^1_{T_{i-1}^{1,n}}, W^2_{T_i^{2,n}} - W^2_{T_{i-1}^{2,n}})_{i=1,\cdots,n}),$$

si $T_0^{1,n} = T_0^{2,n} = 0$ et pour chaque $k = 1, \cdots, n$:

$$T_k^{1,n} = \sum_{i=1}^{k} \tau_i^{1,n},\ T_k^{2,n} = \sum_{i=1}^{k} \tau_i^{2,n}(W^{1,1}_{\tau_1^{1,n}}, \cdots, W^{1,i}_{\tau_i^{1,n}}).$$

En conséquence, nous avons:

$$\mathcal{L}((W^1_{T_k^{1,n}}, W^2_{T_k^{2,n}})_{k=1,\cdots,n}) = \mathcal{L}((\frac{S_k}{\sqrt{n}})_{k=1,\cdots,n}).$$

Les relations sont des conséquences faciles du théorème 4 et de la construction de $T_k^{1,n}$ et $T_k^{2,n}$. D'après le théorème 4 et la construction précédente, il existe pour chaque $p \geq 1$ une constante c telle que pour tout $n \geq 1$, $k = 1, \cdots, n$ et $x_1, \cdots, x_k \in C_n$:

$$E[\tau_k^{2,n}(x_1, \cdots, x_k)^p] \leq cE[(U_k^2)^{2p} | U_k^1 = x_k].$$

Alors, puisque U_k^1 ne prend ses valeurs que dans l'ensemble C_n:

$$\begin{aligned} E[\tau_k^{2,n}(x_1, \cdots, x_k)^p] \leq & \frac{cE[(U_k^2)^{2p} I_{(U_k^1 = x_k)}]}{\min_{k\geq 1} P(X_k^1 = \sqrt{n}x_k)} \\ \leq & \frac{cE[(U_k^2)^{2p}]}{\min_{k\geq 1} P(X_k^1 = \sqrt{n}x_k)} \\ \leq & \frac{c}{n^p} \frac{\max(|x|^{2p} : P(X_1^2 = x) > 0)}{\min_{x: P(X_1^1 = x) > 0} P(X_1^1 = x)} \end{aligned}$$

d'où l'inégalité. La deuxième inégalité concernant le temps d'arrêt $\tau_k^{1,n}$ s'obtient de la même manière.

Pour les égalités, remarquons que pour chaque $n \geq 1$, $k = 1, \cdots, n$ et $x_1, \cdots, x_k \in C_n$, d'après le théorème 4 et la construction précédente:

$$E[(X_1^2)^2]E[\tau_k^{2,n}(x_1, \cdots, x_k)] = E[(U_k^2)^2 | U_1^1 = x_1, \cdots, U_k^1 = x_k].$$

Alors, puisque $\tau_k^{2,n}(x_1, \cdots, x_k)$ est adapté à la filtration $\mathcal{F}^{W^{2,k}(x_1, \cdots, x_{k-1})}$, et cette filtration est indépendante des filtrations $\mathcal{F}^{W^{1,1}}, \cdots, \mathcal{F}^{W^{1,k}}$, on a

$$\mathcal{L}(\tau_k^{2,n}(x_1, \cdots, x_k)) = \mathcal{L}(\tau_k^{2,n}(W^{1,1}_{\tau_1^{1,n}}, \cdots, W^{1,k}_{\tau_k^{1,n}}) | W^{1,1}_{\tau_1^{1,n}} = x_1, \cdots, W^{1,k}_{\tau_k^{1,n}} = x_k)$$

et donc

$$E[(X_1^2)^2]E[\tau_k^{2,n}(W^{1,1}_{\tau_1^{1,n}}, \cdots, W^{1,k}_{\tau_k^{1,n}})] = E[(U_k^2)^2] = \frac{E[(X_1^2)^2]}{n}.$$

Enfin, l'égalité concernant $\tau_k^{2,n}$ est une conséquence directe du théorème 4 ∎

Nous pouvons maintenant prouver le lemme 1.

Preuve du lemme 1 Soit pour chaque $n \geq 1$, $(T^n_{[nt]})_{t \leq 1} = (T^{1,n}_{[nt]}, T^{2,n}_{[nt]})_{t \leq 1}$ la suite construite dans le lemme 3. Elle vérifie l'égalité en loi du lemme 1. Il reste donc à montrer le résultat de convergence. Par abus de notation, appelons comme dans l'énonçé du lemme 1 $(W_{T^n_{[nt]}})_{t \leq 1}$ le processus $(W^1_{T^{1,n}_{[nt]}}, W^2_{T^{2,n}_{[nt]}})_{t \leq 1}$.

Soit pour chaque $n \geq 1$, $\epsilon_n = 1/\log n$ et $\lambda_n = \log n/\sqrt{n}$. On a, si $n \geq 1$:

$$\begin{aligned} &E[\sup_{t \leq 1} |W_{T^n_{[nt]}} - W_t|] \\ &\leq E[\sup_{t \leq 1} |W_{T^n_{[nt]}} - W_t| I_{(\sup_{t \leq 1} |W_{T^n_{[nt]}} - W_t| < \epsilon_n)}] \\ &\quad + E[\sup_{t \leq 1} |W_{T^n_{[nt]}} - W_t| I_{(\sup_{t \leq 1} |W_{T^n_{[nt]}} - W_t| \geq \epsilon_n)}] \\ &\leq \epsilon_n + E[\sup_{t \leq 1} (W_{T^n_{[nt]}} - W_t)^2]^{1/2} P(\sup_{t \leq 1} |W_{T^n_{[nt]}} - W_t| \geq \epsilon_n)^{1/2}. \end{aligned}$$

Or, puisque la mesure Q est à support compact, il est clair d'après le lemme 1 que

$$\sup_{n \geq 1} E[\sup_{t \leq 1} (W_{T^n_{[nt]}} - W_t)^2] < \infty,$$

et il suffit donc de prouver que

$$P(\sup_{t \leq 1} |W_{T^n_{[nt]}} - W_t| > \varepsilon_n) \xrightarrow[n \to \infty]{} 0$$

pour obtenir le résultat du lemme 1. Notant pour chaque $t \leq 1$, $\vec{t} = (t, t)$, on a si $n \geq 1$:

$$\begin{aligned} &P(\sup_{t \leq 1} |W_{T^n_{[nt]}} - W_t| > \varepsilon_n) \\ &\leq P(\sup_{t \leq 1} |W_{T^n_{[nt]}} - W_t| > \varepsilon_n, \sup_{t \leq 1} |T^n_{[nt]} - \vec{t}| < \lambda_n) \\ &\quad + P(\sup_{t \leq 1} |T^n_{[nt]} - \vec{t}| \geq \lambda_n) \\ &\leq P(\sup_{|u-v| \leq \lambda_n} |W_u - W_v| \geq \varepsilon_n) + P(\sup_{t \leq 1} |T^n_{[nt]} - \vec{t}| \geq \lambda_n). \end{aligned} \tag{1}$$

D'après l'inégalité exponentielle donnant le module de continuité d'un mouvement brownien, il existe deux constantes c_1 et c_2 telles que, si $0 < \lambda_n < 1$:

$$P(\sup_{|u-v|\leq\lambda_n,\, u\leq 1+\lambda_n}|W_u - W_v| \geq \varepsilon_n) \leq c_1 \frac{1}{\sqrt{\lambda_n}\varepsilon_n}\exp(-\frac{\varepsilon_n^2}{c_2\lambda_n}).$$

Puisque pour chaque $n \geq 1$, $0 < \lambda_n < 1$, nous obtenons la convergence vers 0 du premier terme de l'inégalité (1). Nous montrons maintenant que le deuxième terme de (1) tend lui aussi vers 0. Puisque d'après le lemme 3, $E[T^{i,n}_{[nt]}] = [nt]/n$ si $n \geq 1$, $t \leq 1$ et $i = 1, 2$, on a pour chaque $n \geq 1$:

$$\begin{aligned}P(\sup_{t\leq 1}|T^n_{[nt]} - \vec{t}| \geq \lambda_n) &\leq P(\sup_{t\leq 1}|T^n_{[nt]} - E[T^n_{[nt]}]| \geq \frac{\lambda_n}{2}) \\ &\quad + P(\sup_{t\leq 1}|\frac{[nt]}{n} - t| \geq \frac{\lambda_n}{2\sqrt{2}}) \\ &\leq P(\sup_{t\leq 1}|T^n_{[nt]} - E[T^n_{[nt]}]| \geq \frac{\lambda_n}{2}),\end{aligned}$$

si n est tel que $\lambda_n > 2\sqrt{2}/n$. Or, si $n \geq 1$:

$$P(\sup_{t\leq 1}|T^n_{[nt]} - E[T^n_{[nt]}]| \geq \frac{\lambda_n}{2}) \leq \sum_{j=1}^{2} P(\sup_{t\leq 1}|T^{j,n}_{[nt]} - E[T^{j,n}_{[nt]}]| \geq \frac{\lambda_n}{4}).$$

Majorons l'expression correspondant à $j = 2$ à l'intérieur de cette somme (le cas $j = 1$ se traitant d'une façon plus simple). Utilisant la définition de $T^{2,n}_k$ (si $n \geq 1$ et $k = 1, \cdots, n$) donnée dans le lemme 3, on obtient:

$$P(\max_{k\leq n}|T^{2,n}_k - E[T^{2,n}_k]| \geq \frac{\lambda_n}{4}) = P(\max_{k\leq n}|\sum_{i=1}^{k}\{\tau^{2,n}_i(W^{1,1}_{\tau^{1,n}_1}, \cdots, W^{1,i}_{\tau^{1,n}_i})\}| \geq \frac{\lambda_n}{4})$$

$$= \sum P(\max_{k\leq n}|\sum_{i=1}^{k}\{\tau^{2,n}_i(x_1, \cdots, x_i)\}| \geq \frac{\lambda_n}{4})P(W^{1,1}_{\tau^{1,n}_1} = x_1)\cdots P(W^{1,n}_{\tau^{1,n}_n} = x_n)$$

(ou l'on somme sur tous les éléments $x_1, \cdots, x_n \in C_n$) car, d'après le lemme 3, il y indépendance entre $(\tau^{2,n}_i(x_1, \cdots, x_i))_{i\leq n}$ et $(W^{1,i}_{\tau^{1,n}_i})_{i\leq n}$, et donc

$$\mathcal{L}((\tau^{2,n}_i(W^{1,1}_{\tau^{1,n}_1}, \cdots, W^{1,i}_{\tau^{1,n}_i}))_{i\leq n}) = \mathcal{L}((\tau^{2,n}_i(x_1, \cdots, x_i))_{i\leq n}|\bigcap_{i=1}^{n}[W^{1,i}_{\tau^{1,n}_i} = x_i]).$$

Le processus $(\sum_{i=1}^{k}\{\tau^{2,n}_i(x_1, \cdots, x_i)\})_k$ étant une martingale, on obtient d'après l'inégalité de Doob et l'inégalité du lemme 3 appliquée avec $p = 2$:

$$\begin{aligned}P(\max_{k\leq n}|\sum_{i=1}^{k}\{\tau^{2,n}_i(x_1, \cdots, x_i)\}| \geq \frac{\lambda_n}{4}) &\leq \frac{16}{\lambda_n^2}E[(\sum_{i=1}^{n}\{\tau^{2,n}_i(x_1, \cdots, x_i)\})^2] \\ &\leq \frac{32}{\lambda_n^2}\sum_{i=1}^{n}E[(\tau^{2,n}_i(x_1, \cdots, x_i))^2] \\ &\leq \frac{c}{\lambda_n^2}\frac{n}{n^2} = \frac{c}{\lambda_n^2 n},\end{aligned}$$

qui converge vers 0 d'après notre choix pour la suite $(\lambda_n)_{n\geq 1}$ ∎

Avant de prouver le lemme 2, nous énonçons un lemme, établi par Westwater ([23], lemme 5):

Lemme 4 *Soit $(U_i)_i$ une suite de variables aléatoires réelles indépendantes. Supposons qu'il existe deux constantes k_1 et k_2 telles que pour tout $i \geq 1$ et p pair : $||U_i||_{L^p} \leq k_2 p^{k_1}$. Alors, pour chaque $n \geq 1$:*

$$||\sum_{i=1}^{n}\{U_i\}||_{L^p} \leq 2k_2 p^{k_1+1/2}\sqrt{n}.$$

Preuve du lemme 2 Dans la suite de cette preuve, c désignera une constante dont la valeur pourra varier de ligne en ligne. On peut supposer que s et t sont tels que pour chaque $n \geq 1$, $[nt] \geq [ns] + 1$.

Il est clair que pour chaque $n \geq 1$, si

$$E_n^{s,t} = \{(i,j) \in \mathbb{N}^2 \ : \ i = [ns], \cdots, [nt]-1; \ j = i+1, \cdots, [nt]\},$$

on a:

$$\gamma_n(z, D_{[nt]}) - \gamma_n(z, D_{[ns]}) = \gamma_n(z, \{0, \cdots, [ns]-1\} \times \{[ns]+1, \cdots, [nt]\}) + \gamma_n(z, E_n^{s,t}).$$

Or, d'après le lemme 2 iii), pour chaque entier pair p:

$$\begin{aligned} ||\gamma_n(z, \{0, \cdots, [ns]-1\} \times \{[ns]+1, \cdots, [nt]\})||_{L^p} &\leq c\frac{\sqrt{[ns]([nt]-[ns])}}{n} \\ &\leq c\sqrt{\frac{[nt]-[ns]}{n}}, \end{aligned}$$

et il reste donc à montrer que $||\gamma_n(z, E_n^{s,t})||_{L^p}$ vérifie aussi une inégalité du même type. Dans ce but, on introduit l'équivalent discret de la partition utilisée par Le Gall [10]. Si $\kappa = \min(i \ : \ 2^i > [nt] - [ns])$, on note pour chaque $\xi = 0, \cdots, \kappa - 1$ et $\eta = 0, \cdots, 2^{\xi} - 1$:

$$A_\eta^\xi = \{(i,j) \in D_{[nt]-[ns]} \ : \ \exists i_0, j_0 < 2^{\kappa-\xi-1}; \ i = \eta 2^{\kappa-\xi} + i_0, \ j = (\eta + \frac{1}{2})2^{\kappa-\xi} + j_0\}.$$

Les $(A_\eta^\xi)_{\xi=0,\cdots,\kappa-1,\eta=0,\cdots,2^\xi-1}$ sont disjoints et

$$D_{[nt]-[ns]} = \bigcup_{\xi=0}^{\kappa-1}\bigcup_{\eta=0}^{2^\xi-1} A_\eta^\xi.$$

Notons de plus, pour $\xi = 0, \cdots, \kappa - 1$ et $\eta = 0, \cdots, 2^\xi - 1$, $P(\eta, \xi)$ l'ensemble des $(i,j) \in \mathbb{N}^2$ tels que :

$$\begin{aligned} i &= [ns] + \eta 2^{\kappa-\xi}, \cdots, [ns] + \eta 2^{\kappa-\xi} + 2^{\kappa-\xi-1} - 1 \\ j &= [ns] + \eta 2^{\kappa-\xi} + 2^{\kappa-\xi-1}, \cdots, [ns] + \eta 2^{\kappa-\xi} + 2^{\kappa-\xi} - 1. \end{aligned}$$

Il vient, avec un décomposition évidente:

$$\gamma_n(z, E_n^{s,t}) = \frac{1}{n}\sum_{\xi=0}^{\kappa-1}\sum_{\eta=0}^{2^\xi-1}\sum_{(i,j)\in A_\eta^\xi}\{I_{\{S_{j+[ns]}-S_{i+[ns]}=z\}}\} = \sum_{\xi=0}^{\kappa-1}\sum_{\eta=0}^{2^\xi-1}\gamma_n(z, P(\eta,\xi)).$$

Or, d'après le lemme 2 iii), pour chaque $\xi = 0, \cdots, \kappa-1$ et $\eta = 0, \cdots, 2^\xi - 1$:

$$||\frac{1}{n}\sum_{(i,j)\in P(\eta,\xi)} I_{(S_j - S_i = z)}||_{L^p} \leq c\frac{2^{\kappa-\xi}}{n} \leq c\frac{[nt]-[ns]}{n}2^{-\xi}.$$

Par indépendance des variables $(\gamma_n(z, P(\eta,\xi)))_{\eta=0,\cdots,2^\xi-1}$, le lemme 4 implique alors que pour chaque $\xi = 0, \cdots, \kappa-1$:

$$||\sum_{\eta=0}^{2^\xi-1}\gamma_n(z, P(\eta,\xi))||_{L^p} \leq c\frac{[nt]-[ns]}{n}2^{-\frac{\xi}{2}}.$$

Par suite, d'après l'inégalité triangulaire:

$$||\gamma_n(z, E_n^{s,t})||_{L^p} \leq \sum_{\xi=0}^{\kappa-1}||\sum_{\eta=0}^{2^\xi-1}\gamma_n(z, P(\eta,\xi))||_{L^p} \leq c\frac{[nt]-[ns]}{n}\sum_{\xi=0}^{\kappa-1}2^{-\frac{\xi}{2}} \leq c\frac{[nt]-[ns]}{n},$$

d'où l'inégalité recherchée ∎

IV- Preuve du Corollaire 1

Le corollaire 1 sera une conséquence du lemme suivant:

Lemme 5 *Pour tout $g \geq 0$, il existe une constante $c(g)$ telle que*

$$\sup_{n\geq 1} E[\exp(-g\gamma_n(0, D_n))] \leq c(g).$$

Remarque D'après le théorème 1, ce lemme entraîne que $\gamma(0, \Delta_1)$ admet des moments exponentiels. A ce sujet, un résultat plus général a été obtenu par Le Gall [13] et Pitman-Yor [14], à l'aide de méthodes très différentes.

Preuve du corollaire 1 Soit f une fonction continue bornée. Nous voulons prouver que $\mu_g^n(f) \to \mu_g(f)$ si $n \to \infty$. Or, pour chaque $n \geq 1$,

$$\mu_g^n(f) = \frac{1}{\overline{L}_g^n} E[f((\frac{S_{[nt]}}{\sqrt{n}})_{t\leq 1})\exp(-g\gamma_n(0, D_n))],$$

si $\overline{L_g^n} = L_g^n \exp(-g/nE[\sum_{i<j\leq n} I_{(S_i=S_j)}])$. D'après le lemme 5 et le fait que f est borné, la suite $(f((S_{[nt]}/\sqrt{n})_{t\leq 1}) \exp(-g\gamma_n(0, D_n)))_{n\geq 1}$ est évidemment uniformément intégrable et donc, d'après le théorème 1,

$$E[f((\frac{S_{[nt]}}{\sqrt{n}})_{t\leq 1}) \exp(-g\gamma_n(0, D_n))] \xrightarrow[n\to\infty]{} E[f(W) \exp(-g\gamma(0, \Delta_1))].$$

De même, $\overline{L_g^n}$ converge vers L_g et donc $\mu_g^n(f)$ converge vers $\mu_g(f)$ si n tend vers l'infini ∎

Preuve du lemme 5 Il s'agit en fait d'une version discrète de la méthode habituellement utilisée pour montrer que $\gamma(0, \Delta_1)$ admet des moments exponentiels (voir Varadhan [21]).

Soit $t \in I\!R_+$. Si pour chaque $n \geq 1$, $\kappa = \min(i : 2^i > [nt])$, on introduit de nouveau pour chaque $\xi = 0, \cdots, \kappa - 1$ et $\eta = 0, \cdots, 2^\xi - 1$:

$$A_\eta^\xi = \{(i,j) \in D_{[nt]} : \exists i_0, j_0 < 2^{\kappa-\xi-1},\ i = \eta 2^{\kappa-\xi} + i_0,\ j = (\eta + \frac{1}{2})2^{\kappa-\xi} + j_0\}.$$

De plus, notons pour chaque $n \geq 1$ et $\xi = 0, \cdots, \kappa - 1$, $\beta_n(\xi)$ la variable aléatoire

$$\beta_n(\xi) = \frac{1}{n} \sum_{\eta=0}^{2^\xi - 1} \sum_{(i,j)\in A_\eta^\xi} I_{(S_i=S_j)}.$$

Enfin, pour chaque $n \geq 1$ et $0 \leq u < v$, appelons

$$\gamma_n^{u,v} = \frac{1}{n} \sum_{i=[nu]}^{[nv]-1} \sum_{j=i+1}^{[nv]} \{I_{(S_i=S_j)}\}.$$

Remarquons que $\gamma_n^{0,t} = \gamma_n(0, D_{[nt]}) = \sum_{\xi=0}^{\kappa-1} \{\beta_n(\xi)\}$.

D'après le lemme 2 iii), il existe une constante c_1 telle que pour chaque $n \geq 1$ et $\xi = 0, \cdots, \kappa - 1$:

$$E[\beta_n(\xi)] \leq c_1 \sum_{\eta=0}^{2^\xi - 1} \frac{2^{\kappa-\xi-1}}{n} \leq c_1 t.$$

Alors, pour tout $k = 1, \cdots, \kappa - 1$, d'après l'inégalité de Tchebytchev:

$$\begin{aligned} P(\gamma_n^{0,t} \leq -2c_1 tk) &\leq P(\sum_{\xi=k}^{\kappa-1} \{\beta_n(\xi)\} \leq -2c_1 tk + \sum_{\xi=0}^{k-1} E[\beta_n(\xi)]) \\ &\leq \frac{E[(\sum_{\xi=k}^{\kappa-1} \{\beta_n(\xi)\})^2]}{(c_1 tk)^2} \leq \frac{(\sum_{\xi=k}^{\kappa-1} \|\{\beta_n(\xi)\}\|_{L^2})^2}{(c_1 tk)^2}. \end{aligned}$$

Or, pour chaque $n \geq 1$, $\xi = 0, \cdots, \kappa - 1$ et $\eta = 0, \cdots, 2^\xi - 1$, on a d'après le lemme 2 iii):

$$\|\sum_{(i,j)\in A_\eta^\xi} I_{(S_i=S_j)}\|_{L^2} \leq c_1 t 2^{-\xi}.$$

Par suite, d'après le lemme 5, $\|\{\beta_n(\xi)\}\|_{L^2} \leq c_2 t 2^{-\xi}$, pour une certaine constante c_2. Ainsi, pour chaque $k = 0, \cdots, \kappa - 1$:

$$P(\gamma_n^{0,t} \leq -2c_1 tk) \leq \frac{c_2^2}{2^k k^2 (c_1(1-2^{-1/2}))^2}.$$

Posons $c_3 = c_2^2(c_1(1-2^{-1/2}))^{-2}$. En utilisant le fait que pour chaque $n \geq 1$,

$$\Omega = \bigcup_{k\geq 1} \{-2c_1 t(k+1) < \gamma_n^{0,t} \leq -2c_1 tk\} \bigcup \{\gamma_n^{0,t} > -2c_1 t\},$$

nous obtenons pour chaque $n \geq 1$:

$$\begin{aligned} E[\exp(-g\gamma_n^{0,t})] &\leq \exp(2gtc_1)(1 + \sum_{k\geq 1} \exp(2gc_1 tk) P(\gamma_n^{0,t} \leq -2c_1 tk)) \\ &\leq \exp(2gtc_1)(1 + c_3 \sum_{k\geq 1} \frac{\exp(2gc_1 t - \log 2)k}{k^2}), \end{aligned}$$

et donc $\sup_{n\geq 1} E[\exp(-g\gamma_n^{0,t}] < \infty$ pour tout couple (g,t) tel que $gt \leq c_4 := \log 2/(2c_1)$. De même, il est clair que la preuve précédente permet de prouver que, si $0 \leq u < v$ sont tels que $g(v-u) \leq c_4$, alors

$$\sup_{n\geq 1} E[\exp(-g\gamma_n^{u,v})] < \infty. \tag{2}$$

Soit alors $g > 0$ tel que $g \leq 2c_4$. Nous avons pour chaque $n \geq 1$:

$$\begin{aligned} \gamma_n(0, D_n) = \gamma_n^{0,1} &= \gamma_n^{0,1/2} + \gamma_n^{1/2,1} + \gamma_n(0, \{0, \cdots, [\frac{n}{2}] - 1\} \times \{[\frac{n}{2}] + 1, \cdots, n\}) \\ &\geq \gamma_n^{0,1/2} + \gamma_n^{1/2,1} - \frac{1}{n} E[\sum_{i=0}^{[n/2]-1} \sum_{j=[n/2]+1}^{n} I_{(S_i = S_j)}] \\ &\geq \gamma_n^{0,1/2} + \gamma_n^{1/2,1} - c_5, \end{aligned}$$

pour une certaine constante c_5, d'après le lemme 2 iii). Alors, par indépendance de $\gamma_n^{0,1/2}$ et $\gamma_n^{1/2,1}$:

$$E[\exp(-g\gamma_n(0, D_n))] \leq \exp(gc_5) E[\exp(-g\gamma_n^{0,1/2})]\, E[\exp(-g\gamma_n^{1/2,1})],$$

qui est fini d'après (2) et le fait que $g \leq 2c_4$.

Lorsque g est quelconque, on peut trouver $k \geq 1$ tel que $g \leq 2^k c_4$. On itère alors le procédé précédent, ce qui donne le résultat ■

Références

[1] R.F.Bass, D.Khoshnevisan - *Intersection Local Times and Tanaka Formulas*, Ann. Inst. Henri Poincaré, vol. 29, no. 3, p.391-418, (1993).

[2] P.Billingsley - **Convergence of Probability Measures**, Wiley and Sons, New-York, (1968).

[3] A.N.Borodin - *On the asymptotic behavior of Local Times of recurrent Random Walks with finite variance*, Theory Prob. Appl., vol. XXVI no 4, p.758-772, (1981).

[4] D.C.Brydges, G.Slade - *The diffusive phase of a model of self-interacting walks*, Probab. Theory Relat. Fields, 103, p.285-315, (1995).

[5] B.Cadre - *Etudes de convergences en loi de fonctionnelles de processus : Formes quadratiques ou multilinéaires aléatoires, Temps locaux d'intersection de marches aléatoires, Théorème central limite presque sûr*, Thèse de l'Université de Rennes I, (1995).

[6] E.Csàki, P.Révész - *Strong invariance for Local Times*, Z. Wahrs. verw Gebiete, vol. 62, p.263-278, (1983).

[7] C.Domb, G.S.Joyce - *Cluster expansion for a Polymer Chain*, J. Phys. C5, p.956-976, (1975).

[8] D.Geman, J.Horowitz, J.Rosen - *The Local Time of intersection for Brownian Paths in the Plane*, Ann. Prob., vol. 12, p.86-107, (1984).

[9] E.Haeusler - *An exact rate of convergence in the Functional Limit Theorem for special Martingale difference array*, Z. Wahrs. verw Gebiete, vol. 65, p.523-534, (1984).

[10] J.F.Le Gall - *Sur le temps local d'intersection du mouvement brownien plan, et la méthode de renormalisation de Varadhan*, Sém. Prob. XIX, Lect. Notes in Math., vol. 1123, Springer, Berlin, p.314-331, (1985).

[11] J.F.Le Gall - *Some properties of Planar Brownian Motion*, Ecole d'été de Saint-Flour XX, Lect. Notes in Math., vol. 1527, Springer, Berlin, (1992).

[12] J.F.Le Gall - *Marches aléatoires auto-évitantes et modèles de polymères*, non publié.

[13] J.F.Le Gall - *Exponential moments for the renormalized self-intersection local time of Planar Brownian Motion*, Sém. Prob. XXVIII, Lect. Notes in Math., vol. 1583, Springer, Berlin, p.172-180, (1994).

[14] J.W.Pitman, M.Yor - Appendice 1 de *Quelques identités en loi pour les processus de Bessel*, Société Mathématique de France, Astérisque, vol. 236, p.249-276, (1996).

[15] D.Revuz, M.Yor - **Continuous Martingales and Brownian Motion**, Springer-Verlag, Berlin, (1991).

[16] J.Rosen - *A Local Time approach to the self-intersection of Brownian Motion Paths in Space*, Comm. Math. Phys., vol. 88, p.327-338, (1983).

[17] J.Rosen - *A renormalized Local Time for multiple intersection of Planar Brownian Motion*, Sém. Prob. XX, Lect. Notes in Math., vol. 1204, Springer, Berlin, p.515-531, (1986).

[18] J.Rosen - *Random Walks and intersection Local Time*, Ann. Prob., vol. 18 no 3, p.959-977, (1990).

[19] F.Spitzer - **Principle of Random Walks**, Van Nostrand, Princeton, New-York, (1964).

[20] A.Stoll - *Invariance Principles for Brownian intersection Local Time and Polymer Measures*, Math. Scand., vol. 64, p.133-160, (1989).

[21] S.R.S.Varadhan - Appendix to **Euclidean Quantum Field Theory**, by K.Symanzik, in **Local Quantum Theory**, R.Jost (Ed.), Academic Press, New-York, (1969).

[22] W.Werner - *Sur les singularités des temps locaux d'intersection du mouvement brownien plan*, Ann. Inst. Henri Poincaré, vol. 29, no. 3, p.419-451, (1993).

[23] J.Westwater - *On Edward's Model for long Polymer Chain*, Comm. Math. Phys., vol. 72, p.131-174, (1980).

[24] M.Yor - *Sur la représentation comme intégrale stochastique du temps d'occupation du mouvement brownien dans* $\mathbb{R}^d$, Sém. Prob. XX, Lect. Notes in Math., vol. 1204, Springer, Berlin, p.543-552, (1986).

Marches aléatoires auto-évitantes et mesures de polymère

Jean-François Le Gall

1. Introduction.

L'objet de cette note est de montrer que la loi d'une marche aléatoire plane faiblement auto-évitante, considérée sur un long intervalle de temps et convenablement changée d'échelle, se rapproche de la mesure de polymère en dimension deux. Les mesures de polymère ont été introduites formellement par Edwards [4], et une définition mathématique rigoureuse en dimension deux a été rendue possible par le travail de Varadhan [11]. La mesure de polymère s'interprète comme la loi d'un mouvement brownien faiblement auto-évitant, et notre résultat est donc un analogue auto-évitant du classique théorème d'invariance de Donsker. Le théorème principal du présent travail a déjà été obtenu par Stoll [10], sous des hypothèses cependant plus restrictives et à l'aide de techniques d'analyse non-standard. Tout récemment, Cadre [3] a développé une autre approche de ce résultat, sous des hypothèses voisines de celles de Stoll et en utilisant une méthode originale de plongement de marches aléatoires planes dans le mouvement brownien. Pour la marche aléatoire simple, une discussion plus générale, s'appliquant aussi aux modèles "auto-attractifs", est donnée dans le travail de Brydges et Slade [2]. Signalons enfin que le problème beaucoup plus difficile de l'approximation de la mesure de polymère en dimension trois par des marches aléatoires faiblement auto-évitantes vient d'être résolu par Albeverio, Bolthausen et Zhou [1]. Le but de cette note est donc surtout pédagogique, et son intérêt réside dans la simplicité des techniques utilisées, qui ont déjà été appliquées à d'autres problèmes, tels que l'étude asymptotique du nombre de sites visités par une marche aléatoire plane [5] ou l'existence de moments exponentiels pour le temps local d'intersection brownien renormalisé [7]. Nous espérons aussi que les estimations du présent travail pourront rendre quelque service dans l'étude des nombreuses questions ouvertes concernant les mesures de polymère.

Ce travail est la rédaction d'un exposé donné dans le cadre du Cours Peccot au Collège de France en 1989. Je remercie Marc Yor de m'avoir donné la possibilité de le publier dans le Séminaire de Probabilités.

2. Hypothèses et énoncé du théorème principal.

Nous considérons une marche aléatoire $X = (X_n, n \in \mathbb{N})$ à valeurs dans $\mathbb{Z}^2$, issue de 0 sous la probabilité P. On a donc $X_0 = 0$ et pour tout $n \geq 1$,

$$X_n = \sum_{i=1}^{n} Y_i$$

où les variables $Y_i, i = 1, 2, \ldots$ sont indépendantes et équidistribuées à valeurs dans $\mathbb{Z}^2$.

Nous supposerons toujours que les trois hypothèses suivantes sont satisfaites :

(H1) La marche aléatoire est centrée et a des moments d'ordre deux :

$$E[|X_1|^2] < \infty, \qquad E[X_1] = 0.$$

(H2) La marche aléatoire X est adaptée, au sens où la loi de X_1 n'est pas portée par un sous-groupe strict de $\mathbb{Z}^2$.

(H3) La marche aléatoire est isotrope, au sens où la matrice de covariance de X_1 s'écrit

$$\text{cov}(X_1) = \sigma^2 \,\text{Id}$$

où $\sigma > 0$ et Id est la matrice identité en dimension deux.

L'hypothèse importante est (H1). L'hypothèse (H3) a pour seul but de simplifier les énoncés qui suivent, en évitant l'introduction de mouvements browniens "non-isotropes".

Pour tout entier $N \geq 1$, pour $0 \leq t \leq 1$, on pose

$$X_t^{(N)} = \frac{1}{\sigma\sqrt{N}} X_{[Nt]}$$

où $[Nt]$ désigne la partie entière de Nt. Soit $Q^{(N)}$ la loi de $(X_t^{(N)}, 0 \leq t \leq 1)$, qui est une mesure de probabilité sur l'espace de Skorokhod $\mathbb{D}([0,1], \mathbb{R}^2)$. D'après le théorème de Donsker,

$$Q^{(N)} \xrightarrow[N\to\infty]{\text{(e)}} W$$

où la notation $\xrightarrow{\text{(e)}}$ indique la convergence étroite, et W est la loi sur $\mathbb{D}([0,1], \mathbb{R}^2)$ de $(B_t, 0 \leq t \leq 1)$, si B désigne un mouvement brownien plan issu de 0.

Introduisons maintenant les lois des processus auto-évitants. Pour tous $\lambda \geq 0$, $N \geq 1$ on pose

$$L_\lambda^{(N)} = E\big[\exp\big(-\lambda \sum_{0\leq i<j\leq N} I(X_i = X_j)\big)\big]$$

et on définit alors $Q_\lambda^{(N)}$ comme la loi de $(X_t^{(N)}, 0 \leq t \leq 1)$ sous la probabilité

$$(L_\lambda^{(N)})^{-1} \exp\big(-\lambda \sum_{0\leq i<j\leq N} I(X_i = X_j)\big) \cdot P \quad .$$

L'idée est d'attribuer un poids plus faible, d'autant plus faible que λ est grand, aux trajectoires qui présentent beaucoup d'auto-intersections.

Il reste à introduire la loi du mouvement brownien auto-évitant, c'est-à-dire la mesure de polymère en dimension deux. On utilise pour cela les temps locaux d'auto-intersection du mouvement brownien plan B (voir [8] ou [6], Chapitre VIII). Soit $\Delta = \{(s,t), 0 \leq s < t \leq 1\}$. Il existe p.s. une unique famille $(\alpha_x, x \in \mathbb{R}^2)$ de mesures de Radon sur Δ telle que :

(i) L'application $x \to \alpha_x$ est continue pour la topologie de la convergence vague.

(ii) Pour toute partie borélienne H de Δ et pour toute fonction h mesurable positive sur $\mathbb{R}^2$,

$$\int_H h(B_s - B_t)\, ds\, dt = \int_{\mathbb{R}^2} h(x)\, \alpha_x(H)\, dx.$$

En prenant pour h une approximation de la mesure de Dirac en 0, on obtient l'expression formelle

$$\alpha_0(H) = \int_H \delta_0(B_s - B_t)\, ds\, dt.$$

On vérifie aisément que $\alpha_x(\Delta) < \infty$ si $x \neq 0$, p.s. et que $\alpha_0(\Delta) = \infty$ p.s. On peut néanmoins "renormaliser" $\alpha_0(\Delta)$ de la manière suivante (voir par exemple [6], Chapitre VIII). Pour tous entiers $p \geq 1$, $k \in \{1, \ldots, 2^{p-1}\}$, on pose

$$A_k^p = [(2k-2)2^{-p}, (2k-1)2^{-p}[\times](2k-1)2^{-p}, 2k2^{-p}] \subset \Delta.$$

Des arguments simples de changement d'échelle montrent que la série

$$\sum_{p=1}^{\infty} \Big(\sum_{k=1}^{2^{p-1}} (\alpha_0(A_k^p) - E[\alpha_0(A_k^p)]) \Big)$$

converge dans L^2 et p.s. La somme de cette série, notée γ, est le temps local d'intersection renormalisé de B sur l'intervalle $[0,1]$. On montre que, pour tout $\lambda > 0$,

$$L_\lambda = E[\exp(-\lambda\, \gamma)] < \infty$$

(voir [7], p.178, pour un argument simple, ce résultat étant dû à Varadhan [11] dans un cadre un peu différent).

La mesure de polymère W_λ est par définition la loi de $(B_t, 0 \leq t \leq 1)$ sous la probabilité

$$(L_\lambda)^{-1} \exp(-\lambda\, \gamma) \cdot P.$$

Nous sommes maintenant en mesure d'énoncer le résultat principal de ce travail.

Théorème 1. *Pour tout* $\lambda > 0$,

$$Q_{\lambda/N}^{(N)} \xrightarrow[N\to\infty]{(e)} W_{\sigma^{-2}\lambda}\, .$$

De manière équivalente, pour toute fonction continue bornée F sur $\mathbb{D}([0,1], \mathbb{R}^2)$,

$$\lim_{N\to\infty} Q_{\lambda/N}^{(N)}[F] = W_{\sigma^{-2}\lambda}[F].$$

Or, par définition,

$$Q_{\lambda/N}^{(N)}[F] = (L_{\lambda/N}^{(N)})^{-1} E[\exp(-\lambda J_N) F(X_s^{(N)}, 0 \leq s \leq 1)]$$

où

$$J_N = \frac{1}{N} \sum_{0 \leq i < j \leq N} I(X_i = X_j).$$

Pour toute variable aléatoire intégrable U, notons $\{U\} = U - E[U]$. Remarquons qu'on peut aussi écrire

$$Q_{\lambda/N}^{(N)}[F] = (\tilde{L}_{\lambda/N}^{(N)})^{-1}\, E[\exp(-\lambda\{J_N\})\, F(X_s^{(N)}, 0 \leq s \leq 1)],$$

à condition de poser

$$\tilde{L}_{\lambda/N}^{(N)} = E[\exp(-\lambda\{J_N\})].$$

On voit alors que le Théorème 1 est une conséquence de la définition de W_λ et des deux propositions suivantes.

Proposition 2. *On a*

$$\big(\{J_N\}, (X_s^{(N)}, 0 \le s \le 1)\big) \xrightarrow[N\to\infty]{\text{(loi)}} \big(\sigma^{-2}\gamma, (B_s, 0 \le s \le 1)\big).$$

Proposition 3. *Pour tout* $\lambda > 0$,

$$\sup_{N\ge 1} E[\exp(-\lambda\,\{J_N\})] < \infty.$$

En effet, supposons démontrées les Propositions 2 et 3. Alors la suite

$$U_N = \exp(-\lambda\{J_N\})\,F(X_s^{(N)}, 0 \le s \le 1)$$

converge en loi vers $U = \exp(-\sigma^{-2}\lambda\,\gamma)\,F(B_s, 0 \le s \le 1)$. De plus, la Proposition 3 montre que la suite (U_N) est bornée dans L^2. On conclut alors que $E[U_N]$ converge vers $E[U]$, et en prenant $F = 1$ on voit de même que $\tilde{L}^{(N)}_{\lambda/N}$ converge vers $L_{\sigma^{-2}\lambda}$.

La Proposition 2 est très proche d'un résultat de Rosen [9], qui suppose cependant la marche aléatoire X apériodique. La convergence conjointe de $\{J_N\}$ et $X^{(N)}$ n'est pas énoncée par Rosen mais découle de la méthode qu'il utilise. Nous donnons dans la partie 3 une démonstration de la Proposition 2 un peu différente de celle de Rosen, reposant sur des estimations que nous utiliserons aussi dans la preuve de la Proposition 3. Cette dernière proposition est démontrée dans la partie 4.

3. Etude asymptotique des nombres d'intersection.

Nous commençons par un résultat relatif au nombre de couples d'intersection de deux marches aléatoires indépendantes. Nous considérons une seconde marche aléatoire plane X' issue de 0 indépendante de X. Nous supposons que X' satisfait les mêmes hypothèses (H1),(H2),(H3) que X avec la même constante σ (cependant X et X' n'ont pas nécessairement même loi). Pour tout $N \ge 1$, on pose

$$I_N = \frac{1}{N}\sum_{i=0}^{N}\sum_{j=0}^{N} I(X_i = X'_j)$$

et on définit $X'^{(N)}$ de la même manière que $X^{(N)}$.

Lemme 4. *On a*

$$\big(I_N, (X_s^{(N)}, X_s'^{(N)}; 0 \le s \le 1)\big) \xrightarrow[N\to\infty]{\text{(loi)}} \big(\sigma^{-2}\beta([0,1]^2), (B_s, B'_s; 0 \le s \le 1)\big),$$

où B' est un mouvement brownien plan indépendant de B issu de 0, et

$$\beta([0,1]^2) = \int_0^1\int_0^1 \delta_{(0)}(B_s - B'_t)\,ds\,dt$$

est le temps local d'intersection de B et B' sur $[0,1]^2$ *(voir par exemple* [6], *Chapitre VIII). De plus, il existe une constante $C_1 < \infty$ telle que*

$$\sup_{N\ge 1} E[(I_N)^2] \le C_1. \tag{1}$$

Démonstration. Nous reprenons les arguments de Rosen [9], en supposant d'abord que X et X' sont apériodiques (i.e. si ϕ est la fonction caractéristique de X_1, ou de X'_1, les conditions $|\phi(\xi)| = 1$ et $\xi \in]-\pi, \pi]^2$ entraînent $\xi = 0$). Pour $\varepsilon > 0$, on pose

$$I_N^\varepsilon = \sigma^{-2} \int_0^1 \int_0^1 p_\varepsilon(X_s^{(N)}, X_t'^{(N)})\, ds\, dt,$$

où $p_\varepsilon(x, y) = (2\pi\varepsilon)^{-1} \exp(-|y-x|^2/(2\varepsilon))$. D'après le Lemme 1 et la formule (2.6) de Rosen [9], il existe deux constantes $C > 0$, $\delta > 0$ telles que, pour tout $\varepsilon \in]0, 1[$,

$$\limsup_{N\to\infty} E[(I_N^\varepsilon - I_N)^2]^{1/2} \leq C\, \varepsilon^\delta \tag{2}$$

(Rosen traite le cas $\sigma = 1$, mais des modifications triviales de son argument donnent le résultat pour σ quelconque). Pour $\varepsilon > 0$ fixé, I_N^ε est une fonction continue bornée du couple $(X^{(N)}, X'^{(N)})$. Le théorème de Donsker entraîne alors

$$\big(I_N^\varepsilon, (X_s^{(N)}, X_s'^{(N)}; 0 \leq s \leq 1)\big) \xrightarrow[N\to\infty]{\text{(loi)}} \big(\sigma^{-2} \int_0^1 \int_0^1 p_\varepsilon(B_s, B'_t) ds dt, (B_s, B'_s; 0 \leq s \leq 1)\big).$$

D'autre part,

$$\lim_{\varepsilon\to 0} \int_0^1 \int_0^1 p_\varepsilon(B_s, B'_t)\, ds\, dt = \beta([0,1]^2), \quad \text{p.s.}$$

grâce à la formule de densité de temps d'occupation pour le temps local d'intersection (voir par exemple [6], Chapitre VIII). La première assertion du lemme découle alors de (2), et il en va de même pour la seconde, puisque, pour ε fixé, les variables I_N^ε, $N \geq 1$ sont bornées uniformément.

Il reste à s'affranchir de l'hypothèse d'apériodicité. Pour cela on introduit une suite (ε_n) de variables aléatoires indépendantes équidistribuées, indépendantes du couple (X, X') et telles que $P[\varepsilon_n = 1] = 1 - P[\varepsilon_n = 0] = \rho \in]0, 1[$. On définit $S_n = \varepsilon_1 + \cdots + \varepsilon_n$ puis $\tilde{X}_n = X_{S_n}$, de sorte que $\tilde{X}$ est une marche aléatoire apériodique vérifiant les mêmes hypothèses que X (la constante σ^2 est remplacée par $\rho\sigma^2$). On construit de même $\tilde{X}'$ à partir de X' et d'une autre suite (ε'_n) de même loi que (ε_n) et indépendante du triplet $(\varepsilon_n, X_n, X'_n; n \geq 0)$. On définit alors $\tilde{I}_N$ comme I_N en remplaçant le couple (X, X') par $(\tilde{X}, \tilde{X}')$, et de même $(\tilde{X}^{(N)}, \tilde{X}'^{(N)})$. On voit facilement que, pour tout $\eta > 0$, $I_N \leq \tilde{I}_{[(1+\eta)N]}$ sur l'ensemble $\{S_{(1+\eta)N} \geq N, S'_{(1+\eta)N} \geq N\}$, indépendant du couple (X, X') et dont la probabilité tend vers 1, uniformément en N, lorsque ρ croît vers 1. En appliquant à $\tilde{I}_N$ la majoration (1) on obtient aussitôt que cette majoration est aussi vraie dans le cas général.

De même, pour obtenir la première partie de la proposition, on remarque d'abord que

$$(X_t^{(N)}, X_t'^{(N)}, \tilde{X}_t^{(N)}, \tilde{X}_t'^{(N)}; 0 \leq t \leq 1) \xrightarrow[N\to\infty]{\text{(loi)}} (B_t, B'_t, \frac{1}{\sqrt{\rho}} B_{\rho t}, \frac{1}{\sqrt{\rho}} B'_{\rho t}; 0 \leq t \leq 1).$$

En utilisant le cas apériodique et la majoration ci-dessus de I_N en fonction de $\tilde{I}_{[(1+\eta)N]}$, et en faisant tendre ρ vers 1, on en déduit que toute valeur d'adhérence de la suite $(X^{(N)}, X'^{(N)}, I_N)$ doit être de la forme (B, B', I_∞) avec $I_\infty \leq \sigma^{-2}\beta([0,1]^2)$. D'autre part, on vérifie immédiatement que pour tout N, $E[I_N] \geq (1-\rho)^4 E[\tilde{I}_N]$, et donc en faisant à nouveau tendre ρ vers 1, on voit qu'on a nécessairement $E[I_\infty] = E[\sigma^{-2}\beta([0,1]^2)]$ ce qui force l'égalité $I_\infty = \sigma^{-2}\beta([0,1]^2)$ et complète la preuve. □

Lemme 5. *Il existe une constante C_2 telle que, pour tout $N \geq 1$,*

$$E[\{J_N\}^2] \leq C_2.$$

Démonstration. On a pour tout $N \geq 2$,

$$\begin{aligned} J_N &= \frac{1}{N} \sum_{0 \leq i < j \leq N/2} I(X_i = X_j) + \frac{1}{N} \sum_{N/2 \leq i < j \leq N} I(X_i = X_j) \\ &+ \frac{1}{N} \sum_{0 \leq i < N/2 < j \leq N} I(X_i = X_j) \\ &= \frac{[N/2]}{N} J_{[N/2]} + \frac{[N/2]}{N} \tilde{J}_{[N/2]} + L_N, \end{aligned}$$

où, d'une part $\tilde{J}_{[N/2]}$ est indépendante de $J_{[N/2]}$ et a même loi que $J_{[N/2]}$, et d'autre part,

$$L_N \leq \frac{1}{N} \sum_{i=0}^{[N/2]} \sum_{j=0}^{[N/2]+1} I(X_{[N/2]-i} = X_{[N/2]+j}).$$

On peut appliquer la majoration (1) aux marches aléatoires $X'_i = X_{[N/2]-i}$, $X''_j = X_{[N/2]+j}$, ce qui conduit à

$$E[L_N^2] \leq C_1.$$

Ensuite, en soustrayant les espérances et en appliquant l'inégalité triangulaire,

$$\begin{aligned} E[\{J_N\}^2]^{1/2} &\leq E\Big[\Big(\frac{[N/2]}{N}\{J_{[N/2]}\} + \frac{[N/2]}{N}\{\tilde{J}_{[N/2]}\}\Big)^2\Big]^{1/2} + (C_1)^{1/2} \\ &= 2^{1/2}\frac{[N/2]}{N} E[\{J_{[N/2]}\}^2]^{1/2} + (C_1)^{1/2} \end{aligned}$$

Si $a_k = \sup\{E[\{J_N\}^2]^{1/2}; 2^k \leq N \leq 2^{k+1}\}$, l'inégalité précédente montre que, pour tout $\rho \in]2^{-1/2}, 1[$, on a dès que k est assez grand

$$a_{k+1} \leq \rho\, a_k + (C_1)^{1/2}.$$

On conclut que la suite (a_k) est bornée. □

Démonstration de la Proposition 2. Reprenons les notations de la preuve du Lemme 5, en supposant N pair :

$$J_N = \frac{1}{2} J_{N/2} + \frac{1}{2} \tilde{J}_{N/2} + L_N$$

avec

$$L_N = \frac{1}{N} \sum_{i=1}^{N/2} \sum_{j=1}^{N/2} I(X'_i = X''_j).$$

Notons comme précédemment $X_t'^{(N)} = (\sigma N)^{-1/2} X'_{[Nt]}$, $X_t''^{(N)} = (\sigma N)^{-1/2} X''_{[Nt]}$ pour $0 \leq t \leq 1/2$. Le Lemme 5 donne

$$((X_t'^{(N)}, X_t''^{(N)}; 0 \leq t \leq 1/2), L_N) \xrightarrow[N\to\infty, N \text{ pair}]{\text{(loi)}} ((B'_t, B''_t; 0 \leq t \leq 1/2), \beta([0, 1/2]^2))$$

où B', B'' sont deux mouvements browniens plans issus de 0 indépendants, et

$$\beta([0,1/2]^2) = \int_0^{1/2}\int_0^{1/2} \delta_0(B'_s - B''_t)\, ds\, dt.$$

D'autre part, on a

$$(X'^{(N)}, X''^{(N)}, X^{(N)}) \xrightarrow[N\to\infty]{\text{(loi)}} (B', B'', B)$$

avec $B'_t = B_{1/2-t} - B_{1/2}$, $B''_t = B_{1/2+t} - B_{1/2}$ pour $0 \le t \le 1/2$. Avec cette définition de B' et B'' on a l'égalité (formellement évidente)

$$\int_0^{1/2}\int_0^{1/2} \delta_0(B'_s - B''_t)\, ds\, dt = \int_{[0,1/2[\times]1/2,1]} \delta_0(B_s - B_t)\, ds\, dt\,.$$

En combinant ce qui précède, on obtient donc que

$$(X^{(N)}, L_N) \xrightarrow[N\to\infty, N \text{ pair}]{\text{(loi)}} (B, \alpha_0([0,1/2[\times]1/2,1])).$$

Soit ensuite $m \geq 1$. On obtient par récurrence la formule

$$J_N = 2^{-m} \sum_{p=1}^{2^m} J^{(p)}_{2^{-m}N} + \sum_{k=1}^{p}\sum_{p=1}^{2^{k-1}} L_N^{k,p}, \tag{3}$$

avec

$$J^{(p)}_{2^{-m}N} = \frac{2^m}{N} \sum_{(p-1)2^{-m}N \le i < j \le p2^{-m}N} I(X_i = X_j)$$

$$L_N^{k,p} = \frac{1}{N} \sum_{(2p-2)2^{-k}N \le i < (2p-1)2^{-k}N < j \le 2p2^{-k}N} I(X_i = X_j).$$

Restreignons-nous aux valeurs de N multiples de 2^m. Cette restriction est sans importance à cause de la propriété

$$\lim_{N\to\infty} E[(J_N - J_{2^m[2^{-m}N]})^2] = 0$$

qui est très facile à vérifier. Alors, les variables $J^{(p)}_{2^{-m}N}$, $p = 1, \ldots, 2^m$ sont indépendantes et de même loi que $J_{2^{-m}N}$. Le Lemme 4 entraîne

$$E\big[\big(2^{-m}\sum_{p=1}^{2^m}\{J^{(p)}_{2^{-m}N}\}\big)^2\big] = 2^{-2m}\, 2^m\, E[\{J_{2^{-m}N}\}^2] \le C_2\, 2^{-m}. \tag{4}$$

D'autre part, le même raisonnement que ci-dessus pour l'étude de $L_N = L_N^{1,1}$ montre que, pour tous k, p,

$$(X^{(N)}, \{L_N^{k,p}\}) \xrightarrow{\text{(loi)}} (B, \{\alpha_0(A_p^k)\})$$

où l'ensemble A_p^k a été défini dans la partie 2. Un argument simple de tension montre que cette convergence a lieu conjointement pour tous k, p. En particulier,

$$(X^{(N)}, \sum_{k=1}^{m}\sum_{p=1}^{2^{k-1}}\{L_N^{k,p}\}) \xrightarrow{\text{(loi)}} (B, \sum_{k=1}^{m}\sum_{p=1}^{2^{k-1}}\{\alpha_0(A_p^k)\}).$$

Il est maintenant facile d'en déduire la Proposition 2 : on utilise la formule (3), en remarquant que pour m grand,

$$\{2^{-m}\sum_{p=1}^{2^m} J^{(p)}_{2^{-m}N}\}$$

est petit en norme L^2, uniformément en N, d'après (4), cependant que

$$\sum_{k=1}^{m}\sum_{p=1}^{2^{k-1}}\{\alpha_0(A^k_p)\}$$

est proche de γ, par la définition même de γ.

4. Preuve de la Proposition 3.

Nous reprenons les notations de la preuve de la Proposition 2, sans supposer N multiple de 2^m. On pose pour $m \geq 1$,

$$J^m_N = J_N - \frac{1}{N}\sum_{k=1}^{2^m}\sum_{(k-1)2^{-m}N\leq i<j\leq k2^{-m}N} I(X_i = X_j).$$

Comme on a aussi

$$J^m_N = \sum_{k=1}^{m}\sum_{p=1}^{2^{k-1}} L^{k,p}_N,$$

la majoration (1) entraîne que pour tout m fixé,

$$C^m_\lambda := \sup_{N\geq 1} E[\exp(-\lambda\{J^m_N\})] \leq \sup_{N\geq 1}\exp(\lambda E[J^m_N]) < \infty.$$

Puisque $J_N = J^m_N$ dès que $N < 2^m$, il suffit pour établir la Proposition 3 de montrer que $\sup_{m\geq 1} C^m_\lambda < \infty$. Pour cela, on va majorer C^{m+1}_λ en fonction de C^m_λ. Partons de l'égalité

$$J^{m+1}_N = J^m_N + \sum_{p=1}^{2^m} L^{m+1,p}_N.$$

Soient λ', λ'' tels que $\frac{\lambda}{\lambda'} + \frac{\lambda}{\lambda''} = 1$. Alors,

$$\begin{aligned} E[\exp(-\lambda\{J^{m+1}_N\})] &\leq E[\exp(-\lambda'\{J^m_N\})]^{\lambda/\lambda'} E[\exp(-\lambda''\sum_{p=1}^{2^m}\{L^{m+1,p}_N\})]^{\lambda/\lambda''} \\ &= E[\exp(-\lambda'\{J^m_N\})]^{\lambda/\lambda'}\prod_{p=1}^{2^m} E[\exp(-\lambda''\{L^{m+1,p}_N\}]^{\lambda/\lambda''}. \end{aligned}$$

L'inégalité $e^{-u} - 1 + u \leq u^2$ pour $u \geq 0$ montre cependant que

$$\left|E[\exp(-\lambda'' L^{m+1,p}_N)] - (1 - \lambda'' E[L^{m+1,p}_N])\right| \leq (\lambda'')^2 E[(L^{m+1,p}_N)^2] \leq C_1(\lambda'')^2 2^{-2m},$$

d'après le Lemme 4. De façon plus précise, les arguments de la partie 3 montrent que $N\,L^{m+1,p}_N$ est majoré par le nombre de couples d'intersection de deux marches aléatoires indépendantes sur l'intervalle $\{0, 1, \ldots, [2^{-(m+1)}N] + 1\}$, et le résultat énoncé découle de la majoration (1) si $2^{-m}N \geq 1$. Si $2^{-m}N < 1$, on voit aisément que $L_N^{m+1,p} = 0$ et la majoration est triviale.

On a ensuite, toujours d'après la majoration (1),

$$\begin{aligned}|\exp(\lambda'' E[L_N^{m+1,p}]) - (1 + \lambda'' E[L_N^{m+1,p}])| &\leq (\lambda'')^2 E[L_N^{m+1,p}]^2 \exp(\lambda'' E[L_N^{m+1,p}]) \\ &\leq C_1(\lambda'')^2\, 2^{-2m} \exp((C_1)^{1/2}\lambda'' 2^{-m}).\end{aligned}$$

Finalement en combinant les deux majorations obtenues on arrive facilement à

$$\left|E[\exp(-\lambda''\{L_N^{m+1,p}\})] - 1\right| \leq 3\,C_1(\lambda'')^2 2^{-2m} \exp((C_1)^{1/2}\lambda'' 2^{-m}).$$

En reportant cette majoration dans les calculs précédents, on trouve

$$\begin{aligned}C_\lambda^{m+1} &\leq (C_{\lambda'}^m)^{\lambda/\lambda'}\Big(1 + 3C_1(\lambda'')^2 2^{-2m} \exp((C_1)^{1/2}\lambda'' 2^{-m})\Big)^{2^m\lambda/\lambda''} \\ &\leq (C_{\lambda'}^m)^{\lambda/\lambda'} \exp\Big(3C_1\lambda\lambda''\, 2^{-m} \exp((C_1)^{1/2}\lambda'' 2^{-m})\Big).\end{aligned}$$

Il est facile de déduire le résultat recherché de cette inégalité. Fixons $\rho > 0$ et pour tout $m \geq 2$ posons

$$u_m = \prod_{k=m}^{\infty} \frac{1}{1 - k^{-2}}, \qquad \lambda_m = \rho u_m$$

de sorte que $\lambda_{m+1}/\lambda_m = 1 - m^{-2}$. On applique alors la majoration précédente avec $\lambda = \lambda_{m+1}$, $\lambda' = \lambda_m$, $\lambda'' = m^2\lambda_{m+1}$:

$$C_{\lambda_{m+1}}^{m+1} \leq C_{\lambda_m}^m \exp\Big(3C_1\lambda_{m+1}^2 m^2\, 2^{-m} \exp((C_1)^{1/2}\lambda_{m+1} m^2 2^{-m})\Big)$$

(noter que $C_\lambda^m \geq 1$). Comme la suite (λ_m) est bornée l'inégalité précédente entraîne

$$\sup_{m\geq 1} C_{\lambda_m}^m < \infty.$$

Pour conclure, on remarque que d'après l'inégalité de Jensen,

$$C_\rho^m \leq (C_{\lambda_m}^m)^{\rho/\lambda_m} \leq C_{\lambda_m}^m,$$

et donc $\sup_{m\geq 1} C_\rho^m < \infty$ ce qui termine la preuve puisque ρ était arbitraire.

Bibliographie.

[1] S. Albeverio, E. Bolthausen, X.Y. Zhou : On the discrete Edwards model in three dimensions. Preprint (1996)

[2] D.C. Brydges, G. Slade : The diffusive model of self-avoiding walks. *Probab. Th. Rel. Fields* **103**, 285-315 (1995)

[3] B. Cadre : Une preuve standard du principe d'invariance de Stoll. Dans ce volume.

[4] S.F. Edwards : The statistical mechanics of polymers with excluded volume. *Proc. Phys. Sci.* **85**, 613-624 (1965)

[5] J.F. Le Gall : Propriétés d'intersection des marches aléatoires I. Convergence vers le temps local d'intersection. *Comm. Math. Phys.* **104**, 471-507 (1986)

[6] J.F. Le Gall : Some properties of planar Brownian motion. *Lecture Notes Math.* **1527**, pp. 111-234. Springer (1992)

[7] J.F. Le Gall : Exponential moments for the renormalized self-intersection local time of planar Brownian motion. *Séminaire de Probabilités XXVIII. Lecture Notes Math.* **1583**, pp. 172-180. Springer (1994)

[8] J. Rosen : Self-intersections of random fields. *Ann. Probab.* **12**, 108-119 (1984)

[9] J. Rosen : Random walks and intersection local time. *Ann. Probab.* **18**, 959-977 (1990)

[10] A. Stoll : Invariance principles for Brownian intersection local time and polymer measures. *Math. Scand.* **64**, 133-160 (1989)

[11] S.R.S. Varadhan : Appendix to "Euclidean quantum field theory" by K. Symanzik. In : *Local Quantum Theory* (R. Jost ed.). Academic Press (1969)

Laboratoire de Probabilités, Université Pierre et Marie Curie
4, Place Jussieu, F-75252 PARIS Cedex 05

On the tails of the supremum and the quadratic variation of strictly local martingales

K.D. Elworthy[(1)] *, X.M. Li*[(1)] *, M. Yor*[(2)]

(1) Mathematics Department, University of Warwick, Coventry CV4 7AL, UK

(2) Laboratoire de Probabilités, Université Pierre et Marie Curie, Tour 56, 3ème Etage, 4, Place Jussieu, F - 75252 PARIS CEDEX 05

Introduction and main results.

In this paper, we study some properties of continuous strictly local martingales, i.e : local martingales which are not martingales. Our interest for this class of local martingales stems from the fact, under some mild additional conditions on such a process $(M_t, t \geq 0)$, the tails of the distributions of $\sup_{t \geq 0} M_t$ and $\langle M \rangle_\infty^{1/2}$ are equivalent to $\frac{c}{x}$, as $x \longrightarrow \infty$, for two related constants c_1 and c_2 (depending on M). Precisely, one of our main results, which has a number of applications, is the

Theorem 1 : *Let* $(M_t, t \geq 0)$ *be a continuous local martingale taking its values in* $\mathbb{R}_+$ *, and satisfying* $E[M_0] < \infty$.

Then, both :

$$\ell = \lim_{x \to \infty}\left\{x\, P\left(\sup_{t \geq 0} M_t \geq x\right)\right\} \quad \text{and} \quad \sigma = \lim_{y \to \infty}\left\{y\, P\left(\langle M \rangle_\infty^{1/2} \geq y\right)\right\}$$

exist in $\mathbb{R}_+$ *, and satisfy :*

$$(1) \qquad \ell = \sqrt{\frac{\pi}{2}}\,\sigma = E[M_0 - M_\infty].$$

It is particularly easy to prove this theorem if $M_\infty = 0$, and $M_0 = c > 0$, for simplicity. In this case, using the Dubins-Schwarz representation of $(M_t, t \geq 0)$ as :

$M_t = \beta_{\langle M \rangle_t}$, where $(\beta_u, u \geq 0)$ denotes a Brownian motion starting from c, we obtain :

$$\sup_{t\geq 0} M_t = \sup_{u\leq T_0} \beta_u \quad \text{and} \quad \langle M\rangle_\infty = T_0 \equiv \inf\{u : \beta_u = 0\}.$$

It is now easy to show that :

$$\sup_{t\geq 0} M_t \overset{(law)}{=} \frac{c}{U}, \text{ and } \langle M\rangle_\infty^{1/2} \overset{(law)}{=} \frac{c}{|N|},$$

where U is uniform on $[0,1]$, and N is a standard reduced gaussian r.v. The double equality *(1)* now follows easily.

In fact, in the first paragraph below, we shall prove a more general result than that of Theorem 1 ; indeed, we shall consider a general $\mathbb{R}$-valued continuous local martingale M and we shall prove the following

Theorem 1' : *Let* $(M_t, t \geq 0)$ *be a continuous local martingale, with* $M_0 = 0$. *Assume that :*

(i) $\{M_V^-$; V *finite stopping time*$\}$ *is uniformly integrable.*

Then, $\{M_t, t \longrightarrow \infty\}$ *converges a.s. ; we denote this limit by* M_∞.

Assume furthermore that :

(ii) *there exists* $\varepsilon > 0$ *such that :* $E[\exp(\varepsilon M_\infty^-)] < \infty$.

Then both :

$$\ell = \lim_{x\to\infty}\left\{x\, P\left(\sup_{t\geq 0} M_t \geq x\right)\right\} \quad \text{and} \quad \sigma = \lim_{y\to\infty}\left\{y\, P(\langle M\rangle_\infty^{1/2} \geq y)\right\} \quad \textit{exist in } \mathbb{R}_+,$$

and satisfy :

(2)
$$\ell = \sqrt{\frac{\pi}{2}}\,\sigma = -E(M_\infty).$$

In our second paragraph, we apply Theorem 1' to transient diffusions, in particular Bessel processes, and we show how the identity *(2)* translates into some remarkable identities involving Bessel functions.
A more general discussion of strictly local martingales and their relations with strong completeness of stochastic flows is made in [9].

Acknowledgment and priority :
The proof of Theorem 1', concerning σ, uses essentially the Tauberian theorem ; after writing a first draft of this paper in May 1995, we learnt that Galtchouk - Novikov [10] already went through a similar discussion.

Ron Doney (Manchester) and J. Warren (Bath) also convinced us that the argument, if not the result, was "well-known" (to some...).

1. Proof of Theorem 1'.

(1.1) We first show that $\{M_t, t \longrightarrow \infty\}$ converges a.s. ; indeed, we remark that (i) implies, from Fatou's lemma, that : $E[L_\infty] < \infty$, where $(L_t, t \geq 0)$ is the local time at 0 of M.

But, it is well-known that the sets : $\{M_t \xrightarrow[t\to\infty]{} \cdot\}$, $\{L_\infty < \infty\}$, and $\{\langle M\rangle_\infty < \infty\}$ are all a.s. equal ; in our situation, they all have probability 1.

(1.2) We first show that ℓ exists, and satisfies :

(1.a) $$\ell = E[-M_\infty].$$

To prove this (fairly well-known result), we apply the optional stopping theorem to $\tilde{M} = (M_{t\wedge T_x} ; t \geq 0)$, for $x > 0$; from (i) , $\tilde{M}$ is uniformly integrable ; hence, we obtain :

$$0 = E[M_{T_x}] = E[M_\infty 1_{(T_x=\infty)}] + x P(T_x < \infty)$$

Consequently :

$$x P\left(\sup_{t\geq 0} M_t \geq x\right) = E\left[(-M_\infty) 1_{(T_x=\infty)}\right].$$

The right-hand side converges, as $x \longrightarrow \infty$, to : $E[-M_\infty]$, thanks to the dominated convergence theorem, since $E[|M_\infty|] < \infty$. This integrability property follows from the equality : $E[M^+_{T_x}] = E[M^-_{T_x}]$, our hypothesis (i), and Fatou's lemma.

(1.3) The proof that σ exists, and satisfies :

(1.b) $$\sqrt{\frac{\pi}{2}}\, \sigma = E(-M_\infty)$$

hinges essentially on the following variant of the Tauberian theorem.

Lemma 1 **(Feller [0], XIII.5 : Tauberian theorems, Example (c)).**

Let X *be an* $\mathbb{R}_+$*-valued random variable, and* $L : \mathbb{R}_+ \longrightarrow \mathbb{R}_+$ *be a slowly* *varying function at* ∞ *; finally, let* $0 < \alpha < 1$.

The following properties are equivalent :

i) $$\frac{1}{\lambda^{\alpha}}(1-E[\exp(-\lambda X)]) \underset{\lambda\to 0}{\sim} L(\tfrac{1}{\lambda})$$

ii) $$x^{\alpha} P(X \geq x) \underset{x\to\infty}{\sim} \frac{1}{\Gamma(1-\alpha)} L(x).$$

Proof of (1.b) : We write :

$$\frac{1}{\nu} E[1 - \exp(-\frac{\nu^2}{2} \langle M\rangle_\infty)]$$

(•) $$= \frac{1}{\nu} E[\exp(\nu(-M_\infty) - \frac{\nu^2}{2} \langle M\rangle_\infty) - \exp(-\frac{\nu^2}{2} \langle M\rangle_\infty)]$$

$$= E\left[\exp\left(-\frac{\nu^2}{2} \langle M\rangle_\infty\right)\left(\frac{\exp(\nu(-M_\infty))-1}{\nu}\right)\right]$$

It is then easily shown that, thanks to the hypothesis (ii) in Theorem 1', the last written expectation converges towards : $E(-M_\infty)$ (precisely, we use dominated convergence, and the elementary fact :

$$\left|\frac{1}{\nu}(\exp(\nu x) - 1)\right| \leq \begin{cases} \exp(\nu x), & \text{if } x \geq 0 ; \\ x^-, & \text{if } x \leq 0 ; \end{cases}$$)

Thus, we see that Lemma 1 applies with $\lambda = \frac{\nu^2}{2}$, or equivalently : $\nu = \sqrt{2\lambda}$, $X = \langle M\rangle_\infty$, and $L(\mu) \equiv \sqrt{2}E(-M_\infty)$. □

To be complete, we add the following justification of (•) : we need to show that :

(••) $$E[\exp(\nu(-M_\infty) - \frac{\nu^2}{2} \langle M\rangle_\infty)] = 1,$$

which also follows from the hypotheses (i) and (ii) ; indeed, they imply that :

$$\exp(\nu M_V^-) \leq E[\exp(\nu M_\infty^-)|\mathcal{F}_V] \text{ , for } \nu \leq \varepsilon$$

hence the uniform integrability of

$$\{\exp(\nu M_V^-) ; \text{ V finite stopping time, } \nu \leq \varepsilon\}$$

which yields (••).

(1.4) We now make some comments about the hypotheses and the conclusion of Theorem 1' :

- first, remark that (i) and (ii) imply, using both Jensen's and Doob's L^p inequalities, that :

$$\text{for } \varepsilon' < \varepsilon\,, \quad E[\exp(\varepsilon' \sup_{t\geq 0} M_t^-)] < \infty$$

- consequently, ℓ is also equal to :

$$\ell^* = \lim_{x\to\infty}\left\{x\, P\left(\sup_{t\geq 0} |M_t| \geq x\right)\right\} ;$$

likewise, σ is also equal to

$$\sigma^+ \stackrel{\text{def}}{=} \lim_{y\to\infty}\{y\, P((\langle M^+\rangle_\infty)^{1/2} \geq y)\}$$

$\left(\text{recall that : } \langle M^+\rangle_t = \int_0^t 1_{(M_s>0)} d\langle M\rangle_s\right)$,

since, from (ii), $\langle M^-\rangle_\infty$ is integrable, and in fact admits moments of all orders.

To summarize, starting from an asymmetric hypothesis about a local martingale M, the conclusion of Theorem 1' may be presented in "symmetric" terms (i.e. involving only $|M|$).

The following variant of Theorem 1' seems to have a wider domain of applicability.

Theorem 1" : *Let* $X_t = M_t + A_t$ *be an* $\mathbb{R}_+$*-valued continuous local sub-martingale such that its increasing process* $(A_t, t \geq 0)$ *satisfies :*

(+) $$E[\exp(\varepsilon A_\infty)] < \infty, \text{ for some } \varepsilon > 0.$$

Then, the following limits exist in $\mathbb{R}_+$ *:*

$$\ell = \lim_{x\to\infty}\left\{x\, P\left(\sup_{t\geq 0} X_t > x\right)\right\} \quad \text{and} \quad \sigma = \lim_{y\to\infty}\left\{y\, P(\langle X\rangle_\infty^{1/2} \geq y)\right\}$$

and satisfy :

(3) $$\ell = \sqrt{\frac{\pi}{2}}\,\sigma = E[M_o - M_\infty].$$

The proof of Theorem 1" is quite similar to that of Theorem 1' ; hence, it is left to the reader.

As an illustration, we remark that Theorem 1" applies to :

$$X_t^{(1)} = B^+_{t\wedge\tau_\lambda} , \text{ and } X_t^{(2)} = |B_{t\wedge\tau_\lambda}| , \text{ for some } \lambda > 0,$$

where : $\tau_\lambda = \inf\{u : \ell_u > \lambda\}$, with (ℓ_u) the local time at 0 of the Brownian motion B ; note that Theorem 1' does not apply to $(B_{t\wedge\tau_\lambda}, t \geq 0)$.

For these examples, *(3)* becomes :

$$\ell^{(1)} = \sqrt{\frac{\pi}{2}}\,\sigma^{(1)} = \frac{\lambda}{2} \quad\text{and}\quad \ell^{(2)} = \sqrt{\frac{\pi}{2}}\,\sigma^{(2)} = \lambda.$$

These results may also be checked directly, since it is well-known that :

$$\sup_{t\leq\tau_1} B_t \overset{(law)}{=} 1/_{2e} \quad\text{and}\quad \sup_{t\leq\tau_1} |B_t| \overset{(law)}{=} 1/_e ,$$

whereas : $\displaystyle\int_0^{\tau_1} ds\, 1_{(B_s>0)} \overset{(law)}{=} \frac{1}{4}\tau_1 \overset{(law)}{=} \frac{1}{4} T_1$, with

$T_1 = \inf\{t : B_t = 1\}$, and e is an exponential variable with mean 1.

Finally, concerning possible further generalizations, it would be most interesting to know whether one can avoid the Tauberian argument, and weaken the hypothesis (+) in Theorem 1".

2. Strictly local martingales, transient diffusions and some remarkable identities.

(2.1) Our examples will take place in the framework considered by Pitman-Yor ([3], [4] ; (1981, 1982)) and Le Gall ([1] ; (1986)) of a regular diffusion $(R_t , 0 \leq t < \zeta, P_r , 0 < r < \infty)$ on the interval $]0,\infty[$ of $\mathbb{R}$; let $s(\cdot)$ denote a scale function for the diffusion, and m the speed measure normalized so that the infinitesimal generator is $\frac{1}{2}\frac{d}{dm}\frac{d}{ds}$. We assume :

(i) $\zeta = \inf\{t > 0 : X_{t-} = 0 \text{ or } \infty\}$

(ii) $s(0) = -\infty$, $s(\infty) < \infty$

(iii) 0 is an entrance point for the diffusion R.

In the sequel, we shall always take $s(\infty) = 0$.

We still need to introduce, for $0 < \rho < \infty$, the last passage time $L_\rho = \sup\{t \geq 0 : R_t = \rho\}$, and the expression of the semi-group

$$Q_t(r,d\rho) = p_t^{*}(r,\rho)m(d\rho).$$

Then, we have the following

Theorem 2 : *1. For* $0 \le r < \rho < \infty$,

(4) $$P_r(L_\rho \in dt) = -\frac{1}{2s(\rho)}\, p_t^{*}(r,\rho)dt$$

2. For every $r > 0$, *and* $t \ge 0$, *the limit*

(5) $$\lim_{\alpha\to\infty}\left\{\alpha\, P_r\left(\sup_{u\le t}\left(-\,s(R_u)\right) \ge \alpha\right)\right\}$$

exists, and is equal to :

(6) $$E_r\left[s(R_t) - s(r)\right] = \frac{1}{2}\int_0^t du\; p_u^{*}(0,r).$$

In particular, $(s(R_t), t \ge 0)$ *is a strictly local martingale.*

Proof : a) For the first statement, see Pitman-Yor (1981).

b) For the second statement, the existence of the limit and its equality to the left-hand side of *(6)* follows from *(1)* in Theorem 1, whereas Le Gall ([1] (1986) ; Theorem 1.1, p. 1222)) expresses the limit *(5)* as the right- hand side of *(6)*. □

(2.2) The most standard example of a diffusion which satisfies the above hypothesis is the Bessel process with dimension $d = 2(1+\nu) > 2$, i.e : $\nu > 0$. We then have : $s(\rho) = -\dfrac{1}{\rho^{d-2}}$,

and the identity *(4)*, taken for $r = 0$, becomes :

(7) $$P_o(L_\rho \in dt) = \left(\frac{\rho^2}{2}\right)^{\nu}\frac{dt}{\Gamma(\nu)t^{\nu+1}}\exp\left(-\frac{\rho^2}{2t}\right).$$

Now, the identity *(6)* may be written as :

(6') $$E_r\left[\frac{s(R_t)}{s(r)}\right] = 1 - \left(\frac{-1}{2s(r)}\int_0^t du\; p_u^{*}(0,r)\right) = P_o(L_r \ge t)$$

which, as a consequence of *(7)*, becomes :

(8) $$\int_0^\infty \frac{d\rho}{\rho^{2\nu}} p_t(r,\rho) = \frac{1}{2^\nu\Gamma(\nu)} \int_t^\infty \frac{du}{u^{1+\nu}} \exp(-\frac{r^2}{2u}),$$

where $$p_t(r,\rho) = \frac{1}{t}\,\rho(\frac{\rho}{r})^\nu \exp -\left(\frac{r^2+\rho^2}{2t}\right) I_\nu\left(\frac{r\rho}{t}\right)$$

is the density of the semi-group $Q_t(r,d\rho)$ with respect to $d\rho$.
Easy changes of variables then show that *(8)* is equivalent to

(8') $$\int_0^\infty \frac{\xi d\xi}{\xi^\nu a^{\nu-1}} \exp\left(-\frac{a^2}{2}(1+\xi^2)\right) I_\nu(a\xi) = \frac{1}{2^\nu\Gamma(\nu)} \int_0^1 \frac{du}{u}\, u^\nu \exp\left(-\frac{a^2 u}{2}\right)$$

and also to :

(8") $$\int_0^\infty \eta\, d\eta \exp\left(-\frac{\eta^2}{2}\right) \frac{I_\nu(\sqrt{a}\,\eta)}{(\sqrt{a}\,\eta)^\nu} = \frac{1}{2^\nu\Gamma(\nu)} \int_0^1 dv(1-v)^{\nu-1} \exp\left(\frac{a}{2}\,v\right).$$

This identity *(8")* may be verified by developing both sides as a series expansion in powers of a with the help, for the left-hand side, of the classical formula :

(9) $$I_\nu(z) = (\frac{z}{2})^\nu \sum_{n=0}^\infty \frac{1}{n!\Gamma(\nu+n+1)} (\frac{z}{2})^{2n}.$$

In any case, the identity *(8")* is a particular case of the Lipschitz-Hankel integrals ; see, e.g., chap. XIII of Watson [6], formula 3, p. 394, which gives a formula for :

$$\int_0^\infty dt\, t^{\mu-1} e^{-p^2t^2} J_\nu(at)\,,$$

with the help of the ${}_1F_1$ hypergeometric functions ; such formulae are also found in Lebedev ([2], p. 278, Exercise 12).

For clarity and future reference, we write again the equalities *(2)* and *(6)* in the particular case where $M_t = \frac{1}{R_t^{d-2}}$, under $P_r^{(\nu)}$, the law of R, starting from $r > 0$.

Proposition 1 : *The 4 following quantities are equal :*

$$\lim_{\alpha\to\infty}\left\{\alpha P_r^{(\nu)}\left(\frac{1}{(\inf_{s\le t} R_s^{d-2})} \ge \alpha\right)\right\}$$

$$\lim_{y\to\infty}\left\{\sqrt{\frac{\pi}{2}}\, y\, P_r^{(\nu)}\left((d-2)\left(\int_0^t \frac{ds}{R_s^{2(d-1)}}\right) \ge y\right)\right\}$$

(11)
$$E_r^{(\nu)}\left(\frac{1}{r^{d-2}} - \frac{1}{R_t^{d-2}}\right) = \frac{1}{2^\nu \Gamma(\nu)} \int_0^t \frac{du}{u^{1+\nu}} \exp\left(-\frac{r^2}{2u}\right).$$

(2.3) Associated with the 2-dimensional Bessel process $(R_t, t \ge 0)$, starting from $r > 0$, there is also the strictly local martingale $M_t = \log \frac{1}{R_t}$, which satisfies the hypothesis of our Theorem 1'.

In order to obtain the corresponding values of the quantities in *(1')* in the present case, it suffices to divide both sides of *(11)* by (2ν), and to let $\nu \to 0$.

Thus, we obtain the following

Proposition 2 : *Let* $P_r^{(o)}$ *be the law of* $(R_t, t \ge 0)$, *the 2-dimensional Bessel process starting from* $r > 0$. *Then, the following 4 quantities are equal :*

$$\lim_{\alpha\to\infty}\left\{\alpha\, P_r^{(o)}\left(\log\frac{1}{(\inf_{s\le t} R_s)} \ge \alpha\right)\right\} ; \quad \lim_{y\to\infty}\left\{\sqrt{\frac{\pi}{2}}\, y\, P_r^{(o)}\left(\left(\int_0^t \frac{ds}{R_s^2}\right)^{1/2} \ge y\right)\right\}$$

(12)
$$E_r^{(o)}\left[\log R_t - \log r\right] = \frac{1}{2}\int_0^t \frac{du}{u}\, e^{-\frac{r^2}{2u}}.$$

We again remark that the identity *(12)*, may be expressed as an integral identity involving the Bessel function I_o , i.e. see *(2.4)* below.

On the other hand, if we particularize our argument in the proof of Theorem 1' concerning the quadratic variation of M, we obtain, in the present case :

(13)
$$\frac{1}{\nu} E_r^{(o)}\left\{1 - \exp\left(-\frac{\nu^2}{2}\int_0^t \frac{ds}{R_s^2}\right)\right\} \xrightarrow[\nu\to 0]{} E_r^{(o)}[\log R_t - \log r]$$

But, the following formula is known (see, e.g., Yor [8]) :

(14)
$$E_r^{(o)}\left\{\exp\left(-\frac{\nu^2}{2}\int_0^t \frac{ds}{R_s^2}\right) \Big| R_t\right\} = \frac{I_\nu}{I_o}\left(\frac{rR_t}{t}\right).$$

Hence, it is deduced from *(13)* that :

$$(15)\qquad E_r^{(o)}\left\{-\frac{\partial}{\partial\nu}\Big|_{\nu=0}\left(\frac{I_\nu}{I_o}\right)\left(\frac{rR_t}{t}\right)\right\} = E_r^{(o)}\left[\log R_t - \log r\right].$$

It now follows from a classical integral representation of $I_\nu(\xi)$ that :

$$-\frac{\partial}{\partial\lambda}\Big|_{\lambda=0} I_\lambda(\xi) = \int_0^\infty du\, e^{-\xi(\cosh u)} \equiv K_o(\xi)\ ^{(1)}.$$

Hence, we deduce from *(15)* and the explicit formula for the semi-group $Q_t^{(o)}(r,d\rho)$ that :

$$(16)\qquad \frac{1}{t}\int_0^\infty d\rho\,\rho\, e^{-\frac{r^2+\rho^2}{2t}}\int_0^\infty du\, e^{-\frac{r\rho}{t}(\cosh u)} = \frac{1}{2}\int_0^t \frac{du}{u}\, e^{-\frac{r^2}{2u}}.$$

(2.4) <u>Some remarks following Proposition 2, and formulae *(13)* through *(16)*.</u>

i) We can write formula *(12)* in the form :

$$(12')\qquad \int_0^\infty \frac{d\rho\,\rho}{t}\exp\left(-\frac{r^2+\rho^2}{2t}\right) I_o\left(\frac{r\rho}{t}\right)(\log\rho - \log r) = \frac{1}{2}\int_0^t \frac{du}{u}\, e^{-(r^2/2u)}$$

and, as above with *(11)* and *(12)* may deduce this identity *(12')* from the corresponding identity involving I_ν, and deduced from *(11)*.

ii) From *(14)*, we can obtain a result similar to, but deeper than, *(13)*, namely :

$$\lim_{\nu\to 0}\frac{1}{\nu} E_r^{(o)}\left[1-\exp\left(-\frac{\nu^2}{2}\int_0^t \frac{ds}{R_s^2}\right)\Big| R_t = \rho\right] = -\frac{\partial}{\partial\nu}\Big|_{\nu=0+}\frac{I_\nu(\xi)}{I_o(\xi)} = \frac{K_o(\xi)}{I_o(\xi)}$$

---

(1) In fact, the equality : $K_o(\xi) = -\frac{\partial}{\partial\lambda}\big|_{\lambda=0}(I_\lambda(\xi))$ also follows immediately from the formula : $K_\lambda(\xi) = \frac{\pi}{2}\,\frac{I_{-\lambda}(\xi)-I_\lambda(\xi)}{\sin(\pi\lambda)}$ (see, e.g., formula (5.7.2), p. 108 in Lebedev [2]). Also, the formula : $K_o(\xi) = \int_0^\infty du\, e^{-\xi(\cosh u)}$ is a particular case of : $K_\nu(\xi) = \int_0^\infty du\, e^{-\xi(\cosh u)}\cosh(\nu u)$.

where $\xi = \frac{r\rho}{t}$, on one hand, and, on the other hand, thanks to the Tauberian theorem, we find that the same quantity is equal to :

$$\sqrt{\frac{\pi}{2}}\ \lim_{y\to\infty}\left\{y\ P_r^o\left(\left(\int_0^t \frac{ds}{R_s^2}\right)^{1/2} \geq y \,\middle|\, R_t = \rho\right)\right\}$$

(2.5) An important part of the results presented in Propositions 1 and 2 is well-known ; in fact, some of these results, namely those concerning

$$\lim_{\alpha\to\infty}\left\{\alpha\ P\left(\sup_{s\leq T} M_s \geq \alpha\right)\right\}$$

in their applications to Bessel processes, form the core of the arguments of the proof of the main asymptotics of the Wiener sausage, i.e : Le Gall [1], Theorem 1.1, and, in part, Spitzer [5].

The results about the asymptotics of $\sqrt{\frac{\pi}{2}}\ y\ P(\langle M\rangle_T^{1/2} \geq y)$ are perhaps less known, although they also appear in Werner [7].

An interesting consequence of Proposition 2 is the following

Corollary : *Let* $(\theta_u, u \geq 0)$ *be a continuous determination of the argument around* 0 *of the* 2*-dimensional Brownian motion* $(Z_u, u \geq 0)$, *starting from* $z_o \neq 0$. *Then, we have :*

$$\lim_{\alpha\to\infty}\{\alpha P(|\theta_t| \geq \alpha)\} = \lim_{\alpha\to\infty}\left\{\alpha\ P\left(\sup_{s\leq t} \theta_s \geq \alpha\right)\right\} = \frac{1}{\pi}\int_0^t \frac{du}{u}\exp\left(-\frac{r^2}{2u}\right)$$

$$\lim_{\alpha\to\infty}\left\{\alpha\ P\left(\sup_{s\leq t} |\theta_s| \geq \alpha\right)\right\} = \frac{1}{2}\int_0^t \frac{du}{u}\exp\left(-\frac{r^2}{2u}\right).$$

Proof : Thanks to the skew-product representation of Z, there exists a 1-dimensional Brownian motion $(\gamma_t, t \geq 0)$, independent of $(R_u, u \geq 0)$ such that :

$$\theta_t = \gamma_{\left(\int_0^t \frac{ds}{R_s^2}\right)}.$$

Hence, we have : $|\theta_t| \overset{(law)}{=} \sup_{s\leq t} \theta_s \overset{(law)}{=} \left(\int_0^t \frac{ds}{R_s^2}\right)^{1/2}\left(\sup_{u\leq 1} \gamma_u\right) \overset{(law)}{=} \left(\int_0^t \frac{ds}{R_s^2}\right)^{1/2} |\gamma_1|$

thanks to the reflection principle. Thus, we have :

$$\alpha P(|\theta_t| \geq \alpha) = \alpha P\left(\sup_{s\leq t} \theta_s \geq \alpha\right) = \alpha P\left(\left(\int_0^t \frac{ds}{R_s^2}\right)^{1/2} \geq \frac{\alpha}{|\gamma_1|}\right),$$

so that, by dominated convergence, we find :

$$\lim_{\alpha\to\infty} \alpha P\left(\sup_{s\leq t} \theta_s \geq \alpha\right) = E(|\gamma_1|) \lim_{\beta\to\infty} \beta P\left(\left(\int_0^t \frac{ds}{R_s^2}\right)^{1/2} \geq \beta\right)$$

and, since : $E(|\gamma_1|) = \sqrt{\frac{2}{\pi}}$, we obtain, from Proposition 2, that :

$$\lim_{\alpha\to\infty} \alpha P\left(\sup_{s\leq t} \theta_s \geq \alpha\right) = \frac{2}{\pi}\left(\frac{1}{2}\int_0^t \frac{du}{u}\, e^{-(r^2/2u)}\right).$$

Likewise, we obtain :

$$\lim_{\alpha\to\infty}\left\{\alpha P\left(\sup_{s\leq t} |\theta_s| \geq \alpha\right)\right\} = E[\gamma_1^*] \lim_{\beta\to\infty}\left\{\beta P\left(\left(\int_0^t \frac{ds}{R_s^2}\right)^{1/2} \geq \beta\right)\right\}$$

$$= \left(E[\gamma_1^*]\sqrt{\frac{2}{\pi}}\right)\left(\frac{1}{2}\int_0^t \frac{du}{u}\, e^{-r^2/2u}\right)$$

and the desired result follows from the next

Lemma 2 : *Define* $\gamma_1^* = \sup_{s\leq 1} |\gamma_s|$. *Then, one has :* $E[\gamma_1^*] = \sqrt{\frac{\pi}{2}}$.

Proof : Define $\hat{T}_1 = \inf\{t : |\gamma_t| = 1\}$. Then, from the scaling property of Brownian motion, we deduce : $\gamma_1^* \overset{(law)}{=} 1/(\hat{T}_1)^{1/2}$, so that :

$$E[\gamma_1^*] = E\left[\frac{1}{(\hat{T}_1)^{1/2}}\right] = \frac{1}{\Gamma(\frac{1}{2})}\int_0^\infty du\, u^{-1/2}\, E\left[e^{-u\hat{T}_1}\right]$$

$$= \sqrt{\frac{2}{\pi}}\int_0^\infty dv\, E\left[e^{-\frac{v^2}{2}\hat{T}_1}\right] = \sqrt{\frac{2}{\pi}}\int_0^\infty \frac{dv}{(\operatorname{ch} v)}$$

Now, we have : $\displaystyle\int_0^\infty \frac{dv}{(\operatorname{ch} v)} = 2\int_0^\infty e^{-v}\frac{dv}{(1+e^{-2v})} = 2\int_0^1 \frac{dx}{1+x^2} = 2\,\mathrm{Arctg}(1) = \frac{\pi}{2}$, so that, finally : $E[\gamma_1^{\bullet}] = \sqrt{\frac{\pi}{2}}$. □

References

[0] W. Feller : An Introduction to probability and its Applications. *Volume 2 - Wiley (1970).*

[1] J.F. Le Gall : Sur la saucisse de Wiener et les points multiples du mouvement brownien. *The Annals of Proba. 14, (4), p. 1219-1244 (1986).*

[2] N.N. Lebedev : Special Functions and their Applications. *Dover Publications (1972).*

[3] J.W. Pitman, M. Yor : Bessel processes and infinitely divisible laws. *In : Stochastic Integrals, ed : D. Williams, Durham Conference 1980. Lect. Notes in Maths. 851, p. 285-370. Springer (1981).*

[4] J.W. Pitman, M. Yor : A decomposition of Bessel bridges. *Zeit. für Wahr., 59, p. 425-457 (1982).*

[5] F. Spitzer : Some theorems about 2-dimensional Brownian motion. *Trans. Amer Math. Soc. 87, p. 187-197 (1958).*

[6] G.N. Watson : A treatise of the theory of Bessel functions. *Cambridge Univ. Press (1966).*

[7] W. Werner : Sur l'ensemble des points autour desquels le mouvement brownien tourne beaucoup. *Prob. Th. Rel. Fields 99,* p. 111-142 (1994).

[8] M. Yor : Loi de l'indice du lacet brownien et distribution de Hartman-Watson. *Zeit. für Wahr., 53, p. 71-95 (1980).*

[9] K.D. Elworthy, X.M. Li, M. Yor : The importance of strictly local martingales ; applications to radial Ornstein-Uhlenbeck process. *Preprint (1996).*

[10] L.I. Galtchouk, A.A. Novikov : Wald equation ; discrete time case. *Preprint (March 1994). Strasbourg University.*

ON WALD'S EQUATION. DISCRETE TIME CASE

Leonid I. GALTCHOUK† and Alexandre A. NOVIKOV‡

† Institut de Recherche Mathématique Avancée
Université Louis Pasteur et C.N.R.S., 7 rue René-Descartes
67084 Strasbourg Cedex, France.
e-mail:galtchou@math.u-strasbg.fr.

‡ Steklov Mathematical Institute
42, Vavilova, 117333, Moscow, Russia.
e-mail: alex@novikov.mian.su.

0. Summary

Let m_t be a square integrable martingale, $m_0 = 0$, such that there exists $\lim_{t\to\infty} m_t = m_\infty$ a.s. We study a minimal possible sufficient condition for the validity of Wald's equation $Em_\infty = 0$ in terms of the tail behavior of a square characteristic $S(m)_\infty$ of m_t.

1. The background for the problem and the main result

Let $(\Omega, \mathbf{F}, \mathbf{F}_t, P), t \in Z^+ = \{0, 1, \ldots\}$ be a stochastic basis with the filtration $\mathbf{F}_t$. We *always* consider in this paper the process m_t as a square integrable martingale , $m_0 = 0$, that is a square integrable process such that $Em_\tau = Em_0 = 0$ for any bounded (by a constant) stopping time τ (of course, with respect to $\mathbf{F}_t$) . Denote a square characteristic of m_t by $S(m)_t := \sum_{k=1}^t E_k X_k^2$, where we used notations for the martingale-difference $X_k := m_k - m_{k-1}$ and for conditional expectations $E_t(\,) := E\{(\,)|\mathbf{F}_{t-1}\}$.

It is well known (see Meyer (1972,Theorem 64), Liptser and Shiryaev (1991)) that if $S(m) = S(m)_\infty < \infty$ a.s. then there exists $\lim_t m_t = m_\infty$ a.s. (all limits over t are considered as $t \to \infty$). For many applications , for example in sequential analysis, it is of interest to know under what minimal conditions the equation

$$Em_\infty = 0$$

still holds.

The prehistory of this question goes back to classical Wald monograph (1945) in which this equality was used to establish some general properties of sequential tests. The first results obtained by Wald (in modern form) is the following:

AMS 1991 classification : 60G42, 60G40.

Key words, phrases :local martingale, Wald's equation, uniform integrability, tauberian theorem.

$$ES(m) < \infty \implies Em_\infty = 0.$$

Later Burkholder and Gundy (1970) proved the maximal inequality

$$E \sup_t |m_t| \leq CE(S(m))^{1/2}$$

which implies uniforme integrability of m if

$$E(S(m))^{1/2} < \infty,$$

and, of course, Wald's equation:

$$Em_\infty = 0. \tag{1}$$

(we denote all constants whose values are not important for this exposition by C).

Note (1) is valid also for continuous time martingale and it seems the first result in this direction was obtained by Novikov (1971) for stochastic integrals with respect to a brownian motion (the paper of Novikov (1971) was presented for publishing at the same time as Burkholder and Gundy (1970)).

Azema, Gundy and Yor (1979) discussed a problem of uniform integrability (U.I.) of continuous martingales ($m_t \in \mathbf{M}^c$) and, particularly, they showed that if $m_t \in \mathbf{M}^c$ and $\sup_t E|m_t| < \infty$ then

$$\lim_t P\{S(m) > t\}t^{1/2} = 0 \Longleftrightarrow m_t \text{ is U.I.} \implies Em_\infty = 0. \tag{2}$$

The similar result as in (2) for discrete time case was obtained by Gundy (1981) but for a special case of martingales satisfying the following conditions:

$$\sup_t E|m_t| < \infty\,,\ \lim_t P\{S(m) > t\}t^{1/2} = 0,$$

$$X_t = V_t D_t, V_t \textit{ is } \mathbf{F}_{t-1}\text{-measurable}, E_t X_t = 0, E_t|X_t| > C > 0, E_t X_t^2 = 1 \tag{3}$$

(all inequalities for random variables in our paper hold with probability one).

In the present paper we prove that under some different bounds for conditional moments of the martingale-difference X_t a weaker condition on $S(m)$ instead of that one in (3) may be used and ever more detailed information concerning the asymptotic behaviour of $P\{S(m) > t\}$ may be obtained (see Lemma 1 and Remark 1 below).

To formulate the basic result introduce the following class of deterministic functions

$$G = \{g(x) > 0,\ g(x) \uparrow,\ \int_1^\infty x^{-3/2} g(x)dx = \infty\}.$$

Theorem 1. *Let $S(m) < \infty, E|m_\infty| < \infty$, $|X_t| < C$. Then*

$$(\textit{there exists } g(x) \in G : Eg(S(m)) < \infty) \implies Em_\infty = 0. \tag{4}$$

To see that condition of (4) is a less restrictive that the condition

$$\lim_t P\{S(m) > t\}t^{1/2} = 0$$

one may take the function $g(x)$ with the step-wise derivative

$$g'(x) = \sum_{k=1}^{\infty}(klogk)^{-1}(x_{k+1}^{1/2} - x_k^{1/2})I\{x_k \le x < x_{k+1}\}$$

where $I\{\ \}$ is an indicator function, $x_1 = 1, x_{k+1} > x_k+1$. As for any nondecreasing positive function $f(x)$

$$\int_1^{\infty} x^{-3/2}f(x)dx = \infty \iff \int_1^{\infty} x^{-1/2}df(x) = \infty$$

then $g(x) \in G$.

Take now

$$x_{k+1} = \inf\{x \ge x_k + 1 : \sup_{t \ge x} P\{S(m) > t\}t^{1/2} \le 1/k\} .$$

As

$$Eg(S(m)) < \infty \iff \int_1^{\infty} P(S(m) > x)\sqrt{x}\frac{dg(x)}{\sqrt{x}} < \infty,$$

then it is easy to see that $Eg(S(m)) < \infty$. Note that, of course,

$$(\text{there exists } g(x) \in G : Eg(S(m)) < \infty) \implies \liminf_t P\{S(m) > t\}t^{1/2} = 0.$$

It should be noted that unlike the case of nonnegative martingales the validity of Wald's equation $Em_\infty = 0$, generally speaking, does not imply Wald's identity , that is the equality $Em_\tau = 0$ for any stopping time τ : consider for example, sums of Rademacher' variables (with jumps 1 and -1) stopped at moment of the first hitting zero after first passaging of the level +1. But if one assumes that m_t^+ is U.I. then (as remarked by Vallois (1991)) Wald's equation is equivalent to U.I. of m_t .

The technique used in the present paper is based on exponential martingales and tauberian theorem (see Feller (1966)) and it is very different from one used in Burkholder and Gundy (1970), Azema, Gundy and Yor (1979), Gundy (1981) and related papers of Kinderman (1980), Klass (1988), de la Pena (1993) (all these papers exploited so-called "good-lambda" inequality first appeared in Burkholder and Gundy (1970)).

We note that the idea of using exponential supermartingales was used earlier by Meyer (1972, th. 71) for obtaining some asymptotic results for martingales.

Our method can be easily extended to the case of continuous time martingale (results for quasi left-continuous martingales was reported by the authors to Probability seminar at Strasbourg university, February , 1994) but the authors plan to consider in a separate paper a more general case of so-called optional martingales (that is, without standard condition on right-continuity of $\mathbf{F}_t$ (see Galtchouk (1980)).

Note that the result of Theorem 1 for a special case of stopped processes with independent increments was proved in Novikov (1981a,1982).

2. Two lemmas

Lemma 1. *Suppose $S(m) < \infty$ and there exists $\lambda_+ > 0$ such that for all* $\lambda \in [0, \lambda_+)$

$$E_t|X_t|^3 exp(\lambda X_t) \leq C E_t|X_t|^2 \tag{5}$$

and

$$\sup_t Eexp(\lambda m_t) < \infty . \tag{6}$$

Then

$$0 \leq Em_\infty < \infty, \tag{7}$$

and

$$\lim_t P\{S(m) > t\} t^{1/2} = (2/\pi)^{1/2} Em_\infty . \tag{8}$$

Proof. By condition (5) the following predictable function

$$\psi_t(\lambda) = logE_t exp(\lambda X_t),\ 0 \leq \lambda < \lambda_+,$$

is finite and it is non negative due to Jensen's inequality and the condition $E_t X_t = 0$.

Below we exploit the following well-known facts: the process

$$Z_t(\lambda) = exp\{\lambda m_t - \sum_1^t \psi_k(\lambda)\},\ 0 \leq \lambda < \lambda_+,$$

is a non negative martingale and there exists

$$\lim_t Z_t(\lambda) = Z_\infty(\lambda).$$

The limit m_∞ exists thanks the condition $S(m) < \infty$. By (6) and Fatou's lemma we have $Eexp(\lambda m_\infty) < \infty$ and by the dominated convergence theorem the following equality (Wald's exponential identity) holds

$$EZ_\infty(\lambda) = 1,\ \ 0 \leq \lambda < \lambda_+ . \tag{9}$$

Assumption (6) implies uniform integrability of $m_t^+ = max(m_t, 0)$ by Vallée-Poussin's theorem.

As $Em_t^+ = Em_t^-, (m_t^- = max(-m_t, 0))$ then by Fatou's lemma

$$Em_\infty^+ = lim_t Em_t^+ = lim_t Em_t^- \geq Em_\infty^-.$$

So we have $0 \leq Em_\infty < \infty$ (this type of arguments was used by Novikov (1981) and a recent paper of Vallois (1991)).

The condition (6) implies

$$E|m_\infty|exp(\lambda m_\infty) < \infty. \tag{10}$$

Indeed

$$E|m_\infty|exp(\lambda m_\infty) = -Em_\infty exp(\lambda m_\infty)I_{m_\infty<0} + Em_\infty exp(\lambda m_\infty)I_{m_\infty\geq 0}. \tag{11}$$

The first right hand term is finite by (7).

Further, for all $\lambda \in [0, \lambda_+)$ there exists a such $\epsilon > 0$ that $\lambda + \epsilon < \lambda_+$. For all $\epsilon > 0$ there exists a such constant $K = K_\epsilon$ that

$$exp((\lambda + \epsilon)m_\infty) > m_\infty exp(\lambda m_\infty)$$

if $m_\infty > K_\epsilon$.

Then for second right hand term of (11) we have

$$Em_\infty exp(\lambda m_\infty)I_{m_\infty\geq 0} = Em_\infty exp(\lambda m_\infty)I_{0\leq m_\infty\leq K_\epsilon} + Em_\infty exp(\lambda m_\infty)I_{m_\infty > K_\epsilon}$$

$$\leq K_\epsilon e^{\lambda K_\epsilon} + Ee^{(\lambda+\epsilon)m_\infty} < \infty.$$

This inequality and (11) imply (10).

Since $E|m_\infty|exp(\lambda m_\infty) < \infty$ then by the dominated convergence theorem from (9) it follows:

$$1 - Eexp\{-\sum_1^\infty \psi_t(\lambda)\} = EZ_\infty(\lambda) - Eexp\{-\sum_1^\infty \psi_t(\lambda)\} =$$

$$\lambda Em_\infty + o(\lambda), \ \ \lambda \to 0.$$

Below, in Lemma 2, we shall prove the following relation

$$Eexp\{-1/2\ \lambda^2 S(m)\} - Eexp\{-\sum_1^\infty \psi_t(\lambda)\} = o(\lambda), \ \ \lambda \to 0, \tag{12}$$

That gives us

$$1 - Eexp\{-\lambda^2/2S(m)\} = \lambda Em_\infty + o(\lambda), \ \ \lambda \to 0.$$

This equation is equivalent to (8) by following tauberian theorem.

Theorem 2 (Feller W.(1971) Ch.XIII,Example (c)). *Let Y be an R_+-valued random variable, and $L : R_+ \to R_+$ be a slowly varying function at ∞. Let $0 \leq \rho < \infty$.*

Then following relations are equivalent:

$$i)\ (1 - E\exp(-\nu Y)) \sim \nu^{1-\rho} L(\frac{1}{\nu}), \ \ \nu \to 0,$$

$$ii)\ x^{1-\rho}P(Y \geq x) \sim \frac{1}{\Gamma(\rho)}L(x),\ x \to \infty.$$

This theorem applies with $Y = S(m), L(x) = Em_\infty, \nu = \frac{\lambda^2}{2}$, or $\lambda = \sqrt{2\nu}$.

So to finish the proof of Lemma 1 we need only to prove equation (12). We shall use the following

Lemma 2. *Under conditions of Lemma 1*

$$(\lambda^2/2 - \lambda^3 C)S(m) \leq \sum_1^\infty \psi_k(\lambda) \leq (\lambda^2/2 + \lambda^3 C)S(m), 0 \leq \lambda < \lambda_+ .$$

Proof of Lemma 2. From the definition of $\psi_t(\lambda)$ it follows that for for all $\lambda \in [0, \lambda_+)$

$$\frac{\partial \psi_t(\lambda)}{\partial \lambda} := \psi_t(\lambda)' = E_t(X_t exp\{\lambda X_t - \psi_t(\lambda)\}),$$

$$0 \leq \psi_t(\lambda)'' = E_t(X_t - \psi_t(\lambda)')^2 exp\{\lambda X_t - \psi_t(\lambda)\}) =$$
$$E_t(X_t^2 exp\{\lambda X_t - \psi_t(\lambda)\}) - (\psi_t(\lambda)')^2.$$

Integrating the last inequality with respect to λ and applying the inequality $exp(x) - 1 \leq x^+ exp(x)$ we get

$$\psi_t(\lambda)' \leq \lambda E_t X_t^2 + \frac{\lambda^2}{2} E_t(X_t^+)^3 exp\{\lambda X_t\}. \qquad (13)$$

As by (5) $E_t(X_t^+)^3 exp\{\lambda X_t\} \leq C E_t X_t^2$ we get the upper bound in Lemma 2.
To prove the lower bound, let us note that the same arguments give

$$\psi_t(\lambda)' \geq \int_0^\lambda E_t(X_t^2 exp\{uX_t - \psi_t(u)\} du - \lambda(\psi_t(\lambda)')^2 \geq$$

$$\lambda(E_t(X_t^2) exp\{-\psi_t(\lambda)\} - \int_0^\lambda E_t X_t^2 (1 - exp\{-uX_t^-) du - (\psi_t(\lambda)')^2 \geq$$
$$\lambda exp\{-\psi_t(\lambda)\} E_t X_t^2 - \lambda^2 E_t(X_t^-)^3/2 - \lambda(\psi_k(\lambda)')^2 .$$

Finally, integrating again, by (5) and by the bound (13) we have

$$\psi_t(\lambda) \geq \lambda^2/2\, E_t X_t^2 - \lambda^3 C E_t X_t^2.$$

The proof of Lemma 2 is completed.

Now to complete the proof of Lemma 1 let us note that due to the upper bound from Lemma 2 and the inequality $1 - exp(-x) \leq x^+$

$$Eexp\{-\lambda^2/2\ S(m)\} - Eexp\{-\sum_1^\infty \psi_k(\lambda)\} \leq$$

$$Eexp\{-\frac{\lambda^2}{2}S(m)\}(1 - exp\{-\lambda^3 CS(m)\}) \leq$$

$$C\lambda E(exp\{-\lambda^2/2S(m)\}\lambda^2 S(m)) = o(\lambda)\ \lambda \to 0\,,$$

(by the dominated convergence theorem).

On the other side, by the lower bound from Lemma 2 and the same as above arguments

$$Eexp\{-\lambda^2/2\ S(m)\} - Eexp\{-\sum_1^\infty \psi_k(\lambda)\} \geq$$

$$C\lambda E(exp\{-\lambda^2/2\ (1 - \lambda C)S(m)\}\lambda^2 S(m)) = o(\lambda),\ \ \lambda \to 0$$

That completes the proof of Lemma 1.

3. Proof of Theorem 1.

Introduce the stopping time

$$\tau(A) = inf\{t : m_t > A - g(S(m)_t)\},\ (inf\{\phi\} = \infty\})$$

where a parameter A is positive, $g \in G$, and consider the stopped martingale

$$m_t^A := m_{t\wedge\tau(A)}.$$

It is easy to see that all conditions of Lemma 1 are fulfilled for m_t^A (condition (5) is fulfilled by the boundness of jumps of m_t and (6) by the definition of $\tau(A)$). So $0 \leq Em_\infty^A < \infty$ and

$$\lim_t P\{S(m^A) > \ t\}t^{1/2} = (2/\pi)^{1/2} Em_\infty^A\,. \tag{14}$$

Now show that

$$Em_\infty^A = 0\,.$$

Indeed, due to the relation $E|m_\infty^A| < \infty$ and by $|X_t| < C$ it follows that $Eg(S(m^A)) < \infty$ or, equivalently (integrating by parts),

$$\int_1^\infty P\{S(m^A) > t\}dg(t) < \infty\,.$$

But $g \in G$ and so by (14) we have now $Em_\infty^A = 0$ or, equivalently ,

$$EI\{\tau(A)=\infty\}m_\infty + EI\{\tau(A)<\infty\}m_{\tau(A)} = 0$$

Finally, note that since $\tau(A)\to\infty$, *as* $A\to\infty$, then by the assumption of (4)

$$EI\{\tau(A)<\infty\}m_{\tau(A)} \geq -EI\{\tau(A)<\infty\}g(S(m)) \ \to 0, \quad A\to\infty\,.$$

As

$$EI\{\tau(A)=\infty\}m_\infty \to Em_\infty$$

we get the lower bound

$$Em_\infty \geq 0\,.$$

Repeating the same arguments for the martingale $(-m_t)$ we obtain the upper bound $Em_\infty \leq 0$.

Proof of Theorem 1 is completed.

4. Remarks

1. The arguments used in proof of Lemma 1 entail the following result which may be known but we have no references.

Proposition. *Let* (m_t) *be a local martingale,* $m_0=0$, *such that* $m_t > -Z$ *for any* t, *where* Z *is a positive integrable r.v.*

Then (m_t) *is a martingale.*

Proof.

Let (τ_l) be a localization sequence of stopping times for the local martingale (m_t).

Then by the martingale property

$$Em_{\tau_l\wedge\tau} \ = 0,$$

where τ is an arbitrary stopping time less than T =const. Hence

$$Em^+_{\tau_l\wedge\tau} = Em^-_{\tau_l\wedge\tau}.$$

Taking a limit as l tends to infinity we get by Fatou's lemma

$$Em^+_\tau = Em^-_\tau < \infty.$$

So m_τ is an integrable r.v. and the sequence $(m^+_{\tau_l\wedge\tau})$, $l=1,2,\dots$ is uniformly integrable. Hence we have for any bounded τ

$$Em^+_\tau = Em^-_\tau.$$

This fact means that m_t is a martingale.

2. It seems the conditions of Theorem 1 and Lemma 1 concerning boundness of jumps of a martingale m_t may be essentially weekned. That is true , at any rate,

for the case when $m_t = Y_{t\wedge\tau}, Y_t = \sum_1^t X_k, X_k$ are iid, $EX_k = 0, EX_k^2 = \sigma^2 > 0$. In this case $S(m) = \sigma^2\tau$ and so by Lemma 1 under additional conditions (5) and (6) we have

$$\lim_t P\{\sigma^2\tau > t\}t^{1/2} = (2/\pi)^{1/2}EY_\tau .$$

For the special case of stopping time $\tau = inf\{t : Y_t > f(t)\}$), $f(1) > 0$, which was studied by Novikov (1981b), more stronger results can be obtained. In particular, Novikov (1981b) proved the following result: if a function $f(t)$ is increasing and convex, or $f(t)$ is decreasing, concave and additionally, $E\exp(\lambda X_1) < \infty$ for some $\lambda > 0$ then

$$0 < EY_\tau < \infty \Longleftrightarrow \int_1^\infty \mid f(t) \mid t^{-3/2}dt < \infty.$$

The authors express their gratitude to M.Emery, M.Lifshits, J.Memin, P.-A.Meyer for stimulating conversations about the results.

The paper was completed while the second author was visiting Department of Mathematics of Strasbourg University and he thanks all staff for hospitality.

References

Azema(J.), Gundy(R.F.), Yor(M.) (1979). Sur l'intégrabilitée uniforme des martingales continues. Séminaire de Probabilités XIV, Lecture Notes in Mathematics, 784, 53-61, Springer-Verlag, Berlin.

Burkholder (D.L.), Gundy(R.F.) (1970). Extrapolation and interpolation of quasilinear operators on martingales. Acta Math., 124, 249-304.

Galtchouk (L.I.) (1980). Optional martingales. Mathematical Sbornik, 112 (154), N 4 (8), 483 - 521 (English translation :(1981) Vol.40, N4, 435-468).

Gundy (R.F.) (1981). On a theorem of F. and M.Riesz and an equation of A.Wald. Indiana Univ.Math.Journal, 30(4), 589-605.

Feller, W. (1971). An Introduction to Probability Theory and Its Applications, vol. 2, Wiley, New York.

Kinderman (R.P.) (1980). Asymptotic comparisons of functionals of Brownian Motion and Random Walk. Ann. Prob., 8,N6)), 1135-1147.

Klass (M.J.) (1988). A best possible improvement of Wald's equation. Ann.prob., 16, N2, 840-853.

Liptser (R.Sh.), Shiryaev (A.N.) (1986). Theory of Martingales, Kluwer Academic Publ.

Meyer (P.-A.) (1972). Martingales and Stochastic Integrals I. Lecture Notes in Mathematics, 284, Springer-Verlag.

Novikov (A.A.) (1971). On the moment of stopping of a Wiener process. Teor. Veroythn. Primen., 16, N3, 458-465 (English translation : pp.449- 456).

Novikov (A.A.) (1981a). A martingale approach to first passage problems and a new condition for Wald's identity. Proc.of the 3rd IFIP-WG 7/1 Working Conf.,Visegrad 1980,Lecture Notes in Control and Inf.Sci. 36, 146-156.

Novikov (A.A.) (1981b). Martingale approach to first passage problems of nonlinear boundaries. Proc.Steklov Inst. 158, 130-158.

Novikov (A.A.) (1982). On the time of crossing of a one-sided nonlinear boundary. Theor.Prob. Appl., 27, N4, 668 - 702 (English translation).

de la Pena (V.H.) (1993). Inequalities for tails of adapted processes with an application to Wald's lemma. J. of Theoretic Prob., 6, N2, 285-302.

Vallois (P.) (1991). Sur la loi du maximum et du temps local d'une martingale continue uniformément intégrable. Preprint, Université de Paris VI.

Wald (A.) (1947). Sequential Analysis, Wiley, New York;

Remarques sur l'hypercontractivité et l'évolution de l'entropie pour des chaînes de Markov finies

Laurent Miclo

Université Paul Sabatier de Toulouse

Summary : We will show how, in the discrete times setting of finite (irreducible and aperiodic) Markov chains, one can still use some logarithmic-Sobolev inequalities to study hypercontractivity and the evolution of entropy. As an application, we will give a new simple proof of a criterion of Hwang and Sheu for the strong ergodicity in law of the generalised simulated annealing algorithms in discrete times.

Résumé : Nous allons montrer comment, dans le cadre du temps discret des chaînes de Markov finies irréductibles et apériodiques, on peut encore utiliser certaines inégalités de Sobolev-logarithmiques pour obtenir des résultats sur l'hypercontractivité et l'évolution de l'entropie. On illustrera ces techniques de semi-groupes en retrouvant simplement un critère de Hwang et Sheu pour l'ergodicité en loi forte des algorithmes de recuit généralisés à temps discret.

Abbreviated title : Hypercontractivité des chaînes de Markov finies.

American Mathematical Society 1991 subject classifications : Primary 60J10 ; secondary 47A50 and 47N30.

Key words and phrases : Irreducible and aperiodique finite Markov chains, logarithmic-Sobolev inequalities, hypercontractivity, entropy evolution, strong ergodicity in law for generalised simulated annealing.

1 Introduction

Les inégalités de Sobolev-logarithmiques furent introduites par Gross dans [8] pour traiter notamment de l'hypercontractivité des processus d'Ornstein-Uhlenbeck en dimension infinie. Mais elles se sont vite révélées intéressantes pour d'autres types de processus (cf. par exemple Rothaus [18] et Holley et Stroock [9]) et jusque dans un cadre général de théorie des semi-groupes (voir Bakry [1] et les références qui y sont données). Même dans le contexte relativement plus simple (en apparence!) des processus de Markov finis homogènes, elles permirent de faire des progrès dans la compréhension des vitesses de convergence vers l'équilibre (cf. Diaconis et Saloff-Coste [4], nous reprendrons d'ailleurs ici les notations de cet article). Cependant leur domaine d'application semblait restreint aux cas de processus indicés par un temps continu, car l'intérêt de ces inégalités était souvent de permettre une majoration des dérivées temporelles de certaines quantités naturellement associées aux semi-groupes. Notre but ici est de montrer comment on peut également les utiliser pour étudier des chaînes de Markov finies à temps discret. Précisons tout de suite que de supposer l'espace des états finis est un peu frustrant, mais nous a été imposé par les inégalités de Sobolev-logarithmiques présentées dans la section suivante, qui ne sont pertinentes que dans ce cadre fini (les autres calculs étant sinon souvent valables dans une plus grande généralité).

Soit donc S un ensemble fini (non réduit à un singleton), muni d'un noyau $p = (p(x,y))_{x,y\in S}$ de probabilités de transitions que l'on supposera irréductible: pour tous $x, y \in S$, il existe $n \in \mathbb{N}$ et une suite finie $x = x_0, x_1, \cdots, x_n = y$ d'éléments de S telle que pour tout $1 \leq i \leq n$, $p(x_{i-1}, x_i) > 0$ (une telle suite sera appelée un chemin de longueur n, de points de départ x et d'arrivée y), i.e. il n'y a qu'une classe de récurrence et pas de points transients. Il existe donc une unique probabilité invariante μ pour p, qui est caractérisée par

$$\forall\, x \in S, \qquad \sum_{y\in S} \mu(y)p(y,x) = \sum_{y\in S} \mu(x)p(x,y) = \mu(x)$$

et elle charge tous les points.

Comme d'habitude, on associe à p un opérateur P qui agit d'une part sur les fonctions réelles définies sur S (leur ensemble sera désormais noté $\mathcal{F}(S)$) par

$$\forall\, f \in \mathcal{F}(S),\ \forall\, x \in S, \qquad Pf(x) = \sum_{y\in S} p(x,y)f(y)$$

et d'autre part sur les probabilités sur S (dont l'ensemble sera désigné par $\mathcal{P}(S)$) par

$$\forall\, m \in \mathcal{P}(S),\ \forall\, x \in S, \qquad mP(x) = \sum_{y\in S} m(y)p(y,x)$$

l'invariance de μ s'écrivant alors $\mu P = \mu$.

Ces opérations ont clairement les interprétations probabilistes suivantes: si $X = (X_n)_{n\in\mathbb{N}}$ est une chaîne de Markov sur S dont les probabilités de transitions sont données par p et qui est issue de $x \in S$, alors pour tout $f \in \mathcal{F}(S)$, $Pf(x) = \mathbb{E}_x[f(X_1)]$. Si par contre on suppose X de loi initiale $m \in \mathcal{P}(S)$, alors mP est la loi de X_1.

Pour $n \in \mathbb{N}$, on désignera aussi par P^n les itérés n fois de P (auxquelles correspondent les matrices p^n), en convenant que $P^0 = I$, où I est l'application identité, agissant suivant les cas sur $\mathcal{F}(S)$ ou sur $\mathcal{P}(S)$.

La première question que l'on se pose est de savoir si le semi-groupe $(P^n)_{n\in\mathbb{N}}$ est hypercontractif: si $2 \leq q_0 < +\infty$ est donné, existe-t-il des $q > q_0$ tels que pour tout $f \in \mathcal{F}(S)$,

$$\|Pf\|_q \leq \|f\|_{q_0} \tag{1}$$

où pour tout $1 \leq r \leq \infty$, $\|\cdot\|_r$ désignera la norme usuelle de $L^r(S,\mu)$.

La seconde question concerne l'évolution de l'entropie le long des itérées de P agissant sur $\mathcal{P}(S)$, et plus particulièrement de décrire quantitativement sa décroissance, pour pouvoir dans les situations apériodiques donner sa vitesse de convergence vers 0 en temps grand. Rappelons que l'entropie par rapport à μ d'une mesure $m \in \mathcal{P}(S)$ est définie par

$$\mathrm{Ent}(m) = \sum_{x\in S} \ln\left(\frac{m(x)}{\mu(x)}\right) m(x)$$

(par la suite, dans le cas de processus inhomogènes, plusieurs mesures invariantes instantanées apparaîtront et cette expression sera alors plutôt notée $\mathrm{Ent}(m|\mu)$ pour éviter les confusions), et que c'est une quantité qui mesure d'une certaine manière un écart à la probabilité μ, car on a par exemple

$$\|m-\mu\|_{vt} \leq \sqrt{2}\sqrt{\mathrm{Ent}(m)} \tag{2}$$

où $\|\cdot\|_{vt}$ représente la variation totale (cf. Stroock [19] formule (1.12)).

Dans la section suivante, on introduira certaines inégalités de Sobolev-logarithmiques dont les constantes associées permettront d'apporter des réponses aux problèmes précédents, respectivement dans les sections 3 et 4. Enfin dans une dernière section, on illustrera ces techniques en retrouvant, relativement facilement, un critère d'ergodicité en loi des algorithmes de recuit généralisés finis à temps discret.

2 Inégalités de Sobolev-logarithmiques

Les inégalités de Sobolev-logarithmiques consistent à comparer une certaine fonctionnelle $\mathcal{L}$, qui ressemble à l'entropie mais qui est définie sur $\mathcal{F}(S)$ par

$$\forall\, f \in \mathcal{F}(S), \qquad \mathcal{L}(f) = \int f^2 \ln\left(\frac{f^2}{\|f\|_2^2}\right) d\mu$$

(cette quantité sera aussi notée $\mathcal{L}_\mu(f)$ quand la probabilité μ ne sera plus sous-entendue), à une forme de Dirichlet. Quand les processus étudiés sont à temps continu, la forme intervenant est celle associée à l'opposé du générateur (ou de manière équivalente, à l'opposé du symétrisé additif de cet opérateur dans $L^2(\mu)$), mais quand le temps est discret, c'est suivant les cas celle associée à $I - P^*P$ ou à $I - PP^*$ qui apparaît naturellement (voir aussi Fill [5] et Diaconis et Saloff-Coste [3], où P^*P est appelé le symétrisé multiplicatif pour le distinguer du symétrisé additif $(P+P^*)/2$).

On a noté ci-dessus P^* l'adjoint de P dans $L^2(\mu)$, et pour tout $f \in \mathcal{F}(S)$, soit

$$\mathcal{E}(f,f) = \int (I-P^*P)(f)f\,d\mu = \int f^2 - (Pf)^2\,d\mu = \|f\|_2^2 - \|Pf\|_2^2$$

(quand on voudra préciser les opérateurs intervenant, cette expression sera aussi notée $\mathcal{E}_{I-P^*P}(f,f)$). Du fait que P est une contraction dans $L^2(\mu)$, il est clair que $\mathcal{E}(f,f) \geq 0$. On calcule immédiatement que P^* agit sur $\mathcal{F}(S)$ par

$$\forall\, f \in \mathcal{F}(S),\ \forall\, x \in S, \qquad P^*f(x) = \sum_{y\in S} p^*(x,y)f(y)$$

avec pour tous $x, y \in S$, $p^*(x,y) = \mu(y)p(y,x)/\mu(x)$.

Le fait que μ est invariante pour P implique que P^* est un noyau markovien

$$\forall\, x \in S, \qquad \sum_{y\in S} p^*(x,y) \;=\; \frac{1}{\mu(x)}\sum_{y\in S}\mu(y)p(y,x) = 1$$

qui de plus admet aussi μ pour probabilité invariante

$$\forall\, x \in S, \qquad \sum_{y\in S}\mu(x)p^*(x,y) = \sum_{y\in S}\mu(y)p(y,x) = \mu(x)$$

Notons qu'en posant pour $x, y \in S$,

$$q(x,y) = (p^*p)(x,y) = \mu(x)^{-1}\sum_{z\in S}\mu(z)p(z,x)p(z,y)$$

on peut expliciter un peu plus la forme de Dirichlet,

$$(3) \qquad \forall\, f \in \mathcal{F}(S), \qquad \mathcal{E}(f,f) = \frac{1}{2}\sum_{x,y\in S}\mu(x)q(x,y)(f(y)-f(x))^2$$

Définition :

On dit que $I - P^*P$ satisfait une inégalité de Sobolev-logarithmique s'il existe une constante $\alpha > 0$ telle que pour tout $f \in \mathcal{F}(S)$,

$$\mathcal{L}(f) \leq \alpha^{-1}\mathcal{E}(f,f)$$

et $\alpha(I - P^*P)$ désignera alors la plus grande constante (appelée constante de Sobolev-logarithmique) telle que toutes ces inégalités soient satisfaites.

Remarque :

Une telle inégalité implique que pour tout $f \in \mathcal{F}(S)$, $\mathcal{L}(f) \leq \alpha^{-1}\|f\|_2^2$, ce qui à son tour impose que S est fini (du moins que μ est une combinaison convexe d'un nombre fini de masses de Dirac, ce qui nous ramène à ce cas), comme on en s'en rend compte en considérant sinon des indicatrices d'ensembles de mesure de plus en plus petites pour μ (voir aussi les contre-exemples de la fin de cette section).

En fait dans le contexte des processus de Markov finis, il est bien connu qu'une inégalité de Sobolev-logarithmique du type précédent est équivalente à l'irréductiblité de p^*p: en effet pour la condition nécessaire, il suffit de remarquer que si on pose $f_\epsilon = \mathbb{I} + \epsilon h$, avec $\mathbb{I}$ la fonction prenant toujours la valeur 1 et $h \in \mathcal{F}(S)$ fixé, on a $\mathcal{E}(f_\epsilon, f_\epsilon) = \epsilon^2\mathcal{E}(h,h)$ et pour ϵ tendant vers 0, $\mathcal{L}(f_\epsilon) \sim 2\epsilon^2\mu((f_\epsilon - \mu(f_\epsilon))^2)$, d'où l'existence d'un trou spectral (d'au moins $2\alpha(I - P^*P)$) pour l'opérateur symétrique $I - P^*P$ (pour plus de détails, on renvoit à Rothaus [18] ou à Diaconis et Saloff-Coste

[4]), ce qui est aussi équivalent à l'irréductibilité de p^*p (ce dernier point se montrant à partir de l'expression (3), voir par exemple la preuve de l'inégalité de Poincaré faite par Holley et Stroock [10]). Quant à la condition suffisante, puisque $\mathcal{L}$ et $\mathcal{E}$ sont tous deux homogènes d'ordre 2, il suffit de voir que

$$\inf_{f\in\mathcal{F}(S)\,/\,\|f\|_2=1,\,f\notin\mathrm{Vect}(\mathbb{I})} \frac{\mathcal{E}(f,f)}{\mathcal{L}(f)} > 0$$

et même que, puisque pour tout $f \in \mathcal{F}(S)$, $\mathcal{E}(f,f) \geq \mathcal{E}(|f|,|f|)$,

$$\inf_{f\in\mathcal{F}(S)\,/\,\|f\|_2=1,\,f\neq\mathbb{I},\,f\geq 0} \frac{\mathcal{E}(f,f)}{\mathcal{L}(f)} > 0$$

mais ceci découle, outre du trou spectral et d'un développement du type précédent au voisinage de $\mathbb{I}$, du fait que l'application $f \mapsto \mathcal{E}(f,f)/\mathcal{L}(f)$ est continue (on aura noté que par l'inégalité de Jensen et la stricte convexité de $\mathbb{R}_+ \ni t \mapsto t\ln(t)$, $\mathcal{L}(f)$ est nul si et seulement si f^2 est constante) sur le compact formé de l'ensemble $\{f \in \mathcal{F}(S)\,/\,\|f\|_2 = 1, f \neq \mathbb{I}, f \geq 0\}$ privé de son intersection avec une petite boule (pour $\|\cdot\|_2$) autour de $\mathbb{I}$, elle y atteint donc son minimum qui ne peut être nul car $\mathcal{E}(f,f) = 0$ équivaut de par la formule (3) et l'irréductibilité de p^*p à f constant.

Pour ceci et pour des estimations de $\alpha(I - P^*P)$ (qui constituent le point crucial dans les applications, voir par exemple la dernière section) on renvoit aussi à Diaconis et Saloff-Coste [4].

Notons que le trou spectral de $I - P^*P$ est toujours majoré par 1, ce qui montre que d'une manière générale, on a $\alpha(I - P^*P) \leq 1/2$.

Le résultat très simple suivant va nous permettre de voir quelles sont les situations pour lesquelles des inégalités de Sobolev-logarithmiques peuvent être utiles. On y suppose seulement que le noyau p admet μ pour mesure invariante. L'adjoint P^* est alors toujours bien défini dans $L^2(\mu)$, même si la matrice p^* n'est plus unique, il suffit de poser $p^*(x,y) = \mu(y)p(y,x)/\mu(x)$ si $\mu(x) > 0$ et de prendre pour $p^*(x,\cdot)$ une probabilité quelconque sur S si $\mu(x) = 0$ (pour permettre au noyau p^* de rester markovien et d'admettre également μ pour mesure invariante). Le résultat ci-dessous ne dépend pas du choix éventuel de p^*.

Proposition 1

On a équivalence entre
(*i*) Il existe un $k \geq 1$ tel que $p^{k*}p^k$ est irréductible.
(*ii*) p est irréductible et apériodique.

Démonstration :

Supposons (*ii*), il est bien connu qu'alors il existe un $k \geq 1$ tel que pour tous $x, y \in S$, $p^k(x,y) > 0$, or il est clair à partir des égalités

$$p^{k*} = (p^k)^* = (p^*)^k$$

que ceci implique aussi que pour tous $x, y \in S$, $p^{k*}(y,x) > 0$, puis que

$$\forall\, x,y \in S, \qquad (p^{k*}p^k)(x,y) > 0$$

et donc notamment l'irréductibilité de $p^{k*}p^k$.

Réciproquement, supposons d'abord que p admette une classe de récurrence $C \neq S$. Il existe une telle classe satisfaisant pour tout $x \in C$, $\mu(x) > 0$. Or pour $x \in C$ et $y \in S$, l'inégalité $p^*(x,y) > 0$ implique, puisque $\mu(x) > 0$, d'une part que $\mu(y) > 0$, et donc notamment que y ne peut pas être un point transient, et d'autre part que $p(y,x) > 0$. Ces deux informations montrent que $y \in C$ et on aboutit à la même conclusion si $p(x,y) > 0$. Ainsi pour tous $k \geq 1$, $x \in C$ et $y \in S$, l'inégalité $(p^{k*}p^k)(x,y) > 0$ impose que y appartient aussi à la classe C et il en découle que $p^{k*}p^k$ a une classe de récurrence dans C, ce qui est incompatible avec l'irréductibilité de $p^{k*}p^k$. Supposons maintenant que p irréductible admette plusieurs classes de périodicité, disons $C_0, \cdots, C_{d-1}$ ($d > 1$ est alors la période, et on suppose que $C_0, \cdots, C_{d-1}$ sont rangés consécutivement). En considérant leurs indices comme des éléments de $\mathbb{Z}/d\mathbb{Z}$, pour tous $k \geq 1$, $i \in \mathbb{Z}/d\mathbb{Z}$, $x \in C_i$ et $y \in S$, on a l'implication $p^k(x,y) > 0 \Rightarrow y \in C_{i+k}$ (respectivement $p^{k*}(x,y) > 0 \Rightarrow y \in C_{i-k}$), ce qui prouve que $p^{k*}p^k(x,y) > 0$ assure que x et y sont dans une même classe de périodicité et donc que $p^{k*}p^k$ ne peut être irréductible.

□

Si p est irréductible et apériodique, notons

$$k(p) = \min\{k \in \mathbb{N}^* \,/\, p^{k*}p^k \text{ est irréductible}\} \tag{4}$$

Pour appliquer les résultats des deux sections suivantes, qui seront satisfaits sous l'hypothèse d'existence d'inégalités de Sobolev-logarithmiques, il faudra au moins remplacer la matrice p par $p^{k(p)}$, et même souvent il sera intéressant de considérer plutôt p^k avec un certain $k > k(p)$ pour accroître fortement la constante de Sobolev-logarithmique (voir l'exemple de la remarque (b) de la fin de la section 5). L'argument précédent montre que

$$k(p) \leq \min\{k \geq 1 \,/\, \forall\, x,y \in S,\ p^k(x,y) > 0\}$$

mais on peut avoir une inégalité stricte comme le montre l'exemple donné par la matrice

$$p = \frac{1}{2}\begin{pmatrix} 0 & 1 & 1 \\ 1 & 0 & 1 \\ 0 & 2 & 0 \end{pmatrix}$$

pour laquelle $k(p) = 1$. Par ailleurs pour tout $n \in \mathbb{N}$, $n \geq 2$, la matrice p définie sur $\mathbb{Z}/(n\mathbb{Z})$ par

$$\forall\, i,j \in \mathbb{Z}/(n\mathbb{Z}), \qquad p(i,j) = \begin{cases} 1 & , \text{si } i \neq 0 \text{ et } j = i+1 \\ 1/2 & , \text{si } i = 0 \text{ et } j = 1 \\ 1/2 & , \text{si } i = 0 = j \\ 0 & , \text{sinon} \end{cases}$$

fournit un exemple très simple pour lequel $k(p) = n - 1$.

Revenons au cas général, à partir des inégalités de Cauchy-Schwarz,

$$\forall\, k \geq 1, \forall\, f \in \mathcal{F}(S),\ \forall\, x \in S, \qquad (P^{k+1}f(x))^2 \leq P((P^k f)^2)(x)$$

et de l'invariance de μ par rapport à P, on remarque que

$$\forall\, k \geq 1, \forall\, f \in \mathcal{F}(S), \qquad \mathcal{E}_{I-P^{k+1*}P^{k+1}}(f,f) \geq \mathcal{E}_{I-P^{k*}P^{k}}(f,f)$$

ce qui montre que

$$\mathbb{N}^* \ni k \mapsto \alpha(I - P^{k*}P^k)$$

est une application croissante (en convenant que $\alpha(I - P^{k*}P^k) = 0$ si $p^{k*}p^k$ n'est pas irréductible). Celle-ci admet d'ailleurs $\alpha(I - E_\mu^* E_\mu) = \alpha(I - E_\mu)$ comme limite en l'infini, où E_μ est la matrice correspondant à l'espérance par rapport à μ et qui est donnée par

$$\forall\, x, y \in S, \qquad E_\mu(x,y) = \mu(y)$$

Indiquons que la valeur de $\alpha(I - E_\mu)$ a été astucieusement calculée par Diaconis et Saloff-Coste (cf. le théorème 5.1 de l'appendice de [4]) et qu'elle vaut $(1-2\underline{\mu})/\ln(1/\underline{\mu}-1)$, où $\underline{\mu} = \min_{x\in S} \mu(x)$.

Il apparaît également que pour tout $k \geq k(p)$, $p^{k*}p^k$ sera irréductible, puisque ceci revient à dire que $\alpha(I - P^{k*}P^k) > 0$, inégalité elle-même équivalente à

$$\forall\, f \in \mathcal{F}(S), \qquad \mathcal{E}_{I-P^{k*}P^{k}}(f,f) = 0 \Rightarrow f \text{ est constant}$$

Remarques :

a) Dans le même ordre d'idées, notons pour tout $n \in \mathbb{N}$,

$$V_n = \{f \in \mathcal{F}(S) \,/\, \|P^n f\|_2 = \|f\|_2\}$$

Puisque pour tout $x \in S$, on a $(P^n f(x))^2 \leq P^n(f^2)(x)$ avec égalité si et seulement si f est constante sur l'ensemble $\{y \in S \,/\, p^n(x,y) > 0\}$, il apparaît que V_n est un sous-espace vectoriel de $\mathcal{F}(S)$. De plus, en utilisant les inégalités $\|P^{n+1}f\|_2 \leq \|P^n f\|_2 \leq \cdots \leq \|Pf\|_2 \leq \|f\|_2$ valables pour tout $f \in \mathcal{F}(S)$, on montre que pour tout $n \in \mathbb{N}$,

$$f \in V_{n+1} \iff Pf \in V_n \text{ et } f \in V_1$$

De ceci il découle facilement que si pour un $n \in \mathbb{N}$ on a $V_{n+1} = V_n$, alors pour tout $r \in \mathbb{N}$, $V_{n+r} = V_n$. Or en supposant p irréductible apériodique, on a $V_{k(p)} = \text{Vect}(\mathbb{1})$, ainsi (V_n) est une suite de sous-espaces vectoriels qui commence par être strictement décroissante ($V_0 = \mathcal{F}(S)$) puis qui finit par valoir $\text{Vect}(\mathbb{1})$. On en déduit que l'on a toujours $k(p) \leq \text{card}(S) - 1$ (l'égalité étant possible comme on l'a déjà vu dans l'exemple ci-dessus sur $\mathbb{Z}/(n\mathbb{Z})$). Ceci nous donne un résultat sur les chemins qui n'est pas facile à obtenir directement : en effet, $k(p)$ est le plus petit $k \in \mathbb{N}$ tel que pour tous $x, y \in S$, on puisse trouver deux suites finies $(q^{(i)})_{0\leq i\leq n}$ et $(\check{q}^{(i)})_{0\leq i\leq n}$ de chemins (pour p) de longueur k satisfaisant $q_0^{(0)} = x$, $\check{q}_0^{(n)} = y$, $q_k^{(i)} = \check{q}_k^{(i)}$ pour tout $0 \leq i \leq n$ et $q_0^{(i+1)} = \check{q}_0^{(i)}$ pour tout $0 \leq i < n$. Les résultats précédents montrent donc que l'on peut trouver de telles suites avec $k = \text{card}(S) - 1$.

b) D'autre part, remarquons que P^*P est diagonalisable et n'admet que des valeurs propres positives, la plus grande étant 1. Puisque cette diagonalisation est orthogonale dans $L^2(\mu)$, $\|Pf\|_2 = \|f\|_2$ équivaut à l'appartenance de f à l'espace propre associé à la valeur propre 1 et l'irréductibilité de p^*p est donc équivalente au fait que celle-ci est de multiplicité 1. Or il est bien connu que P^*P et PP^* ont même spectre, ainsi l'irréductibilité de p^*p est équivalente à celle de pp^*, on a notamment $k(p^*) = k(p)$ et $I - P^*P$ satisfait une inégalité de Sobolev-logarithmique si et seulement si il en est de même pour $I - PP^*$. Cependant il n'est pas clair en général, mis à part dans les cas où p est normal (notamment si p est réversible, car on a alors $P^* = P$), que les constantes $\alpha(I - P^*P)$ et $\alpha(I - PP^*)$ soient égales. Or dans la section suivante c'est la première qui interviendra, alors que dans les sections 4 et 5 ce sera la seconde.

c) Si p est irréductible apériodique et réversible, notons que $k(p) = 1$, car s'il existait $f \in \mathcal{F}(S)$ non constante telle que $\mathcal{E}_{I-P^*P}(f,f) = 0$, ceci impliquerait que l'espace propre associé à la valeur propre 1 pour $P^2 = P^*P$ serait au moins de dimension 2, i.e. soit l'espace propre associé à 1 pour P est au moins de dimension 2, ce qui est exclus par l'irréductibilité, soit -1 est valeur propre de P, ce qui est en contradiction avec l'apériodicité.

On peut en déduire un résultat un peu plus général : supposons que p soit tel que

$$\forall\, x,y \in S, \qquad p(x,y) > 0 \Longleftrightarrow p(y,x) > 0 \tag{5}$$

alors on a aussi $k(p) = 1$.

En effet, $k(p)$ ne dépend que de la matrice d'incidence $i(p)$ associée à p, qui est donnée par

$$\forall\, x,y \in S, \qquad i(p)(x,y) = \mathbb{I}_{p(x,y)>0}$$

ainsi $k(p) = k(\widetilde{p})$, où $\widetilde{p}$ est la matrice markovienne définie par

$$\forall\, x,y \in S, \qquad \widetilde{p}(x,y) = \begin{cases} 1/\mathrm{card}\{z \in S \,/\, p(x,z) > 0\} & \text{, si } p(x,y) > 0 \\ 0 & \text{, sinon} \end{cases}$$

or cette dernière est réversible si et seulement si p satisfait (5).

3 Hypercontractivité à temps discret

Traditionnellement dans le cas des processus de Markov finis, une inégalité d'hypercontractivité du type (1) se prouve pour $P = \exp(t_0 L)$ où $t_0 > 0$ et L est un générateur irréductible sur S. Plus précisément, on montre que toute une famille d'inégalités est satisfaite :

$$\forall\, t \geq 0, \forall\, f \in \mathcal{F}(S), \qquad \|\exp(tL)f\|_{q(t)} \leq \|f\|_{q_0}$$

où $q : \mathbb{R}_+ \to [2, +\infty[$ est une certaine application telle que $q(0) = q_0 \geq 2$. On les prouve en les dérivant par rapport à $t \geq 0$ et en imposant aux dérivées d'être négatives grâce à des inégalités de Sobolev-logarithmiques (cf. par exemple Gross [8] ou Diaconis et Saloff-Coste [4]). Ainsi, si on veut montrer (1), on peut commencer par chercher s'il n'existerait pas un générateur irréductible L sur S tel que $P = \exp(L)$. Cependant ceci est rarement satisfait, ne serait-ce que parce que cela implique déjà que pour tous $x, y \in S$, $p(x,y) > 0$. Nous allons plutôt essayer d'estimer directement la différence $\|Pf\|_q - \|f\|_{q_0}$ et voir comment les inégalités de Sobolev-logarithmiques de la section précédente peuvent permettre de la rendre négative.

On supposera donc que $I - P^*P$ satisfait une inégalité de Sobolev-logarithmique avec constante $\alpha(I - P^*P) > 0$, c'est-à-dire que P^*P est irréductible, ce qui entraîne notamment l'irréductibilité et l'apériodicité de P. Notons également

$$\nu(P) = \max\{1/p(x,y) \,/\, x,y \in S,\, p(x,y) > 0\} - 1$$

(on a donc $\nu(P) \geq 1$ par apériodicité de P), et si $q \geq 2$ et $\nu > 0$ sont donnés, on pose

$$g(q,\nu) = \frac{(1+\nu)^q - 1 - q\nu}{((1+\nu)^{q/2}-1)^2}$$

Si $q = 2$, on a $g(2,\cdot) \equiv 1$, mais si $q > 2$ est fixé, on vérifie aisément que l'application $\mathbb{R}_+^* \ni \nu \mapsto g(q,\nu)$ est strictement décroissante et satisfait

$$g(q,0_+) = 2\frac{q-1}{q}, \quad g(q,+\infty) = 1$$

D'autre part, on peut aussi montrer qu'à $\nu > 0$ fixé, $[2,+\infty[\ni q \mapsto g(q,\nu)$ commence par être croissante puis décroît, les limites étant $g(2,\nu) = 1 = g(+\infty,\nu)$.

Avec ces notations, le principal résultat de cette section s'énonce alors,

Proposition 2

Pour tous $q_0 \geq 2$ et $f \in \mathcal{F}(S)$, on a

$$\|Pf\|_q \leq \|f\|_{q_0}$$

dès que q satisfait $q \leq [1 + g(q,\nu(P))\alpha(I - P^*P)]q_0$, et donc notamment si $q = [1 + \alpha(I - P^*P)]q_0$

La preuve de cette proposition est basée sur deux lemmes très simples :

Lemme 3

Fixons $f \in \mathcal{F}(S)$, l'application

$$[0,1] \ni t \mapsto \|f\|_{1/t}$$

est convexe. Notamment pour tous $q \geq q_0 \geq 1$,

$$\|f\|_q - \|f\|_{q_0} \leq \frac{q-q_0}{qq_0} \|f\|_q^{1-q} \mathcal{L}(f^{q/2})$$

Démonstration :

Soit $0 < \theta < 1$ et $0 < s < t < 1$, on veut estimer $\|f\|_{1/(\theta s+(1-\theta)t)}$. Pour ceci soit $0 < r < 1$ tel que

$$\frac{1}{\theta s + (1-\theta)t} = \frac{r}{s} + \frac{1-r}{t}$$

on calcule que $r = \theta s/(\theta s + (1-\theta)t)$.

Par l'inégalité de Hölder on a

$$\begin{aligned}
\|f\|_{1/(\theta s+(1-\theta)t)} &\leq \|f\|_{1/s}^{r(\theta s+(1-\theta)t)/s} \|f\|_{1/t}^{(1-r)(\theta s+(1-\theta)t)/t} \\
&= \|f\|_{1/s}^{\theta} \|f\|_{1/t}^{1-\theta} \\
&\leq \theta \|f\|_{1/s} + (1-\theta)\|f\|_{1/t}
\end{aligned}$$

par convexité de l'application exponentielle. Pour la seconde affirmation, il reste à vérifier que

$$\begin{aligned}\frac{d}{dt}\|f\|_{1/t} &= -\|f\|_{1/t}^{1-1/t}\int \ln\left(\frac{f^{1/t}}{\|f\|_{1/t}^{1/t}}\right) f^{1/t}\,d\mu \\ &= -\|f\|_{1/t}^{1-1/t}\,\mathcal{L}(f^{\frac{1}{2t}})\end{aligned}$$

□

Le second lemme concerne une petite inégalité qui précise la stricte convexité de l'application $\mathbb{R}_+ \ni t \mapsto t^q$, pour $q \geq 2$. C'est elle qui a déterminé la fonction $g(\cdot,\cdot)$ précédente.

Lemme 4

> Soit $\nu \geq 0$ et $q \geq 2$ fixés. Pour tous $t \geq 0$ et $-t \leq s \leq \nu t$, on a
>
> $$(t+s)^q \geq t^q + qt^{q-1}s + g(q,\nu)((t+s)^{q/2} - t^{q/2})^2$$

Démonstration :

En effet, $t > 0$ étant fixé, notons pour $u = s/t$, $h(u)$ le membre de gauche moins le membre de droite le tout divisé par t^q. En dérivant deux fois, il apparaît que h a le comportement suivant :

u	-1		0		$u(q,\nu)$		ν
h	$h(-1)$	$\searrow$	0	$\nearrow$	$h(u(q,\nu))$	$\searrow$	0

pour une certaine valeur $0 < u(q,\nu) < \nu$, ce qui permet de conclure au résultat annoncé.

□

Démonstration de la proposition 2 :

Soit $q \geq q_0 \geq 2$ a priori quelconques, et comme d'habitude, on commence par se ramener à ne considérer que des fonctions f positives. On écrit ensuite

$$\|Pf\|_q - \|f\|_{q_0} = \|Pf\|_q - \|f\|_q + \|f\|_q - \|f\|_{q_0}$$

et on majore l'expression $\|f\|_q - \|f\|_{q_0}$ comme dans lemme 3.

Pour l'autre terme, on utilise l'inégalité de concavité $a^{1/q} - b^{1/q} \leq \frac{1}{q}b^{1/q-1}(a-b)$, valable pour tous réels $a, b \geq 0$, pour obtenir

$$\|Pf\|_q - \|f\|_q \leq \frac{1}{q}\|f\|_q^{1-q}(\|Pf\|_q^q - \|f\|_q^q)$$

ce qui nous ramène à estimer $\int (Pf)^q\,d\mu - \int f^q\,d\mu$. Pour ceci on se sert du lemme 4 appliqué avec $\nu = \nu(P)$, $t = Pf(x)$ et $t + s = f(y)$, où $x, y \in S$ sont tels que $p(x,y) > 0$.

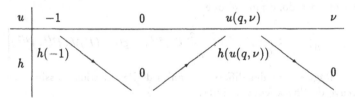

Mais $x \in S$ étant fixé, on intègre ces inégalités en y par rapport à $p(x,y)$ pour obtenir

$$P(f^q)(x) \geq (P(f)(x))^q + g(q,\nu(P)) \sum_{y\in S} p(x,y)(f^{q/2}(y) - (Pf)^{q/2}(x))^2$$

Cependant, toujours à $x \in S$ fixé, on a

$$\begin{aligned}\sum_{y\in G} p(x,y)\left(f^{q/2}(y) - (Pf)^{q/2}(x)\right)^2 &\geq \min_{c\in\mathbb{R}} \sum_{y\in G} p(x,y)\left(f^{q/2}(y) - c\right)^2 \\ &= \sum_{y\in G} p(x,y)\left(f^{q/2}(y) - P(f^{q/2})(x)\right)^2 \\ &= P(f^q)(x) - \left(P(f^{q/2})(x)\right)^2\end{aligned}$$

d'où

$$(6) \qquad P(f^q)(x) \geq (P(f)(x))^q + g(q,\nu(P))\left(P(f^q)(x) - \left(P(f^{q/2})(x)\right)^2\right)$$

Il reste alors à intégrer ces inégalités en x par rapport à μ et à utiliser l'invariance de μ par P pour faire apparaître que

$$\|Pf\|_q^q - \|f\|_q^q \leq -g(q,\nu(P))\mathcal{E}(f^{q/2}, f^{q/2})$$

Finalement on a donc montré que

$$(7) \qquad \|Pf\|_q - \|f\|_{q_0} \leq \frac{1}{q}\|f\|_q^{1-q}\left[\frac{q-q_0}{q_0}\mathcal{L}(f^{q/2}) - g(q,\nu(P))\mathcal{E}(f^{q/2}, f^{q/2})\right]$$

qui est un analogue pour des différences finies de l'estimation classique de la dérivée dans la preuve de l'hypercontractivité.

Si maintenant $q \leq [1 + g(q,\nu(P))\alpha(I - P^*P)]q_0$, on peut conclure par l'inégalité de Sobolev-logarithmique

$$\alpha(I - P^*P)\mathcal{L}(f^{q/2}) \leq \mathcal{E}(f^{q/2}, f^{q/2})$$

□

<u>Remarques :</u>

Dans les applications, on a souvent $\nu(P)$ très grand, et on peut alors se contenter de prendre $q = (1 + \alpha(I - P^*P))q_0$. Notons d'ailleurs que pour la preuve du résultat d'hypercontractivité avec ce q, le lemme 4 est inutile, car on a besoin de l'inégalité (6) qu'avec $g(q,\nu(P))$ remplacé par 1, et dans ce cas elle se réduit à

$$\left(P(f^{q/2})(x)\right)^2 \geq (P(f)(x))^q$$

qui est trivialement satisfait pour $q \geq 2$.

Néanmoins nous avons tenu à présenter le lemme 4, car c'est une inégalité du même genre qui interviendra dans la section suivante, et curieusement la différence entre $g(q, 0_+)$ et $g(q, +\infty)$ est la "même" que celle qui intervient dans le lemme

2.6 de Diaconis et Saloff-Coste [4] entre les situations réversibles et non réversibles. Malheureusement, nous ne sommes pas arrivé à exploiter ceci pour montrer par exemple que si P est réversible, alors on peut obtenir un meilleur résultat que la proposition 2.

Dans le même ordre de considérations, notons que (7), où on remplace $g(q,\nu(P))$ par 1, permet de retrouver le lemme 2.6 de Diaconis et Saloff-Coste [4] dans les cas non réversibles : soit L un générateur markovien sur S, matriciellement $L = (L(x,y))_{x,y\in S}$ avec pour tous $x \neq y \in S$, $L(x,y) \geq 0$, et pour tout $x \in S$, $\sum_{y\in S} L(x,y) = 0$. Supposons L irréductible et soit $P = \exp(tL)$ pour un $t > 0$. Si $q : \mathbb{R}_+ \to \mathbb{R}_+$ est une application dérivable telle que $q(0) = q_0 \geq 2$, écrivons que

$$\begin{aligned} &\frac{\|\exp(tL)f\|_{q(t)} - \|f\|_{q_0}}{t} \\ &\leq \frac{1}{q(t)}\|f\|_{q(t)}^{1-q(t)}\left[\frac{q(t)-q_0}{t}\frac{\mathcal{L}(f^{q(t)/2})}{q_0} - \frac{\mathcal{E}_{I-\exp(tL^*)\exp(tL)}(f^{q(t)/2}, f^{q(t)/2})}{t}\right] \end{aligned}$$

Or on sait (voir la preuve du théorème 3.5 de Diaconis et Saloff-Coste [4]) que

$$\left.\frac{d}{dt}\|\exp(tL)f\|_{q(t)}\right|_{t=0} = \|f\|_{q_0}^{1-q_0}\left[\left.\frac{dq(t)}{dt}\right|_{t=0}\frac{\mathcal{L}(f^{q_0/2})}{q_0^2} - \mathcal{E}_{-L}(f, f^{q_0-1})\right]$$

et d'autre part il est clair que

$$\lim_{t\to 0_+}\frac{\mathcal{E}_{I-\exp(tL^*)\exp(tL)}(f^{q(t)/2}, f^{q(t)/2})}{t} = \mathcal{E}_{-L^*-L}(f^{q_0/2}, f^{q_0/2}) = 2\mathcal{E}_{-L}(f^{q_0/2}, f^{q_0/2})$$

Ainsi en passant à la limite en t petit, on retrouve bien que pour tout $q_0 \geq 2$ et tout $f \in \mathcal{F}(S)$,

$$\mathcal{E}_{-L}(f^{q_0/2}, f^{q_0/2}) \leq \frac{q_0}{2}\mathcal{E}_{-L}(f, f^{q_0-1})$$

4 Evolution de l'entropie à temps discret

Un des moyens de montrer la convergence en loi vers l'équilibre des processus de Markov finis irréductibles et homogènes (à temps continu), est d'étudier l'évolution de l'entropie, par rapport à la mesure invariante μ du système, de la loi à un instant donné. En effet, il n'est pas très difficile de majorer la dérivée par rapport au temps de cette quantité par l'opposé de deux fois la forme de Dirichlet associée au symétrisé additif de l'opposé du générateur (si le processus est réversible, on peut même faire intervenir quatre fois cette forme de Dirichlet, voir Diaconis et Saloff-Coste [4]). Par le biais d'une inégalité de Sobolev-logarithmique, ceci permet d'obtenir une inégalité différentielle simple satisfaite par l'entropie, qui s'intègre pour prouver que cette quantité converge exponentiellement vite vers 0, la vitesse étant donnée par la constante de Sobolev-logarithmique considérée.

Nous allons voir comment ceci peut également s'adapter pour traiter de l'ergodicité des chaînes de Markov finies homogènes irréductibles et apériodiques.

Comme dans la section précédente, on a besoin d'une inégalité qui quantifie d'une certaine manière la stricte convexité de l'application $\mathbb{R}_+ \ni u \mapsto u\ln(u)$.

Lemme 5

Pour tous $t \geq 0$ et $s \geq -t$, on a

$$(t+s)\ln(t+s) \geq t\ln(t) + (1+\ln(t))s + (\sqrt{t+s} - \sqrt{t})^2$$

Démonstration :

A $t > 0$ fixé, soit pour $u = s/t$, $h(u)$ le membre de gauche moins celui de droite, le tout divisé par t. En dérivant deux fois on montre que h a le tableau de variations suivant :

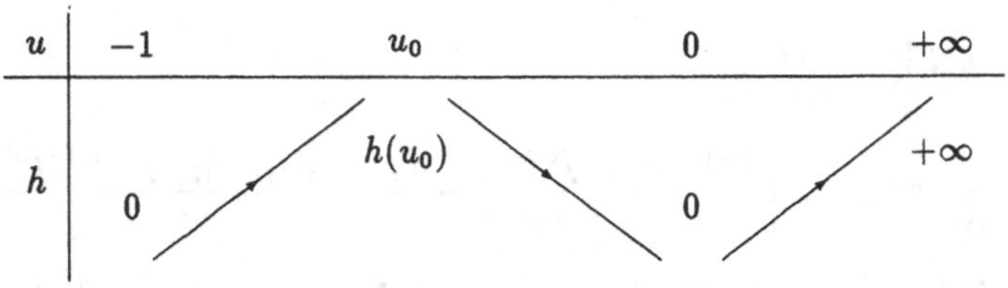

pour une certaine valeur $-1 < u_0 < 0$, caractérisée par $\ln(1+u_0) - 1 + (1+u_0)^{-1/2} = 0$.

□

En supposant que $I - PP^*$ satisfait une inégalité de Sobolev-logarithmique avec constante $\alpha(I - PP^*) > 0$, on en déduit le résultat suivant :

Proposition 6

Pour tout $m \in \mathcal{P}(S)$, on a

$$\mathrm{Ent}(mP) \leq (1 - \alpha(I - PP^*))\mathrm{Ent}(m)$$

Démonstration :

Désignons par f la densité de m par rapport à μ. On vérifie immédiatement que la densité de mP par rapport à μ est alors donnée par P^*f.

En appliquant le lemme 5 avec $t = P^*f(x)$ et $t + s = f(y)$, pour $x, y \in S$, puis en multipliant cette inégalité par $p^*(x,y)$ et en sommant sur $y \in S$, on obtient que pour tout $x \in S$,

$$P^*(f\ln(f))(x) \geq P^*f(x)\ln(P^*f(x)) + \sum_{y \in S} p^*(x,y)\left(\sqrt{f(y)} - \sqrt{P^*f(x)}\right)^2$$

Cependant, toujours à $x \in S$ fixé, on a comme dans la section précédente

$$\sum_{y \in S} p^*(x,y)\left(\sqrt{f(y)} - \sqrt{P^*f(x)}\right)^2 \quad \geq \quad P^*f(x) - \left(P^*(\sqrt{f})(x)\right)^2$$

ainsi en intégrant par rapport à μ en x et en utilisant l'invariance de μ par rapport à P^*, on fait apparaître que

$$(8) \quad \mathrm{Ent}(mP) = \sum_{x \in S} P^*f(x)\ln(P^*f(x))\,\mu(x) \quad \leq \quad \mathrm{Ent}(m) - \mathcal{E}_{I-PP^*}(\sqrt{f}, \sqrt{f})$$

Il reste à utiliser l'inégalité de Sobolev-logarithmique

$$\mathrm{Ent}(m) = \mathcal{L}(\sqrt{f}) \leq \frac{1}{\alpha(I - PP^*)}\mathcal{E}_{I-PP^*}(\sqrt{f}, \sqrt{f})$$

pour conclure au résultat escompté.

□

Remarques :

a) Pour $0 < \nu \leq 1$, soit $g(\nu) = ((1-\nu)\ln(1-\nu) + \nu)/(\sqrt{1-\nu} - 1)^2$, on vérifie sans difficulté que pour tous $t \geq 0$ et $s \geq -\nu t$, on a

$$(9) \qquad (t+s)\ln(t+s) \geq t\ln(t) + (1 + \ln(t))s + g(\nu)(\sqrt{t+s} - \sqrt{t})^2$$

Remarquons que g est strictement décroissante et que $g(0_+) = 2$ et $g(1) = 1$. Une nouvelle coïncidence veut que la différence entre ces valeurs est la même que celle qui intervient dans le lemme 2.7 de Diaconis et Saloff-Coste [4] entre les situations réversibles et non réversibles. Notons aussi que l'on peut se servir d'une inégalité du type (9) avec un $\nu > 0$ dépendant de m et p, si l'on sait a priori que m charge tous les points. Puis quand on finit par savoir plus précisément que f est assez proche de $\mathbb{1}$, on peut utiliser plutôt une variante de l'inégalité de trou spectral à la place d'une véritable inégalité de Sobolev-logarithmique (cf. les inégalités énergie-entropie de Bakry [1], que l'on écrit ici sous la forme

$$\forall\, f \in \mathcal{F}(S),\ \|f\|_2 = 1, \qquad \mathcal{L}(f) \leq \alpha_{\mathcal{L}(f)}(I - PP^*)\mathcal{E}(f, f)$$

avec pour tout $h > 0$, $\alpha_h(I - PP^*) = h^{-1}\inf_{f\in\mathcal{F}(S)\setminus\mathrm{Vect}(\mathbb{1}),\,\|f\|_2=1,\,\mathcal{L}(f)=h} \mathcal{E}(f,f) \geq \alpha(I - PP^*)$, et on se sert du fait que $\lim_{h\to 0_+} \alpha(h)$ est égal à la moitié du trou spectral).

b) Par des arguments identiques à ceux de la fin de la section précédente, en prenant $P = \exp(tL)$ avec t petit et L générateur irréductible, on montre que (8) permet de retrouver le lemme 2.7 de Diaconis et Saloff-Coste [4] : pour tout $f \in \mathcal{F}(S)$, $f \geq 0$,

$$\mathcal{E}_{-L}(\sqrt{f}, \sqrt{f}) \leq \frac{1}{2}\mathcal{E}_{-L}(\ln(f), f)$$

(car il suffit de l'avoir pour toutes ces fonctions qui satisfont $\int f\, d\mu = 1$).

c) Le choix précédent de $t = P^*f(x)$ est commode et est suggéré par la preuve de l'inégalité de Jensen pour la fonction convexe $\mathbb{R}_+ \ni t \mapsto t\ln(t)$ et la probabilité donnée par $p^*(x, \cdot)$ à $x \in S$ fixé (qui permet de voir que l'entropie est toujours décroissante le long des itérés du semi-groupe agissant sur $\mathcal{P}(S)$), mais n'est pas optimal en général : en effectuant les mêmes calculs, on obtient pour tout $t \geq 0$,

$$\begin{aligned} P^*(f\ln(f))(x) &- \left(P^*(f)(x) - (P^*(\sqrt{f})(x))^2\right) \\ &\geq t\ln(t) + (1 + \ln(t))(P^*(f)(x) - t) + \left(P^*(\sqrt{f})(x) - \sqrt{t}\right)^2 \end{aligned}$$

et à $x \in S$ fixé, on est donc amené à optimiser le membre de droite en t. On vérifie facilement que le maximum est atteint pour

$$t = \left(\frac{P^*(f)(x)}{P^*(\sqrt{f})(x)}\right)^2$$

si $P^*(\sqrt{f})(x) \neq 0$ (sinon de toute façon, l'inégalité précédente s'écrit $0 \geq 0$ pour tout $t > 0$). En remplaçant t par cette valeur, on obtient

$$\begin{aligned} &P^*(f\ln(f))(x) - \left(P^*(f)(x) - (P^*(\sqrt{f})(x))^2\right) \\ &\geq \; -P^*(f)(x) + (P^*(\sqrt{f})(x))^2 + 2\ln\left(\frac{P^*(f)(x)}{P^*(\sqrt{f})(x)}\right) P^*(f)(x) \end{aligned}$$

c'est-à-dire

$$P^*(f\ln(f))(x) - \ln(P^*(f)(x))P^*(f)(x) \geq \ln\left(\frac{P^*(f)(x)}{(P^*(\sqrt{f})(x))^2}\right) P^*(f)(x)$$

En intégrant par rapport à μ en x, il apparaît donc que

$$\mathrm{Ent}(m) - \mathrm{Ent}(mP) \geq \int [\ln(P^*(f)) - \ln((P^*(\sqrt{f}))^2)]P^*(f)\,d\mu$$

ainsi en utilisant l'inégalité de concavité

$$\forall\; a,b > 0, \qquad \ln(a) - \ln(b) \geq \frac{1}{a}(a-b)$$

on retrouve la majoration (8).

De manière similaire le choix de $t = Pf(x)$ dans la section précédente n'est pas a priori le meilleur, cependant le problème d'optimisation en t n'est plus alors aussi trivial (et dépend de l'exposant q considéré).

Comme application immédiate de la proposition 6, on obtient une vitesse de convergence vers l'équilibre des chaînes de Markov irréductibles et apériodiques :

Corollaire 7

Soit p une matrice markovienne irréductible et apériodique. Alors pour toute probabilité initiale $m \in \mathcal{P}(S)$, tout $k \geq k(p)$ et tout $n \in \mathbb{N}$, on a

$$\mathrm{Ent}(mP^n) \leq (1 - \alpha(I - P^k P^{k*}))^{\lfloor n/k \rfloor}\mathrm{Ent}(m)$$

où $\lfloor \cdot \rfloor$ représente la partie entière.

En fait on aurait pu se servir de ce résultat pour montrer que $(i) \Rightarrow (ii)$ dans la proposition 1, du moins si l'on y suppose p sans point transient, car il permet de voir que si pour un $k \geq 1$, $\alpha(I - P^k P^{k*}) > 0$, alors on a unicité de la mesure invariante.

Notons également que contrairement aux processus de Markov à temps continu, ici l'entropie ne décroît pas nécessairement strictement en dehors de l'équilibre μ. Plus précisément, par le cas d'égalité dans l'inégalité de Jensen pour l'application strictement convexe $\mathbb{R}_+ \ni x \mapsto x\ln(x)$, il apparaît que $\mathrm{Ent}(mP) = \mathrm{Ent}(m)$ si et seulement si pour tout $x \in S$ fixé, la densité $f = dm/d\mu$ est constante sur l'ensemble $\{y \in S \,/\, p^*(x,y) > 0\}$ (ce qui équivaut aussi à $\mathcal{E}_{I-PP^*}(\sqrt{f},\sqrt{f}) = 0$), ainsi dans l'exemple sur $\mathbb{Z}/n\mathbb{Z}$ de la section 2, si on part d'une masse de Dirac en 1, il faut attendre le temps $n-1$ pour que l'entropie commence à décroître strictement à la prochaine transition (car $\mathrm{Ent}(mP) = \mathrm{Ent}(m)$ équivaut ici à $m(0) = 2m(n-1)$).

Si p est une matrice apériodique irréductible telle que $k(p) = 1$, on pourrait conjecturer que

$$\mathbb{N} \ni n \mapsto \frac{\mathrm{Ent}(mP^{n+1}) - \mathrm{Ent}(mP^n)}{\mathrm{Ent}(mP^n)}$$

est une application décroissante, ce qui permettrait de sortir du cadre d'ensembles d'états S finis et obtenir en général une décroissance exponentielle de l'entropie, mais avec une vitesse qui dépend de la loi initiale m par l'intermédiaire de la constante $(\mathrm{Ent}(mP) - \mathrm{Ent}(m))/\mathrm{Ent}(m) < 0$. Cependant ceci est faux : considérons à nouveau l'exemple ci-dessus sur $\mathbb{Z}/n\mathbb{Z}$ avec $n \geq 3$ et prenons $m = \frac{1}{5}\delta_{n-2} + \frac{4}{5}\delta_0$, on vérifie immédiatement que $\mathrm{Ent}(mP^2) = \mathrm{Ent}(mP) < \mathrm{Ent}(m)$. Quitte à considérer pour un $\epsilon > 0$ assez petit la matrice donnée par

$$\forall\, x, y \in S, \qquad p_\epsilon(x,y) = \frac{p(x,y) + \epsilon}{1 + n\epsilon}$$

on obtient alors un contre-exemple à la conjecture précédente.

5 Application aux algorithmes de recuit généralisés

L'ergodicité forte en loi des algorithmes de recuit généralisés à temps discret, si la température décroît suffisamment lentement, est bien connue (cf. les articles de Hwang et Sheu [11] et [12]). Dans [14], nous avons présenté une preuve très courte de ce résultat pour les équivalents de ces processus à temps continu (voir aussi la simplification donnée ultérieurement dans la section 2 de [16]), tout en retrouvant la constante critique qui y apparaît, en étudiant l'évolution de l'entropie à l'aide du comportement à basse température, donné par Holley et Stroock dans [10], de certaines constantes de Sobolev-logarithmiques. Nous allons voir ici comment les techniques décrites dans la section précédente étendent cette démonstration aux cas d'algorithmes à temps discret. Le résultat n'étant pas original, la justification de cette section se trouve plutôt dans l'espoir que la relative simplicité de la preuve lui permettra de s'adapter à des situations plus complexes que l'on rencontre dans la pratique.

On considère toujours un ensemble fini S muni d'un noyau de probabilités de transitions p irréductible (pas nécessairement apériodique), mais on suppose que l'on dispose en plus d'une fonction de coût V, définie sur $S \times S$ privé de sa diagonale et à valeurs dans $\overline{\mathbb{R}_+}$, compatible avec (S, p) (qui dans ce contexte est appelé noyau de communication a priori) :

$$\forall\, x \neq y \in S, \qquad p(x,y) = 0 \iff V(x,y) = +\infty$$

Pour $\beta \geq 0$ donné (représentant l'inverse de la température), soit p_β le noyau irréductible défini par

$$\forall\, x, y \in S, \qquad p_\beta(x,y) = \begin{cases} \exp(-\beta V(x,y))p(x,y) & \text{, si } x \neq y \\ 1 - \sum_{z \neq x} p_\beta(x,z) & \text{, sinon.} \end{cases}$$

Pour ne pas considérer des situations homogènes, on suppose que V n'est pas seulement à valeurs dans $\{0, +\infty\}$. Ceci assure que pour tout $\beta > 0$, il existe un $x_0 \in S$ tel que $p_\beta(x_0, x_0) > 0$ et donc que p_β est de plus apériodique.

Si $m_0 \in \mathcal{P}(S)$ et une évolution de l'inverse de la température $\beta : \mathbb{N} \to \mathbb{R}_+$ sont donnés, on notera $X = (X_n)_{n\in\mathbb{N}}$ une chaîne de Markov inhomogène de loi initiale m_0 et dont le noyau de probabilités de transitions à tout instant $n \in \mathbb{N}$ est p_{β_n}. Quand l'évolution β est telle que $\lim_{n\to\infty} \beta_n = +\infty$, le processus précédent est appelé un algorithme de recuit généralisé.

Pour donner des conditions d'ergodicité forte en loi de ces chaînes, il apparaît une constante $c \geq 0$ que nous allons maintenant décrire, mais il faut pour cela faire quelques rappels sur les probabilités invariantes μ_β associées aux p_β pour $\beta \geq 0$: il existe (voir le lemme 3.1 p. 177 de Freidlin et Wentzell [6]) deux applications $\rho : S \to \mathbb{R}_+^*$ et $U : S \to \mathbb{R}_+$ (appelée parfois énergie virtuelle ou quasi-potentiel associé à (S, p, V)), telles que pour tout $x \in S$ fixé, on ait pour β grand,

$$\mu_\beta(x) \sim \rho(x) \exp(-\beta U(x))$$

Ceci montre notamment que quand β devient grand, les probabilités μ_β convergent vers une certaine mesure μ_∞ dont le support est l'ensemble des minima globaux de U.

Pour $x, y \in S$, soit $\mathcal{C}_{x,y}$ l'ensemble des chemins (dont les transitions sont permises par p) allant de x à y. Si $q = (q_i)_{0\leq i\leq n}$ est un tel chemin, son élévation est le nombre

$$e(q) = \begin{cases} \max_{0\leq i<n}(U(x_i) + V(x_i, x_{i+1})) & \text{, si } n \geq 1 \\ U(x) & \text{, si } n = 0 \text{ (et donc } x = y) \end{cases}$$

(où on a convenu que pour tout $z \in S$, $V(z, z) = 0$ si $p(z, z) > 0$), et on pose

$$H(x, y) = \min_{q\in\mathcal{C}_{x,y}} e(q)$$

cette quantité étant appelée la hauteur de communication de x à y (pour le triplet (S, p, V)). Enfin on note

$$c = \max_{x,y\in S} H(x, y) - U(x) - U(y) \geq 0$$

Le résultat classique que nous voulons retrouver s'énonce alors,

Proposition 8

Supposons que l'évolution β satisfasse

$$\begin{aligned} \lim_{n\to+\infty} \beta_n &= +\infty \\ \limsup_{n\to+\infty} \beta_n / \ln(n) &< 1/c \\ \limsup_{n\to+\infty} n\, |\beta_{n+1} - \beta_n| &< +\infty \end{aligned}$$

alors la loi de X_n converge pour n grand vers μ_∞.

Bien que la proposition reste vérifiée dans le cas où $c = 0$, cette situation n'est pas très intéressante en théorie du recuit simulé et on se restreindra désormais à $c > 0$.

Exemple :

Une évolution typique de l'inverse de la température qui satisfait ces conditions est celle donnée par

$$\forall\, n \in \mathbb{N}, \qquad \beta_n = b^{-1}\ln(1+n)$$

avec $b > c$. Profitons-en pour préciser que ces conditions ne sont pas optimales, on pourrait d'ailleurs les étendre un peu en reprenant plus soigneusement la preuve qui suit, cependant il semblerait qu'aucune de ces généralisations ne puissent permettre de traiter le cas où pour tout $n \in \mathbb{N}$, $\beta_n = c^{-1}\ln(1+n)$, or il est connu que cette évolution assure également la convergence en loi vers μ_∞ (cf. par exemple [17], mais il est à noter que la démonstration est alors beaucoup plus complexe et passe par une évaluation précise des temps et positions de sortie de certains ensembles, car nous ne sommes pas arrivé à la déduire de l'étude de l'évolution de fonctionnelles (comme l'entropie) définies sur $\mathcal{P}(S)$, et bien que la preuve soit donnée pour des processus à temps continu, elle peut s'adapter au cas du temps discret). Néanmoins, notons que si l'on prend pour tout $n \in \mathbb{N}$, $\beta_n = b^{-1}\ln(1+n)$, avec $0 < b < c$, alors la conclusion de la proposition 8 est fausse, car on sait (cf. Hwang et Sheu [12] ou [15] dont les résultats sont donnés pour des algorithmes de recuit simulé classiques à temps continu, mais qui peuvent également s'étendre à la situation d'algorithmes généralisés à temps discret) que dans ce cas la loi limite dépend effectivement de la loi initiale, et en ce sens, les conditions de cette proposition ne sont pas si mauvaises.

La preuve de cette proposition va se faire en plusieurs étapes. Tout d'abord, en reprenant les notations de la section 2, remarquons que $k(p_\beta)$ ne dépend pas de $\beta > 0$, on notera désormais cette quantité $\underline{k}$. Pour $l \geq \underline{k}$, soit $\lambda(I - P_\beta^l P_\beta^{l*})$ la plus petite valeur propre non nulle de l'opérateur $I - P_\beta^l P_\beta^{l*}$, qui est auto-adjoint dans $L^2(\mu_\beta)$, c'est-à-dire la plus grande constante $\lambda > 0$ telle que

$$(10) \qquad \forall\, f \in \mathcal{F}(S), \qquad \lambda \int (f - \mu_\beta(f))^2\, d\mu_\beta \leq \int f(I - P_\beta^l P_\beta^{l*}) f\, d\mu_\beta$$

Commençons par rappeler le comportement pour β grand de ce trou spectral, et pour ceci introduisons les quantités suivantes :

$$\forall\, x \neq y \in S, \qquad W_l(x,y) = -\lim_{\beta \to +\infty} \beta^{-1}\ln((p_\beta^l p_\beta^{l*})(x,y))$$

puis

$$c(U, W_l) = \max_{x,y \in S}\ \min_{q=(q_i)_{0\leq i\leq n} \in \mathcal{S}_{x,y}}\ \max_{0 \leq i < n} (U(q_i) + W_l(q_i, q_{i+1})) \wedge (U(q_{i+1}) + W_l(q_{i+1}, q_i))$$

où $\mathcal{S}_{x,y}$ représente l'ensemble des suites finies de S dont le premier élément est x et le dernier y.

Lemme 9

Il existe une certaine constante $K > 0$ assez grande, telle que pour tout $\beta \geq 0$,

$$K^{-1}\exp(-c(U, W_l)\beta) \leq \lambda(I - P_\beta^l P_\beta^{l*}) \leq K\exp(-c(U, W_l)\beta)$$

La majoration est donnée à titre indicatif, car elle n'est pas nécessaire à la preuve de la proposition 8.

Démonstration :

Il s'agit là d'une estimation classique, dès que l'on a remarqué que U est la fonctionnelle des grandes déviations satisfaites pour β grand par la mesure invariante de $p_\beta^l p_\beta^{l*}$, car celle-ci n'est autre que μ_β.

Pour la minoration, on renvoit à la preuve présentée dans [14], qui adapte celle de Holley et Stroock [10] donnée dans le cas d'algorithmes de recuit classiques réversibles et basée sur des inégalités de type Poincaré (plus précisément, on utilise ici que pour tous $x \neq y \in S$ tels que $W_l(x,y) < +\infty$, il existe $\widehat{\rho}_l(x,y) > 0$ tel que pour β grand, $p_\beta^l p_\beta^{l*}(x,y) \sim \widehat{\rho}_l(x,y)\exp(-\beta W_l(x,y))$).

Pour la majoration, on peut également reprendre la démonstration de Holley et Stroock [10] qui est basée sur le choix d'une bonne fonction f dans (10) : soit C un cycle relatif à la fonction de coût W_l (voir Hwang et Sheu [12] pour une définition récursive des cycles, ou la section 3 de [17] qui les fait apparaître comme les traces sur S des cycles usuels relatifs à la structure de Hajek virtuelle associée à $(S, \widehat{\rho}_l, W_l)$, qui est une notion définie un peu plus loin), dont le complémentaire contient au moins un des minima globaux de U et qui est de hauteur de sortie maximale parmi ceux-ci. Cette hauteur de sortie de C vaut alors $c(U, W_l)$ et il suffit de prendre pour f l'indicatrice de C dans (10) pour obtenir la majoration escomptée.

□

Pour tous $x \neq y \in S$, notons

$$H_{U,W_l}(x,y) = \min_{q=(q_i)_{0\leq i\leq n}\in S_{x,y}} \max_{0\leq i\leq n-1} U(q_i) + W_l(q_i, q_{i+1})$$

et remarquons que

$$(11)\quad \min_{q=(q_i)_{0\leq i\leq n}\in S_{x,y}} \max_{0\leq i<n} (U(q_i) + W_l(q_i, q_{i+1})) \wedge (U(q_{i+1}) + W_l(q_{i+1}, q_i)) = H_{U,W_l}(x,y)$$

car puisque μ_β est réversible pour le noyau $p_\beta^l p_\beta^{l*}$, on a pour tous $z, z' \in S$, $U(z) + W_l(z,z') = U(z') + W_l(z',z)$ (néanmoins il est connu que l'égalité (11) ci-dessus est toujours satisfaite, même si μ_β n'avait été qu'invariante).

L'étape suivante est cruciale et consiste à trouver un $k \geq 1$ tel que

$$c(W_k, U) = c$$

Pour décrire cet entier, on va faire quelques rappels sur la structure du paysage d'énergie associé à la fonction de coût V.

On dit que $x \in S$ est un minimum local, si et seulement si pour tout $y \in S$ tel que $U(y) < U(x)$, on a $H(x,y) > U(x)$, ainsi notamment les minima globaux de U, qui sont ceux qui annulent cette application, en sont. Il ne s'agit pas là d'une notion de minima locaux pour U (bien que ceux-ci en soient tous), mais par rapport à une fonction $\widetilde{U}$ qui prolonge naturellement U sur un graphe $(\widetilde{S}, \widetilde{p})$ qui lui même étend canoniquement le graphe (S, p), comme on l'a déjà présenté dans [17] : Pour tous $x \neq y \in S$ tels que $p(x,y) > 0$, soit $x \cdot y$ un nouveau point n'appartenant pas à S. On prend alors

$$\widetilde{S} = S \sqcup \{x \cdot y \,/\, x \neq y \in S \text{ tels que } p(x,y) > 0\}$$

et

$$\forall\, z,z' \in \widetilde{S}, \qquad \widetilde{p}(z,z') = \begin{cases} p(z,z') & \text{, si } z = z' \in S \\ p(z,z') & \text{, si } z \in S \text{ et } z' = z\cdot y \text{, pour un } y \in S \\ 1 & \text{, si } z = x \cdot z' \text{ et } z' \in S \text{, pour un } x \in S \\ 0 & \text{, sinon} \end{cases}$$

puis

$$\forall\, z \in \widetilde{S}, \qquad \widetilde{U}(z) = \begin{cases} U(z) & \text{, si } z \in S \\ U(x) + V(x,y) & \text{, si } z = x\cdot y \text{, avec } x,y \in S \end{cases}$$

Cependant, si V dérivait déjà d'un potentiel $\widetilde{U}$ satisfaisant la condition de réversibilité faible de Hajek, c'est-à-dire si pour tous $x \neq y \in S$ tels que $p(x,y) > 0$, on a $V(x,y) = (\widetilde{U}(y) - \widetilde{U}(x))_+$, et si en posant pour tous $x,y \in S$,

$$H_{\widetilde{U}}(x,y) = \min_{q=(q_i)_{0\leq i\leq n}\in \mathcal{C}_{x,y}} \max_{0\leq i\leq n} \widetilde{U}(q_i)$$

on a que $H_{\widetilde{U}}$ est symétrique en ses deux arguments, alors cette construction est en fait inutile, car son but est de ramener d'une certaine manière toutes les situations à celle-ci.

Revenons au cas général, il apparaît que H correspond à la hauteur de communication classique relative au potentiel $\widetilde{U}$ (i.e. la restriction à $S \times S$ de $H_{\widetilde{U}}$, qui elle est définie sur $\widetilde{S} \times \widetilde{S}$), et on a vu dans [17] que cette application $\widetilde{U}$ satisfaisait une condition de réversibilité faible de Hajek (sur $(\widetilde{S},\widetilde{p})$), ce qui permit de retrouver que H est symétrique en ses deux arguments (ce résultat est dû initialement à Trouvé, cf. le lemme 1.33 p. 34 de [20], qui l'a prouvé par une méthode différente). En conséquence, $(\widetilde{S},\widetilde{p},\widetilde{U})$ sera appelé la structure de Hajek virtuelle associée à (S,p,V).

Notons L l'ensemble des minima locaux (qui est donc la trace sur S de l'ensemble des véritables minima locaux de $\widetilde{U}$ sur $\widetilde{S}$), sur lequel on introduit une relation d'équivalence $\bowtie$ par

$$\forall\, x,y \in L, \qquad x \bowtie y \iff H(x,y) = U(x) = U(y)$$

Appelons $C_1, \cdots, C_s$ les classes d'équivalence correspondantes. Du fait de l'hypothèse faite sur V, pour tout $1 \leq i \leq s$, il existe au moins un $x \in C_i$ tel que pour tout $\beta > 0$, $p_\beta(x,x) > 0$. On notera $\widehat{C}_i$ leur ensemble.

Soit $q = (q_i)_{0\leq i\leq n}$ un chemin (on désignera désormais par chemin, une suite finie d'éléments de S dont toutes les transitions sont permises par les p_β, pour $\beta > 0$, et non pas seulement par p, c'est-à-dire que l'on admet éventuellement certaines transitions supplémentaires qui restent en un même point, et pour tout $x \in S$, on conviendra d'ailleurs que $V(x,x) = 0$ ou $V(x,x) = +\infty$, suivant que $p_1(x,x) > 0$ ou $p_1(x,x) = 0$), on dira que q est descendant (respectivement montant) si

$$\forall\, 0 < i < n, \qquad V(q_i, q_{i+1}) = 0$$

(resp. si pour tout $0 \leq i < n-1$, $U(q_i) + V(q_i,q_{i+1}) = U(q_{i+1})$). Notons, en utilisant que pour tous $x \neq y \in S$ tels que $p(x,y) > 0$, on a $\widetilde{U}(x \cdot y) \geq U(x) \vee U(y)$, que les chemins qui leur sont canoniquement associés dans $(\widetilde{S},\widetilde{p}_\beta)$ sont effectivement descendants (resp. montants) pour $\widetilde{U}$, sauf éventuellement la première (resp. dernière) transition. Si V dérivait déjà d'un potentiel de Hajek, descendant (resp. montant) signifiera véritablement décroissant (resp. croissant) pour le potentiel.

Il est clair (en travaillant par exemple sur $(\widetilde{S},\widetilde{p}_\beta)$), que pour tout $x \in S$, il existe au moins un chemin descendant (resp. montant) issu de x et aboutissant à un élément de $\widehat{C}_1 \sqcup \cdots \sqcup \widehat{C}_s$ (resp. partant de $\widehat{C}_1 \sqcup \cdots \sqcup \widehat{C}_s$ et aboutissant en x). Notons $D(x)$ (resp. $M(x)$) l'ensemble des éléments de $\widehat{C}_1 \sqcup \cdots \sqcup \widehat{C}_s$ pour lesquels il existe un tel chemin descendant (resp. montant). Sur l'exemple simple suivant on voit qu'il se peut que $D(x)$ et $M(x)$ soient différents :

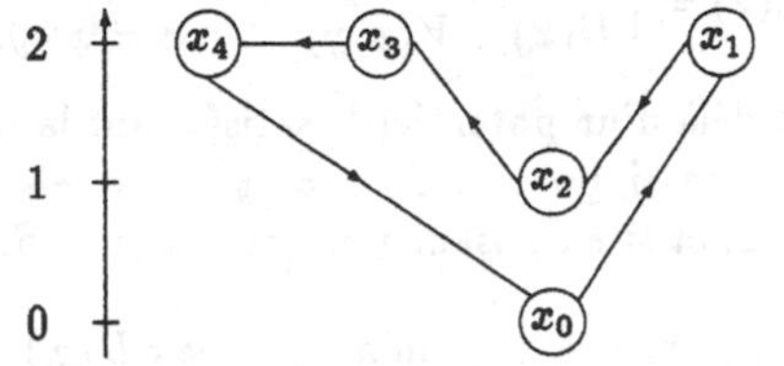

On a $S = \{x_0, x_1, x_2, x_3, x_4\}$ et les flèches désignent les transitions strictement positives pour p (toutes ici de valeur 1), les autres transitions étant nulles. De plus, la fonction de coût V dérive du potentiel de Hajek dont les valeurs sont indiquées en ordonnée. Avec ces définitions il apparaît que $D(x_4) = \{x_0\}$ et que $M(x_4) = \{x_2\}$. On pourrait d'ailleurs remplacer x_3 et x_4 par un seul point, mais on les a distingués pour que cet exemple puisse aussi servir dans la remarque (b) de la fin de cette section.

Revenons au cas général, pour tous $x \in S$ et $y \in D(x)$ (resp. $y \in M(x)$) fixés, considérons $\widehat{\mathcal{C}}^{(\mathrm{d})}_{x,y}$ (resp. $\widehat{\mathcal{C}}^{(\mathrm{m})}_{y,x}$) l'ensemble des chemins descendants (resp. montants) allant de x à y (resp. de y à x) qui sont d'élévation minimale parmi de tels chemins. Rappelons par ailleurs que la longueur d'un chemin $q = (q_i)_{0\leq i\leq n}$ est définie par $l(q) = n$ et notons

$$\begin{aligned} d(x,y) &= \min_{q\in\widehat{\mathcal{C}}^{(\mathrm{d})}_{x,y}} l(q) \\ m(y,x) &= \min_{q\in\widehat{\mathcal{C}}^{(\mathrm{m})}_{y,x}} l(q) \end{aligned}$$

On pose ensuite provisoirement

$$k = \max_{x\in S,\, y\in D(x)} d(x,y)$$

<u>Proposition 10</u>

On a $c(W_k, U) = c$, ce qui implique que $k \geq \underline{k}$, et ainsi il existe une constante $K \geq 1$ telle que

$$\forall\, \beta \geq 0, \qquad K^{-1}\exp(-c\beta) \leq \lambda(I - P_\beta^k P_\beta^{k*}) \leq K\exp(-c\beta)$$

Démonstration :

Pour tous $x \in S$ et $y \in D(x)$, il existe au moins un chemin de $\widehat{\mathcal{C}}^{(\mathrm{d})}_{x,y}$ de longueur k , quitte à obliger le chemin à rester plusieurs pas en y. Choisissons un tel chemin que l'on notera désormais $q^{(\mathrm{d})}(x,y) = (q_i^{(\mathrm{d})}(x,y))_{0\leq i\leq k}$. Il sera commode aussi de faire intervenir son retourné $\check{q}^{(\mathrm{d})}(y,x) = (q^{(\mathrm{d})}_{k-i}(x,y))_{0\leq i\leq k}$, qui n'est plus nécessairement un chemin au sens précédent.

Rappelons également que si $q = (q_i)_{0\leq i\leq k}$ est une suite de k éléments de S, sa valuation par rapport à p_β (resp. p^*_β) est la quantité

$$v_\beta(q) = \prod_{i=0}^{k-1} p_\beta(q_i, q_{i+1}) \leq p^k_\beta(q_0, q_k)$$

(resp. $v^*_\beta(q) = \prod_{i=0}^{k-1} p^*_\beta(q_i, q_{i+1}) \leq p^{*k}_\beta(q_0, q_k)$).

Ainsi pour les valuations $v_\beta(q^{(\mathrm{d})}(x,y))$ et $v^*_\beta(\check{q}^{(\mathrm{d})}(y,x))$, on a le comportement suivant pour β grand :

$$\begin{aligned} v_\beta(q^{(\mathrm{d})}(x,y)) &\sim \exp(-\beta V(x, q^{(\mathrm{d})}_0(x,y)))r^{(\mathrm{d})}(x,y) \\ &= \exp(-\beta[e(q^{(\mathrm{d})}(x,y)) - U(x)])r^{(\mathrm{d})}(x,y) \\ v^*_\beta(\check{q}^{(\mathrm{d})}(y,x)) &= \mu_\beta(x)v_\beta(q^{(\mathrm{d})}(x,y))/\mu_\beta(y) \\ &\sim \exp(-\beta[e(q^{(\mathrm{d})}(x,y)) - U(y)])\check{r}^{(\mathrm{d})}(y,x) \end{aligned}$$

où $r^{(\mathrm{d})}(x,y)$ et $\check{r}^{(\mathrm{d})}(y,x)$ sont des quantités strictement positives.

Ceci nous amène à considérer une nouvelle fonction de coût $\check{W}$ (non nécessairement compatible avec (S,p)) dont l'intérêt apparaîtra par la suite et qui est définie par

$$\forall\, x \neq y \in S, \qquad \check{W}(x,y) = \begin{cases} e(q^{(\mathrm{d})}(x,y)) - U(x) & \text{, si } y \in D(x) \\ e(q^{(\mathrm{d})}(y,x)) - U(x) & \text{, si } x \in D(y) \\ +\infty & \text{, sinon} \end{cases}$$

que l'on prolonge sur la diagonale de $S\times S$ par $W(x,x) = 0$ ou $W(x,x) = +\infty$ suivant que $x \in \widehat{C}_1 \sqcup \cdots \sqcup \widehat{C}_s$ ou $x \notin \widehat{C}_1 \sqcup \cdots \sqcup \widehat{C}_s$.

Par ailleurs, pour tous $x \neq y$, soit

$$\begin{aligned} \widehat{W}(x,y) &= -\lim_{\beta\to+\infty} \beta^{-1}\ln(p^k_\beta(x,y)) \\ &= \min_{x_1,\cdots,x_{k-1}\in S} V(x,x_1) + V(x_1,x_2) + \cdots + V(x_{k-1},y) \end{aligned}$$

Du fait que pour tous $z, z' \in S$, $U(z) + V(z,z') \geq U(z')$, on s'aperçoit sur cette dernière expression que

$$\forall\, x \neq y \in S, \qquad \widehat{W}(x,y) \geq H(x,y) - U(x)$$

Remarquons également que pour tous $x \neq y$,

$$W(x,y) = \min_{z\in S} \widehat{W}(x,z) + U(y) + \widehat{W}(y,z) - U(z)$$

où désormais W désigne W_k.

En utilisant que pour tout $z \in S$ fixé, on a

$$\begin{aligned} \widehat{W}(x,z) + U(y) + \widehat{W}(y,z) - U(z) &\geq H(x,z) - U(x) + H(y,z) - U(z) \\ &= H(x,z) + H(z,y) - U(z) - U(x) \\ &\geq H(x,y) - U(x) \end{aligned}$$

il apparaît que l'on a aussi

(12) $$\forall\, x,y \in S, \qquad W(x,y) \geq H(x,y) - U(x)$$

D'après les inégalités $H(x,y) \leq H(x,z) \vee H(z,y)$, valables pour tous $x,y,z \in S$, et (12), il est clair que $H_{U,W} \geq H$ (en convenant comme d'habitude que pour tout $x \in S$, $H_{U,W}(x,x) = U(x)$). Mais vérifions qu'en fait on a une égalité

$$H_{U,W} = H$$

En utilisant que $\lim_{\beta\to+\infty} \beta^{-1} \ln[v_\beta(q^{(\mathrm{d})}(y,y))] = 0 = \lim_{\beta\to+\infty} \beta^{-1} \ln[v_\beta^*(\check{q}^{(\mathrm{d})}(y,y))]$ pour tout $y \in \widehat{C}_1 \sqcup \cdots \sqcup \widehat{C}_s$, il apparaît facilement que $W \leq \check{W}$, d'où $H_{U,W} \leq H_{U,\check{W}}$ (cette dernière application étant définie de manière similaire, avec $\check{W}$ remplaçant W), et on peut donc se contenter de montrer que $H_{U,\check{W}} \leq H$.

Par construction des quantités précédentes et en se ramenant sur $(\widetilde{S},\widetilde{p})$, il suffit de voir que si une fonction de coût V dérive d'un potentiel U satisfaisant la condition de Hajek sur un graphe irréductible (S,p), alors pour tous $x \neq y \in S$, il existe une suite finie $q = (q_i)_{1\leq i\leq n} \in \mathcal{S}_{x,y}$ qui soit une succession de chemins véritablement descendants à des minima locaux (pour U) et de retournés de tels chemins à partir de ces minima locaux et telle que

$$\max_{0\leq i\leq n} U(q_i) = H(x,y)$$

Pour ceci, notons pour tout $1 \leq i \leq s$, A_i le domaine d'attraction de $\widehat{C}_i$, c'est-à-dire l'ensemble des $x \in S$ tels que $D(x) \cap \widehat{C}_i \neq \emptyset$. Soit $\widetilde{q} = (\widetilde{q}_j)_{0\leq j\leq m} \in \mathcal{C}_{x,y}$ un véritable chemin tel que $e(\widetilde{q}) = H(x,y)$. On construit q à partir de $\widetilde{q}$ de la manière suivante : soit $1 \leq i_0 \leq s$ tel que $x \in A_{i_0}$ et soit $m_0 = \max\{0 \leq j \leq m \,/\, \forall\, 0 \leq l \leq j,\ \widetilde{q}_l \in A_{i_0}\}$. On peut alors remplacer le tronçon $(\widetilde{q}_i)_{0\leq i\leq m_0}$ par une descente à partir de x en $\widehat{C}_{i_0}$ suivi du retourné d'une descente de $\widetilde{q}_{m_0}$ en $\widehat{C}_{i_0}$. Ensuite si $m_0 = m$ c'est gagné, et sinon il existe $1 \leq i_1 \leq s$ tel que $\widetilde{q}_{m_0} \in A_{i_1}$ et tel qu'en posant $m_1 = \max\{m_0 \leq j \leq m \,/\, \forall\, m_0 \leq l \leq j,\ \widetilde{q}_l \in A_{i_0}\}$, on ait $m_1 > m_0$ (ceci provenant du fait que l'on a nécessairement $U(\widetilde{q}_{m_0}) > U(\widetilde{q}_{m_0+1})$). On remplace alors le trajet $(\widetilde{q}_j)_{m_0\leq j\leq m_1}$ par une descente en $\widehat{C}_{i_1}$ suivi du retourné d'une autre telle descente. En procédant ainsi, on finit par arriver en y, d'où l'existence de q qui est la concaténation de ces suites finies de descentes et de retournées de descentes.

L'identité $H_{U,W} = H$ implique alors par le biais de (11) que $c(U,W) = c$.

□

Si l'on avait posé

$$k = \max_{x\in S,\, y\in M(x)} m(y,x)$$

en travaillant plutôt à partir des chemins montants et de leur retournés (ce qui amène à introduire pour $x \in S$ et $y \in M(x)$, des $q^{(\mathrm{m})}(y,x)$ et des $\check{q}^{(\mathrm{m})}(x,y)$) et des domaines de répulsion des $\widehat{C}_i$, $R_i = \{x \in S, / M(x) \cap \widehat{C}_i \neq \emptyset\}$ (en fait plutôt ceux associés à la structure de Hajek virtuelle), on montre d'une manière similaire que pour une certaine constante $K > 1$,

$$\forall\, \beta \geq 0, \qquad K^{-1} \exp(-c\beta) \leq \lambda(I - P_\beta^{k*} P_\beta^k) \leq K \exp(-c\beta)$$

Il reste à utiliser que $\lambda(I - P_\beta^k P_\beta^{k*}) = \lambda(I - P_\beta^{k*} P_\beta^k)$ pour conclure que

$$\text{(13)} \qquad \forall\, \beta \geq 0, \qquad K^{-1}\exp(-c\beta) \leq \lambda(I - P_\beta^k P_\beta^{k*}) \leq K\exp(-c\beta)$$

est toujours vérifié avec k défini par

$$\text{(14)} \qquad k = \left(\max_{x\in S,\, y\in D(x)} d(x,y)\right) \wedge \left(\max_{x\in S,\, y\in M(x)} m(y,x)\right)$$

ce qui admet aussi pour conséquence que $k \geq \underline{k}$.

Soit $\alpha(I - P_\beta^k P_\beta^{k*})$ la constante de Sobolev-logarithmique associée à l'opérateur $I - P_\beta^k P_\beta^{k*}$, c'est-à-dire la plus grande constante $\alpha > 0$ telle que

$$\forall\, f \in \mathcal{F}(S), \qquad \alpha \int f^2 \ln(f^2)\, d\mu_\beta \leq \int f(I - P_\beta^k P_\beta^{k*}) f\, d\mu_\beta$$

Un calcul de Holley et Stroock (cf. le théorème 3.21 p. 568 de [10]) permet de déduire de la minoration de $\lambda(I - P_\beta^k P_\beta^{k*})$ dans (13) et du comportement pour β grand de μ_β, qu'il existe une constante $K > 0$ assez grande, telle que

$$\forall\, \beta \geq 1, \qquad \alpha(I - P_\beta^k P_\beta^{k*}) \geq K^{-1}\beta^{-1}\exp(-c\beta)$$

(plus simplement encore et d'une manière peut-être plus explicite, ceci découle aussi du corollaire 5.4 de l'appendice de Diaconis et Saloff-Coste [4]).

D'autre part, la constante de Sobolev-logarithmique étant toujours majorée par la moitié du trou spectral de l'opérateur symétrique considéré (voir par exemple Diaconis et Saloff-Coste [4]), on a en fin de compte un encadrement de la forme

$$\text{(15)} \qquad \forall\, \beta \geq 1, \qquad K^{-1}\beta^{-1}\exp(-c\beta) \leq \alpha(I - P_\beta^k P_\beta^{k*}) \leq K\exp(-c\beta)$$

(notons que l'appendice de Diaconis et Saloff-Coste [4], qui donne une expression exacte pour la constante de Sobolev-logarithmique dans le cas où S n'est constitué que de deux points, permet de voir que l'on ne peut pas espérer avoir en général une estimation du type (13) pour $\alpha(I - P_\beta^k P_\beta^{k*})$, mais que la minoration (15) est du bon ordre).

On a donc prouvé l'estimation suivante :

Proposition 11

Si k est l'entier donné par (14), on a

$$\lim_{\beta\to+\infty} \beta^{-1}\ln(\alpha(I - P_\beta^k P_\beta^{k*})) = -c$$

Remarquons que de la même manière, on montre que $\lim_{\beta\to+\infty} \beta^{-1}\ln(\alpha(I - P_\beta^{k*} P_\beta^k)) = -c$, ce qui aurait pu être intéressant si on avait cherché à utiliser plutôt l'hypercontractivité, dans l'esprit de Concordet [2], ce qui permet plus précisément de montrer que la densité converge vers $\mathbf{I}$ et de donner des estimées de la vitesse de convergence.

Des arguments standards de comparaisons permettent d'étendre ce résultat, mais introduisons d'abord l'ensemble suivant :

$$R = \{x \in S \,/\, \lim_{\beta\to+\infty} p_\beta(x,x) > 0\}$$

(on a aussi, puisque pour tout $x \in S$ l'application $\mathbb{R}_+ \ni \beta \mapsto p_\beta(x,x)$ est soit constante soit strictement croissante, $R = \{x \in S \,/\, \exists\, \beta \geq 0 \text{ avec } p_\beta(x,x) > 0\} = \{x \in S \,/\, \forall\, \beta > 0, p_\beta(x,x) > 0\}$).

Supposons maintenant que l'on dispose de k (qui désormais désignera la constante définie ci-dessus) matrices markoviennes $p(1,t), \cdots, p(k,t)$ dépendant d'un paramètre $t \geq 0$ telles que pour une certaine fonction $\beta : \mathbb{R}_+ \to \mathbb{R}_+$ vérifiant $\lim_{t\to+\infty} \beta_t = +\infty$, on ait les limites suivantes pour tout $1 \leq i \leq k$,

$$(16)\begin{cases} \forall\, x \neq y \in S, & \lim_{t\to+\infty} \beta_t^{-1} \ln(p(i,t)(x,y)) = -V(x,y) \\ \forall\, x \in S, & \lim_{t\to+\infty} \beta_t^{-1} \ln(p(i,t)(x,x)) = \begin{cases} 0 & \text{, si } x \in R \\ -\infty & \text{, si } x \notin R \end{cases} \end{cases}$$

Posons

$$\alpha[I - P(1,t)\cdots P(k,t)(P(1,t)\cdots P(k,t))^*] = \inf_{f\in\mathcal{F}\setminus\mathrm{Vect}(\mathbb{1})} \frac{\mathcal{E}_{I-P(1,t)\cdots P(k,t)(P(1,t)\cdots P(k,t))^*}(f,f)}{\mathcal{L}_{\widehat{\mu}_t}(f)}$$

avec $\widehat{\mu}_t$ la probabilité invariante de $p(1,t)\cdots p(k,t)$ (qui intervient aussi dans la forme de Dirichlet au numérateur comme mesure d'intégration et dans la définition de l'adjoint qui est compris ici dans $L^2(\widehat{\mu}_t)$), qui est bien définie du moins pour t assez grand.

Le résultat suivant se base sur la proposition 11 pour la généraliser.

Proposition 12

Les constantes de Sobolev-logarithmique précédentes satisfont

$$\lim_{t\to+\infty} \beta_t^{-1} \ln\left(\alpha[I - P(1,t)\cdots P(k,t)(P(1,t)\cdots P(k,t))^*]\right) = -c$$

Démonstration :

Il suffit de vérifier que

$$\lim_{t\to+\infty} \beta_t^{-1} \ln\left(\frac{\alpha[I - P(1,t)\cdots P(k,t)(P(1,t)\cdots P(k,t))^*]}{\alpha[I - P_{\beta_t}^k P_{\beta_t}^{k*}]}\right) = 0$$

Cependant la condition (16) permet de voir que

$$\forall\, x,y \in S, \qquad \lim_{t\to+\infty} \beta_t^{-1} \ln((p(1,t)\cdots p(k,t))(x,y)) = \lim_{t\to+\infty} \beta_t^{-1} \ln((p_{\beta_t}^k)(x,y))$$

Par la formule de Freidlin et Wentzell pour la mesure invariante, il apparaît alors que

$$(17) \qquad \forall\, x \in S, \qquad \lim_{t\to+\infty} \beta_t^{-1} \ln(\widehat{\mu}_t) = \lim_{t\to+\infty} \beta_t^{-1} \ln(\mu_{\beta_t}) = -U(x)$$

En conséquence, on a donc également pour tous $x, y \in S$,

$$\lim_{t\to+\infty} \beta_t^{-1} \ln((p(1,t)\cdots p(k,t))^*(x,y)) = \lim_{t\to+\infty} \beta_t^{-1} \ln(p_{\beta_t}^{k*}(x,y))$$

et

$$(18) \qquad \lim_{t\to+\infty} \beta_t^{-1} \ln([p(1,t)\cdots p(k,t)(p(1,t)\cdots p(k,t))^*](x,y)) = \lim_{t\to+\infty} \beta_t^{-1} \ln([p_{\beta_t}^k p_{\beta_t}^{k*}](x,y))$$

Grâce à (17) et (18), via une application de l'argument de comparaison du lemme 3.3 de Diaconis et Saloff-Coste [4], on obtient le résultat annoncé.

□

Nous pouvons maintenant appliquer les calculs de la section précédente. On considère donc une chaîne de Markov inhomogène X vérifiant les hypothèses de la proposition 8 et soit $0 \leq r < k$ fixé.

Pour simplifier les notations, convenons de poser pour tout $n \in \mathbb{N}$, $\widehat{m}_n$ la loi de X_{kn+r}, $\widehat{p}_n$ la matrice produit $p_{\beta_{kn+r}} \cdots p_{\beta_{kn+k+r-1}}$ et $\widehat{\mu}_n$ la probabilité invariante qui lui est associée. Notons que pour n grand, on a aussi

$$\forall\, x \in S, \qquad \widehat{\mu}_n(x) \sim \rho(x)\exp(-\beta_{kn+r}U(x))$$

En effet, il est facile de voir, d'après la troisième condition de la proposition 8, que les entrées des matrices $\widehat{p}_n$ et $p^k_{\beta_{kn+r}}$ ne diffèrent que de facteurs de la forme $(1+\mathcal{O}(1/n))$ (du moins dès que $\beta_{kn+r} > 0$, car sinon, si de plus $\beta_{kn+k+r-1} > 0$, il se peut que pour un $x \in S$, on ait $(p^k_0)(x,x) = 0$ et $\widehat{p}_n(x,x) > 0$), ainsi par la formule explicite de la mesure invariante donnée par Freidlin et Wentzell, il apparaît, en considérant $\mu_{\beta_{kn+r}}$ comme la mesure invariante de $p^k_{\beta_{kn+r}}$, que pour n grand, on a

$$\text{(19)} \qquad \forall\, x \in S, \qquad \widehat{\mu}_n(x) = \mu_{\beta_{kn+r}}(x)(1+\mathcal{O}(1/n))$$

d'où l'équivalence annoncée, et d'après la première condition de la proposition 8, le fait que

$$\text{(20)} \qquad \forall\, x \in S, \qquad \lim_{n\to\infty} \widehat{\mu}_n(x) = \mu_\infty = \begin{cases} \rho(x) & \text{, si } U(x) = 0 \\ 0 & \text{, sinon.} \end{cases}$$

Mais on peut déduire de (19) un autre résultat qui nous sera fort utile : en effet, de la même manière, on prouve que pour n grand,

$$\forall\, x \in S, \qquad \mu_{\beta_{k(n+1)+r}}(x) = \mu_{\beta_{kn+r}}(x)(1+\mathcal{O}(1/n))$$

et puisque pour tout $x \in S$, $\widehat{\mu}_{n+1}(x) = \mu_{\beta_{k(n+1)+r}}(x)(1+\mathcal{O}(1/n))$, on obtient que

$$\forall\, x \in S, \qquad \frac{\widehat{\mu}_{n+1}(x)}{\widehat{\mu}_n(x)} = 1 + \mathcal{O}(1/n)$$

puis que

$$\left\| \ln\left(\frac{\widehat{\mu}_{n+1}}{\widehat{\mu}_n}\right) \right\|_\infty = \mathcal{O}(1/n)$$

où $\|\cdot\|_\infty$ représente la norme uniforme sur $\mathcal{F}(S)$.

D'autre part, pour $n \in \mathbb{N}$, appelons $\widehat{\alpha}_n$ la constante de Sobolev-logarithmique associée à l'opérateur $I - \widehat{P}_n\widehat{P}_n^*$:

$$\widehat{\alpha}_n = \inf_{f\in\mathcal{F}\setminus\mathrm{Vect}(\mathbb{1})} \frac{\mathcal{E}_{I-\widehat{P}_n\widehat{P}_n^*}(f,f)}{\mathcal{L}_{\widehat{\mu}_n}(f)}$$

Une application de (12) montre que

$$\text{(21)} \qquad \lim_{n\to\infty} \beta_{kn+r}^{-1}\ln(\widehat{\alpha}_n) = -c$$

ce qui combiné avec la seconde hypothèse de la proposition 8, donne

$$\liminf_{n\to\infty} \frac{\ln(\widehat{\alpha}_n)}{\ln(n)} = \liminf_{n\to\infty} \frac{\ln(kn+r)}{\ln(n)} \frac{\beta_{kn+r}}{\ln(kn+r)} \frac{\ln(\widehat{\alpha}_n)}{\beta_{kn+r}} > -1$$

En notant $\gamma = -(\liminf_{n\to\infty} \ln(\widehat{\alpha}_n)/\ln(n) - 1)/2 < 1$, il existe donc deux constantes $n_0 \geq 0$ et $K > 0$ telles que pour tout $n \geq n_0$, $\widehat{\alpha}_n \geq K/(n+1)^\gamma$. On peut d'ailleurs prendre $n_0 = \max\{n \in \mathbb{N} \,/\, \beta_{kn+r} = 0\} + 1$ (ou $n_0 = 0$ si cet ensemble est vide) et trouver un $K > 0$ correspondant, et comme on peut supposer β à valeurs dans $\mathbb{R}_+^*$ pour ce qui nous intéresse, on prendra ci-dessous $n_0 = 0$.

Nous disposons désormais de toutes les estimations nécessaires à l'étude de l'évolution de l'entropie $\mathrm{Ent}(\widehat{m}_n|\widehat{\mu}_n)$. On aura remarqué que pour prouver la proposition 8, il suffit, par le biais de la majoration (2) et de (20), de voir que $\lim_{n\to\infty} \mathrm{Ent}(\widehat{m}_n|\widehat{\mu}_n) = 0$ (pour tout $1 \leq r < k$ fixé). Pour ceci, on écrit

$$\begin{aligned}\mathrm{Ent}(\widehat{m}_{n+1}|\widehat{\mu}_{n+1}) - \mathrm{Ent}(\widehat{m}_n|\widehat{\mu}_n) &= \mathrm{Ent}(\widehat{m}_{n+1}|\widehat{\mu}_{n+1}) - \mathrm{Ent}(\widehat{m}_{n+1}|\widehat{\mu}_n)\\ &\quad + \mathrm{Ent}(\widehat{m}_{n+1}|\widehat{\mu}_n) - \mathrm{Ent}(\widehat{m}_n|\widehat{\mu}_n)\end{aligned}$$

Pour évaluer la première différence du membre de droite, notons que

$$\begin{aligned} a_n &\stackrel{\text{déf}}{=} |\mathrm{Ent}(\widehat{m}_{n+1}|\widehat{\mu}_{n+1}) - \mathrm{Ent}(\widehat{m}_{n+1}|\widehat{\mu}_n)|\\ &= \left|\int \ln(\widehat{\mu}_{n+1}/\widehat{\mu}_n)\, d\widehat{m}_{n+1}\right|\\ &\leq \left\|\ln\left(\frac{\widehat{\mu}_{n+1}}{\widehat{\mu}_n}\right)\right\|_\infty\\ &\leq \frac{K'}{n+2}\end{aligned}$$

pour tout $n \in \mathbb{N}$, pour un bon choix de $K' > 0$.

Quant à la seconde différence, d'après la section 4, on a

$$\mathrm{Ent}(\widehat{m}_{n+1}|\widehat{\mu}_n) - \mathrm{Ent}(\widehat{m}_n|\widehat{\mu}_n) \leq -\widehat{\alpha}_n \mathrm{Ent}(\widehat{m}_n|\widehat{\mu}_n)$$

d'où en fin de compte, en posant pour tout $n \in \mathbb{N}$, $b_n = 1 - \widehat{\alpha}_n \geq 0$,

$$\begin{aligned}&\mathrm{Ent}(\widehat{m}_{n+1}|\widehat{\mu}_{n+1})\\ &\leq b_n \mathrm{Ent}(\widehat{m}_n|\widehat{\mu}_n) + a_n\\ &\leq a_n + b_n a_{n-1} + b_n b_{n-1} a_{n-2} + \cdots + b_n b_{n-1} \cdots b_1 a_0 + b_n b_{n-1} \cdots b_0 \mathrm{Ent}(\widehat{m}_0|\widehat{\mu}_0)\end{aligned}$$

Il reste donc à voir que cette dernière expression tend vers 0 pour n grand. On utilise pour ceci les inégalités

$$b_n = 1 - \widehat{\alpha}_n \leq \exp(-\widehat{\alpha}_n) \leq \exp(-K/(n+1)^\gamma)$$

qui permettent de voir pour tout $0 \leq j \leq n$,

$$b_n \cdots b_j \leq \exp(-K((n+1)^{-\gamma} + \cdots + (j+1)^{-\gamma})) \leq \exp(-K(1-\gamma)^{-1}((n+2)^{1-\gamma} - (j+1)^{1-\gamma})$$

On est donc ramené à voir que

$$\lim_{n\to\infty} \exp(-K(1-\gamma)^{-1}(n+2)^{1-\gamma}) \left[\sum_{j=0}^{n} \frac{1}{j+2} \exp(K(1-\gamma)^{-1}(j+2)^{1-\gamma})\right] = 0$$

et pour cela on majore la somme entre crochet par l'intégrale suivante (car l'application $\mathbb{R}_+^* \ni t \mapsto \exp(K(1-\gamma)^{-1}t^{1-\gamma})/t$ est décroissante puis croissante), que l'on intègre par parties

$$\begin{aligned}
&\int_1^{n+3} \frac{1}{t}\exp(K(1-\gamma)^{-1}t^{1-\gamma})\,dt\\
&= \int_1^{n+3} \frac{1}{t^{1-\gamma}}\frac{1}{t^\gamma}\exp(K(1-\gamma)^{-1}t^{1-\gamma})\,dt\\
&= K^{-1}([t^{\gamma-1}\exp(K(1-\gamma)^{-1}t^{1-\gamma})]_1^{n+3} + (1-\gamma)\int_1^{n+3} t^{\gamma-2}\exp(K(1-\gamma)^{-1}t^{1-\gamma})\,dt)
\end{aligned}$$

Choisissons $A \geq 1$ tel que $K(1-\gamma)^{-1}A^{1-\gamma} \geq 2$ et écrivons que

$$\begin{aligned}
&K^{-1}(1-\gamma)\int_1^{n+3} \frac{1}{t^{2-\gamma}}\exp(K(1-\gamma)^{-1}t^{1-\gamma})\,dt\\
&\leq K^{-1}(1-\gamma)\int_1^{A} \frac{1}{t^{2-\gamma}}\exp(K(1-\gamma)^{-1}t^{1-\gamma})\,dt + \frac{1}{2}\int_1^{n+3}\frac{1}{t}\exp(K(1-\gamma)^{-1}t^{1-\gamma})\,dt
\end{aligned}$$

ce qui fait apparaître que pour une certaine constante $K'' > 0$,

$$\int_1^{n+3} \frac{1}{t}\exp(K(1-\gamma)^{-1}t^{1-\gamma})\,dt \leq K'' + 2K^{-1}\frac{1}{(n+3)^{1-\gamma}}\exp(K(1-\gamma)^{-1}(n+3)^{1-\gamma})$$

puis que pour n grand,

$$\exp(-K(1-\gamma)^{-1}(n+2)^{1-\gamma})\left[\sum_{j=0}^{n}\frac{1}{j+2}\exp(K(1-\gamma)^{-1}(j+2)^{1-\gamma})\right] = \mathcal{O}\left(\frac{1}{n^{1-\gamma}}\right)$$

(et en étant un peu plus soigneux, il est possible de voir que le membre de gauche multiplié par $n^{1-\gamma}$ tend pour n grand vers K^{-1}), d'où le résultat annoncé, et une majoration de la vitesse de convergence.

Remarques :

a) Rappelons que les algorithmes de recuit simulé classiques sont ceux pour lesquels il existe une mesure μ telle que le noyau a priori p soit réversible par rapport à μ et pour lesquels V dérive d'un potentiel U.

Dans cette situation un encadrement de la forme (15) est en fait valable pour $\alpha(I - P_\beta P_\beta^*)$. En effet, comme d'habitude par les techniques de Holley et Stroock, on se ramène à une estimation du trou spectral λ_β de $I - P_\beta P_\beta^*$.

Or ici, il apparaît que la mesure invariante μ_β est la mesure de Gibbs donnée par

$$\forall\, x \in S, \qquad \mu_\beta(x) = \exp(-\beta U(x))\mu(x)$$

et que p_β est réversible par rapport à cette mesure, montrant ainsi que $P_\beta^* = P_\beta$.

On en déduit que $1 - \lambda_\beta$ est la plus grande valeur propre de P_β^2 différente de 1. Or d'après les calculs qui précèdent, il existe $k \in \mathbb{N}^*$ tel que la plus petite valeur propre non nulle $\widetilde{\lambda}_\beta$ de $I - P_\beta^k P_\beta^{k*} = I - P_\beta^{2k}$ satisfasse pour une certaine constante $K \geq 1$, un encadrement de la forme

$$\forall\, \beta \geq 1, \qquad K^{-1}\exp(-c\beta) \leq \widetilde{\lambda}_\beta \leq K\exp(-c\beta)$$

Mais en utilisant que toutes les valeurs propres de P_β^2 sont positives, il apparaît que pour β grand,

$$\widetilde{\lambda}_\beta = 1 - (1 - \lambda_\beta)^k \sim k\lambda_\beta$$

et il en découle qu'il existe une autre constante $K > 1$ telle que

$$(22) \qquad \forall\, \beta \geq 1, \qquad K^{-1}\exp(-c\beta) \leq \lambda_\beta \leq K\exp(-c\beta)$$

Frigerio et Grillo (voir les équations (2.12) et (2.13) de [7]) affirment qu'il est aussi possible d'obtenir directement (22) à partir des calculs classiques de Holley et Stroock. Cependant remarquons que si l'on note pour $\beta \geq 0$, $-1 \leq \lambda_1(\beta) \leq \cdots \leq \lambda_{\mathrm{card}(S)-1}(\beta) < \lambda_{\mathrm{card}(S)}(\beta) = 1$ les valeurs propres de p_β (Vect($\mathbb{I}$) étant l'espace propre associé à $1 = \lambda_{\mathrm{card}(S)}(\beta)$), les estimations de [10] portent sur $1 - \lambda_{\mathrm{card}(S)-1}(\beta)$, or on a $\lambda_\beta = 1 - (\lambda^2_{\mathrm{card}(S)-1}(\beta) \vee \lambda_1^2(\beta))$, et il faudrait encore connaître le comportement de $1 + \lambda_1(\beta)$.

Mais montrons que celui-ci n'est en fait pas gênant, car

$$\lim_{\beta \to +\infty} 1 + \lambda_1(\beta) > 0$$

En effet, soit p_∞ la matrice markovienne limite pour β grand des p_β, par un résultat usuel de perturbation (cf. Kato [13], théorème 5.1 p. 107), $\lambda_1(\beta)$ converge vers la plus petite valeur propre de p_∞ pour β grand (même si p_∞ n'est plus nécessairement diagonalisable, contrairement ici aux p_β, on aura noté que les valeurs propres de p_∞ restent réelles). Il suffit donc de voir que -1 n'est pas valeur propre de p_∞. Pour ceci considérons par l'absurde un vecteur propre $f \in \mathcal{F}(S) \setminus \{0\}$ associé. On a donc pour tout $n \in \mathbb{N}$,

$$(23) \qquad P_\infty^n f = (-1)^n f$$

Or remarquons que les classes de récurrences de p_∞ sont exactement les classes d'équivalence pour la relation $\bowtie$ dans L, i.e. les $C_1, \cdots, C_s$. Posons pour tout $1 \leq i \leq s$, $p_\infty^{(i)}$ la restriction de p_∞ à C_i, il s'agit de matrices markoviennes irréductibles et apériodiques (car pour tout $x \in \widehat{C}_i$, $p_\infty(x,x) > 0$). On notera $\mu_\infty^{(i)}$ les probabilités invariantes associées (sur C_i). Alors, pour tout $x \in S$ fixé, on a

$$\lim_{n \to +\infty} P_\infty^n f(x) = \sum_{i=1}^{s} \gamma_i(x)\mu_\infty^{(i)}(f)$$

où pour tout $1 \leq i \leq s$, les limites suivantes existent bien

$$\gamma_i(x) = \lim_{n \to +\infty} \uparrow P_\infty^n(\mathbb{I}_{C_i})(x)$$

Ainsi en vertu de (23), $(-1)^n f$ doit converger pour n grand, c'est-à-dire que $f \equiv 0$, ce qui est la contradiction recherchée.

b) Souvent on peut aussi prendre un k plus petit que celui donné par (14) tout en étant assuré d'avoir encore (15). Par exemple, pour $x \in S$, $y \in D(x)$ et $q \in \widehat{\mathcal{C}}_{x,y}^{(\mathrm{d})}$, soit $j(q)$ la plus grande longueur d'un sous-chemin de q qui, en dehors de ses points de départ et d'arrivée, ne rencontre pas $R = \{z \in S \,/\, p_1(z,z) > 0\}$, posons ensuite

$$d(x,y) = \min_{q \in \mathcal{C}_{x,y}^{(\mathrm{d})}} j(q)$$

et définissons de manière similaire $m(y,x)$ pour $x \in S$ et $y \in M(x)$. On peut alors prendre dans les preuves précédentes le k construit comme dans (14) mais à partir de ces nouvelles quantités. Néanmoins même celui-ci n'est pas minimal en général, comme on peut le voir sur les algorithmes de recuit classiques.

On pourrait conjecturer que l'on a toujours pour une certaine constante $K \geq 1$,

$$\forall\, \beta \geq 0, \qquad K^{-1}\exp(-c\beta) \leq \lambda(I - P_\beta^{\underline{k}} P_\beta^{\underline{k}*}) \leq K\exp(-c\beta)$$

mais c'est faux :

En effet, considérons un triplet (S,p,V) tel que $\underline{k} > 1$, par exemple celui donné par une figure précédente. On rajoute à S un point $\overline{x}$ qui n'y appartenait pas, et on pose $\overline{S} = S \sqcup \{\overline{x}\}$. Soit $x_0 \in S$ tel que $U(x_0) = 0$ et tel qu'il existe $x_1 \in S$ avec $0 < V(x_0,x_1) < +\infty$. On définit un noyau de probabilités de transitions $\overline{p}$ par

$$\forall\, x,y \in \overline{S}, \qquad \overline{p}(x,y) = \begin{cases} p(x,y) & \text{, si } x,y \in S \text{, avec } x \neq x_0 \\ p(x,y)/2 & \text{, si } x = x_0 \text{ et } y \in S \\ 1/2 & \text{, si } x = x_0 \text{ et } y = \overline{x} \\ 1/\mathrm{card}(S) & \text{, si } x = \overline{x} \text{ et si } y \in S \\ 0 & \text{, sinon.} \end{cases}$$

Pour $K \geq 0$, on munit $(\overline{S},\overline{p})$ de la fonction de coût $\overline{V_K}$ définie par

$$\forall\, x \neq y \in \overline{S}, \qquad \overline{V_K}(x,y) = \begin{cases} V(x,y) & \text{, si } x,y \in S \\ K & \text{, si } x = x_0 \text{ et } y = \overline{x} \\ 0 & \text{, si } x = \overline{x} \text{ et si } y \in S \\ +\infty & \text{, sinon.} \end{cases}$$

Par la formule de Freidlin et Wentzell, il apparaît que si K est assez grand, alors la restriction à S du quasi-potentiel $\overline{U}$ de $(\overline{S},\overline{p},\overline{V_K})$ n'est autre que U et que $\overline{U}(\overline{x}) = K$. Par ailleurs, sur l'expression de c, il est clair que cette constante sera aussi celle associée à $(\overline{S},\overline{p},\overline{V_K})$, quitte à prendre K encore plus grand. Or on se convainc facilement, puisque $\underline{k} > 1$, que pour tout $C > c$ assez grand, on peut choisir K toujours plus grand (et plus précisément, pour C grand, K et C seront équivalents) de manière à ce que $\lambda(I - \overline{P_\beta}\overline{P_\beta^*})$ soit de l'ordre de $\exp(-C\beta)$ (au sens d'une minoration et d'une majoration du type (22)) et non pas de l'ordre de $\exp(-c\beta)$, ce qui fournit le contre-exemple cherché.

Pour d'autres remarques sur les preuves de cette section et tout particulièrement sur les manières d'affaiblir la seconde condition de (16) tout en gardant le bénéfice de la proposition 12, on renvoit à la prépublication 07-96 du Laboratoire de Statistique et Probabilités de l'Université Toulouse III qui correspond à cet article.

Références

[1] D. Bakry. L'hypercontractivité et son utilisation en théorie des semigroupes. In P. Bernard, editor, *Lectures on Probability Theory. Ecole d'Eté de Probabilités de Saint-Flour XXII-1992*, Lecture Notes in Mathematics 1581. Springer-Verlag, 1994.

[2] D. Concordet. Estimation de la densité du recuit simulé. *Annales de l'Institut Henri Poincaré*, 30(2):265–302, 1994.

[3] P. Diaconis and L. Saloff-Coste. Nash inequalities for finite Markov chains. A paraître dans *Journal of Theoretical Probability*, 1992.

[4] P. Diaconis and L. Saloff-Coste. Logarithmic Sobolev inequalities for finite Markov chains. Préprint, Octobre 1995.

[5] J.A. Fill. Eigenvalue bounds on convergence to stationarity for nonreversible Markov chains, with an application to the exclusion process. *The Annals of Applied Probability*, 1(1):62–87, 1991.

[6] M.I. Freidlin and A.D. Wentzell. *Random Perturbations of Dynamical Systems*. A Series of Comprehensive Studies in Mathematics 260. Springer-Verlag, 1984.

[7] A. Frigerio and G. Grillo. Simulated annealing with time-dependent energy function. *Mathematische Zeitschrift*, 213:97–116, 1993.

[8] L. Gross. Logarithmic Sobolev inequalities. *American Journal of Mathematics*, 97(4):1061–1083, 1976.

[9] R. Holley and D. Stroock. Logarithmic Sobolev inequalities and stochastic Ising models. *Journal of Statistical Physics*, 46:1159–1194, 1987.

[10] R. Holley and D. Stroock. Simulated annealing via Sobolev inequalities. *Communications in Mathematical Physics*, 115:553–569, 1988.

[11] C.R. Hwang and S.J. Sheu. Large-time behavior of perturbed diffusion Markov processes with applications to the second eigenvalue problem for Fokker-Planck operators and simulated annealing. *Acta Applicandae Mathematicae*, 19:253–295, 1990.

[12] C.R. Hwang and S.J. Sheu. Singular perturbed Markov chains and exact behaviors of simulated annealing processes. *Journal of Theoretical Probability*, 5(2):223–249, 1992.

[13] T. Kato. *Perturbation Theory for Linear Operators*. Classics in Mathematics. Springer, 1980.

[14] L. Miclo. Recuit simulé sans potentiel sur un ensemble fini. In J. Azéma, P.A. Meyer, and M. Yor, editors, *Séminaire de Probabilités XXVI*, Lecture Notes in Mathematics 1526, pages 47–60. Springer-Verlag, 1992.

[15] L. Miclo. Une étude des algorithmes de recuit simulé sous-admissibles. *Annales de la Faculté des sciences de Toulouse*, 4:819–877, 1995.

[16] L. Miclo. Sur les problèmes de sortie discrets inhomogènes. Préprint à paraître dans *The Annals of Applied Probability*, 1995.

[17] L. Miclo. Sur les temps d'occupations des processus de Markov finis inhomogènes à basse température. Préprint, 1995.

[18] O.S. Rothaus. Diffusion on compact Riemannian manifolds and logarithmic Sobolev inequalities. *Journal of Functional Analysis*, 42:102–109, 1981.

[19] D.W. Stroock. Logarithmic Sobolev inequalities for Gibbs states. In G. Dell'Antonio and U. Mosco, editors, *Dirichlet Forms*, Lecture Notes in Mathematics 1563, pages 194–228. Springer-Verlag, 1993.

[20] A. Trouvé. *Parallélisation massive du recuit simulé.* PhD thesis, Université Paris 11, Janvier 1993. Thèse de doctorat.

Laboratoire de Statistique et Probabilités
Université Paul Sabatier et CNRS
118, route de Narbonne
31062 Toulouse cedex, France

Comportement des temps d'atteinte d'une diffusion fortement rentrante

Mădălina DEACONU, Sophie WANTZ

INRIA-Institut Elie Cartan

Université Henri Poincaré , B.P 239

54506 Vandoeuvre-les-Nancy

Résumé - Soit $y \in \mathbb{R}$ et $(X_t^y; t \geq 0)$ la solution de l'EDS unidimensionelle : $X_t^y = y + B_t - \frac{1}{2}\int_0^t u(X_s^y)ds$, où la dérive $-u$ est "fortement rentrante" (cf. H_1 et H_2 ci-dessous). Nous étudions le comportement asymptotique de $E(exp\ \alpha T_x^y)$, lorsque $y \to \infty$ avec $\alpha \geq 0$, $y \geq x \geq 0$ et $T_x^y = \inf\{t \geq 0; X_t^y = x\}$.

1. Introduction

Soit $(X_t^y; t \geq 0)$ la solution de l'EDS unidimensionelle

$$X_t^y = y + B_t - \frac{1}{2}\int_0^t u(X_s^y)\,ds \qquad (E)$$

où $(B_t; t \geq 0)$ est un mouvement brownien linéaire issu de zéro. En fait, nous ne nous intéressons au processus $(X_t^y; t \geq 0)$ que lorsqu'il séjourne dans $\mathbb{R}_+$, si bien que les hypothèses suivantes ne portent que sur les valeurs prises par u sur $\mathbb{R}_+$:

(H_1) • $u : \mathbb{R}^+ \to \mathbb{R}^+$ est de classe C^2, u et u' tendent vers l'infini à l'infini, u'' et $\frac{u'^2}{u}$ sont des $o(u)$ à l'infini et $u'' \leq 2\frac{u'^2}{u}$ en dehors d'un compact .

(H_2) • $\int_.^{+\infty} \frac{1}{u} < +\infty$

Un exemple simple d'une telle fonction est $u(x) = x^\gamma$ avec $\gamma > 1$.

Quitte à modifier u sur $\mathbb{R}_-$ pour obtenir le caractère localement lipschitzien et rentrant de $-u$, l'équation (E) possède une unique solution forte. Soit : $T_x^y = \inf\{t \geq 0; X_t^y = x\}$. Notre résultat est :

Théorème 1.1 . — *Il existe un compact K de $\mathbb{R}_+$ et deux constantes $0 < C_1 < C_2 < \infty$ tels que pour tout $x \notin K$, $x \geq 0$:*

i) Pour tout $\alpha > 0$, tel que $u(x) \geq C_2\alpha^{\frac{1}{2}}$, on a : $\forall\ y \geq x$, $E(exp\ \alpha T_x^y) < \infty$.
De plus : $\sup_{y \geq x} E(exp\ \alpha T_x^y) < \infty$ et $\lim_{y\to\infty} E(exp\ \alpha T_x^y)$ existe .

ii) Pour tout $\alpha > 0$, tel que $u(x) \leq C_1\alpha^{\frac{1}{2}}$, on a : $\forall\ y \geq x$, $E(exp\ \alpha T_x^y) = \infty$.

iii) Pour tout $\alpha > 0$ on a :

$$1 \geq \limsup_{y\to\infty} E(exp\ -\alpha T_x^y) = \liminf_{y\to\infty} E(exp\ -\alpha T_x^y) = k(\alpha, x) > 0\ .$$

C_2 (resp. C_1) peut être choisie arbitrairement proche et plus grande que $2\sqrt{2}$ (resp. arbitrairement proche et plus petite que $\sqrt{2}$) .

Remarque 1.2 . — *Si X est le processus d'Ornstein-Uhlenbeck (c'est à dire $u(x) = k\,x$, $k > 0$), un tel résultat est faux . Bien que, pour un $\alpha > 0$ assez petit par rapport à k, on ait : $E(exp\ \alpha T_x^y) < \infty$ $(y \geq x)$, nous avons : $\sup_{y \geq x} E(exp\ \alpha T_x^y) = +\infty$ pour tout $\alpha > 0$ et tout $x \geq 0$ (cf [GNRS], p.402, formule 12). Ainsi, l'hypothèse (H_2) est-elle tout à fait essentielle à notre résultat .*

Le but de ce travail est de démontrer le Théorème 1.1 . Heuristiquement, ce résultat s'explique par le caractère fortement rentrant de $-u$.

2. Démonstration du théorème 1.1

On cherche à estimer la vitesse de retour en un point x partant de y, $y \geq x$: T_x^y . Afin de calculer $E(exp\ \alpha T_x^y)$, nous aurons besoin des fonctions propres du générateur infinitésimal de X :

$$L := \tfrac{1}{2}\Big(\frac{\partial^2}{\partial x^2} - u\frac{\partial}{\partial x}\Big).$$

On rappelle que, pour toute fonction $f : \mathbb{R} \to \mathbb{R}_+$ de classe C^2,

$$M_t^{y,f} := f(X_t^y)\ exp\Big(-\int_0^t \frac{Lf}{f}(X_s^y)ds\Big)\ ; \quad t \geq 0, \tag{1}$$

est une martingale locale (cf [RY] p. 277) . Aussi est-il classique que l'étude de $E(exp\ \alpha T_x^y)$, pour x fixé assez grand, passe par celle des fonctions propres de L, donc des solutions de :

$$f''(z) - u(z)f'(z) = \alpha f(z)\ (\alpha \in \mathbb{R}, z \geq x) \tag{P_α}$$

(La constante $\frac{1}{2}$ a disparu mais elle est prise en compte dans les constantes intervenant dans notre résultat.)

Nous allons chercher des solutions de (P_α) sous la forme :

$$f_\alpha(z) = \Big(exp \int_x^z \frac{\varphi_\alpha(y)}{u(y)}dy\Big)\ \frac{exp\ \int_x^z u(y)dy}{u(z)} \tag{2}$$

Dès que φ_α est bornée, l'hypothèse (H_2) assure l'existence d'une limite finie de $\displaystyle\int_x^z \frac{\varphi_\alpha}{u}(y)dy$ quand $z \to +\infty$.

Le lemme suivant résulte de calculs élémentaires . Nous en omettons la démonstration .

Lemme 2.1 . — *f_α est solution de (P_α) si et seulement si φ_α est solution de l'équation différentielle ordinaire :*

$$S_\alpha(\varphi_\alpha, x) = \varphi'_\alpha(x) \tag{D_α}$$

avec : $$S_\alpha(g,x) = 3g\frac{u'}{u}(x) + u''(x) - \frac{2u'^2}{u}(x) - g^2\frac{1}{u}(x) + (\alpha - g)u(x)\,.$$

Afin de déterminer l'allure de la fonction φ_α qui nous renseignera sur le fonction propre f_α, résolvons l'équation $S_\alpha\ (y,x) = 0$. En la multipliant par u, on trouve une équation en y du second degré : $y^2 + y(u^2 - 3u') + 2u'^2 - u''u - \alpha u^2 = 0.$

Pour que cette équation ait des racines réelles, on doit imposer :

$$\Delta = u^4\Big(1 - 6\frac{u'}{u^2} + \frac{u'^2}{u^4} + 4\frac{u''}{u^3} + 4\frac{\alpha}{u^2}\Big) \geq 0.$$

Par l'hypothèse (H_1), $\frac{u'^2}{u} = o(u)$ et $u'' = o(u)$, donc Δ se comporte comme

$$\Delta \simeq u^4\Big(1 - \varepsilon + \frac{4\alpha}{u^2}\Big).$$

On en déduit la condition suivante pour avoir des racines :

$$\begin{cases} \bullet \quad \alpha \geq 0 \textit{ et } x \textit{ assez grand} \\ ou \\ \bullet \quad \alpha < 0 \textit{ et } x \textit{ est tel que } u(x) \geq \overline{C_2}|\alpha|^{\frac{1}{2}}, \textit{ avec } \overline{C_2} > 2 \end{cases} \qquad (C)$$

Nous verrons que cette condition (C), selon qu'elle est ou non vérifiée, détermine deux classes de fonctions propres .

Désignons par Γ_α^+ (resp. $\Gamma_\alpha^-, \Gamma_\alpha^0$) le sous-ensemble de $\mathbb{R}_+ \times \mathbb{R}$ défini par :

$$\Gamma_\alpha^+ = \{(x,y); x \geq 0,\ S_\alpha(y,x) > 0\ (resp. < 0, = 0)\}.$$

Désignons par Δ_x la droite verticale de $\mathbb{R}_+ \times \mathbb{R}$ passant par $(x, 0)$. On a alors les faits suivants :

• $\Delta_x \cap \Gamma_\alpha^+$ est soit vide, soit un compact non vide . Cette dichotomie nous amènera à distinguer deux sortes de fonctions propres (voir plus loin).

• $\Delta_x \cap \Gamma_\alpha^0$ est, pour x assez grand, constitué de deux points : $\widetilde{\Gamma_x^0}$ et $\widetilde{\widetilde{\Gamma_x^0}}$, avec $\widetilde{\Gamma_x^0} > \widetilde{\widetilde{\Gamma_x^0}}$. $\widetilde{\Gamma_x^0}$ est la plus grande racine de l'équation $S_\alpha(y,x) = 0$. Elle est de la forme : $\widetilde{\Gamma_x^0} = \alpha - \varepsilon_\alpha(x)$ avec $\varepsilon_\alpha(x) > 0$ et $\varepsilon_\alpha(x) \to_{x\to\infty} 0$.

• Pour x assez grand, $\Gamma_\alpha^+ \cap \{(z,y); z \geq x, y \in \mathbb{R}\} =: \Gamma_\alpha^{+,x}$ est absorbant pour l'équation (D_α), i.e. : si φ est une solution de (D_α), avec $\varphi(x) = s$ et $(x,s) \in \Gamma_\alpha^{+,x}$, alors φ est définie sur $[x, \infty[$ et $(z, \varphi(z)) \in \Gamma_\alpha^{+,x}$ pour tout $z \geq x$.

Nous considérons à présent les deux cas induits par la condition (C) .

a) Plaçons nous ici dans le cas où la condition (C) est vérifiée . Cela se traduit par : $\Delta_x \cap \Gamma_\alpha^+ \neq \emptyset$. On a alors le résultat suivant, que nous montrerons dans le paragraphe suivant :

THÉORÈME 2.2 . — *Dans le cas où la condition (C) est vérifiée, définissons une fonction f par (pour $z \geq x$) :*

$$f(z) := \frac{exp \int_x^z u(y)dy}{u(z)} \Big[exp\Big(\int_x^\infty \frac{\varphi_1}{u} + \int_x^z \frac{\varphi_2}{u}\Big) - exp\Big(\int_x^\infty \frac{\varphi_2}{u} + \int_x^z \frac{\varphi_1}{u}\Big)\Big], \quad (3)$$

avec φ_1 et φ_2 deux solutions de D_α sur l'intervalle $[x, +\infty[$, telles que, pour $s > \alpha$ et $s' > \alpha$ fixés, $\varphi_1(x) = s$, $\varphi_2(x) = s'$ et $\varphi_1 \geq \varphi_2$, cf. figure 1 . Alors f est une fonction propre de L et $\lim_{t\to+\infty} f(t) > 0$.

Nous sommes maintenant en mesure d'établir la formule de la vitesse de rappel. On note T_x^y le temps de retour en x partant de y, $x \leq y$.

Théorème 2.3 . — *Lorsque (C) est vérifiée, on a pour tout $y > x$:*

$$E_y\big(e^{-\alpha T_x^y}\big) = \frac{f(y)}{f(x)}. \tag{4}$$

Démonstration du Théorème 2.3 :

(i) Soient $\alpha < 0$ et $u(x) \geq \overline{C_2}|\alpha|^{\frac{1}{2}}$ et soit z tel que $x < y < z$. On a, en notant $\mathbb{1}$ la fonction indicatrice :

$$f(y) = E_y\big(f(X_t)e^{|\alpha|t}\mathbb{1}_{t<T_x^y\wedge T_z^y}\big)+f(x)E_y\big(e^{|\alpha|T_x^y}\mathbb{1}_{T_x^y<t\wedge T_z^y}\big)+f(z)E_y\big(e^{|\alpha|T_z^y}\mathbb{1}_{T_z^y<t\wedge T_x^y}\big) \tag{5}$$

en appliquant le fait que $M_t^{y,f}$ donnée par (1) est une martingale locale .
Quand $t \to \infty$, d'après le théorème de convergence monotone, on a pour les deux derniers termes de (5) :

$$\lim_{t\to\infty} f(x)E_y\big(e^{|\alpha|T_x^y}\mathbb{1}_{T_x^y<t\wedge T_z^y}\big) = f(x)E_y\big(e^{|\alpha|T_x^y}\mathbb{1}_{T_x^y<T_z^y}\big)$$
$$\lim_{t\to\infty} f(z)E_y\big(e^{|\alpha|T_z^y}\mathbb{1}_{T_z^y<t\wedge T_x^y}\big) = f(z)E_y\big(e^{|\alpha|T_z^y}\mathbb{1}_{T_z^y<T_x^y}\big).$$

Pour le premier terme du membre de droite de (5), on sait que :

$$\lim_{t\to\infty} f(X_t)e^{|\alpha|t}\mathbb{1}_{t<T_x^y\wedge T_z^y} = 0 \qquad p.s.$$

Montrons que cette famille est équiintégrable en t. Pour cela, il suffit de voir qu'elle est bornée dans L^p, avec $p > 1$. En écrivant la relation (5) pour une fonction propre g de valeur propre $\alpha' < 0$, avec $\alpha' < \alpha$, on en déduit que :

$$g(y) \geq E_y\big(g(X_t)\,e^{|\alpha'|t}\,\mathbb{1}_{t<T_x^y\wedge T_z^y}\big) \geq C\,E_y\big(e^{|\alpha'|t}\,\mathbb{1}_{t<T_x^y\wedge T_z^y}\big)$$

puisque g est minorée. D'où :

$$\begin{aligned} E_y\big(f^{\frac{\alpha'}{\alpha}}(X_t)\,e^{|\alpha'|t}\,\mathbb{1}_{t<T_x^y\wedge T_z^y}\big) &\leq C'\,E_y\big(e^{|\alpha'|t}\,\mathbb{1}_{t<T_x^y\wedge T_z^y}\big) \\ &\leq C''\,g(y). \end{aligned}$$

Donc la famille $\big(f(X_t)\,e^{|\alpha|t}\,\mathbb{1}_{t<T_x^y\wedge T_z^y}\big)_t$ est bornée dans $L^{\frac{\alpha'}{\alpha}}$ et est donc équiintégrable. Ainsi :

$$\lim_{t\to\infty} E_y\big(f(X_t)e^{|\alpha|t}\mathbb{1}_{t<T_x^y\wedge T_z^y}\big) = 0.$$

Ce qui nous donne, pour $t \to \infty$ dans (5) :

$$f(y) = f(x)E_y\big(e^{|\alpha|T_x^y}\mathbb{1}_{T_x^y<T_z^y}\big)+f(z)E_y\big(e^{|\alpha|T_z^y}\mathbb{1}_{T_z^y<T_x^y}\big).$$

On fait à présent tendre z vers $+\infty$, et par la même méthode (l'équiintégrabilité), on montre que :

$$\lim_{z\to\infty} E_y\big(e^{|\alpha|T_z^y}\mathbb{1}_{T_z^y<T_x^y}\big) = 0.$$

On conclut par (4) dans le cas (i) :

$$E_y(e^{|\alpha|T_x^y}) = \frac{f(y)}{f(x)}.$$

(ii) Soient $\alpha > 0$ et x assez grand . Ce cas est beaucoup plus simple car $f(X_{t\wedge T_x^y} e^{-\alpha(t\wedge T_x^y)})$ est une martingale bornée . Il suffit donc d'appliquer le théorème d'arrêt et de faire tendre t vers $+\infty$. Ceci achève la démonstration du théorème 2.3.

On a alors : $\lim_{y\to+\infty} E_y(e^{-\alpha T_x^y}) = \dfrac{C(\alpha)}{f(x)}$. Les points (i) et (iii) du théorème 1.1 en découlent aisément .

COROLLAIRE 2.4 . — *Pour tout $x \notin K$, il existe un $\alpha > 0$ tel que, pour tout $r > 0$ et uniformément en $y \geq x$, on a :*

$$P(T_x^y > r) \leq C(x)e^{-\alpha r}.$$

COROLLAIRE 2.5 . — *Pour $\alpha > 0$ et x en dehors d'un compact K et pour tout $\alpha < 0$ tel que $u(x) \geq \overline{C_2}|\alpha|^{\frac{1}{2}}$, les lois des v.a. T_x^y convergent étroitement, quand $y \to +\infty$, vers une loi de probabilité ν telle que $\int_0^\infty e^{-\alpha z}\,\nu(dz) < \infty$.*

Idée de la démonstration du Corollaire 2.5 : On se place dans le cas $\alpha > 0$ et x assez grand . Soit Q_x^y la loi de la *v.a.* T_x^y . Alors, on montre que la famille $(Q_x^y)_y$ est tendue et monotone, et on en déduit qu'il existe une loi de probabilité ν telle que $\displaystyle\int_0^\infty e^{-\alpha z}\,\nu(dz) < +\infty$ et Q_x^y converge vers ν quand $y \to +\infty$. Ceci établit notre corollaire .

b) On se place à présent en dehors du domaine d'application de la condition (C), c'est-à-dire lorsque $\alpha < 0$ et $|\alpha| > 1$. Alors :

THÉORÈME 2.6 . — *Lorsque la condition (C) n'est pas vérifiée, une fonction propre de L peut être écrite sous la forme :*

$$f(z) := \left(exp \int_{x_1}^{z} \frac{\varphi}{u}\right) \frac{exp\int_{x_1}^{z} u(y)dy}{u(z)}, \tag{6}$$

avec x_1 tel que $u(x_1) \leq \overline{C}_1|\alpha|^{\frac{1}{2}}$, et φ telle que :

- *pour $z \geq x_1$, $\varphi(z) \leq \rho|\alpha| - u^2(z)$* $\qquad (\rho < 1)$
- *pour $z \leq x_1$, $\varphi(z) \geq \frac{|\alpha|}{z - x_0}$, voir figure 2 .*

Par les mêmes arguments d'équiintégrabilité que dans la démonstration du théorème 2.3, on en déduit que : $E(exp\ \alpha T_{x_0}^y) = +\infty$, avec $x_0 < C_1|\alpha|^{\frac{1}{2}}$, ce qui est le point (ii) du théorème 1.1 avec $C_1 = \sqrt{2}\,\overline{C_1}$. Notre principal résultat est donc établi .

REMARQUE 2.7 . — *L'hypothèse $\int_{\cdot}^{+\infty} \frac{1}{u} < +\infty$, sous laquelle notre théorème 1.1 est vrai, est précisement celle qui assure l'ultracontractivité du semigroupe associé à X_t^y par rapport à la mesure de probabilité $\nu(dx) = Ce^{-v(x)}dx$ avec v une primitive de u (cf. [KKR]) .*

Nous allons achever la démonstration du théorème 1.1 en prouvant les théorèmes 2.2 et 2.6 .

3. Démonstration du théorème 2.2 (première classe de fonctions propres : cas où (C) est satisfaite)

Supposons que $\alpha < 0$ et $u(x) \geq \overline{C_2}|\alpha|^{\frac{1}{2}}$ (le cas $\alpha > 0$ et x assez grand se traite de manière analogue).
On remarque que la fonction constante α est une sursolution de (D_α) pour x assez grand (i.e. $-S_\alpha(\alpha, x) \geq 0$) .
Soit $s > \alpha$. Nous désignons par φ la solution de (D_α) définie sur l'intervalle $[x, \infty[$ et telle que $\varphi(x) = s$. Il est alors clair que :

$$\sup_{z \geq x} |\varphi(z)| \leq k \qquad \text{et} \qquad \lim_{z \to \infty} \varphi(z) = \alpha.$$

Fig. 1.

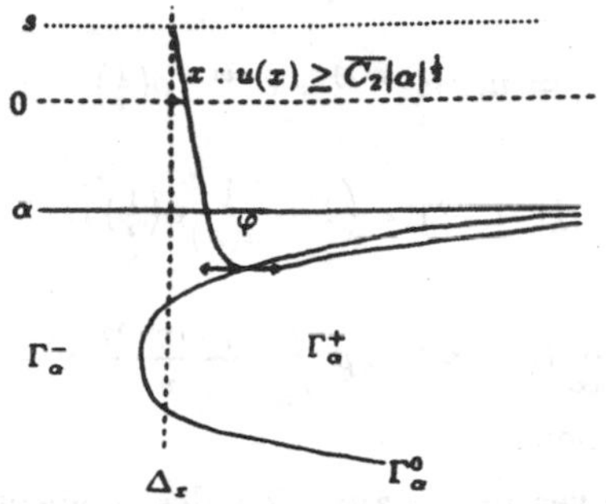

Montrons alors que la fonction f définie dans l'énoncé du théorème 2.2 par (3) est une solution positive de P_α, et que sa limite à l'infini existe et est strictement positive . On note :

$$v(z) := \int_x^z u(y)dy \quad \text{et} \quad h(z) := \int_x^z \frac{\varphi_1 + \varphi_2}{u}(y)dy\,.$$

On voit facilement que : $(\varphi_1 - \varphi_2)(z) = Cu^3(z)exp(-v(z) - h(z))$.
f étant une combinaison linéaire de deux fonctions propres de la forme (2), de valeur propre α, est elle-même une fonction propre de valeur propre α. Ecrivons f sous la forme :

$$f(z) = \Big(exp \int_x^\infty \frac{\varphi_1}{u} + \int_x^z \frac{\varphi_2}{u}\Big)\Big(1 - exp \int_z^\infty \frac{\varphi_2 - \varphi_1}{u}\Big)\frac{e^{v(z)}}{u(z)} \tag{7}$$

Puisque $(\varphi_2 - \varphi_1) \leq 0$ on a $f(z) \geq 0$ pour tout $z \geq x$.
Notons :

$$\gamma(z) := \Big(exp \int_x^\infty \frac{\varphi_1}{u} + \int_x^z \frac{\varphi_2}{u}\Big), \qquad g(z) := 1 - exp \int_z^\infty \frac{\varphi_2 - \varphi_1}{u}.$$

Il est evident que $\gamma(z)$ a une limite quand z tend vers l'infini, égale à $exp \int_x^\infty \frac{\varphi_1 + \varphi_2}{u}$.

D'autre part :

$$g(z) = \Big(\int_z^\infty \frac{\varphi_1 - \varphi_2}{u} + o\Big(\int_z^\infty \frac{\varphi_1 - \varphi_2}{u}\Big)\Big)$$

car $(\varphi_1 - \varphi_2)(z) \to_{z\to\infty} 0$ et donc on peut encadrer l'intégrale par deux fonctions tendant vers zéro, ce qui nous autorise à faire de développement limité de l'exponentielle au voisinage de l'origine . On obtient :

$$\int_z^\infty \frac{\varphi_1 - \varphi_2}{u}(y)dy = \int_z^\infty \widetilde{C}(y)u^2(y)e^{-v(y)}dy \quad \text{avec } \lim_{y\to\infty} \widetilde{C}(y) = 1.$$

D'autre part : $\displaystyle\int_z^\infty u^2(y)e^{-v(y)}dy = u(z)e^{-v(z)} + e^{-v(z)}o(\tfrac{1}{z})$.

D'où : $g(z) = u(z)e^{-v(z)} + e^{-v(z)}o(\frac{1}{z})$;

et finalement, d'après (7) :

$$f(z) = \gamma(z)\Big(1 + \frac{1}{u(z)}o(\tfrac{1}{z})\Big).$$

On en déduit que :

$$\lim_{z\to\infty} f(z) = C exp \int_x^\infty \frac{\varphi_1 + \varphi_2}{u}(y)dy.$$

Ceci achève la démonstration .

De plus, on peut trouver une majoration des moments d'ordre p des temps d'atteinte ; et pour $p = 1$, on a une estimation optimale .

Théorème 3.1 . — *i) Pour tout* $p \geq 1$, *il existe une constante* $C_p < \infty$ *telle que :*

$$\limsup_{x\to\infty} u(x)\, E\big[(T_x^{x+h})^p\big] \leq h\, C_p,$$

ii) Pour $p = 1$, *cette estimation est optimale : il existe deux constantes* $C < \infty$ *et* $C' < \infty$ *telles que :*

$$h\, C' \leq \liminf_{x\to\infty} u(x)E\big(T_x^{x+h}\big) \leq \limsup_{x\to\infty} u(x)E\big(T_x^{x+h}\big) \leq h\, C.$$

4. Démonstration du théorème 2.6 (seconde classe de fonctions propres : cas où (C) n'est pas satisfaite)

Rappelons que dans ce cas, la condition (C) n'est pas vérifiée, ce qui équivaut à : $\alpha < 0$ et $|\alpha| > 1$. Faisons trois remarques :

i) La fonction ψ_1, définie par : $\psi_1(z) := \rho|\alpha| - u^2(z)$, avec $\rho < 1$, est une sursolution de (D_α) pour z en dehors d'un compact, i.e. :

$$\psi_1'(z) - S_\alpha(\psi_1, z) \geq 0\,.$$

ii) Soient $\overline{C_1} < 1$, x_1 tel que $u(x_1) \leq \overline{C_1}|\alpha|^{\frac{1}{2}}$ et $x_0 = x_1 - k$, $\Big(k > \dfrac{1}{\rho - \overline{C_1}^2}\Big)$.

Soit ψ_2 définie sur $]x_0, x_1]$ par :

$$\psi_2(z) := \frac{|\alpha|}{z - x_0}.$$

Alors, ψ_2 est une sursolution de (D_α) . En effet, on vérifie aisément que $\psi_2'(z) - S_\alpha(\psi_2, z) \geq 0$.

iii) On peut choisir $\overline{C_1}$ (aussi proche de 1 que l'on veut) telle que : $\psi_2(x_1) \leq \psi_1(x_1)$, c'est-à-dire :

$$\frac{|\alpha|}{x_1 - x_0} \leq \rho|\alpha| - u^2(x_1)\ , \qquad (\rho - \overline{C_1}^2) \geq \frac{1}{x_1 - x_0} = \frac{1}{k}.$$

Ceci justifie notre choix de la constante k $\left(k > \dfrac{1}{\rho - \overline{C_1}^2}\right)$.

Soit maintenant s tel que $\psi_2(x_1) \leq s \leq \psi_1(x_1)$. Et soit φ la solution de (D_α) vérifiant $\varphi(x_1) = s$. Définissons φ sur son intervalle maximal de définition .

Fig. 2.

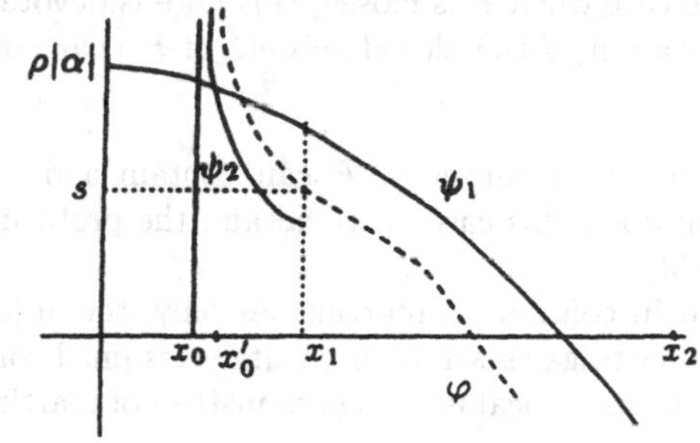

On a donc :

- pour $z \geq x_1$ $\varphi(z) \leq \rho|\alpha| - u^2(z)$ (8)
- pour $z \leq x_1$ $\varphi(z) \geq \dfrac{|\alpha|}{z - x_0}$ (9)

Soit alors

$$f(z) := \left(exp \int_{x_1}^{z} \frac{\varphi}{u}\right) \frac{exp \int_{x_1}^{z} u(y)dy}{u(z)}.$$

Ainsi, il est clair, d'après (8) et (9), que f est solution de (P_α) et :

• f est définie sur $[x_0', x_2]$ (avec x_0' proche de x_0, $x_0 < x_0' < x_1$ et x_2 éventuellement égal à $+\infty$), de classe C^2 sur $]x_0', x_2[$ et telle que :

• $f(x_0') = f(x_2) = 0 \quad (x_0 \leq x_0' \leq x_1 < x_2 \leq \infty)$.

Références

[**GNRS**] V. Giorno, A.G. Nobile, L.M. Ricciardi, L. Sacerdote *Some remarks on the Rayleigh process*, J. Appl. Prob. 23 , 398-408 (1986)

[**KKR**] O. Kavian, G. Kerkyacharian, B. Roynette *Quelques remarques sur l'ultra-contractivité*, Journal of Functional Analysis, 111 (1993)

[**RY**] D. Revuz, M. Yor *Continuous Martingales and Brownian Motion*, Springer Verlag

CLOSED SETS SUPPORTING A CONTINUOUS DIVERGENT MARTINGALE

by M. Émery[1]

Let $(X_t)_{t\geqslant 0}$ be a continuous martingale with values in a finite-dimensional affine space E. We shall call X *divergent* if almost surely $\lim_{t\to\infty} X_t$ does not exist in E. For which subsets F of E does there exist in E a divergent, continuous martingale with values in F? Unable to answer this question in general, we shall restrict ourselves to the case when F is closed; this note is devoted to giving a non-probabilistic characterization of the closed subsets of E that contain a divergent, continuous martingale.

When $\dim E = 1$, no strict subset of E can contain a divergent, continuous martingale, but E itself does; this case is trivial and the problem is interesting for $\dim E \geqslant 2$ only (if at all!).

As we are interested in continuous martingales only, the adjective 'continuous' will be omitted and all martingales will be implicitely assumed continuous. By time-change, considering continuous local martingales instead of martingales would make no difference.

Our statements will involve only the affine structure on E (and the associated topology); but in some proofs, E will be endowed with an additional Euclidean structure: the distance will be denoted by d, the open balls will be called $B(x,r)$, the closed ones $\overline{B}(x,r)$, the spheres $S(x,r)$, orthogonality will be used, etc.

1. Prominent points and humpless kernel of a closed set

If X is a topological space and if $A \subset B \subset X$, $\mathrm{i}_B\, A$ and $\partial_B A$ will respectively denote the interior and the boundary of A in the topological space B (endowed with the topology inherited from X, of course). One always has $(\mathring{A} =)\, \mathrm{i}_X\, A \subset \mathrm{i}_B\, A$, for $\mathrm{i}_X\, A$ is an open subset of X included in A, hence also an open subset of B included in A. The reverse inclusion may fail (for instance when $A = B$ and A is not open).

LEMMA 1. — *Let A, B and C be three subsets of a topological space.*

a) *If $A \subset B$, one has $\partial_A(A\cap C) \subset \partial_B(B\cap C)$.*

b) *If $A \subset B\cap C$ and if A is both open and closed in $B\cap C$, then $\partial_B A \subset \partial C$.*

PROOF. — a) Let $x \in \partial_A(A\cap C)$. If V is a neighbourhood of x in B, $V\cap A$ is a neighbourhood of x in A and must meet $A\cap C$ and $A\cap C^{\mathrm{c}}$; a fortiori, V itself meets $B\cap C$ and $B\cap C^{\mathrm{c}}$. As V is arbitrary, x is in $\partial_B(B\cap C)$.

1. This note originates from enjoyable conversations with Chris Burdzy.

b) Calling X the ambient topological space, a) yields

$$\partial_B(B \cap C) \subset \partial_X(X \cap C) = \partial C ;$$

so it suffices to verify that $\partial_B A \subset \partial_B(B \cap C)$. Setting $D = B \cap C$ and taking B as the new ambient topological space, it now suffices to verify that *if A is a closed and open subset of D, then $\partial A \subset \partial D$.*

Let $x \in \partial A$. We must show that any neighbourhood V of x meets both D and D^c. We already know that it meets A, hence also D; it remains to see that it meets D^c. Consider first the case when $x \in A$. Write $A = D \cap O$ with O open; $V \cap O$ is a neighbourhood of x, so it meets $A^c = D^c \cup O^c$, hence also D^c, and we are done. Now the other case: $x \notin A$. Write A as $D \cap F$ with F closed; from $A \subset F$ one gets $\partial A \subset F$ and $x \in F$; since x does not belong to $A = D \cap F$, it must be in D^c, and V meets D^c at point x. ∎

DEFINITIONS. — *Let F be a closed subset of the affine space E and x a point of F. One says that x is a* prominent point of F *if there exist an affine hyperplane H in E not containing x and a compact K included in F, containing x and with boundary $\partial_F K$ included in H.*

The set of all points of F that are not prominent points of F will be called the non-prominence *of F and abbreviated* np(F).

Very roughly, x is a prominent point of F if a plane blade can cut off a bounded part of F containing x.

We shall see below (Lemma 5) a seemingly stronger but equivalent definition of the prominent points of a closed set: generality is not restricted by demanding, in the above definition, that the compact K be also open in the closed set $F \cap D$, where D is the closed half-space with boundary H and containing x.

The figure below shows a closed set F_1 in the plane (in gray; it consists of a half-plane, minus a square, plus a disk and a triangle) and its non-prominence np(F_1). The prominent points of F_1 are the points of the triangle minus its base and the points of the disk minus its horizontal diameter. This can be checked directly from the definitions, or, more easily, by using Proposition 4 a) below. Notice on this exemple that the requirement $H \not\ni x$ in the definition of prominent points cannot be weakened to $x \notin \partial_F K$: the points of the horizontal diameter of the disk are not prominent, but would become so after this modification (take $K =$ the disk, so that $\partial_{F_1} K$ consists of two points).

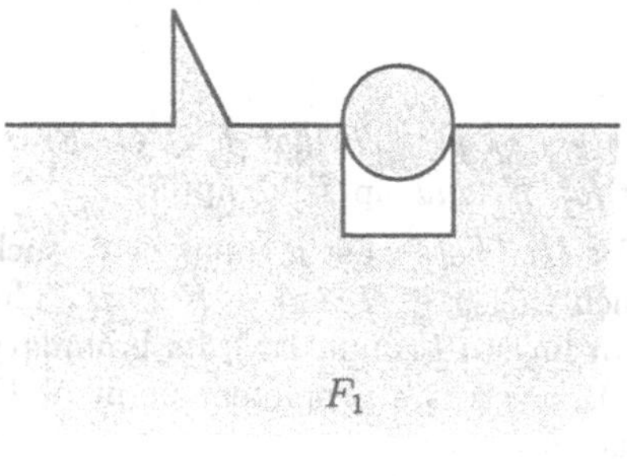

F_1

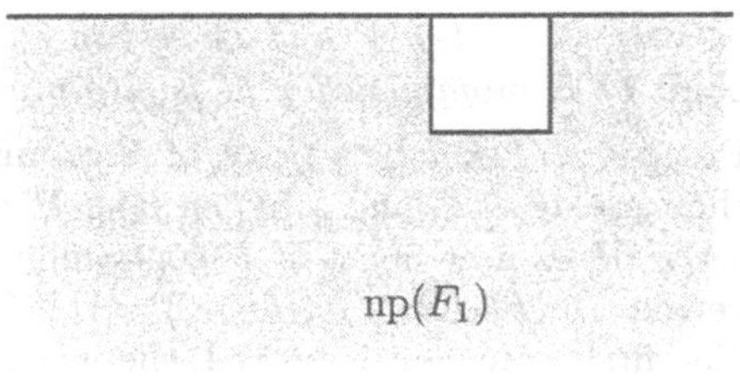

np(F_1)

If F is compact, all its points are prominent and $\mathrm{np}(F)$ is empty (take $K = F$, so the boundary $\partial_F K$ is empty, and call H any hyperplane not meeting F). More generally, for the same reason, every compact open in F (for instance, every connected component of F that is isolated in F and bounded) consists of prominent points of F.

REMARK. — The definition of prominent points seems to involve the reference space E, via the constraint that H must be a hyperplane. But actually it does not: if E' is an affine sub-space of E and F a closed subset of E', the prominent points of F are the same, whether defined with respect to E or E'. Indeed, every hyperplane H of E' is the trace on E' of some hyperplane H of E; conversely, if a hyperplane H of E does not contain a given point of E' (here, the prominent point), $H \cap E'$ is either empty or a hyperplane of E'.

LEMMA 2. — *Let F be a closed subset of E. The set* $\mathrm{np}(F)$ *is closed too.*

PROOF. — Let x be a prominent point of F; there exist a hyperplane H and a compact K such that $x \notin H$, $x \in K \subset F$ and $\partial_F K \subset H$. One has $x \notin \partial_F K$, whence $x \in \mathrm{i}_F K$. All points of F close enough to x are also in the open subset $H^c \cap \mathrm{i}_F K$ of F, hence they are also prominent. So the set of all prominent points of F is open in F, and $\mathrm{np}(F)$ is closed in F, hence also closed in E. ∎

F_1 in the preceding picture is such that $\mathrm{np}(F_1)$ has no prominent point. This is not a general rule: the figure below shows a closed set F_2 made of all the edges of an infinite hexagonal lattice except one (call this one e); $\mathrm{np}(F_2)$ consists of all the edges except the neighbours of e, and the set of all prominent points of $\mathrm{np}(F_2)$ is the union of all the edges that are second-order neighbours of e.

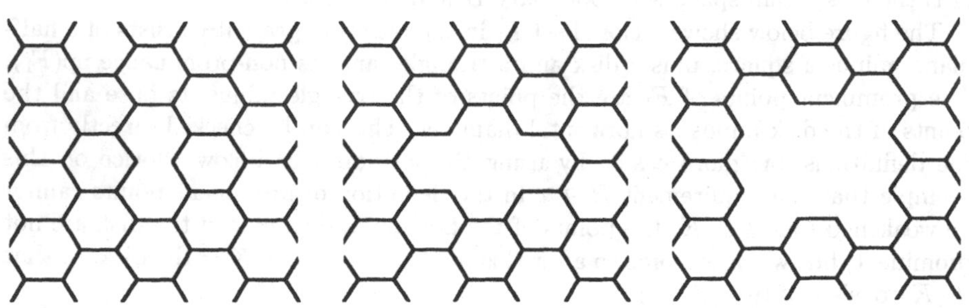

F_2 $\mathrm{np}(F_2)$ $\mathrm{np}(\mathrm{np}(F_2))$

LEMMA 3. — *Let F and G be two closed subsets of E such that $F \subset G$. Every point of F prominent for G is also prominent for F, and* $\mathrm{np}(F) \subset \mathrm{np}(G)$.

PROOF. — Let x be a point of F prominent for G. There exist a compact K such that $x \in K \subset G$ and a hyperplane H of E such that $x \notin H$ and $\partial_G K \subset H$. The set $F \cap K$ is a compact of F containing x; according to Lemma 1 a), its boundary verifies $\partial_F(F \cap K) \subset \partial_G(G \cap K) = \partial_G K \subset H$. Hence x is a prominent point of F. The first statement is proved, the inclusion follows. ∎

DEFINITION. — *A closed subset of* E *is* humpless *if it has no prominent point.*

LEMME 4. — *Let* $\mathcal{F}$ *be a set of humpless closed subsets of* E *and* $G = \bigcup_{F \in \mathcal{F}} F$ *the union of this set. The closure* $\overline{G}$ *of* G *is humpless.*

PROOF. — The prominent points of $\overline{G}$ form an open subset of $\overline{G}$ (Lemma 2). If this open set were not empty, it would meet G, for G is dense in $\overline{G}$; hence there would exist some $F \in \mathcal{F}$ containing some prominent point x of $\overline{G}$; x would also be a prominent point of F (Lemma 3); this would contradict the humplessness of F. So $\overline{G}$ has no prominent point. ∎

PROPOSITION 1 and DEFINITION. — *Let* F *be a closed subset of* E. *All humpless closed subsets of* F *are included in one of them.*

This biggest humpless closed subset of F *will be called the* humpless kernel *of* F *and denoted by* $\check{F}$.

PROOF. — Apply Lemma 4 to the closure of the union of all humpless closed subsets of F. ∎

Proposition 1 can also be proved by transfinite induction, using only the fact that the mapping np from the set of all closed subsets of E to itself is a *derivation* (that is, it is increasing, and it verifies $\mathrm{np}(F) \subset F$ for all F). This makes it easy to construct, for every ordinal α, the transfinite iterate $\mathrm{np}^{\alpha}(F)$; the so-obtained transfinite sequence is decreasing, hence stationary, and it is not difficult to see that its limit is the biggest fixed point of np included in F.

Such a transfinite induction is not necessary here, and ordinary induction will suffice: Proposition 2 will show that the limit is reached at or before the first infinite ordinal.

REMARK. — One has always $\check{F} \subset \mathrm{np}(F)$. In other words, no prominent point of F can belong to the humpless kernel $\check{F}$; indeed, according to Lemma 3, such a point should be prominent for $\check{F}$ too, but this is impossible since $\check{F}$ is humpless. Hence the humpless kernel is made of non prominent points only.

But the reverse inclusion is false: F may have points that are neither prominent, nor in the humpless kernel $\check{F}$. Consider for instance F_2 drawn on the preceding page; the iterate $\mathrm{np}^n(F_2)$ is obtained by deleting from F_2 all the edges that are neighbours of order $\leqslant n$ of the edge e; this can be checked by induction using Proposition 4 b). Consequently, the humpless kernel $\check{F}_2$, included in each iterate $\mathrm{np}^n(F_2)$, is empty, though $\mathrm{np}(F_2)$ is not.

LEMMA 5. — *Let* x *be a prominent point of a closed set* F. *There exist a closed half-space* D *such that* $x \in \mathring{D}$, *and a compact open subset* K *of* $F \cap D$, *such that* $x \in K$ *and* $\partial_F K \subset \partial D$.

PROOF. — By hypothesis there exist a hyperplane $H \not\ni x$ and a compact L such that $x \in L \subset F$ and $\partial_F L \subset H$. Let H' be a hyperplane parallel to H and separating x and H; call D the closed half-space with boundary H' and containing x. $K = L \cap D$ is a compact containing x. Lemma 1 a) gives $\partial_{F \cap D} K = \partial_{F \cap D}\big((F \cap D) \cap L\big) \subset \partial_F (F \cap L) = \partial_F L \subset H$; but $\partial_{F \cap D} K$ is a subset of $F \cap D$, hence also of D, which does not meet H. Consequently, $\partial_{F \cap D} K = \varnothing$, and the compact K of $F \cap D$ is also open in $F \cap D$. Last, inclusion $\partial_F K \subset \partial D$ follows immediately from Lemma 1 b). ∎

LEMMA 6. — *Let $(F_n)_{n\in\mathbb{N}}$ be a decreasing sequence of closed subsets of E; call F_∞ the limit of this sequence.*

a) *If K_∞ is a compact open subset of F_∞, then for every n large enough there exists a compact open subset K_n of F_n such that $K_n \supset K_\infty$.*

b) *Each prominent point of F_∞ is a prominent point of all F_n's but finitely many.*

c) *The closed sets $\mathrm{np}(F_n)$ form a decreasing sequence with limit $\mathrm{np}(F_\infty)$.*

PROOF. — a) There exists an open subset U of E such that $K_\infty = F_\infty \cap U$; there exists a compact L such that $K_\infty \subset \mathring{L} \subset L \subset U$. The compacts $F_n \cap \partial L$ are decreasing with limit $F_\infty \cap \partial L = F_\infty \cap (U \cap \partial L) = (F_\infty \cap U) \cap \partial L = K_\infty \cap \partial L = \varnothing$; so, for n large enough, $F_n \cap \partial L$ is empty and $F_n \cap L = F_n \cap \mathring{L}$ is a compact and open subset of F_n containing K_∞.

b) Let x be a prominent point of F_∞. Lemma 5 gives a closed half-space D such that $x \in \mathring{D}$ and a compact K open in $F_\infty \cap D$ such that $x \in K$. Apply a) to the decreasing closed sets $F_n \cap D$ with limit $F_\infty \cap D$: for n large enough, there is a compact K_n open in $F_n \cap D$ and containing K, hence also x; Lemma 1 b) yields the inclusion $\partial_{F_n} K_n \subset \partial D$, showing that x is a prominent point of these F_n's.

c) Inclusions $\mathrm{np}(F_{n+1}) \subset \mathrm{np}(F_n)$ and $\mathrm{np}(F_\infty) \subset \bigcap_n \mathrm{np}(F_n)$ are straightforward from Lemma 3. Conversely, a point belonging to all the $\mathrm{np}(F_n)$'s is in each F_n hence in F_∞; but it cannot be prominent in F_∞ because of b); so it belongs to $\mathrm{np}(F_\infty)$. ∎

PROPOSITION 2. — *Let F be closed in E. The decreasing sequence of iterates $\mathrm{np}^n(F)$ of F converges to the humpless kernel $\check{F}$ of F.*

PROOF. — Decreasingness comes from Lemma 3; call F_∞ the limit $\bigcap_n \mathrm{np}^n(F)$. Applied to $F_n = \mathrm{np}^n(F)$, Lemma 6 c) entails that F_{n+1} tends to $\mathrm{np}(F_\infty)$, whence $\mathrm{np}(F_\infty) = F_\infty$ and F_∞ is a humpless closed set contained in F. If G is any humpless closed set included in F, Lemma 3 implies $\mathrm{np}^n(G) \subset \mathrm{np}^n(F)$, that is, $G \subset \mathrm{np}^n(F)$ since G is a fixed point of np. As a result, $G \subset F_\infty$; this shows that F_∞ is the biggest humpless closed set contained in F, that is, $F_\infty = \check{F}$. ∎

2. The case when E is a plane

The case when $\dim E = 1$ is trivial: the only humpless closed sets are E and $\varnothing$ and one has $\mathrm{np}(F) = \check{F} = \varnothing$ if $F \neq E$ and $\mathrm{np}(E) = \check{E} = E$. The simplest non-trivial examples occur when $\dim E = 2$; this section is devoted to describing the non-prominence and the humpless kernel of a planar closed set. But some statements extend to higher dimensions as well; so when we assume E is a plane, we shall mention it explicitely.

Summarized in Proposition 4, the results are quite intuitive; scribbling a few sketchy pictures will convince you much more pleasantly than reading the pedestrian but tedious proofs given below. Experts have shown me how some homological considerations could have saved paper and ink; but I prefer walking the way rather than taking readers in a jet I can hardly pilot. In any case, this section will not be used in the sequel.

In two dimensions, the mapping $F \mapsto \text{np}(F)$ becomes easier to describe when passing to complementaries and dealing with open sets instead of closed ones. If O is open in E, we shall call $\tilde{\text{np}}(O)$ the complementary of the closed set $\text{np}(O^c)$. Lemmas 2 and 3 and Propositions 1 and 2 say that $\tilde{\text{np}}(O)$ is open and contains O, that $\tilde{\text{np}}$ is an increasing mapping from the set of open subsets of E to itself, and that the sequence of iterates $\tilde{\text{np}}^n(O)$ is increasing, with limit the complementary of the humpless kernel of the closed set O^c.

LEMMA 7. — *Let D be a closed half-plane, K a compact in D, V a neighbourhood of K in D and x a point in K. There exists in $V \setminus K$ a continuous curve with endpoints y and z such that x belongs to the segment $[y, z]$.*

PROOF. — Call D' the closed half-plane included in D and whose boundary $\partial D'$ contains x; by replacing D by D', K by $K \cap D'$ and V by $V \cap D'$, we may suppose that x belongs to the boundary $\Delta = \partial D$ of the half-plane D. Without loss of generality, we shall also suppose V open in D and $V \neq D$.

Let $a > 0$ be the distance from the compact K to the closed set $D \setminus V$. Cover K with a finite family $\big(B(c_i, \frac{1}{2}a), i \in I\big)$ of open disks with the same radius $\frac{1}{2}a$ and with centres c_i in K. Choose a number $b \in (\frac{1}{2}a, a)$ meeting the following requirements: any circle $C_i = \partial B(c_i, b)$ with centre c_i and radius b is not tangent to Δ, any two of these circles are not tangent, any three of them have empty intersection, and no intersection point of two of them is on Δ. (This is possible because only finitely many values of b are forbidden, namely the distances $d(c_i, \Delta)$, the half-distances $\frac{1}{2}\, d(c_i, c_j)$, the outradii of the triangles $c_i c_j c_k$, and the distances from c_i to the intersections of Δ with the perpendicular bissectors of the segments $c_i c_j$.) Call F_i the closed disk $\overline{B}(c_i, b)$; its boundary is C_i. The compact $L = \bigcup_i F_i$ verifies $K \subset \bigcup_i B(c_i, b) \subset \mathring{L}$ and $L \cap D \subset V$, hence also $\partial L \cap D \subset V \setminus K$. To prove the lemma, we shall construct in $\partial L \cap D$ a continuous curve whose endpoints are on Δ and encompass x.

Orient the plane, thus defining a counter-clockwise direction on each circle. Orient also the line Δ, in such a way that if a circle C meets Δ at two points y and z, and if y is before z on Δ, z is before y on the arc $C \cap D$ with endpoints y and z.

The intersection $\Delta \cap L = \bigcup_i (\Delta \cap F_i)$ is a finite union of segments with strictly positive lengths, hence also a finite union of disjoint segments with strictly positive lengths; call $s_\alpha = [y_\alpha, z_\alpha]$ these disjoint segments, where α ranges over a finite set A and where y_α is before z_α on the oriented line Δ. Let Y be the set $\{y_\alpha, \alpha \in A\}$ of all left-endpoints of these segments and Z the set $\{z_\alpha, \alpha \in A\}$ of all right-endpoints of these segments. Point x belongs to one of the segments s_α; on Δ, there are before x more points of Y than of Z (exactly one more). We shall construct a one-to-one correspondence between Y and Z, such that any two corresponding points can always be linked by a continuous curve lying in $\partial L \cap D$. At least one point of Y before x will be linked to some point of Z after x, thus proving the lemma.

Remark first that each point of ∂L is in L, hence in one of the closed disks F_i; and it belongs to the closure of the exterior of L and a fortiori to the closure of the exterior of the disk F_i; so it must be on the boundary C_i and this gives $\partial L \subset \bigcup_i C_i$.

Start at a point $z \in Z$. It belongs to ∂L, hence to some C_i, unique owing to the conditions on b. Follow this C_i counter-clockwise until meeting the line Δ or another circle C_j. If Δ is met first, stop; if some C_j (unique owing to b) is met first, leave C_i and follow C_j counter-clockwise until meeting Δ or some C_k ($k \neq j$, but k can be equal to i). If on Δ, stop, else switch to the new circle. Keep doing this as long as possible, that is, indefinitely or until meeting Δ.

To conclude, it suffices to show that, *starting from* $z \in Z$, *the line* Δ *is reached after finitely many steps, at some point* $y \in Y$; that *the path followed from* z *to* y *lies in* $\partial L \cap D$; and that *the so-defined mapping from* Z *to* Y *is one-to-one and onto.*

Consider the reverse algorithm, analogously defined, but with 'clockwise' instead of 'counter-clockwise'. Applied after starting with the direct one, the reverse algorithm follows backwards the same path; this shows that two paths obtained with the direct algorithm cannot merge (that is, coincide after some step without having coincided in the past). In particular, starting from $z \in Z$, it is impossible to pass twice the same point, for the past of the second time should be the same as the past of the first time, and the starting point z should have been met between both times; but meeting Δ terminates the algorithm. As the total number of arcs at our disposal is finite, Δ must be met after finitely many steps, and the algorithm eventually stops.

The starting point z belongs to ∂L; it is on C_i but not in any of the closed disks F_j with $j \neq i$ (it is not on the boundary of those disks because of the condition on b; nor in their interior since it is on ∂L). The first step of the path, until another circle is met, remains in the exterior of all the F_j's, hence on the boundary ∂L; it is also in the half-plane D because of the choice of the orientations. It can be checked inductively that, at each step of the algorithm, the arc of some circle C_k used lies in the exterior of all the other circles C_m, $m \neq k$: indeed, when passing from some circle C_k to another circle C_ℓ, since C_ℓ is reached from the exterior and both motions are counter-clockwise, the path will leave C_k outwards. So the algorithm never leaves ∂L. Similarly, when eventually reaching Δ, the path is counter-clockwise following some C_p while remaining in the exterior of all the other circles. Hence, the point where Δ is met is a point of $\partial L \cap \Delta$ having a right-neighbourhood $F_p \cap \Delta$ included in L; so this meeting point must belong to Y.

This defines a mapping from Z to Y. To see that it is one-to-one and onto, it suffices to exhibit its inverse. The latter is obtained by applying the reverse algorithm starting from the points of Y: by the same argument, one eventually reaches Δ after following the same path backwards. ∎

LEMMA 8. — *Let* O *be a connected open subset of* E. *Any two points of* O *can be linked by a simple curve in* O (that is, with no multiple points).

PROOF. — It suffices to verify that if x is any point of O, the set of endpoints of all simple curves in O started at x is both open and closed in O. Since O is locally convex, it suffices to show that if x, y and z are three points, if c is a simple curve from x to y and if s is the segment $[y, z]$, there exists a simple curve from x to z included in $c \cup s$. Calling t the point of $s \cap c$ closest to z, one gets the required curve by chaining together the (unique) part of c linking x to t and the segment (possibly a singleton) $[t, z]$. ∎

DEFINITION. — *If* A *is a subset of* E, *the union of all segments* $[u, v]$, *where* u *and* v *range over* A, *will be called the* segment-span *of* A.

Clearly, the segment-span of A is included in any convex set containing A, in particular in the convex hull of A. The converse is false in general, but holds in dimension 1 (immediate) and, when A has at most 2 connected components, also in dimension 2. This is a theorem of Fenchel; for references and generalizations, see O. Hanner & H. Rådström [4]. We shall need only the particular case when A is connected:

LEMMA 9. — *Let A be a connected subset of a plane. The convex hull of A is equal to the segment-span of A.*

PROOF. — It suffices to see that the segment-span S of A contains the convex hull of A. Taking $x \in S^c$, we have to show that x is not in the convex hull of A.

Remark first that $S \supset A$ (a point is a segment), so x does not belong to A. Let E be the plane, Γ a circle with centre x and f the mapping from $E \backslash \{x\}$ to Γ such that, for every $y \neq x$, the points y and $f(y)$ are on the same ray emanating from x. Since f is continuous and A connected, the range $f(A)$ is a connected subset of Γ, hence an arc $a \subset \Gamma$. Hypothesis $x \notin S$ entails that this arc never contains both endpoints of a diameter of Γ; so it is either an arc with measure less than π, or a non-closed arc with measure π. In either case, the set $f^{-1}(a) \subset E \backslash \{x\}$ is a convex part of E, containing A, but not x. This prevents the convex hull of A from containing x. ∎

PROPOSITION 3. — *Suppose $\dim E \geqslant 2$; let O be open and F be closed in E.*

a) *Every prominent point of F is in the segment-span of some connected component of the open set F^c.*

b) *The open set $\check{\text{np}}(O)$ is included in the union of the segment-spans of the connected components of O.*

c) *If each connected component of F^c is convex, F is humpless.*

This proposition holds a fortiori if 'segment-span' is replaced by 'convex hull'.

When $\dim E \geqslant 3$, the converse statements to a), b) and c) are false: consider the case when F is a line in E (it is a humpless closed set) and O the complementary of a line (this connected open set segment-spans the whole space).

PROOF OF PROPOSITION 3. — a) Given a prominent point x of F, we have to find in the same connected component of F^c two points y and z such that the segment $[y, z]$ contains x.

By Lemma 5, there exist a closed half-space D' and a compact and open subset K' of $F \cap D'$ such that $x \in K'$ and $\partial_F K' \subset \partial D'$. There exists a V' open in D' such that $K' = F \cap V'$. Let Δ be a line containing x and parallel to the hyperplane $\partial D'$ (this is where the hypothesis $\dim E \geqslant 2$ comes in) and let P be the 2-plane perpendicular to $\partial D'$ and containing Δ. Call D the half-plane $D' \cap P$, K the compact $K' \cap D$ and V the set $V' \cap D$. Since $x \in K$ and V is open in D and contains K, Lemma 7 applies and gives two points y and z of $V \backslash K$ linked by a continuous curve in $V \backslash K$ and such that the segment $[y, z]$ contains x. But $V \backslash K$ is included in $V' \backslash K'$ and hence in F^c; the points y and z, linked by a continuous curve in F^c, are in the same connected component of F^c, and we are done.

b) Observe that the points of $\check{\text{np}}(O)$ are the points of O and the prominent points of O^c and apply a) to $F = O^c$.

c) If the connected components of F^c are convex, each of these components is its own segment-span, and F is humpless according to a). ∎

In two dimensions, the converse statements to a), b) and c) hold true, and prominency and humplessness can be more expressively rephrased.

In the next statement, 'convex hull' can as well be replaced with 'segment-span' since, according to Lemma 9, they are equivalent for planar connected sets.

PROPOSITION 4. — *Suppose* $\dim E = 2$; *let* O *be open and* F *be closed in* E.

a) *A point* $x \in F$ *is prominent if and only if it belongs to the convex hull of some connected component of the open set* F^c.

b) *The open set* $\tilde{\text{np}}(O)$ *is the union of the convex hulls of the connected components of* O.

c) *The closed set* F *is humpless if and only if each connected component of* F^c *is convex.*

PROOF. — a) The necessary condition has been seen in Proposition 3 a); now for the converse. Supposing x in F and in the convex hull of a connected component C of F^c, we shall show it is prominent.

Lemma 9 yields two points y and z of C such that $x \in [y,z]$; Lemma 8 gives the existence of a simple curve c linking y and z in C. Let y' (respectively z') the point of $c \cap [x,y]$ (respectively $c \cap [x,z]$) closest to x and c' the piece of c linking y' and z'. The union $\gamma = c' \cup [y',z']$ is a simple closed curve; by Jordan's theorem, γ^c has two connected components, one of which (call it J) is bounded and verifies $\partial J = \gamma$. Since x is in F, it is not in C and a fortiori not on c'. In a neighbourhood of x, γ is straight, and the oriented normal line to γ at x can be defined: let $\vec{v}$ be a normal vector to $[y',z']$, going from J towards J^c, and of length $r > 0$ small enough for the closed disks $\overline{B}(y',r)$ and $\overline{B}(z',r)$ to be included in C. The translation by $\vec{v}$ transforms y' and z' into y'' and z''; let R be the compact convex rectangle with vertices y', z', z'' and y'', L the compact $\bar{J} \cup R$ and K the compact $L \cap F$.

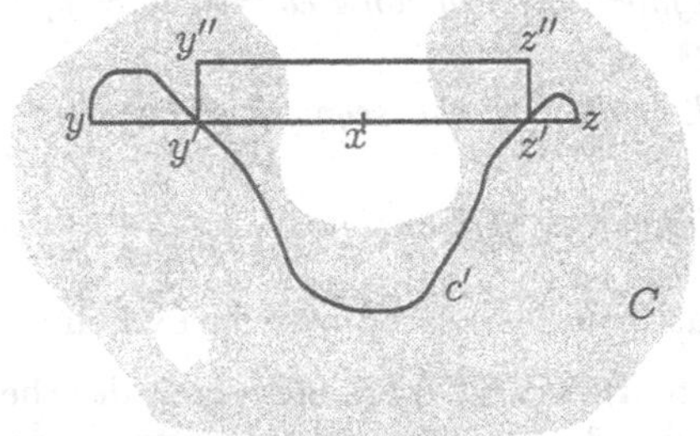

As $x \in K$, to establish that x is prominent, it suffices to show that the boundary $\partial_F K$ is included in the segment $[y'',z'']$. Lemma 1 a) gives

$$\partial_F K = \partial_F(L \cap F) \subset \partial_E(L \cap E) = \partial L = \partial(R \cup \bar{J})$$
$$\subset \partial R \cup \partial \bar{J} = \partial R \cup \gamma = \partial R \cup c' \cup [y',z'] = \partial R \cup c' \ ;$$

but $\partial_F K$ is also included in F, wherefrom

$$\partial_F K \subset (\partial R \cap F) \cup (c' \cap F) = \partial R \cap F \ .$$

Remarking that ∂R consists of four segments, two of which, $[y',y'']$ and $[z',z'']$, are in C, transforms the above inclusion into

$$\partial_F K \subset (y'',z'') \cup ((y',z') \cap F) \ .$$

Now, for every point t of the open interval (y',z'), the segment $[y',z']$ splits a small disk centred at t into two half-disks, one included in $\bar{J}$ and the other one in R; hence L is a neighbourhood of t. Consequently, K is a neighbourhood in F of each point of $(y',z') \cap F$, and those points cannot be on $\partial_K F$; finally $\partial_F K \subset (y'',z'')$ and x is prominent.

b) Apply a) to $F = O^c$, and notice that the points of $\widetilde{\mathrm{np}}(O)$ are the points of O and the prominent points of F.

c) If the connected components of F^c are convex, F is humpless by Proposition 3 c).

If F is humpless, let C be a connected component of F^c. By a), no point of F can belong to the convex hull $\hat{C}$ of C, so $\hat{C}$ is included in F^c. Hence $\hat{C}$ is a connected part of F^c containing C, so $\hat{C} = C$, and C is convex. ∎

3. Martingales, at last!

Prominence and humplessness will be used to describe the closed sets of E that contain a divergent martingale (that is, almost surely not convergent in E when t tends to infinity; recall that we consider only continuous martingales).

The Euclidean structure (balls, distance, etc.) already used several times is also able to measure the length of a curve and its analogue for a martingale: The *Euclidean quadratic variation* of a martingale X in E is the increasing process

$$\langle X, X\rangle_t = \lim_n \textstyle\sum_k d\big(X_{t\wedge k2^{-n}}, X_{t\wedge(k+1)2^{-n}}\big)^2$$

where the limit is in probability; equivalently, it is also the sum $\sum_i \langle X^i, X^i\rangle_t$, where the real martingales X^i are the coordinates of X in an orthonormal affine frame. Recall the equivalence, valid for almost all ω,

$$\lim_{t\to\infty} X_t(\omega) \text{ exists in } E \qquad \Longleftrightarrow \qquad \langle X, X\rangle_\infty(\omega) < \infty\,.$$

LEMMA 10. — *Let x be a prominent point of a closed subset F of E. There exist a number $\alpha > 0$ and a set U open in F such that $x \in U \subset F\setminus \mathrm{np}(F)$ and that, for every F-valued martingale X verifying $X_0 \in U$,*

$$\mathsf{P}\big[\lim_{t\to\infty} X_t \text{ exists in } E\big] \geqslant \alpha\,.$$

PROOF. — There exist a hyperplane H and a compact K such that $x \in K \subset F$, $\partial_F K \subset H$ and $x \notin H$. Let ℓ denote the affine function on E vanishing on H and such that $\ell(x) = 2$; the number $a = \sup_K \ell$ verifies $2 \leqslant a < \infty$. As x is not in $\partial_F K$, it lies in $\mathrm{i}_F K$, and the set $U = \{\ell > 1\} \cap \mathrm{i}_F K$ is open in F and verifies $x \in U \subset K$ and $\ell > 1$ on U. The properties of H and K imply that each point of U is a prominent point of F. If X is an F-valued martingale such that $X_0 \in U$, call T the stopping time $\inf\{t : X_t \in \partial_F K\}$. On $\{T < \infty\}$, $X_T \in \partial_F K \subset H$ and $\ell(X_T) = 0$; on the interval $[\![0, T]\!]$, X is in K for, in F, no continuous curve starting in U can leave K without meeting the boundary $\partial_F K$. Hence, the stopped process $X^{|T}$ is a bounded martingale, and the real process $M = \ell(X^{|T})$ is a bounded real martingale verifying $M_0 > 1$, $M \leqslant a$ and $M_\infty = 0$ on $\{T < \infty\}$. Consequently,

$$a\,\mathsf{P}[T = \infty] \geqslant \mathsf{E}[M_\infty \mathbb{1}_{\{T=\infty\}}] = \mathsf{E}[M_\infty] = \mathsf{E}[M_0] > 1\,,$$

wherefrom $\mathsf{P}[T = \infty] > 1/a$. But on the event $\{T = \infty\}$ the paths of X are in K, hence bounded, hence convergent; so, $\mathsf{P}[X \text{ converges}] > \alpha = 1/a$. ∎

LEMMA 11. — *Let F be closed in E and X be an F-valued divergent martingale. The subset $\{X \notin \mathrm{np}(F)\}$ of $\mathbf{R}_+ \times \Omega$ is evanescent.*

In other words, an F-valued divergent martingale lives in fact in the smaller closed set $\mathrm{np}(F)$.

PROOF. — As the set $O = F \backslash \mathrm{np}(F)$ is open in F, it is a countable union of compacts, and every open covering of O in the topological space F contains a countable subcovering. Now Lemma 10, applied to each point of O, gives a covering of O by open sets $U_x \subset O$, each of them associated with a number $\alpha_x > 0$. Hence there exist a sequence $(V_n)_{n\in\mathbf{N}}$ of open sets of F and a sequence $(\beta_n)_{n\in\mathbf{N}}$ in $(0,\infty)$ such that $\bigcup_n V_n = O$ and that, for all n and all F-valued martingale X verifying $X_0 \in V_n$, the minoration $\mathbf{P}[X \text{ converges}] \geqslant \beta_n$ holds.

Let X be a martingale in F such that the optional set $\{X \notin \mathrm{np}(F)\}$ is not evanescent. This set contains the graph of some stopping time (section theorem); so there are a stopping time T and an n such that the event $\Omega' = \{T < \infty\} \cap \{X_T \in V_n\}$ verifies $\mathbf{P}[\Omega'] > 0$. Applying the above minoration to the martingale $X'_t = X_{T+t}$ (defined on Ω' with the filtration $\mathcal{F}'_t = \mathcal{F}_{T+t}$ and the probability $\mathbf{P}'[A] = \mathbf{P}[A \mid \Omega']$) yields $\mathbf{P}'[X \text{ converges}] \geqslant \beta_n$, whence $\mathbf{P}[X \text{ converges}] \geqslant \beta_n \mathbf{P}[\Omega'] > 0$, and X is not divergent. ∎

PROPOSITION 5. — *Let F be closed in E. Every F-valued divergent martingale takes its values in the humpless kernel $\check{F}$ of F.*

PROOF. — Let X be such a martingale. By induction on n, Lemma 11 shows that each set $\{X \notin \mathrm{np}^n(F)\}$ is evanescent. So is also the union of these sets; now, according to Proposition 2, this union is nothing but $\{X \notin \check{F}\}$. ∎

PROPOSITION 6. — *Let F be a humpless closed set of E and x a point of F. There exists an F-valued divergent martingale X such that $X_0 = x$.*

PROOF. — *Step one.* Given any $a > 0$, we shall construct a Markov kernel N in F (endowed with its Borel σ-field) such that for every $y \in F$, the probability $\varepsilon_y N$ has mass centre y and is carried by the compact $\{z \in F : d(y,z) = a\} = F \cap S(y,a)$.

For $y \in F$, let L_y denote the compact $F \cap S(y,a)$; we shall first show that y is in the convex hull C_y of L_y. If it were false, y and L_y would be separated by a hyperplane H : there would exist a closed half-space D with boundary H such that $y \in \mathring{D}$ and $L_y \subset D^c$. The intersection $F \cap D \cap S(y,a)$ would be empty and the compact $K = F \cap D \cap \overline{B}(y,a)$ would also be equal to $F \cap D \cap B(y,a)$; it would be both closed and open in $F \cap D$. Lemma 1 b) would give $\partial_F K \subset \partial D = H$ and y would be prominent in F. As F is humpless, this is impossible.

Since $y \in C_y$, y is by Carathéodory's theorem the mass centre of a probability N_y carried by $r+1$ points of L_y, where $r = \dim E$. To conclude step one, it suffices to verify that N_y can be chosen mesurable in y. The set of all systems $(y; z_0, \ldots, z_r; \lambda_0, \ldots, \lambda_r)$ verifying $z_i \in L_y$, $0 \leqslant \lambda_i \leqslant 1$, $\sum_{i=0}^r \lambda_i = 1$ and $\sum_{i=0}^r \lambda_i z_i = y$ is closed in $F^{r+2} \times [0,1]^{r+1}$, with non-empty y-sections; so it has a Borel section $y \mapsto (z_0(y), \ldots, z_r(y); \lambda_0(y), \ldots, \lambda_r(y))$ (see for instance Dellacherie [3], page 350). Defining N_y as $\sum_{i=0}^r \lambda_i(y) \varepsilon_{z_i(y)}$ gives the claimed kernel.

Step two. Given $a > 0$, we shall construct a martingale X^a starting at x, with values in the closed set $F^a = \{y \in E : d(y,F) \leqslant a\}$ and with Euclidean quadratic variation $\langle X^a, X^a \rangle_t \equiv t$.

Recall that to each centred probability μ on a vector space V is associated a V-valued *Walsh martingale* (unique in law): denoting by R the absolute value of a real Brownian motion started at the origin, the Walsh martingale is obtained by independently multiplying each excursion of R by a random vector in V with law μ; see for instance [2] for more details. This process is a martingale because μ is centred.

If now y is a point in the affine space E and ν a probability on E centred at y, one can similarly define "the" E-valued Walsh martingale W started at y and associated to ν; if T is the first time when R hits 1, the random variable W_T has law ν.

Take $\nu = \varepsilon_x N$, where x is the given point and N the kernel constructed in step one; since $\varepsilon_x N$ is carried by the sphere $S(x,a)$, the so-obtained Walsh martingale W has Euclidean quadratic variation $\langle W,W\rangle_t = a^2 t$ and its distribution at time $T_1 = \inf\{t \geqslant 0 : W_t \in S(x,a)\}$ is $\varepsilon_x N$. Define a martingale Y^a equal to W on the interval $[\![0,T_1]\!]$; after T_1, start the same construction again independently with x replaced by $Y^a_{T_1}$ (it is in F by construction of N) and $\varepsilon_x N$ by $\varepsilon_{Y^a_{T_1}} N$; and stop at the first hitting time T_2 of $S(Y^a_{T_1},a)$ after T_1, etc. Since the differences $T_{i+1} - T_i$ are i.i.d. (they are distributed as the time needed by R to reach 1), T_i tend to infinity and this construction can be performed step by step, yielding a process Y^a_t well-defined for every $t \geqslant 0$. Moreover, this process is a martingale, with Euclidean quadratic variation $\langle Y^a,Y^a\rangle_t = a^2 t$: this holds on $[\![0,T_i]\!]$ by induction on i (this is where the measurability of N is used). Last, Y^a is in F at times T_i and in the ball $\overline{B}(Y^a_{T_i},a)$ during the interval $[\![T_i,T_{i+1}]\!]$. Consequently, its distance to F remains bounded by a and it lives in F^a. To get X^a as claimed, it suffices to time-change Y^a by a constant factor: $X^a_t = Y^a_{t/a^2}$ is a F^a-valued martingale starting at x and its Euclidean quadratic variation is $\langle X^a,X^a\rangle_t = \langle Y^a,Y^a\rangle_{t/a^2} = t$.

Step three. Construction of a F-valued martingale X started at x, with Euclidean quadratic variation $\langle X,X\rangle_t = t$.

Carrying the construction of the previous step for $a = 1/n$ yields a sequence $(Z^n)_{n\in\mathbb{N}}$ of continuous, E-valued martingales started at x and with the same Euclidean quadratic variation $\langle Z^n,Z^n\rangle_t = t$. Such a sequence has a subsequence convergent in law, whose limit is a martingale X in E verifying also $X_0 = x$ and $\langle X,X\rangle_t = t$ (see Rebolledo [5]). Furthermore, since Z^k is $F^{1/n}$-valued for $k \geqslant n$, so is also X, which lives in each $F^{1/n}$, hence in F. Last, X is divergent since $\langle X,X\rangle_\infty = \infty$ a. s. ∎

COROLLARY 1. — *Let F be closed in E and x be a point in F. There exists in F a divergent martingale starting from x if and only if x is in the humpless kernel $\check{F}$.*

PROOF. — If there exists a divergent martingale in F started at x, it lives in $\check{F}$ according to Proposition 5; consequently its starting point x is in $\check{F}$.

Conversely, if $x \in \check{F}$, Proposition 6 applied to the humpless closed set $\check{F}$ gives the existence of a divergent martingale, started at x, living in $\check{F}$ and a fortiori in F. ∎

COROLLARY 2. — *Let F be closed in E. The following three statements are equivalent :*

(i) *there exists an F-valued divergent martingale;*

(ii) *the humpless kernel $\check{F}$ is not empty;*

(iii) *F contains a non-empty humpless closed set.*

PROOF. — Implication (i) $\Rightarrow$ (ii) stems from Proposition 5, its converse (ii) $\Rightarrow$ (i) from Corollary 1, and equivalence (ii) $\Leftrightarrow$ (iii) from the definition of $\check{F}$. ∎

COROLLARY 3. — *Let F be closed in an affine plane E. There exists an F-valued divergent martingale if and only if there exists an open set U, whose connected components are convex, and such that $F^c \subset U \neq E$.*

PROOF. — This is a restatement of the equivalence (i) $\Leftrightarrow$ (iii) in Corollary 2 using Proposition 4 c). ∎

4. Remarks

a) (Remark by P. A. Meyer.) Prominence with respect to some closed F is far from being a local property: proving that x is prominent requires considering only the intersection of F with some ball centred at x, but this ball can be arbitrarily large, and proving that x is *not* prominent is impossible if you know only a bounded part of F. But humplessness is, in some sense, local. Say that F is *r-humpless* if each point x of F belongs to the convex hull of $F \cap S(x,r)$. *The following are equivalent :*

(i) *F is r_n-humpless for some sequence $(r_n)_{n\in\mathbb{N}}$ with $r_n > 0$ and $r_n \to 0$;*

(ii) *F is r-humpless for every $r > 0$;*

(iii) *F is humpless.*

Indeed, the proof of Proposition 6 first establishes that (iii) $\Rightarrow$ (ii), then uses only the (seemingly) weaker statement (i) to construct in F a martingale started at any given point. So (i) implies the existence of such martingales, and (iii) follows by Corollary 1.

b) Replace now E by a C^2-manifold, endowed with an affine connection. Given a closed set $F \subset E$, do there exist divergent martingales in F? No generalization of humplessness to that case seems to exist. But using some complete Riemannian metric on the manifold (not related to the connection; notice that any two such metrics are comparable on compacts), it is still possible to define r-humplessness and the construction in Proposition 6 carries over to this situation: F contains divergent martingales iff it contains non-empty, closed, r_n-humpless subsets for a sequence $r_n > 0$ tending to 0.

c) What happens if F is no longer supposed closed? Prominent points can still be defined: x is prominent if there are an affine hyperplane $H \subset E$ not containing x and a bounded, closed subset K of F containing x and whose boundary $\partial_F K$ is included in H. As in Lemma 10, it is easily seen that no divergent martingale contained in F can start from a prominent point. But I do not know if Lemma 6 generalizes: if one cuts off the prominent points of F, then the prominent points of the remaining set, and so on, does he eventually get a humpless residue? Or is it necessary to transfinitely iterate this cutting off?

These questions are probably uninteresting since, even if a definition of humpless kernels and Proposition 5 extend, one way or another, to a non-closed F, there is no reason to expect the converse, that is, the existence in any humpless set of a divergent martingale (Proposition 6). The proof given above, constructing the martingale in a slightly larger set and passing to the limit, is clearly doomed to failure for non-closed sets.

Another attempt would be to try to reduce the non-closed case to what we already know. For instance, if a set contains a divergent martingale, does there always exist a smaller closed set containing also a divergent martingale? The answer is no, even for a smooth open set; here is a counter-example.

Call A the open planar set $\{(x,y)\in\mathbf{R}^2 : 0<y<f(x)\}$, where $f:\mathbf{R}\to\mathbf{R}$ is convex, C^2, strictly positive, and has limit 0 when $x\to-\infty$ (for instance $f=\exp$). We shall show that A contains a divergent martingale, but no closed set included in A shares this property.

First, there exists in A a divergent martingale. Let X be a real Brownian motion started at 0 and I be the current infimum of X, given by $I_t=\inf_{0\leqslant s\leqslant t}X_s$. The process $Y=\frac{1}{2}\,f\circ I+\frac{1}{2}\,(X-I)\,f'\circ I$ is a martingale owing to the change of variable formula

$$dY=\tfrac{1}{2}\,f'\circ I\,dX+\tfrac{1}{2}(X-I)\,f''\circ I\,dI=\tfrac{1}{2}\,f'\circ I\,dX\ ,$$

where $(X-I)\,dI=0$ because I varies when $X=I$ only. (This formula extends to the case when f is not smooth: see Azéma & Yor [1], page 92.) As f is increasing, $Y\geqslant\frac{1}{2}\,f\circ I>0$; as $\frac{1}{2}f$ is convex, $Y\leqslant\frac{1}{2}\,f\circ X<f\circ X$. Thus the planar martingale (X,Y) lives in A. And it is divergent, for so is already its projection X on the x-axis.

And yet, every humpless closed set included in A is empty, so no closed set included in A can contain a divergent martingale. To see this, let F be a humpless closed set included in A. Choose any non-empty open ball centred on the x-axis and having no intersection with F. The union of this ball with the x-axis is connected and does not meet F; as F is humpless, Proposition 4 a) says that the convex hull of this union does not meet F either; so F has no point in some strip $\{|y|<\varepsilon\}$. Choose x_0 such that $f(x_0)<\varepsilon$; F cannot meet the line $\{x=x_0\}$, so it does not meet the union of this line with the x-axis, nor the convex hull of this union (same reason as above). As this convex hull is the whole plane, F is empty.

References

[1] J. Azéma & M. Yor. Une solution simple au problème de Skorokhod. *Séminaire de Probabilités XIII*, Lecture Notes in Mathematics 721, Springer 1979.

[2] M. Barlow, J. Pitman & M. Yor. On Walsh's Brownian Motions. *Séminaire de Probabilités XXIII*, Lecture Notes in Mathematics 1372, Springer 1989.

[3] C. Dellacherie. Ensembles analytiques : théorèmes de séparation et applications. *Séminaire de Probabilités IX*, Lecture Notes in Mathematics 465, Springer 1975.

[4] O. Hanner & H. Rådström. A Generalization of a Theorem of Fenchel. *Proc. Amer. Math. Soc. 2*, 1951.

[5] R. Rebolledo. La méthode des martingales appliquée à l'étude de la convergence en loi des processus. *Bull. Soc. math. France*, Mémoire 62, supplément au numéro d'octobre 1979.

Université Louis Pasteur et C.N.R.S.
I.R.M.A.
7 rue René Descartes
67 084 STRASBOURG Cedex

SOME POLAR SETS FOR THE BROWNIAN SHEET

BY
DAVAR KHOSHNEVISAN*
Department of Mathematics
The University of Utah
Salt Lake City. UT. 84112. U.S.A.
davar@math.utah.edu

§1. Introduction. Let $W \triangleq (W(s); s \in \mathbb{R}_+^N)$ denote d-dimensional N-parameter Brownian sheet. That is, W is a centered Gaussian process on $\mathbb{R}^d$ indexed by $\mathbb{R}_+^N$ such that

$$\mathbb{E}W_i(s)W_j(t) = \begin{cases} \prod_{k=1}^N (s_k \wedge t_k), & \text{if } i = j \\ 0, & \text{otherwise} \end{cases}$$

We will write V_i for the i-th coordinate of the k-dimensional vector V and the norm of $V \in \mathbb{R}^k$ is $\|V\| \triangleq \left(\sum_{j=1}^k V_j^2\right)^{1/2}$.

In this article, we are concerned with some interesting sets which are avoided by the path of W. In the language of Markov processes, such sets are said to be *polar*. Let us begin with a result of OREY AND PRUITT [OP] on when singletons are polar.

(1.1) Theorem. **([OP, Theorems 3.3, 3.4])** *For any $a \in \mathbb{R}^d$,*

$$\mathbb{P}\big(W(t) = a, \text{ for some } t \in \mathbb{R}_+^N\big) = \begin{cases} 1, & \text{if } d < 2N \\ 0, & \text{if } d \geq 2N \end{cases}.$$

(1.2) Remark. When the Brownian sheet is non-critical, i.e., $d \neq 2N$, we provide an elementary proof which can be easily extended to show the following: suppose $E \subset \mathbb{R}^d$ is compact and $\liminf_{h \to 0} h \ln(1/h) N_E^{d/2}(h) = 0$ where $N_E(h)$ is the upper (or lower) Kolmogorov entropy of E. Then $\mathbb{P}\big(W(t) \in E, \text{ for some } t \in \mathbb{R}_+^N\big) = 0$. See TAYLOR [T1] for definitions and properties.

The next result concerns k-multiple points. We say that W has k-multiple points, if there exists k distinct times $t^1, \cdots, t^k$, such that $W(t^1) = \cdots = W(t^k)$.

(1.3) Theorem. *The probability that W has k-multiple points is 1 or 0 according as whether $(d - 2N)k < d$ or $(d - 2N)k > d$.*

---

* Research partially supported by the National Science Foundation grant DMS–9503290

Clearly, the above leaves out the critical case, $(d-2N)k=d$. There does not seem to be an elementary way to resolve this problem when $(d-2N)k=d$. However, the problem can be solved. See the forthcoming paper of SALISBURY AND FITZSIMMONS [FS-2]

In Section 2, we prove Theorem (1.1) in the non–critical case, i.e., when $d \neq 2N$. Theorem (1.3) is proved in Section 3.

A historical account of these problems is in order. When $N=1$, W is d–dimensional Brownian motion and the above are amongst the results of DVORETSKY, ERDŐS AND KAKUTANI [DEK1,DEK2] and DVORETSKY, ERDŐS, KAKUTANI AND TAYLOR [DEKT]; see TAYLOR [T1] for a detailed account of this celebrated problem (as well as many other related developments). In this case, (i.e., when $N=1$), much more can be done due to the Markovian structure of the underlying process. For further advances in this area see, for example, BASS, BURDZY AND KHOSHNEVISAN [BBK], BASS AND KHOSHNEVISAN [BK], DYNKIN [D1,D2], FITZSIMMONS AND SALISBURY [FS-1], HAWKES AND PRUITT [HaP], HENDRICKS [He], LE GALL [LG], PERES [P], ROSEN [R1–R3], SALISBURY [S], SHIEH [Sh], TAYLOR [T1–T3], VARADHAN [V], WERNER [W] and YOR [Y], to cite a small sample. When $N>1$ and $k<4N$, the existence of 2–multiple points was dicovered simultaneously and independently by EHM [E] and ROSEN [R2]; see ADLER [A1] and DYNKIN [D1,D2] for improvements and other works. Similar methods to the ones mentioned above (i.e., local time techniques) can be used to show the existence of k–multiple points for any $k \geq 2$ satisfying $(d-2N)k \leq d$; cf. CHEN [C]. (In light of Theorem (1.1) above, the condition $d \geq 2N$ in [C] is superfluous for non-polarity.) For our purposes, the crux of the argument is the proof of the non–existence of k–multiple points. The need to solve this problem was brought to our attention by the review of FRISTEDT [F].

ACKNOWLEDGEMENT. I wish to express my gratitude to R.J. Adler, T. Salisbury, Z. Shi, S.J. Taylor and M. Yor.

§2. The Proof of Theorem (1.1) in the non–critical case. Without loss of much generality, let us only consider the case $a=0$. When $d<2N$, there exists a non–trivial measure which lives on $\{s \in \mathbb{R}_+^N : W(s)=0\}$; see ADLER [A1] and EHM [E]. Consequently, $\mathbb{P}(\,^{\exists}s \in \mathbb{R}_+^N : W(s)=0)=1$. For the sake of completion, we will give a simple Fourier analytic proof of this fact (when $N=1$, this method appears in KAHANE [K], Chapters 16 and 18). Fix a closed cube $I \subset (0,\infty)^N$ and consider the occupation measure, $\nu(A) \triangleq \int_I \mathbf{I}\{W(s) \in A\}ds$. The Fourier transform $\widehat{\nu}$ of ν is $\widehat{\nu}(\xi)=\int_I \exp\big(i\xi \cdot W(s)\big)ds$, where $\xi \in \mathbb{R}^d$ and $\cdot$ denotes the Euclidean dot product. Note that

$$\mathbb{E}|\widehat{\nu}(\xi)|^2 = \mathbb{E}\int_I\int_I \exp\big(i\xi\cdot(W(s)-W(t))\big)ds\,dt$$
$$= \int_I\int_I \exp\Big(-\frac{\|\xi\|^2}{2}\sigma^2(s,t)\Big)ds\,dt,$$

where $\sigma^2(s,t) \triangleq \prod_{j=1}^N s_j + \prod_{j=1}^N t_j - 2\prod_{j=1}^N (s_j \wedge t_j)$ for $s,t \in \mathbb{R}_+^N$. Define, $\sigma^2 \circ \pi(u,v) = \exp(\sum_j u_j) + \exp(\sum_j v_j) - 2\exp\sum_j (u_j \wedge v_j)$. Then by a change of variables.

$$\mathbb{E}|\widehat{\nu}(\xi)|^2 = \int_{\ln(I)}\int_{\ln(I)} \exp\big(-\|\xi\|^2\sigma^2\circ\pi(u,v)/2\big)\exp\sum_j (u_j+v_j)\,du\,dv.$$

For $u,v \in \ln(I)$, let $S = \{1 \le j \le N:\ u_j \le v_j\}$. Recalling that $I \subset (0,\infty)^N$ is a fixed closed cube, consider,

$$\sigma^2\circ\pi(u,v) = \exp\Big(\sum_{j\in S} u_j\Big)\Big[\exp\sum_{j\in S^c} u_j - \exp\sum_{j\in S^c} v_j\Big] +$$
$$+\exp\Big(\sum_{j\in S^c} v_j\Big)\Big[\exp\sum_{j\in S} v_j - \exp\sum_{j\in S} u_j\Big]$$
$$= e^{\sum u_j}\Big[1-\exp\sum_{j\in S^c}|u_j-v_j|\Big] + e^{\sum v_j}\Big[1-\exp\sum_{j\in S}|u_j-v_j|\Big]$$
$$\ge c_0\sum_{j=1}^N |u_j - v_j|,$$

where c_0 depends only on d,N and the size of I. Therefore, for some c_1 depending on d,N and the size of I,

$$\mathbb{E}|\widehat{\nu}(\xi)|^2 \le \int_{\ln(I)}\int_{\ln(I)} \exp\Big(-\frac{c_0\|\xi\|^2\sum_j|u_j-v_j|}{2}\Big)e^{\sum_j(u_j+v_j)}du\,dv$$
$$\le c_1\int_{\ln(I)\ominus\ln(I)} \exp\Big(-c_0\|\xi\|^2\sum_j |w_j|/2\Big)dw,$$

where $A\ominus B \triangleq \{x-y: x\in A, y\in B\}$. By scaling, it follows that for some c_2 (which depends only on d,N and the size of I),

$$\mathbb{E}|\widehat{\nu}(\xi)|^2 \le c_2(\|\xi\|^{-2N}+1).$$

Since $d<2N$, this implies that $\mathbb{E}\int_{\mathbb{R}^d}|\widehat{\nu}(\xi)|^2d\xi<\infty$. In particular, with probability one, $\widehat{\nu}\in L^2(\mathbb{R}^d,d\xi)$. By Parseval's identity, almost surely, $\nu(d\xi) << d\xi$ and the density is a.s. in $L^2(\mathbb{R}^d,d\xi)$. Writing the density as ℓ_I^x, it follows that $\nu(A) = \int_A \ell_I^x dx$. Note that $\mathbb{E}\ell_I^0 = \int_I\big(2\pi\prod_{j=1}^N s_j\big)^{-d/2}ds > 0$. Therefore, $\ell_I^0 > 0$ with positive probability. Since the "measure" $I \mapsto \ell_I^0$ is supported in $W^{-1}(\{0\})$, with positive probability, $I \cap W^{-1}(\{0\}) \neq \varnothing$. An application of Kolmogorov's 0–1 law shows that $W^{-1}(\{0\}) \neq \varnothing$. a.s. .

It remains to investigate the case $d > 2N$: our proof is motivated by the work of KAUFMAN [Ka].

By taking $\eta \to 0$. we see that it suffices to show that for any $\eta \in (0.1)$.

$$\text{(2.1)} \qquad \mathbb{P}(\ {}^{\exists}t \in [\eta, \eta^{-1}]^N\ :\ W(t) = 0) = 0.$$

For any $\varepsilon > 0$ cover $[\eta, \eta^{-1}]^N$ by closed non-overlapping boxes, $B_j(\varepsilon)$, $1 \le j \le n(\varepsilon)$, of side ε. It is easy to see that there exist suitable constants $K_i = K_i(\eta, N)$, $i = 1, 2$, such that

$$\text{(2.2)} \qquad K_1\varepsilon^{-N} \le n(\varepsilon) \le K_2\varepsilon^{-N}.$$

Define the random process N by

$$N(\varepsilon) \triangleq \sum_{j=1}^{n(\varepsilon)} \mathbf{I}\{\ {}^{\exists}s \in B_j(\varepsilon)\ :\ W(s) = 0\},$$

where $\mathbf{I}\{\cdots\}$ is 1 or 0 according to whether or not the event between the braces occurs. Recall the uniform modulus of continuity of W (cf. OREY AND PRUITT [OP] or the proof of ADLER [A2, p.8], for example):

$$\text{(2.3)} \qquad \limsup_{\varepsilon \to 0} \max_{1 \le j \le n(\varepsilon)} \sup_{s,t \in B_j(\varepsilon)} \frac{\|W(s) - W(t)\|}{\sqrt{\varepsilon \ln(1/\varepsilon)}} \le K_3,$$

where $K_3 = K_3(\eta, d, N) \in (0, \infty)$. It follows that for all ε small enough, $N(\varepsilon) \le M(\varepsilon)$, where M is defined by the following:

$$M(\varepsilon) \triangleq \sum_{j=1}^{n(\varepsilon)} \mathbf{I}\{\ {}^{\forall}s \in B_j(\varepsilon)\ :\ \|W(s)\| \le 2K_3\sqrt{\varepsilon \ln(1/\varepsilon)}\}.$$

To finish the proof of the theorem, it suffices to show that with probability one,

$$\liminf_{\varepsilon \to 0} M(\varepsilon) = 0.$$

We will achieve this by proving that

$$\text{(2.4)} \qquad \lim_{\varepsilon \to 0} \mathbb{E}M(\varepsilon) = 0.$$

Note that

$$\mathbf{I}\{\ {}^{\forall}s \in B_j(\varepsilon)\ :\ \|W(s)\| \le 2K_2\sqrt{\varepsilon \ln(1/\varepsilon)}\} \le \mathbf{I}\{\|W(b_j(\varepsilon))\| \le 2K_3\sqrt{\varepsilon \ln(1/\varepsilon)}\},$$

where $b_j(\varepsilon)$ is the center of $B_j(\varepsilon)$, say. Hence,

$$\mathbb{E}M(\varepsilon) \leq \sum_{1\leq j\leq n(\varepsilon)} \mathbb{P}(\|W(b_j(\varepsilon))\| \leq 2K_3\sqrt{\varepsilon \ln(1/\varepsilon)}).$$

For $s \in \mathbb{R}_+^N$ and $a \in \mathbb{R}^d$, let $\varphi_s(a)$ denote the Gaussian density of $W(s)$ at a. From the properties of Gaussian densities, there exist some $K_4 = K_4(\eta, N, d)$ so that

$$\sup_{a\in\mathbb{R}^d} \sup_{s\in[\eta,\eta^{-1}]^N} \varphi_s(a) \leq K_4.$$

Hence, using (2.2), we see that there exists some $K_5 = K_5(\eta, d, N)$ such that

$$\mathbb{E}M(\varepsilon) \leq K_5\varepsilon^{-N+(d/2)}\big(\ln(1/\varepsilon)\big)^{d/2}.$$

Since $d > 2N$, (2.4) and hence the result follow. □

§3. **The Proof of Theorem (1.3).** When $d < 2N$, Theorem (1.3) follows from Theorem (1.1). Suppose $d \geq 2N$. When $(d-2N)k < d$, the existence of k-multiple points follows immediately from CHEN [C]. Equivalently, one can show (as we did for Theorem 1.1) that uniformly in $\varepsilon > 0$, $\varepsilon^{d(1-k)}\widehat{\mu}_\varepsilon \in L^2(\mathbb{R}^d, d\xi)$, where $\mu_\varepsilon(A)$ is given by,

$$\int_{I_1}\cdots\int_{I_k} \mathrm{I}\{W(s^1)\in A\}\prod_{j=2}^{k}\mathrm{I}\{\|W(s^1)-W(s^j)\|\leq\varepsilon\}ds^1\cdots ds^k,$$

and I_j is the box $[2j, 2j+1]^N$, $1 \leq j \leq k$. We will omit the details.

Suppose, next, that $(d-2N)k > d$. Let $\eta \in (0,1)$ be very small and fixed; also fix disjoint boxes $C_1, \cdots, C_k$ such that $C_i \subset [\eta, \eta^{-1}]^N$, $1 \leq i \leq k$ and that if $i \neq j$, $\mathbf{d}(C_i, C_j) \geq \eta$, where $\mathbf{d}$ denotes the usual Euclidean (that is, ℓ^2) distance on $\mathbb{R}^N$. It suffices to show the following:

$$\mathbb{P}(\forall 1 \leq j \leq k,\ \exists t^j \in C_j:\ W(t^1) = \cdots = W(t^k)) = 0. \tag{3.1}$$

Fix any such $\eta \in (0,1)$ and $C_1, \cdots, C_k \subset [\eta, \eta^{-1}]^N$. For any $\varepsilon > 0$ and $j \in \{1, \cdots, k\}$, cover C_j with disjoint boxes $B_{i,j}(\varepsilon)$ of side ε, $1 \leq i \leq n_j(\varepsilon)$. Note that there exists some $K_6 = K_6(\eta, N)$ such that

$$\max_{j\leq k} n_j(\varepsilon) \leq K_6\varepsilon^{-N}. \tag{3.2}$$

Define,

$$N_k(\varepsilon) \triangleq \sum_{i_1=1}^{n_1(\varepsilon)}\sum_{i_2=1}^{n_2(\varepsilon)}\cdots\sum_{i_k=1}^{n_k(\varepsilon)} \mathrm{I}\{\forall 1\leq p\leq k,\ \exists t^p \in B_{i_p,p}(\varepsilon):\ W(t^1)=\cdots=W(t^k)\}.$$

From (2.3), a little thought shows that for all ε small enough, $N_k(\varepsilon) \le M_k(\varepsilon)$, where $M_k(\varepsilon)$ is given by

$$M_k(\varepsilon) \triangleq \sum_{i_1=1}^{n_1(\varepsilon)} \cdots \sum_{i_k=1}^{n_k(\varepsilon)} \mathrm{I}\{\ \forall 1 \le p \le k : \sup_{\substack{s\in B_{i_1,1}(\varepsilon)\\ t\in B_{i_p,p}(\varepsilon)}} \|W(s) - W(t)\| \le 2K_3\sqrt{\varepsilon\ln(1/\varepsilon)}\}.$$

As in §2, Theorem (1.3) follows once we show the following:

$$\lim_{\varepsilon\to 0} \mathbb{E}M_k(\varepsilon) = 0. \tag{3.3}$$

Let $b_{i,j}(\varepsilon)$ denote the center of $B_{i,j}(\varepsilon)$, say. Note that $\mathbb{E}M_k(\varepsilon)$ is bounded above by

$$\sum_{i_1=1}^{n_1(\varepsilon)} \cdots \sum_{i_k=1}^{n_k(\varepsilon)} \mathbb{P}\Big(\ \forall 1 \le p \le k : \|W(b_{i_1,1}(\varepsilon)) - W(b_{i_p,p}(\varepsilon))\| \le 2K_3\sqrt{2\varepsilon\ln(1/\varepsilon)}\Big).$$

However, by the construction of $C_1, \cdots, C_k$, we see that for any $1 < j \le k$, conditional on $\{W(b_{i_{j-1},j-1}(\varepsilon)), \cdots, W(b_{i_1,1}(\varepsilon))\}$, $W(b_{i_j,j}(\varepsilon))$ is a vector of independent normal random variables. Moreover, the (conditional) variance of any of the components of $W(b_{i_j,j}(\varepsilon))$ is bounded below by $K_7\eta$, for some $K_7 = K_7(N)$. By iteration, and since normal distributions are unimodal, the mode being at the mean , we see that

$$\begin{aligned} \mathbb{E}M_k(\varepsilon) &\le K_8 \prod_{j=1}^{k} n_j(\varepsilon) \cdot \Big(\varepsilon\ln(1/\varepsilon)\Big)^{d(k-1)/2} \\ &\le K_9 \varepsilon^{-kN+d(k-1)/2}\big(\ln(1/\varepsilon)\big)^{d(k-1)/2}, \end{aligned} \tag{3.4}$$

by (3.2). Here, $K_8 = K_8(\eta, d)$ and $K_9 \triangleq K_8 \cdot K_6^k$. Recall that we have $(d-2N)k > d$. Equivalently, we have $d(k-1) > 2Nk$. From (3.4) we obtain (3.3) and hence the result. □

REFERENCES.

[A1] R.J. ADLER (1981). *The Geometry of Random Fields*, Wiley, London

[A2] R.J. ADLER (1990). *An Introduction to Continuity, Extrema, and Related Topics for General Gaussian Processes*, Institute of Mathematical Statistics Lecture Notes—Monograph Series, Vol. 12

[BBK] R.F. BASS, K. BURDZY AND D. KHOSHNEVISAN (1994). Intersection local time for points of infinite multiplicity, *Ann. Prob.*, **22**, 566–625

[BK] R.F. BASS AND D. KHOSHNEVISAN (1993). Intersection local times and Tanaka formulas, *Ann. Inst. Henri Poincaré: Prob. et Stat.*, **29**, 419–451

[BG] R. BLUMENTHAL AND R.K. GETOOR (1968). *Markov Processes and Potential Theory*. Academic Press, New York

[C] X. CHEN (1994). Hausdorff dimension of multiple points of the (N, d) Wiener process, *Indiana Univ. Math. J.*, **43**(1), 55–60

[DEK1] A. DVORETSKY, P. ERDŐS AND S. KAKUTANI (1950). Double points of paths of Brownian motion in n–space. *Acta. Sci. Math.* (Szeged), **12**. 74–81

[DEK2] A. DVORETSKY, P. ERDŐS AND S. KAKUTANI (1954). Multiple points of Brownian motion in the plane, *Bull. Res. Council Israel Section F*, **3**, 364–371

[DEKT] A. DVORETSKY, P. ERDŐS, S. KAKUTANI AND S.J. TAYLOR (1957). Triple points of Brownian motion in 3–space. *Proc. Camb. Phil. Soc.*, **53**, 856–862

[D1] E.B. DYNKIN (1988). Self–intersection gauge for random walks and for Brownian motion, *Ann. Prob.*, **16**, 1–57

[D2] E.B. DYNKIN (1985). Random fields associated with multiple points of Brownian motion, *J. Funct. Anal.*, **62**, 397–434

[E] W. EHM (1981). Sample function properties of multiparameter stable processes, *Zeit. Wahr. verw. Geb.*, **56**, 195–228

[E1] S.N. EVANS (1987) Multiple points in the sample paths of a Lévy process, *Prob. Th. Rel. Fields*, **76**, 359–367

[E2] S.N. EVANS (1987) Potential theory for a family of several Markov processes, *Ann. Inst. Henri Poincaré: Prob. et Stat.*, **23**, 499–530

[FS-1] P.J. FITZSIMMONS AND T.S. SALISBURY (1989). Capacity and energy for multi–parameter Markov processes, *Ann. Inst. Henri Poincaré: Prob. et Stat.*, **25**, 325–350

[FS-2] P.J. FITZSIMMONS AND T.S. SALISBURY Forthcoming Manuscript.

[F] B. FRISTEDT (1995). *Math. Reviews*, review 95b:60100, February 1995 issue

[HaP] J. HAWKES AND W.E. PRUITT (1974). Uniform dimension results for processes with independent increments, *Zeit. Wahr. verw. Geb.*, **28**, 277–288

[H] W.J. HENDRICKS (1974). Multiple points for transient symmetric Lévy processes, *Zeit. Wahr. verw. Geb.* **49**, 13–21

[K] J.P. KAHANE (1985). *Some Random Series of Functions*, Cambridge Univ. Press, Cambridge, U.K.

[Ka] R. KAUFMAN (1969). Une propriété métrique du mouvement brownien, *C.R. Acad. Sci. Paris, Sér. A*, **268**, 727–728

[LG] J.F. LEGALL (1990). *Some Properties of Planar Brownian Motion, Ecole d'été de Probabilités de St-Flour* XX, LNM **1527**, 111–235

[OP] S. OREY AND W.E. PRUITT (1973). Sample functions of the N–parameter Wiener process, *Ann. Prob.*, **1**, 138–163

[P] Y. PERES (1995). Intersection–equivalence of Brownian paths and certain branching processes, *Comm. Math. Phys.* (To appear)

[R1] J. ROSEN (1995). Joint continuity of renormalized intersection local times, Preprint

[R2] J. ROSEN (1984). Stochastic integrals and intersections of Brownian sheet. Unpublished manuscript

[R3] J. ROSEN (1984). Self-intersections of random fields, *Ann. Prob.*, **12**. 108–119

[S] T.S. SALISBURY (1995). Energy. and intersections of Markov chains, *Proceedings of the IMA Workshop on Random Discrete Structures* (To appear)

[Sh] N.-R. SHIEH (1991). White noise analysis and Tanaka formulæ for intersections of planar Brownian motion, *Nagoya Math. J.*, **122**, 1–17

[T1] S.J. TAYLOR (1986). The measure theory of random fractals. *Math. Proc. Camb. Phil. Soc.*, **100**. 383–406

[T2] S.J. TAYLOR (19[illegible]). Multiple points for the sample paths of a transient stable process, *J. Math. Mech.*, **16**, 1229–1246

[T3] S.J. TAYLOR (1966). Multiple points for the sample paths of the symmetric stable process, *Zeit. Wahr. verw. Geb.*, **5**, 247–264

[V] S.R.S. VARADHAN (1969). Appendix to "Euclidean Quantum Field Theory", by K. Symanzik. In *Local Quantum Theory* (ed.: R. Jost). Academic Press, New York

[W] W. WERNER (1993). Sur les singularités des temps locaux d'intersection du mouvement brownien plan, *Ann. Inst. Henri. Poincaré: Prob. et Stat.*, **29**, 391–418

[Y] M. YOR (1985). Compléments aux formules de Tanaka–Rosen, *Sém. de Prob.* XIX, LNM **1123**, 332–349

A COUNTER-EXAMPLE CONCERNING A CONDITION OF OGAWA INTEGRABILITY

Pietro Majer * - Maria Elvira Mancino **

* Università di Pisa, Dipartimento di Matematica.
** Università di Firenze, DiMaDEFAS.

Abstract

A counterexample is exibited showing that the condition of Ogawa integrability introduced in [**3**] *is not satisfied by any complete orthonormal system.*

KEY WORDS: Ogawa integral, trace, orthonormal bases.

Introduction

In the past twenty years Itô's integral has been variously generalized by several authors. A very attractive notion has been suggested by Ogawa, who in [**5**] gave a definition of a stochastic integral linked to the results of Itô-Nisio concerning the uniform convergence to the Wiener process of a suitable random walk. To describe it more fully, let $W = (W_t)_{t\in[0\,,1]}$ be a Brownian motion on the probability space $(\Omega, \mathcal{A}, P)$, and let λ be the Lebesgue measure on $[0,\,1]$. A process H belonging to $L^2(\lambda \otimes P)$ is said to be Ogawa integrable, with respect to a given orthonormal system (e_i) of the space $L^2(\lambda)$, if the series

$$\sum_{i=1}^{\infty} \int_0^1 e_i(s)dW_s \int_0^t e_i(s)H_s ds$$

converges in probability for any t in $[0\,,1]$.

Such an integral may depend on the particular orthonormal system chosen. In [**6**] Ogawa studies the integrability of the continuous quasi-martingales and shows their integrability relatively to the trigonometric system. Later on, in [**7**], he proves that the integrability with respect to the trigonometric system implies the integrability with respect to the Haar system. Lastly, in [**8**], the orthonormal systems which make integrable every continuous quasi-martingale are characterized as the ones satisfying the following condition

$$\sup_n \int_0^1 \left(\sum_{i=1}^n e_i(t) \int_0^t e_i(s)ds \right)^2 dt < \infty . \tag{1}$$

The trigonometric system as well as Haar system are easily seen to verify condition (1). However, the problem of deciding if (1) holds for any orthonormal system of $L^2(\lambda)$ is left open (see **[3]**).

More recently some authors investigated Ogawa integrability independently from the base by restricting the class of integrands (e.g., to those processes which are regular in the sense of Malliavin derivative, as in **[3]** **[4]**, or to multiple Itô-Wiener integrals, as in **[9]**). Nevertheless these classes do not contain all continuous quasi-martingales (see the counter-example in **[1]**), so that it is still interesting to decide whether or not condition (1) holds for any orthonormal system. In the present note we give a negative answer to this question by exibiting a counter-example.

In order to construct a complete orthonormal system verifying

$$\sup_n \int_0^1 \left(\sum_{i=1}^n e_i(t) \int_0^t e_i(s)ds \right)^2 dt = \infty , \tag{2}$$

we consider as a starting point the easier problem (Lemma 2) of finding, for a given real number M, a finite orthogonal family of simple functions $u_1, \dots, u_n$ such that

$$\int_0^1 \left(\sum_{i=1}^n u_i(t) \int_0^t u_i(s)ds \right)^2 dt \geq M .$$

In Lemma 1 the latter problem, which has a finite-dimensional nature, turns into the spectral analysis of a suitable matrix. Iterating the above construction on each interval of some countable partition of $[0,1]$ gives rise to an orthonormal system (e_i) of $L^2(\lambda)$ satisfying (2). The system (e_i) can then be completed into an orthonormal base, and, if the latter is conveniently ordered, there results that property (2) still holds.

1. Preliminaries

In this section we shall briefly sketch the proof of Ogawa integrability for continuous quasi-martingales so as to emphasize the necessity of condition (1).

Let H be a continuous quasi-martingale of the form $A + K.W$, where A is an adapted process having bounded variation trajectories, K is a bounded predictable process, and $K.W$ denotes the Itô integral of the process K. Since bounded variation processes are Ogawa integrable with respect to any orthonormal system, as it is readily seen by means of a simple integration by parts, we can restrict ourselves to the processes of the form $H = K.W$.

Given an orthonormal system (e_i) of $L^2(\lambda)$, for any $i \geq 1$ and any t in $[0,1]$, we denote $E_i(t)$ the function $\int_0^t e_i(s)ds$. We have to prove that the series

$$\sum_{i=1}^{\infty} \int_0^1 e_i(s)dW_s \int_0^t H_s e_i(s)ds$$

converges in probability for any t in $[0,1]$.
Due to the integration by parts formula, there holds

$$\int_0^t H_s e_i(s)ds = E_i(t)H_t - \int_0^t E_i(s)H_s dW_s.$$

Applying the Itô formula gives

$$\sum_{i=1}^{n} \int_0^1 e_i(s)dW_s \int_0^t H_s e_i(s)ds = S_1(n) + S_2(n) + S_3(n)$$

where

$$S_1(n) = \sum_{i=1}^{n} \left\{ H_t E_i(t) \int_0^1 e_i(s)dW_s - \int_0^t E_i(s)K_s dW_s \int_0^s e_i(s')dW_{s'} \right\},$$

$$S_2(n) = -\sum_{i=1}^{n} \int_0^1 e_i(s)dW_s \int_0^t E_i(s)K_s dW_s,$$

$$S_3(n) = -\sum_{i=1}^{n} \int_0^t E_i(s)e_i(s)K_s ds.$$

Hence one easily verifies that $S_1(n)$ and $S_3(n)$ converge in probability. Moreover, if the system (e_i) verifies condition (1), then $S_2(n)$ converges in probability to $-\frac{1}{2}\int_0^t K_s ds$.

2. Construction of a counterexample

Theorem. *There exists a complete orthonormal system $\{e_i\}_{i\in\mathbf{N}}$ of $L^2(\lambda)$ such that*

$$\sup_{n\in\mathbf{N}} \int_0^1 \left(\sum_{i=1}^{n} e_i(t) \int_0^t e_i(s)ds \right)^2 dt = \infty \quad .$$

We need some lemmas.

Lemma 1. *Let $n \in \mathbf{N}$ and let V be the $n \times n$ matrix with coefficients*

$$V_{ij} =: \frac{1+(-1)^{n+\max(i,j)}}{2} \qquad 1 \le i,j \le n \quad . \tag{3}$$

Then, letting $\theta_k =: \frac{k\pi}{2n+1}$ for $k = 1,2,...n$, the eigenvalues λ_k of V and the associated eigenvectors $u^k = (u_1^k, \dots, u_n^k)$ are

$$\begin{aligned} \lambda_k &= \frac{(-1)^n}{2}\sec(2\theta_k) \\ u^k &= \rho_k\Big(-\sin\theta_k,\ -\sin(3\theta_k),\ \sin(5\theta_k), \dots, (-1)^{[\frac{n+1}{2}]}\sin((2n-1)\theta_k)\Big) \\ \rho_k &= \left(\frac{n}{2}+\frac{1}{4}\tan\theta_k\right)^{-\frac{1}{2}} \quad . \end{aligned} \tag{4}$$

Proof. First, let us observe that the inverse matrix of V writes

$$\begin{bmatrix} & \cdots & 1 & 0 & 1 \\ \vdots & \ddots & \vdots & \vdots & \vdots \\ 1 & \cdots & 1 & 0 & 1 \\ 0 & \cdots & 0 & 0 & 1 \\ 1 & \cdots & 1 & 1 & 1 \end{bmatrix}^{-1} = (-1)^n \begin{bmatrix} -1 & 1 & & & \\ 1 & 0 & -1 & & \\ & -1 & 0 & & \\ & & & \ddots & \pm 1 \\ & & & \pm 1 & 0 \end{bmatrix}$$

or, in shorter notation, $V^{-1} = (-1)^n DJD$, where D and J are the $n \times n$ matrices with coefficients respectively $D_{ij} =: (-1)^{[\frac{i+1}{2}]}\delta_{ij}$ and $J_{ij} =: -\delta_{i1}\delta_{1j} + \delta_{|i-j|,1}$, that is

$$D = \begin{bmatrix} -1 & & & & \\ & -1 & & & \\ & & +1 & & \\ & & & +1 & \\ & & & & \ddots \end{bmatrix}, \qquad J = \begin{bmatrix} -1 & 1 & & & \\ 1 & 0 & 1 & & \\ & 1 & 0 & & \\ & & & \ddots & 1 \\ & & & 1 & 0 \end{bmatrix}.$$

Therefore, if $u \in \mathbb{R}^n$ is an eigenvector of V corresponding to the eigenvalue λ, then $\mu =: (-1)^n\lambda^{-1}$ is the eigenvalue of J relative to eigenvector $v =: Du$. Now a base of eigenvectors of J, $(v^k)_{1 \le k \le n}$, together with eigenvectors $(\mu_k)_{1 \le k \le n}$, is the one defined by letting $\theta_k =: \frac{k\pi}{2n+1}$ and

$$\begin{aligned} \mu_k &= 2\cos(2\theta_k) \\ v^k &= (\sin\theta_k, \sin(3\theta_k), \sin(5\theta_k), \dots, \sin((2n-1)\theta_k)) \ , \end{aligned} \tag{5}$$

as can be shown by direct computation (or using the argument developed in the next Remark 1). Letting $\rho_k =: \|v_k\|$, one obtains

$$\rho_k^2 = \sum_{j=1}^{n} (\sin(2j-1)\theta_k)^2 = \frac{n}{2} + \frac{1}{4}\tan\theta_k \quad ,$$

whence equations (4) follow letting $\lambda_k = (-1)^n\mu_k^{-1}, \quad u^k = \rho_k D v^k$.

Remark 1. For a more euristic computation of the spectrum of J, let us observe that equation $Jv = 2\xi v$, with v in $\mathbb{R}^n$, ξ in $\mathbb{R}$, is equivalent to:

$$\begin{cases} v = \rho(W_0(\xi), W_1(\xi), \ldots, W_{n-1}(\xi)), & \rho \in \mathbb{R} \\ W_n(\xi) = 0 \ , \end{cases}$$

where the $W_k(x)$, for $k \in \mathbf{N} \cup \{0\}$, are polynomials verifying the linear recurrence formula

$$\begin{cases} W_0(x) = 1 \\ W_1(x) = 2x + 1 \\ W_{k+1}(x) + W_{k-1}(x) = 2xW_k(x) \quad . \end{cases} \tag{6}$$

Hence one recognizes the orthogonal Jacobi polynomials $J_k^{(\alpha,\beta)}(x)$ for $\alpha = \frac{1}{2}$, $\beta = -\frac{1}{2}$; these admit the representation in the following closed formula, which one obtains smoothly from equations (6) (see e.g. [**2**])

$$W_m(x) = \frac{\sin(m + \frac{1}{2})\theta}{\sin(\frac{\theta}{2})}, \qquad x = \cos\theta \tag{7}$$

and one finds again equations (5), taking account that the roots of W_n are, thanks to (7), $x_k = \cos(2\theta_k)$, $1 \leq k \leq n$.

The theorem follows smoothly from a slightly weaker preliminar result, which we now state with the same notations as Lemma 1.

Lemma 2. *Let n be even and, for $1 \leq k \leq n$, let f_k be the simple functions defined by*

$$f_k(t) = \sqrt{n} \sum_{j=1}^{n} u_j^k \, \chi_{[\frac{j-1}{n}, \frac{j}{n}]}(t) \ .$$

Then f_k are orthonormal in $L^2(\lambda)$ and one has

$$\int_0^1 \left(\sum_{1 \leq k \leq \frac{n}{2}} f_k(t) \int_0^t f_k(s) ds \right)^2 dt \geq C_n =: \frac{1}{2} \left(\frac{\log n}{2\pi} \right)^2 \ . \tag{8}$$

Proof. Since the vectors u^k defined by equations (4) are orthonormal in $\mathbb{R}^n$, the functions f_k are immediately seen to be orthonormal in $L^2(\lambda)$. So we are left with the L^2 estimate of the function

$$S(t) = \sum_{1 \leq k \leq \frac{n}{2}} f_k(t) \int_0^t f_k(s) ds \quad .$$

To this end let us consider the test function

$$\phi(t) = \sqrt{2}\sum_{j=1}^{n} \frac{1+(-1)^{n+j}}{2} \chi_{[\frac{j-1}{n},\frac{j}{n}]}(t) \; . \tag{9}$$

There results $\|\phi\|_2 = 1$, whence

$$\begin{aligned} \|S\|_2 \geq \langle S, \phi\rangle &= \int_0^1 \left(\sum_{1\leq k\leq \frac{n}{2}} f_k(t) \int_0^t f_k(s)ds \right) \phi(t)dt \\ &= \sum_{1\leq k\leq \frac{n}{2}} \int_0^1 \int_0^t f_k(t)f_k(s)\phi(t)dsdt \\ &= \frac{1}{2}\sum_{1\leq k\leq \frac{n}{2}} \int_0^1 \int_0^1 f_k(t)f_k(s)\phi(\max(t,s))dtds \; , \end{aligned} \tag{10}$$

where the latter inequality follows from the symmetry of the integrands $f_k(t)f_k(s)\phi(\max(t,s))$ with respect to the pair (t,s).
Let us notice that the function $\phi(\max(t,s))$, using equations (3) and (9), can also be written as

$$\phi(\max(t,s)) = \sqrt{2}\sum_{i,j} V_{ij}\chi_{[\frac{i-1}{n},\frac{i}{n}]}(t)\chi_{[\frac{j-1}{n},\frac{j}{n}]}(s)$$

whence

$$\int_0^1\int_0^1 f_k(t)f_k(s)\phi(\max(t,s))dtds = \frac{\sqrt{2}}{n}\langle Vu^k, u^k\rangle = \frac{\sqrt{2}}{n}\lambda_k \; .$$

Thus equalities (10) yield

$$\begin{aligned} &\frac{1}{\sqrt{2}n}\sum_{1\leq k\leq \frac{n}{2}} \lambda_k = \frac{1}{\sqrt{8}n}\sum_{1\leq k\leq \frac{n}{2}} \frac{1}{\cos\left(\frac{2k\pi}{2n+1}\right)} \\ &= \frac{2n+1}{2\sqrt{8}n\pi}\sum_{1\leq k\leq \frac{n}{2}} \frac{1}{\cos\left(\frac{2k\pi}{2n+1}\right)}\,\frac{2\pi}{2n+1} \geq \frac{2n+1}{2\sqrt{8}n\pi}\int_0^{\frac{n\pi}{2n+1}} \frac{1}{\cos t}dt \end{aligned} \tag{11}$$

Indeed, the sum in the left side of equation (11) is an upper Riemann estimate for the integral in the right side .
On the other hand

$$\begin{aligned} \frac{2n+1}{2\sqrt{8}n\pi}\int_0^{\frac{n\pi}{2n+1}} \frac{1}{\cos t}dt &\geq \frac{1}{\sqrt{8}\pi}\int_0^{\frac{\pi}{2}-\frac{\pi}{4n}} \frac{1}{\cos t}dt \\ &= \frac{1}{\sqrt{8}\pi}\log\cot(\frac{\pi}{8n}) \geq \frac{1}{\sqrt{8}\pi}\log n \; . \end{aligned}$$

Therefore one concludes

$$\|S\|_2^2 \geq \frac{1}{2}\left(\frac{\log n}{2\pi}\right)^2 .$$

Remark 2. In the summation shown in (8) we took only the first $\frac{n}{2}$ functions f_k, that is, the ones corresponding to the positive eigenvalues λ_k of the matrix V. This is due to the fact that, as it is shown in equations (11), the arithmetic mean of the positive eigenvalues of V, $\frac{2}{n}\sum_{1\leq k\leq \frac{n}{2}} \lambda_k$, is of the order (at least) of $\log n$, whereas the mean value extended to the whole spectrum is $\frac{1}{n}\sum_{1\leq k\leq n} \lambda_k = \frac{1}{n}\mathrm{Tr}(V) = \frac{1}{2}$.

Remark 3. For any given subinterval $[a,b]$ of $[0,1]$, one can also choose the functions of Lemma 2 with supports in $[a,b]$. Actually, it is sufficient to consider $\tilde{f}_k(t) =: (b-a)^{-\frac{1}{2}} f_k\left(\frac{t-a}{b-a}\right)$ if $t \in [a,b]$, and $\tilde{f}_k(t) = 0$ if $t \notin [a,b]$. Then inequality (7) holds for $\tilde{f}_k$ relatively to the constant $\widetilde{C}_n =: (b-a)C_n$.

Lemma 3. *Let $g_1, \ldots, g_p$ be orthonormal functions in $L^2(\lambda)$. There holds*

$$\int_0^1 \left(\sum_{i=1}^p g_i(t)\int_0^t g_i(s)ds\right)^2 dt \leq p .$$

Proof. The Fourier coefficients of $\chi_{[0,t]}$ with respect to g_i is $\int_0^t g_i(s)ds$. Then it follows, using Schwarz inequality and Bessel inequality

$$\begin{aligned}\int_0^1 \left(\sum_{i=1}^p g_i(t)\int_0^t g_i(s)ds\right)^2 dt &= \int_0^1 \left(\sum_{i=1}^p g_i(t)\langle g_i, \chi_{[0,t]}\rangle\right)^2 dt \\ &\leq \int_0^1 \left(\sum_{i=1}^p g_i(t)^2\right)\left(\sum_{i=1}^p \langle g_i, \chi_{[0,t]}\rangle^2\right) dt \\ &\leq \int_0^1 \left(\sum_{i=1}^p g_i(t)^2\right) \|\chi_{[0,t]}\|_2^2 dt \\ &\leq \int_0^1 \sum_{i=1}^p g_i(t)^2 dt = p .\end{aligned}$$

Proof of the Theorem. Let $\{I_i\}$ be a countable family of disjoint subintervals of $[0,1]$. For any index $i \in \mathbf{N}$, applying Lemma 2 and Remark 3, we can find n_i simple functions $\{f_{ij}\}_{1\le j\le n_i}$ such that

$$\langle f_{ij}, f_{ik}\rangle = \delta_{jk}, \quad 1 \le j,k \le n_i \tag{12}$$

$$\operatorname{supp}(f_{ij}) \subset I_i \tag{13}$$

$$\int_0^1 \left(\sum_{j=1}^{n_i} f_{ij}(t)\int_0^t f_{ij}(s)ds\right)^2 dt \ge 4 \tag{14}$$

The family $\{f_{ij},\, i \in \mathbf{N},\, 1 \le j \le n_i\}$ is clearly orthonormal, since f_{ij} and f_{lk} verify (12) whenever $i = l$, while have disjoint supports if $i \ne l$, thanks to (13).

Next let us consider an orthonormal base $\{f_{i0}\}_{i\in\mathbf{N}}$ which completes system $\{f_{ij},\, i \in \mathbf{N},\, 1 \le j \le n_i\}$ to an orthonormal base $\{f_{ij},\, i \in \mathbf{N},\, 0 \le j \le n_i\}$; then re-indicize the latter by means of the position

$$e_k = f_{ij} \quad \text{if and only if} \quad k = i + j + \sum_{0\le l<i} n_l \quad \text{and} \quad 0 \le j < n_i\,.$$

(This amounts to give the set of pairs (i,j) its lexicographic order.) For any $p \in \mathbf{N}$, let $m =: p + \sum_{0\le l<p} n_l$. Thus one gets

$$\begin{aligned}\sum_{i=1}^m e_i(t)\int_0^t e_i(s)ds &= \sum_{i=1}^p\sum_{j=0}^{n_i} f_{ij}(t)\int_0^t f_{ij}(s)ds \\ &= \sum_{i=1}^p f_{i0}(t)\int_0^t f_{i0}(s)ds + \sum_{i=1}^p\sum_{j=1}^{n_i} f_{ij}(t)\int_0^t f_{ij}(s)ds\,.\end{aligned}$$

There follows, using the elementary inequality $(a+b)^2 \ge \frac{a^2}{2} - b^2$,

$$\begin{aligned}&\int_0^1 \left(\sum_{i=1}^m e_i(t)\int_0^t e_i(s)ds\right)^2 dt \qquad (15)\\ &\ge \frac{1}{2}\int_0^1 \left(\sum_{i=1}^p\sum_{j=1}^{n_i} f_{ij}(t)\int_0^t f_{ij}(s)ds\right)^2 dt - \int_0^1 \left(\sum_{i=1}^p f_{i0}(t)\int_0^t f_{i0}(s)ds\right)^2 dt \\ &= \frac{1}{2}\sum_{i=1}^p\int_0^1 \left(\sum_{j=1}^{n_i} f_{ij}(t)\int_0^t f_{ij}(s)ds\right)^2 dt - \int_0^1 \left(\sum_{i=1}^p f_{i0}(t)\int_0^t f_{i0}(s)ds\right)^2 dt\,,\end{aligned}$$

where the latter equality is a consequence of the fact that the functions f_{ij} with different indices i have disjoint support, as we remarked above. Due to inequality (14) and to Lemma 3 one has respectively

$$\frac{1}{2}\sum_{i=1}^{p}\int_0^1\left(\sum_{j=1}^{n_i} f_{ij}(t)\int_0^t f_{ij}(s)ds\right)^2 dt \geq 2p$$

$$\int_0^1\left(\sum_{i=1}^{p} f_{i0}(t)\int_0^t f_{i0}(s)ds\right)^2 dt \leq p\,,$$

and we conclude from (15)

$$\int_0^1\left(\sum_{i=1}^{m} e_i(t)\int_0^t e_i(s)ds\right)^2 dt \geq p.$$

The claim follows for p being arbitrary.

References

[1] M.E. MANCINO : *Su alcuni integrali stocastici anticipativi.* PhD Thesis, Univ. di Trento (1995)

[2] I.P. NATANSON : *Constructive Function Theory.* Vol II - Frederick Ungar Publishing Co. NY (1965)

[3] D. NUALART : *Noncausal stochastic integrals and calculus.* L.N.M. 1316

[4] D. NUALART, M. ZAKAI : *Generalized stochastic integrals and the Malliavin calculus.* Prob.Th.Rel.Fields 73 (1986)

[5] S. OGAWA : *Sur le produit direct du bruit blanc par lui-même.* C. R. Acad. Sc. Paris, t.288, Série A-359 (Février 1979)

[6] S. OGAWA : *Quelques propriétés de l'intégrale stochastique du type noncasual.* Japan J.Appl. Math. vol.1 (1984)

[7] S. OGAWA : *Une remarque sur l'approximation de l'intégrale stochastique du type noncasual par une suite d' intégrales de Stieltjes.* Tôhuku Math.J. vol.36 (1984)

[8] S. OGAWA : *The stochastic integral of noncasual type as an extension of the symmetric integrals.* Japan J.Appl. Math. vol.2 (1985)

[9] J. ROSINSKI : *On stochastic integration by series of Wiener integrals.* Appl. Math. Optim. vol.19 (1989)

The Multiplicity of Stochastic Processes

Yukuang Chiu
Department of Mathematics, 0112
University of California, San Diego
La Jolla, CA 92093-0112, USA

This paper studies the multiplicity of non-Gaussian, non-infinitely divisible and non-stationary processes associated with the "chaos" space of N. Wiener [12], and for each positive integer N and for $N = \infty$, constructs a process of multiplicity N. The examination of multiplicity of a process has been of interest to many authors such as H. Cramér [2,3,4], T. Hida [5,6], K. Itô [7] and G. Kallianpur and V. Mandrekar [10].

Our approach here begins with a classical, well known theorem on a separable Hilbert space.

Let U_t, $(t \in R)$ be a one parameter group of unitary operators acting on a separable Hilbert space $\mathbf{H}$, and let E_λ be its spectral measure, i.e.,

$$U_t = \int_{-\infty}^{\infty} e^{it\lambda} dE_\lambda.$$

Then there exists a sequence $\{f_n\}$ of elements in $\mathbf{H}$, which will be referred to as cyclic vectors, such that the Hilbert space $\mathbf{H}$ can be Hellinger-Hahn [9] decomposed into a direct sum

$$\mathbf{H} = \sum_{n \geq 1} \bigoplus \mathbf{H}_n,$$

where

$$\begin{aligned} \mathbf{H}_n &= \left\{ \int_{-\infty}^{\infty} g(\lambda) dE_\lambda f_n;\ g \in L^2(R, \mu_n) \right\} \\ &= \text{linear span of } \{U_t f_n;\ -\infty < t < \infty\}, \end{aligned}$$

which will be referred to as a cyclic subspace of $\mathbf{H}$ with f_n, with the notation

$$d\mu_n(\lambda) = ||dE_\lambda f_n||^2,$$

we further have

$$d\mu \stackrel{\text{def}}{:=} d\mu_1 \gg d\mu_2 \gg \cdots,$$

where $d\mu \gg d\nu$ means that the measure $d\mu$ is absolutely continuous with respect to the measure $d\nu$. The type of the measure sequence $\{d\mu_n\}$ is invariant with respect to the choice of $\{f_n\}'s$. This is to say that if $\mathbf{H} = \sum_{n\geq 1} \oplus \mathbf{H}'_n$ is another decomposition with $\mathbf{H}'_n$, a cyclic subspace with cyclic vector f'_n, then

$$d\mu_n \sim d\mu'_n(\text{ equivalence }),\ n = 1, 2, \cdots,$$

where $d\mu'_n(\lambda) = ||dE_\lambda f'_n||^2$.

Denote the support for $d\mu_n$ by Λ_n. The integer $m(\lambda) = max\{n;\ \lambda \in \Lambda_n\}$ is referred to as the multiplicity of λ, and the pair $\{d\mu, m\}$, the spectral type of U_t. The spectral type of U_t is said to be $\sigma-$Lebesgue if $d\mu$ is equivalent to Lebesgue measure and if $m(\lambda) \equiv \infty$; and that of U_t is said to be simple Lebesgue if $d\mu$ is equivalent to Lebesgue measure and if $m(\lambda) \equiv 1$.

As a further consequence of Hellinger-Hahn decomposition, we have that if U_t and U_t' are one parameter groups of unitary operators acting on $\mathbf{H}$ and $\mathbf{H}'$ respectively, and if they are unitary equivalent, i.e., if there exists an isometry V of $\mathbf{H}$ onto $\mathbf{H}'$ such that $U_t' = VU_tV^{-1}$, then the associated measure sequences $\{d\mu_n\}$ and $\{d\mu_n'\}$ are of the same type. Conversely, if these two sequences are of the same type, then we can construct an isometry between $\mathbf{H}$ and $\mathbf{H}'$ such that $\{U_t\}$ and $\{U_t'\}$ are unitary equivalent. In other words, the sequence $\{d\mu_n\}$ is unitary invariant.

Example Define θ_t to be the transform of $L^2(R)$

$$\theta_t : \begin{array}{ccc} L^2(R) & \to & L^2(R) \\ F(\cdot) & \to & F(\cdot - t). \end{array}$$

Then θ_t consists of a one parameter group of unitary operators on $L^2(R)$, and its spectral type is simple Lebesgue.

To see this, let us write

$$\tau(u) = \begin{cases} e^u, & u < 0 \\ 0, & u \geq 0 \end{cases}.$$

Then

$$\hat{\tau}(\lambda) \stackrel{\text{def}}{:=} \frac{1}{\sqrt{2\pi}} \int_{-\infty}^{0} e^u e^{i\lambda u} du = \frac{1}{\sqrt{2\pi}} \frac{1}{1 + i\lambda}.$$

Since the Fourier transform is topologically isomorphic on $L^2(R)$ by Plancherel's theorem [13], it follows that

$$\text{linear span } \{e^{i\lambda t}\hat{\tau}(\lambda),\, t \in R\} = \text{ linear span } \{\theta_t \tau(\cdot),\, t \in R\} = L^2(R).$$

Thus $L^2(R)$ itself turns out to be a cyclic space with cyclic vector τ. Now

$$\Big(\int_{-\infty}^{\infty} e^{it\lambda} dE_\lambda \tau\Big)(u) = \tau(u - t) = \frac{1}{\sqrt{2\pi}} \int_{-\infty}^{\infty} e^{-i\lambda(u-t)} \hat{\tau}(\lambda) d\lambda,$$

$$\int_{-\infty}^{\infty} e^{it\lambda} ||dE_\lambda \tau||^2 = \int_{-\infty}^{\infty} e^{it\lambda} |\hat{\tau}(\lambda)|^2 d\lambda.$$

Hence

$$d\mu(\lambda) = ||dE_\lambda \tau)||^2 = \frac{1}{2\pi} \frac{1}{1 + \lambda^2} d\lambda.$$

This shows that the measure $d\mu$ is equivalent to Lebesgue measure and $m(\lambda) \equiv 1$, and consequently simple Lebesgue.

In the sequel, let U_t be the one parameter group of unitary operators induced by Brownian motion flow T_t on $L^2(\mathbf{B})$ [5], i.e., the collection of all variables measurable with respect to the σ-field generated by Brownian motion $\mathbf{B}$ with finite variances. First, we look at the spectral type of U_t.

To begin with, let $L^2(\mathbf{B}) = \sum_{n=0}^{\infty} \oplus \mathcal{H}_n$ be the Wiener-Itô decomposition [8, 12] of $L^2(\mathbf{B})$. It is well known that each $\mathcal{H}_n$, which consists of an U_t-invariant subspace, is topologically isomorphic to $\sqrt{n!}\widehat{L}^2(R^n)$ (via $\mathcal{J}$-transformation [5]), where $\widehat{L}^2(R^n)$ denotes all the symmetric functions of $L^2(R^n)$, and that each element in $\mathcal{H}_n$ can be expressed as an n-multiple Wiener integral. Without ambiguity, we still write U_t to be the restriction of U_t on $\mathcal{H}_n$. We then have

Theorem For each $n \geq 2$, the spectral type of U_t on $\mathcal{H}_n$ is σ-Lebesgue.

To prove this, we introduce a unitary isometry V_t of U_t. Since spectral type is unitary invariant, the investigation of spectral type of U_t may be reduced to a search for that of V_t. Now let us put

$$L^2_{nc} = L^2((u_1, u_2, \dots u_n) \in R^n;\ u_1 \leq u_2 \leq \dots \leq u_n)$$

and define $\mathcal{C}$:

$$\mathcal{C}: \begin{array}{ccc} \widehat{L}^2(R^n) & \rightarrow & \sqrt{n!}L^2_{nc} \\ F(u_1, u_2, \dots, u_n) & \rightarrow & F(u_{\pi(1)}, u_{\pi(2)}, \dots, u_{\pi(n)}), \end{array}$$

where π is a permutation of $\{1, 2, \cdots, n\}$ such that $u_{\pi(1)} \leq u_{\pi(2)} \leq \dots, \leq u_{\pi(n)}$. Obviously $\mathcal{C}$ defines an isometric mapping from $\widehat{L}^2(R^n)$ to $\sqrt{n!}L^2_{nc}$. Further let

$$A_n = \begin{pmatrix} \frac{1}{n} & \frac{1}{n} & \cdots & \frac{1}{n} & \frac{1}{n} \\ -1 & 1 & \cdots & 0 & 0 \\ \cdots & \cdots & \cdots & \cdots & \cdots \\ 0 & 0 & \cdots & 1 & 0 \\ 0 & 0 & \cdots & -1 & 1 \end{pmatrix}$$

and define $\mathcal{E}$:

$$\mathcal{E}: \begin{array}{ccc} L^2_{nc} & \rightarrow & L^2(R \times R_+^{n-1}) \\ F(u_1, u_2, \dots u_n) & \rightarrow & G(v_1, v_2, \dots, v_n), \end{array}$$

where $R_+ = [0, \infty)$ and

$$\begin{pmatrix} v_1 \\ v_2 \\ \vdots \\ v_n \end{pmatrix} = A_n \begin{pmatrix} u_1 \\ u_2 \\ \vdots \\ u_n \end{pmatrix}.$$

Then again we verify that $\mathcal{E}$ defines an isometric mapping from ${L^2}_{nc}$ to $L_2(R \times R_+^{n-1})$. Hence if

$$V_t \stackrel{\text{def}}{:=} (\mathcal{E} \cdot \mathcal{C} \cdot \mathcal{J})^{-1} U_t\, (\mathcal{E} \cdot \mathcal{C} \cdot \mathcal{J}),$$

then $\{V_t,\ t \in R\}$ consists of a one parameter group of unitary operators on $L^2(R \times R_+^{n-1})$. As a matter of fact, with the diagram

$$\mathcal{E} \cdot \mathcal{C} \cdot \mathcal{J}: \begin{array}{ccc} \mathcal{H}_n & \xrightarrow{\mathcal{J}} & \sqrt{n!}\ \widehat{L}^2(R^n) \\ \uparrow & & \downarrow \mathcal{C} \\ n!\ L^2(R \times R_+^{n-1}) & \xleftarrow[\mathcal{E}]{} & n!\ {L^2}_{nc} \end{array}$$

in mind, we see that if

$$\mathcal{E} \cdot \mathcal{C} \cdot \mathcal{J}: \begin{array}{ccc} \mathcal{H}_n & \rightarrow & n!\ L^2(R \times R_+^{n-1}) \\ \varphi & \rightarrow & n!\ G(v_1, v_2, \dots, v_n), \end{array}$$

then

$$\begin{array}{lrcl} (1) & U_t\varphi & \rightarrow & n!\ G(v_1 - t, v_2, \dots, v_n) \\ (2) & & = & n!\ (V_tG)(v_1, v_2, \dots, v_n). \end{array}$$

To see the spectral type of V_t on $L^2(R\times R_+^{n-1})$, we decompose $L^2(R\times R_+^{n-1})$ into a direct sum by means of a complete orthonormal basis $\{\eta_n;\ n\geq 0\}$ of $L^2(R_+)$:

$$L^2(R\times R_+^{n-1}) = \sum_{k_2,\cdots,k_n\geq 0} \oplus L_{k_2,\ldots,k_n}, \tag{3}$$

where

$$L_{k_2,\cdots,k_n} = \{f(v_1)\otimes\eta_{k_2}(v_2)\otimes\cdots\otimes\eta_{k_n}(v_n);\ f\in L^2(R)\},$$

and $\otimes$ means tensor product. Such η_n's may be taken, for example as the Laguerre functions. Apparently, the subspace $L_{k_2,\cdots,k_n}$ of $L^2(R\times R_+^{n-1})$ by (1) and (2) is V_t invariant, and the spectral type of V_t on each $L_{k_2,\ldots,k_n}$, as seen in the example, is simple Lebesgue. Combining this with (3), we have proven that the spectral type of V_t on $L^2(R\times R_+^{n-1})$ is σ-Lebesgue.

Here, let us note that if we put

$$X_{k_2,\cdots,k_n}(t) = (\mathcal{E}\cdot\mathcal{C}\cdot\mathcal{J})^{-1}(\tau(v_1-t)\eta_{k_2}(v_2)\cdots\eta_{k_n}(v_n))$$

then $X_{k_2,\cdots,k_n}(t)$ may be expressed as a stochastic integral

$$X_{k_2,\ldots,k_n}(t) = \int_{-\infty}^{t} dB(u_n)\int_{-\infty}^{u_n}\eta_{k_n}(u_n-u_{n-1})dB(u_{n-1})\times\ldots\times$$

$$\int_{-\infty}^{u_3}\eta_{k_3}(u_3-u_2)dB(u_2)\times\int_{-\infty}^{u_2}\tau(\frac{u_1+\ldots+u_n}{n}-t)\eta_{k_2}(u_2-u_1)dB(u_1).$$

Hence if we put

$$\mathcal{H}_n(X_{k_2,\cdots,k_n}) = (\mathcal{E}\cdot\mathcal{C}\cdot\mathcal{J})^{-1}L_{k_2,\ldots,k_n},$$

then

$$\mathcal{H}_n = \sum_{k_2,\cdots,k_n\geq 0}\oplus\mathcal{H}_n(X_{k_2,\cdots,k_n}).$$

This is the decomposition of $\mathcal{H}_n$ corresponding to that of $L^2(R\times R_+^{n-1})$.

Further, if we notice that the expectations in $\mathcal{H}_n$ correspond to the multiple integrations in $L^2(R\times R_+^{n-1})$, then we can immediately compute, for example

$$E[(X_{k_2,\cdots,k_n}(t)-X_{k_2,\cdots,k_n}(s))^2] = \int_{-\infty}^{\max\{t,s\}}(\tau(u-t)-\tau(u-s))^2du,$$

and

$$E[X_{k_2,\cdots,k_n}(t)X_{k_2,\cdots,k_n}(s)] = \frac{1}{2}e^{-|t-s|}. \tag{4}$$

In the case where $n=2$, which is of particular interest, we will write

$$X_n(t) = \int_{-\infty}^{t} dB(u_2)\int_{-\infty}^{u_2}\tau(\frac{u_1+u_2}{2}-t)\eta_n(u_2-u_1)dB(u_1).$$

We now focus on the multiplicity of a process $X(t)\in\mathcal{H}_2$:

$$X(t) \stackrel{\text{def}}{:=} \sum_{n=0}^{\infty}F(t)^n X_n(t),$$

where $F(t)$ on R is an absolutely continuous function with (i) $0<F(t)\leq\delta<1$.

Theorem If $F(t)$ further satisfies the conditions (*ii*) the derivative F' of F is in $L^1(R)$; (*iii*) for any open interval (a,b),

$$\int_a^b F'^2 dt = +\infty,$$

then the multiplicity of $X(t)$ is infinity.

The proof will be done by constructing another process $Y(t)$ which is both canonically represented by Brownian motion and has the same reproducing kernel Hilbert space as that of $X(t)$. Consequently, the determination of the multiplicity for process $X(t)$ may be reduced to that for $Y(t)$.

Before constructing $Y(t)$, let us first find a process $T(t)$ such that $T(t)$ can be canonically represented by Brownian motion, and that $T(t)$ shares the same covariance with $X_n(t)$. Since the covariance of $X_n(t)$ is given by (4), it follows from N. Wiener [11] that such a process must be Ornstein-Uhlenbeck process

$$T(t) = \int_{-\infty}^t e^{-(t-u)} dB(u).$$

Let us prepare a sequence of independent Brownian motions on R: $B_0, B_1, B_2, \cdots$, and let

$$Y_n(t) = \int_{-\infty}^t e^{-(t-u)} dB_n(u).$$

Then a process $Y(t)$ defined as

$$Y(t) = \sum_{n=0}^{\infty} F(t)^n Y_n(t)$$

shares the same reproducing kernel Hilbert space as that of $X(t)$. Hence the multiplicity of $Y(t)$ equals that of $X(t)$.

To say that the multiplicity of $Y(t)$ is infinity, it suffices to show by T. Hida [5,6] that the representation of $Y(t)$ is canonical, i.e., fix $T \in R$, for $n = 0, 1, 2, \cdots$, take $f_n \in L^2((-\infty, T])$ such that

$$\sum_{n=0}^{\infty} \int_{-\infty}^{T} |f_n(t)|^2 dt < \infty$$

and let

$$g_n(t) - \int_{-\infty}^{\min(t,T)} e^{-(t-u)} f_n(u) du.$$

We then have to show that if

$$h_0(t) := \sum_{n=0}^{\infty} F(t)^n g_n(t) = 0,$$

then $f_n = 0$ in $L^2((-\infty, T])$, $n = 0, 1, 2, \cdots$. For this purpose, let

$$\begin{aligned} h_k(t) &= \sum_{n=k}^{\infty} n(n-1)\cdots(n-k+1) F(t)^{n-k} g_n(t), \; k \geq 1 \\ l_k(t) &= \sum_{n=k}^{\infty} n(n-1)\cdots(n-k+1) F(t)^{n-k} g_n'(t), \; k \geq 1 \\ l_0(t) &= \sum_{n=0}^{\infty} F(t)^n g_n'(t). \end{aligned}$$

It is clear that for all k,

$$l_k(t) \in L^2_{loc}(R), \quad h_k(t) \in C(R),$$

where $L^2_{loc}(R)$ and $C(R)$ denote all the locally L^2 integrable functions and all the continuous functions on R respectively. It then follows, by mathematical induction and hypotheses on F that

$$\begin{array}{rcl} h'_0(t) = l_0(t) + F'(t)h_1(t) = 0 & \Longrightarrow & h_1(t) = 0 \\ h'_1(t) = l_1(t) + F'(t)h_2(t) = 0 & \Longrightarrow & h_2(t) = 0 \\ \cdots & & \cdots \cdots \\ h'_k(t) = l_k(t) + F'(t)h_{k+1}(t) = 0 & \Longrightarrow & h_{k+1}(t) = 0 \\ \cdots & & \cdots \cdots \end{array}$$

In matrix form,

$$A_t \cdot \begin{pmatrix} g_0(t) \\ g_1(t) \\ \vdots \\ g_n(t) \\ \vdots \end{pmatrix} = \begin{pmatrix} 0 \\ 0 \\ \vdots \\ 0 \\ \vdots \end{pmatrix},$$

where

$$A_t = \begin{pmatrix} 1 & F(t) & F(t)^2 & F(t)^3 & \cdots & \binom{n}{0}F(t)^n & \cdots & \cdots \\ 0 & 1 & 2F(t) & 3F(t)^2 & \cdots & \binom{n}{1}F(t)^{n-1} & \cdots & \cdots \\ 0 & 0 & 1 & 3F(t) & \cdots & \binom{n}{2}F(t)^{n-2} & \cdots & \cdots \\ 0 & 0 & 0 & 1 & \cdots & \binom{n}{3}F(t)^{n-3} & \cdots & \cdots \\ \cdots & \cdots & \cdots & \cdots & \cdots & \cdots & \cdots & \cdots \\ \cdots & \cdots & \cdots & \cdots & \cdots & \binom{n}{n-1}F(t) & \cdots & \cdots \\ 0 & 0 & 0 & 0 & \cdots & 1 & \cdots & \cdots \\ \cdots & \cdots & \cdots & \cdots & \cdots & \cdots & \cdots & \cdots \end{pmatrix}.$$

On the other hand, if we let

$$B_t = \begin{pmatrix} 1 & -F(t) & F(t)^2 & -F(t)^3 & \cdots & (-1)^n\binom{n}{0}F(t)^n & \cdots & \cdots \\ 0 & 1 & -2F(t) & 3F(t)^2 & \cdots & (-1)^{n-1}\binom{n}{1}F(t)^{n-1} & \cdots & \cdots \\ 0 & 0 & 1 & -3F(t) & \cdots & (-1)^{n-2}\binom{n}{2}F(t)^{n-2} & \cdots & \cdots \\ 0 & 0 & 0 & 1 & \cdots & (-1)^{n-3}\binom{n}{3}F(t)^{n-3} & \cdots & \cdots \\ \cdots & \cdots & \cdots & \cdots & \cdots & \cdots & \cdots & \cdots \\ \cdots & \cdots & \cdots & \cdots & \cdots & -\binom{n}{n-1}F(t) & \cdots & \cdots \\ 0 & 0 & 0 & 0 & \cdots & 1 & \cdots & \cdots \\ \cdots & \cdots & \cdots & \cdots & \cdots & \cdots & \cdots & \cdots \end{pmatrix},$$

then $B_tA_t = A_tB_t$ turns out to be an infinite unit matrix. This results in $g_n(t) = 0$ and hence $f_n = 0, n = 0, 1, 2, \cdots$. The proof of the theorem is thus completed.

As a consequence of the approach, we may easily prove that for each positive integer N, the multiplicity of a process defined as

$$X(t) = \sum_{n=0}^{N-1} F(t)^n X_n(t)$$

is exactly N.

The argument for this follows if, in the proof, we define $Y(t)$ as

$$Y(t) = \sum_{n=0}^{N-1} F(t)^n B_n(t)$$

and A_t as

$$A_t = \begin{pmatrix} 1 & F(t) & F(t)^2 & F(t)^3 & \cdots & \binom{N-1}{0}F(t)^{N-1} \\ 0 & 1 & 2F(t) & 3F(t)^2 & \cdots & \binom{N-1}{1}F(t)^{N-2} \\ 0 & 0 & 1 & 3F(t) & \cdots & \binom{N-1}{2}F(t)^{N-3} \\ 0 & 0 & 0 & 1 & \cdots & \binom{N-1}{3}F(t)^{N-4} \\ \cdots & \cdots & \cdots & \cdots & \cdots & \cdots \\ \cdots & \cdots & \cdots & \cdots & \cdots & \binom{N-1}{N-2}F(t) \\ 0 & 0 & 0 & 0 & \cdots & 1 \end{pmatrix}$$

and B_t as

$$B_t = \begin{pmatrix} 1 & -F(t) & F(t)^2 & -F(t)^3 & \cdots & (-1)^{N-1}\binom{N-1}{0}F(t)^{N-1} \\ 0 & 1 & -2F(t) & 3F(t)^2 & \cdots & (-1)^{N-2}\binom{N-1}{1}F(t)^{N-2} \\ 0 & 0 & 1 & -3F(t) & \cdots & (-1)^{N-3}\binom{N-1}{2}F(t)^{N-3} \\ 0 & 0 & 0 & 1 & \cdots & (-1)^{N-4}\binom{N-1}{3}F(t)^{N-4} \\ \cdots & \cdots & \cdots & \cdots & \cdots & \cdots \\ \cdots & \cdots & \cdots & \cdots & \cdots & -\binom{N-1}{N-2}F(t) \\ 0 & 0 & 0 & 0 & \cdots & 1 \end{pmatrix}.$$

Finally, we need to demonstrate the existence of the function F. The construction will be done by using the Monotone Convergence Theorem.

Notation: Let $f(t)$ be a function locally symmetric at $t = x$ and let $N(x)$ denote the local support of f at x and $|N(x)|$ denote the Lebesgue measure of the support $N(x)$.

We first proceed to construct a sequence of functions $s_n(t), n = 1, 2, \cdots$ as follows. $s_1(t)$: (i) symmetric about y-axis, (ii) locally symmetric at $t = \frac{n}{2}$, $n = 1, 2, \cdots$ and $|N(\frac{n}{2})| \leq \frac{1}{2^2}$, and (iii) $0 < \int_{N(\frac{n}{2})} s_1(t)dt \leq \frac{1}{2}\frac{\delta}{2^{2+n}}$ and $\int_{N(\frac{n}{2})} s_1^2(t)dt = +\infty, n = 1, 2, \cdots$; $s_2(t)$: (i) symmetric about y-axis, (ii) locally symmetric at $t = \frac{n}{2^2}$, $n = 1, 3, 5, \cdots$, and $|N(\frac{n}{2^2})| \leq \frac{1}{2^3}$, and (iii) $0 < \int_{N(\frac{n}{2^2})} s_2(t)dt \leq \frac{1}{2}\frac{\delta}{2^{3+n}}$ and $\int_{N(\frac{n}{2^2})} s_2^2(t)dt = +\infty, n = 1, 2, \cdots$. In general, for $k \geq 3$, we similarly construct $s_k(t)$ as $s_k(t)$: (i) symmetric about y-axis, (ii) locally symmetric at $t = \frac{n}{2^k}$, $n = 1, 3, 5, \cdots$, and $|N(\frac{n}{2^k})| \leq \frac{1}{2^{k+1}}$, and (iii) $0 < \int_{N(\frac{n}{2^k})} s_k(t)dt \leq \frac{1}{2}\frac{\delta}{2^{k+1+n}}$ and $\int_{N(\frac{n}{2^k})} s_k^2(t)dt = +\infty, n = 1, 2, \cdots$.

Now, let us consider the sum

$$S_n(t) \stackrel{\text{def}}{:=} \sum_{k=1}^{n} s_k(t).$$

Since we obviously have

$$0 \leq S_1(t) \leq S_2(t) \leq \cdots,$$

and

$$0 < \lim_{n\to\infty} \int_{-\infty}^{\infty} S_n(t)dt \leq \sum_{k=1}^{\infty} 2^{-k}\delta = \delta,$$

it follows from the Monotone Convergence Theorem that

$$S(t) \stackrel{\text{def}}{:=} \lim_{n\to\infty} S_n(t)$$

exists for almost all t. Now define function F as

$$F(t) = \int_{-\infty}^{t} S(u)du.$$

We may easily verify that the function F satisfies the conditions as in the theorem.

Acknowledgements

I thank Prof. T. Hida of Nagoya University, Japan for his academic supervision. As his 70th birthday anniversary draws close, I wish him in good health. I thank Drs Fred Wright and Chuck Berry at University of California, San Diego for their encouragement and support.

References

[1] N. Aronszajn, Theory of reproducing kernels. Trans. Amer. Math. Soc. Vol. 68, 337-404 (1950).

[2] H. Cramér, Stochastic Processes as Curves in Hilbert Space. Theory Probability Appl., **9**(2), 169-179 (1964).

[3] H. Cramér, A Contribution to the Multiplicity Theory of Stochastic Processes. Proc. Fifth Berkeley Symp. Stat. Appl. Probability, **II**, 215-221 (1965).

[4] H. Cramér, Structural and Statistical Problems for a Class of Stochastic Processes. The First Samuel Stanley Wilks Lecture at Princeton University, March 17 1970, 1-30 (1971).

[5] T. Hida, Brownian Motion. Springer-Verlag (1980).

[6] T. Hida, Canonical Representations of Gaussian Processes and Their Applications. mem. College Sci., Univ. Kyoto, A33 (1), 109-155 (1960).

[7] K. Itô, Spectral Type of The Shift Transformation of Differential Processes With Stationary Increments. Trans. Amer. Math. Soc., Vol. 81, No. 2, 253-263 (1956).

[8] K. Itô, Multiple Wiener integral. J. Math. Soc. Japan 3, 157–169 (1951).

[9] S. Itô, On Hellinger-Hahn's Theorem. (In Japanese) Sūgaku, vol. 5, no. 2, 90-91 (1953).

[10] G. Kallianpur and V. Mandrekar, On the Connection between Multiplicity Theory and O. Hanner's Time Domain Analysis of Weakly Stationary Stochastic Processes. Univ. North Carolina Monograph Ser. Probability Stat., No. 3, 385-396 (1970).

[11] N. Wiener, Time Series. M.I.T. (1949).

[12] N. Wiener, Nonlinear Problems in Random Theory. M.I.T. (1958).

[13] N. Wiener, The Fourier Integral and Certain of Its Applications. Dover Publications. INC., New York (1958).

THEOREMES LIMITES POUR LES TEMPS LOCAUX D'UN PROCESSUS STABLE SYMETRIQUE

Nathalie Eisenbaum

Laboratoire de Probabilités - Université Paris VI - 4, Place Jussieu - Tour 56 - 3ème étage - 75252 Paris Cedex 05

Ceci constitue un complément à l'article [1] intitulé "Une version sans conditionnement du Theoreme d'isomorphisme de Dynkin" paru dans le Séminaire de Probabilités XXIX (p.266-289). Les précisions que nous apportons concernent la partie III traitant des théorèmes limites sur les temps locaux d'un processus stable symétrique.

I - Introduction

Soit $(L_t^x,\ x\in\mathbb{R},\ t\geq 0)$ le processus des temps locaux d'un processus X , à valeurs réelles, issu de 0, stable symétrique d'indice $\beta>1$. En notant $(p_t(x), x\in\mathbb{R}, t\geq 0)$ les densités de transition de X, on pose :

$$c_\beta = \int_0^{+\infty} (p_t(0) - p_t(1))dt.$$

Pour tout γ de [0,1], on appelle drap brownien fractionnaire d'indice γ, un processus gaussien centré continu $(B_t(x)\ ;\ x\in\mathbb{R}\ ,\ t\geq 0)$ de covariance :

$$E(\ B_s(x).B_t(y)\ ) = (s\wedge t).\Gamma^{(\gamma)}(x,y)$$

où $\quad \Gamma^{(\gamma)}(x,y) = \frac{1}{2}\ (\ |x|^\gamma + |y|^\gamma - |x-y|^\gamma\ ).$

Théorème 1 : *Pour* $y_1, y_2, \ldots, y_n$ *n réels distincts, on a :*

$$\left(X\ ,\ (\ \frac{1}{\varepsilon^{\frac{\beta-1}{2}}}\left(L_t^{\varepsilon x+y_k} - L_t^{y_k}\right)\ ;\ x\in\mathbb{R}\ , 1\leq k\leq n,\ t\geq 0)\ \right)$$

$$\xrightarrow[\varepsilon\to 0]{(d)} \left(\ X\ ,\ (\ \sqrt{c_\beta}\ B^{[y_k]}_{2L_t^{y_k}}(x)\ ;\ x\in\mathbb{R},\ 1\leq k\leq n\ ,\ t\geq 0)\ \right)$$

où $\{\ B_t^{[y_k]}(x),\ x\in\mathbb{R},\ 1\leq k\leq n,\ t\geq 0\ \}$ *est un système gaussien indépendant de*

X, *composé de* n *draps browniens fractionnaires d'indice* (β-1), *tous indépendants.*

L'objet de la partie III de [1] était d'établir la convergence du second terme du couple considéré dans le Théorème 1. Nous montrons ici que l'on peut adjoindre X dans le résultat. A la fin de [4], Rosen indique une preuve possible de ce résultat dans le cas n=1. Notre démonstration s'appuie exclusivement sur le Théorème d'isomorphisme de Dynkin qui permet dans un premier temps d'établir le théorème suivant .
On note L_T le processus $(L_T^x \; ; \; x\in\mathbb{R})$.

Théorème 2 : *Pour* $y_1, y_2, \ldots, y_n$ *n réels distincts et tout réel* a *,on a :*

$$\left(\left(L_T \, , \, \left(\frac{1}{\varepsilon^{\frac{\beta-1}{2}}}\left(L_T^{\varepsilon x+y_k} - L_T^{y_k}\right) \; ; \; x\in\mathbb{R} \, , 1\le k\le n\right)\right) \; \middle| \; X_T = a \right)$$

$$\xrightarrow[\varepsilon\to 0]{(d)} \left(\left(L_T \, , \, \left(\sqrt{c_\beta}\, B^{[y_k]}_{2L_T^{y_k}}(x) \; ; \; x\in\mathbb{R}, \; 1\le k\le n \right)\right) \; \middle| \; X_T = a \right)$$

où $\{ B_t^{[y_k]}(x), \; x\in\mathbb{R}, \; 1\le k\le n, \; t\ge 0 \}$ *est un système gaussien indépendant de* X_T *et de* L_T *, composé de* n *draps browniens fractionnaires d'indice* (β-1), *tous indépendants.*

Soulignons le fait que même dans le cas où X est un mouvement brownien et n=1, le Théorème 2 présente un résultat nouveau . En effet dans ce cas particulier Yor [5] a établi la convergence conjointe du Théorème 1. Mais elle ne permet pas à priori d'en déduire la convergence sous conditionnement du Théorème 2.
Nous renvoyons à l'article [1] pour une bibliographie plus étendue.

II - Les arguments

Dans [1] , nous avons prouvé que pour tout y la famille $\left(\frac{1}{\varepsilon^{\frac{\beta-1}{2}}}\left(L_t^{\varepsilon x+y} - L_t^{y}\right) \; ; \; x\in\mathbb{R} \; , \; t\ge 0\right)$ est tendue .Il suffit donc d'établir la convergence fini-dimensionnelle du couple .

1) Démonstration du Théorème 2

Soit T un temps exponentiel de paramètre λ, indépendant de X . Pour tout couple de réels (a,b), soit la probabilité $\tilde{P}_{ab}$ définie par :

$$\tilde{P}_{ab}\Big|_{\mathcal{F}_t} = \frac{g(X_t,b)}{g(a,b)} P_a\Big|_{\mathcal{F}_t} \qquad \text{sur } (t<T)$$

où g est la fonction de Green de X tué en T et $(\mathcal{F}_t)_{t\geq 0}$ désigne la filtration naturelle de X.

Soit $(\phi_x, x\in\mathbb{R})$ un processus gaussien centré indépendant de X, de covariance la fonction de Green de X tué en T. On note < . > l'espérance relativement à ϕ.

Le Théorème d'isomorphisme de Dynkin nous assure alors que pour toute fonctionnelle mesurable F :

$$\text{(I)} \qquad \tilde{P}_{ab} < F(L_T + \frac{\phi^2}{2}) > = < \frac{\phi_a \phi_b}{g(a,b)} F(\frac{\phi^2}{2}) >$$

On note : $\mathbb{A} = \{y_1, y_2, \dots, y_n\}$.

On a montré dans [1] que l'on a la convergence en loi suivante :

$$\Big((\frac{\phi_{\varepsilon x+y} - \phi_y}{\varepsilon^{\frac{\beta-1}{2}}} ; (x,y)\in\mathbb{R}\times\mathbb{A}) , (\frac{\phi_{\varepsilon x+y} + \phi_y}{2} ; (x,y)\in\mathbb{R}\times\mathbb{A}) , (\phi_x ; x\in\mathbb{R}) \Big)$$

$$\Big\downarrow (d) \quad \text{quand } \varepsilon\to 0$$

$$\Big((\sqrt{c_\beta}\, B_1^{[y]}(x) ; (x,y)\in\mathbb{R}\times\mathbb{A}) , (\phi_y ; y\in\mathbb{A}) , (\phi_x ; x\in\mathbb{R}) \Big)$$

où pour tout y , $B^{[y]}$ est un drap brownien fractionnaire d'indice $\beta-1$. Les processus $B^{[y]}$, y variant dans $\mathbb{A}$, étant tous indépendants entre eux et indépendants de ϕ.

En particulier , on a :

$$\Big((\frac{\phi^2_{\varepsilon x+y} - \phi^2_y}{2\,\varepsilon^{\frac{\beta-1}{2}}} ; (x,y)\in\mathbb{R}\times\mathbb{A}) , (\phi_x ; x\in\mathbb{R}) \Big)$$

$$\Big\downarrow (d) \quad \text{quand } \varepsilon\to 0$$

$$\Big((\sqrt{c_\beta}\, B^{[y]}_{\phi_y^2}(x) ; (x,y)\in\mathbb{R}\times\mathbb{A}) , (\phi_x ; x\in\mathbb{R}) \Big).$$

Ce résultat reste vrai sous la mesure $< \frac{\phi_a \phi_b}{g(a,b)}, \ . \ >$.

Notons H_ε l'application linéaire de $\mathcal{C}(\mathbb{R},\mathbb{R})$ dans $\mathcal{C}(\mathbb{R},\mathbb{R}) \times \mathcal{C}(\mathbb{R},\mathbb{R}^n)$ définie par:

$$H_\varepsilon(f) = \left(f,\ \varepsilon^{-(\beta-1)/2}(f(\varepsilon.+y) - f(y))\ ;\ y\in\mathbb{A}\right).$$

En utilisant (I), on obtient alors sous $\tilde{P}_{ab}< \ . \ >$:

$$H_\varepsilon\left(L_T + \frac{\phi^2}{2} \right) \xrightarrow[\varepsilon\to 0]{(d)} \left(L_T + \frac{\phi^2}{2},\ (\sqrt{c_\beta}\, B^{[y]}_{2L_T^y+\phi_y^2}(.)\ ;\ y\in\mathbb{A})\right),$$

ce qu'on peut écrire :

$$H_\varepsilon(L_T) + H_\varepsilon\left(\frac{\phi^2}{2} \right) \xrightarrow[\varepsilon\to 0]{(d)}$$

$$(L_T\ ,\ (\sqrt{c_\beta}\, \tilde{B}^{[y]}_{2L_T^y}(.)\ ;\ y\in\mathbb{A})) \ +\ \left(\frac{\phi^2}{2},\ (\sqrt{c_\beta}\, B^{[y]}_{\phi_y^2}(.)\ ;\ y\in\mathbb{A})\right),$$

où $\tilde{B}^{[y]}, y\in\mathbb{A}$ est une famille de mouvements browniens fractionnaires indépendants, indépendante de L_T, T et de ϕ.
Les transformées de Laplace-Fourier fini-dimensionnelles associées au processus $\left(\frac{\phi^2}{2},\ (\sqrt{c_\beta}\, B^{[y]}_{\phi_y^2}(.)\ ;\ y\in\mathbb{A})\right)$, ne s'annulant pas, on en déduit :

$$H_\varepsilon(L_T) \xrightarrow[\varepsilon\to 0]{(d)} \left[L_T\ ,\ (\ \sqrt{c_\beta}\, B^{[y_k]}_{2L_T^{y_k}}(x)\ ;\ x\in\mathbb{R},\ 1\le k\le n\ ) \right]$$

avec $(B^{[y_i]}\ ;\ 1\le i\le n)$ indépendant de L_T.

Le Théorème 2 s'obtient maintenant grâce à la proposition suivante . $\mathcal{F}_T$ désigne la tribu engendrée par les évènements de la forme $A \cap \{T>t\}$, où $A\in\mathcal{F}_t$.

Proposition 3 : *Les probabilités* $\tilde{P}_{ab}$ *et* $P_a(\ .\ \mid X_T = b)$ *coïncident sur* $\mathcal{F}_T$

<u>Démonstration de la Proposition 3</u> Il suffit de montrer que pour tout borélien B de $\mathbb{R}$ et tout A élément de $\mathcal{F}_t$, on a :

$$\int_B \tilde{P}_{ab}(A \cap \{T>t\}) P_a(X_T \in db) = P_a(A \cap \{T>t\} \cap \{X_T\in B\}) \quad .$$

En remarquant que : $P_a(X_T \in db) = \lambda\, g(a,b)\, db$, on a :

$$\begin{aligned}
\int_B \tilde{P}_{ab}(A \cap \{T>t\}) P_a(X_T \in db) &= \int_B \tilde{P}_{ab}(A \cap \{T>t\}) \lambda\, g(a,b)\, db \\
&= \int_B E_a(A \cap \{T>t\};\, g(X_t,b))\, \lambda\, db \\
&= \int_B E_a(A \cap \{T>t\};\, g(X_t,b))\, \lambda\, db \\
&= \lambda\, e^{-\lambda t} \int_B E_a(A \,;\, g(X_t,b))\, db \\
&= \lambda\, e^{-\lambda t} \int_0^{+\infty} \int_B E_a(A \,;\, p_s(X_t-b))\, e^{-\lambda s} ds\, db \\
&= \lambda\, e^{-\lambda t} \int_0^{+\infty} E_a(A \,;\, X_{s+t}\in B)\, e^{-\lambda s} ds \\
&= \int_t^{+\infty} E_a(A \,;\, X_s\in B)\, \lambda e^{-\lambda s} ds \\
&= P_a(A \cap \{T>t\} \cap \{X_T\in B\}).
\end{aligned}$$

Nous faisons également les deux remarques suivantes :

<u>Remarques</u> :

(i) ϕ étant indépendant de T, on obtient de la même façon :

$$\left(L_T \,,\, \Big(\frac{1}{\varepsilon^{\frac{\beta-1}{2}}} \Big(L_T^{\varepsilon x + (y_k T)^{1/\beta}} - L_T^{y_k T^{1/\beta}} \Big) \,;\, x\in\mathbb{R} \,, 1\le k\le n\Big) \;\Big|\; X_T = b \right)$$

$$\xrightarrow[\varepsilon\to 0]{(d)} \left(L_T \,,\, \Big(\sqrt{c_\beta}\, B^{[y_k]}_{2L_T^{y_k T^{1/\beta}}}(x) \,;\, x\in\mathbb{R},\ 1\le k\le n \Big) \;\Big|\; X_T = b \right)$$

(ii) Par un argument de convergence dominée , on obtient à partir de la remarque (i) :

$$\left(X_T , L_T , \left(\frac{1}{\varepsilon^{\frac{\beta-1}{2}}} \left(L_T^{\varepsilon x+(y_k T^{1/\beta})} - L_T^{y_k T^{1/\beta}} \right) ; x\in\mathbb{R} , 1\le k\le n \right) \right)$$

$$\xrightarrow[\varepsilon\to 0]{(d)} \left(X_T , L_T , \left(\sqrt{c_\beta}\, B^{[y_k]}_{2L_T^{y_k T^{1/\beta}}}(x) ; x\in\mathbb{R}, 1\le k\le n \right) \right)$$

avec $(B^{[y_i]} ; 1\le i\le n)$ indépendant de (L_T, X_T).

2) Passage du temps exponentiel indépendant à un temps déterministe

Montrons maintenant que la convergence obtenue à l'étape 1) permet d'obtenir pour tout $t>0$:

$$\left(X_t , L_t , \left(\frac{1}{\varepsilon^{\frac{\beta-1}{2}}} \left(L_t^{\varepsilon x+y_k} - L_t^{y_k} \right) ; x\in\mathbb{R} , 1\le k\le n \right) \right)$$

$$\xrightarrow[\varepsilon\to 0]{(d)} \left(X_t , L_t , \left(\sqrt{c_\beta}\, B^{[y_k]}_{2L_t^{y_k}}(x) ; x\in\mathbb{R}, 1\le k\le n \right) \right).$$

De façon générale les déductions de ce genre sont fausses . (Par exemple la convergence des transformées de Laplace d'une suite de mesures sur $\mathbb{R}_+$ absolument continues par rapport à la mesure de Lebesgue sur $\mathbb{R}_+$, n'implique pas la convergence des densités respectives). C'est pourquoi nous détaillons notre affirmation. Pour alléger l'écriture nous traitons le cas où X est un mouvement brownien, les autres cas se traitant avec les mêmes arguments.

Nous allons utiliser le lemme suivant qui est une conséquence immédiate de la définition de la convergence en loi :

Lemme 4 : *Soit (X_n) une suite de variables aléatoires à valeurs dans un espace métrisable E, convergeant en loi vers X. Pour toute fonction continue f : $E\to G$ où G est un espace métrisable, la suite $(f(X_n))$ converge en loi vers $f(X)$.*

On remarque que : $T = \int_{\mathbb{R}} L_T^x\, dx$. Soit ϕ la fonction définie sur l'ensemble des fonctions réelles continues à support compact, muni de la topologie de la convergence uniforme sur les compacts , telle que :

$$\phi(f) = (f, \textstyle\int_{\mathbb{R}} f(x)dx).$$

En utilisant le Lemme 4 , on obtient alors à partir de la convergence de la Remarque (ii) :

$$\left(T , X_T , L_T , \left(\frac{1}{\sqrt{\varepsilon}}\left(L_T^{\varepsilon x + y_k\sqrt{T}} - L_T^{y_k\sqrt{T}}\right) ; x\in\mathbb{R} , 1\le k\le n\right) \right)$$

$$\xrightarrow[\varepsilon\to 0]{(d)} \left(T , X_T , L_T , \left(\sqrt{c_2}\, B^{[y_k]}_{2L_T^{y_k\sqrt{T}}}(x) ; x\in\mathbb{R}, 1\le k\le n \right) \right)$$

En utilisant les propriétés de scaling du mouvement brownien, la convergence ci-dessus est équivalente à :

$$\left(T , \sqrt{T}\, X_1 , (\sqrt{T}\, L_1^{x/\sqrt{T}} , x\in\mathbb{R}) , \left(\frac{\sqrt{T}}{\sqrt{\varepsilon}}\left(L_1^{(\varepsilon x/\sqrt{T})+y_k} - L_1^{y_k}\right) ; x\in\mathbb{R} , 1\le k\le n\right) \right)$$

$$\Big\downarrow (d) \quad \text{quand } \varepsilon\to 0$$

$$\left(T , \sqrt{T}\, X_1 , (\sqrt{T}\, L_1^{x/\sqrt{T}} , x\in\mathbb{R}) , \left(\sqrt{c_2}\, B^{[y_k]}_{2\sqrt{T}\, L_1^{y_k}}(x) ; x\in\mathbb{R}, 1\le k\le n \right) \right).$$

Soit G_n la fonction de $\mathbb{R} \times \mathscr{C}(\mathbb{R},\mathbb{R}^n)$ dans $\mathscr{C}(\mathbb{R},\mathbb{R}^n)$ définie par :

$$G_n(a,f)(x) = f(ax).$$

En munissant $\mathscr{C}(\mathbb{R},\mathbb{R}^n)$ de la topologie de la convergence uniforme sur les compacts, G_n est continue.

On considère la fonction F suivante :

$$F : \mathbb{R}_+^* \times \mathbb{R} \times \mathscr{C}(\mathbb{R},\mathbb{R}) \times \mathscr{C}(\mathbb{R},\mathbb{R}^n) \longrightarrow \mathbb{R} \times \mathscr{C}(\mathbb{R},\mathbb{R}) \times \mathscr{C}(\mathbb{R},\mathbb{R}^n)$$

$$F(t,x,f,g) = \left(\frac{x}{\sqrt{t}} , \frac{1}{\sqrt{t}} G_1(\sqrt{t},f) , \frac{1}{\sqrt{t}} G_n(\sqrt{t},g) \right)$$

F étant continue, on utilise le Lemme 4, pour obtenir :

$$F\left(T , \sqrt{T}\, X_1 , (\sqrt{T}\, L_1^{x/\sqrt{T}} , x\in\mathbb{R}) , \left(\frac{\sqrt{T}}{\sqrt{\varepsilon}}\left(L_1^{(\varepsilon x/\sqrt{T})+y_k} - L_1^{y_k}\right) ; x\in\mathbb{R} , 1\le k\le n\right) \right)$$

$$\Big\downarrow (d) \quad \text{quand } \varepsilon\to 0$$

$$F\left(T , \sqrt{T}\, X_1 , (\sqrt{T}\, L_1^{x/\sqrt{T}} , x\in\mathbb{R}) , \left(\sqrt{c_2}\, B^{[y_k]}_{2\sqrt{T}\, L_1^{y_k}}(x) ; x\in\mathbb{R}, 1\le k\le n \right) \right).$$

Ce qui s'écrit également :

$$\left(X_1 , L_1 , \left(\frac{1}{\sqrt{\varepsilon}} \left(L_1^{\varepsilon x+y_k} - L_1^{y_k} \right) ; x\in\mathbb{R} , 1\le k\le n \right) \right)$$

$$\xrightarrow[\varepsilon\to 0]{(d)} \left(X_1 , L_1 , \left(\sqrt{c_2}\, B^{[y_k]}_{2L_1^{y_k}}(x) ; x\in\mathbb{R}, 1\le k\le n \right) \right).$$

Ce résultat s'étend à tout $t>0$ par scaling.

3) Convergence fini-dimensionnelle

Pour établir la convergence fini-dimensionnelle on considère F_1 et F_2 des fonctionnelles définies sur $\mathscr{C}(\mathbb{R},\mathbb{R})\times\mathscr{C}(\mathbb{R},\mathbb{R}^n)$ telles que pour $m = 1,2$:

$$F_m (f_1, f_2) = \exp\{ i \int_{\mathbb{R}} f_1(x)d\mu_m(x) + i \int_{\mathbb{R}} f_2(x)d\nu_m(x) \}$$

où pour tout m, μ_m et ν_m sont des mesures σ-finies sur respectivement $\mathbb{R}$ et $\mathbb{R}^n$.

On considère également g_1 et g_2 deux fonctions bornées éléments de $\mathscr{C}(\mathbb{R},\mathbb{R})$.

On note $Y_t(\varepsilon)$ le processus $\left(\frac{L_t^{\varepsilon x+y_k} - L_t^{y_k}}{\varepsilon^{\frac{\beta-1}{2}}} , x\in\mathbb{R} , 1\le k\le n \right)$ et Z_t le processus $\left(\sqrt{c_\beta}\, B^{[y_k]}_{2L_t^{y_k}}(x) ; x\in\mathbb{R}, 1\le k\le n\right)$. Soient $t,s > 0$. On a :

$$\mathbb{E}\left(F_1(L_t,Y_t(\varepsilon))\, g_1(X_t) \,.\, F_2(L_{t+s},Y_{t+s}(\varepsilon))\, g_2(X_{t+s}) \right)$$

$$= \mathbb{E}\left(F_1F_2(L_t,Y_t(\varepsilon))\, g_1(X_t) \,.\, [F_2(L_s,Y_s(\varepsilon))\, g_2(X_s)] \circ \theta_t \right).$$

Donc le terme de gauche est en fait de la forme :

$$\mathbb{E}\left(F_1F_2(L_t,Y_t(\varepsilon)) .\, \psi_\varepsilon(X_t)\right)$$

avec pour tout $x\in\mathbb{R}$: $\psi_\varepsilon(x) = \mathbb{E}_x\left(F_2(L_s,Y_s(\varepsilon))\, g_2(X_s)\right) . g_1(x)$.

Grâce à l'étape 2), on sait que pour tout x:

$$\psi_\varepsilon(x) \xrightarrow[\varepsilon\downarrow 0]{} \psi(x) = \mathbb{E}_x\left(F_2(L_s,Z_s)\, g_2(X_s)\right) . g_1(x)$$

$$
\begin{aligned}
&\left| \mathbb{E}\Big(F_1(L_t,Y_t(\varepsilon))g_1(X_t).F_2(L_{t+s},Y_{t+s}(\varepsilon))g_2(X_{t+s})\Big) \right.\\
&\qquad \left. - \ \mathbb{E}\Big(F_1(L_t,Z_t)g_1(X_t).F_2(L_{t+s},Z_{t+s})g_2(X_{t+s}) \Big) \right| \\
&= \left| \mathbb{E}\Big(F_1F_2(L_t,Y_t(\varepsilon)).\psi_\varepsilon(X_t)\Big) - \mathbb{E}\Big(F_1F_2(L_t,Z_t).\ \psi(X_t)\Big) \right| \\
&\le \mathbb{E}\Big(|F_1F_2(L_t,Y_t(\varepsilon))|\ |\psi_\varepsilon(X_t) - \psi(X_t)|\Big) \\
&\qquad + \left| \mathbb{E}\Big(F_1F_2(L_t,Y_t(\varepsilon)).\psi(X_t)\Big) - \mathbb{E}\Big(F_1F_2(L_t,Z_t).\psi(X_t)\Big) \right|
\end{aligned}
$$

Dans la somme ci-dessus , quand ε tend vers 0 , le premier terme tend vers 0 par un argument de convergence dominée . Le second tend également vers 0 grâce à la convergence établie à l'étape 2).

On en déduit que :

$$
\Big((X_t,L_t,Y_t(\varepsilon))\ ,\ (X_{t+s},L_{t+s},Y_{t+s}(\varepsilon))\Big) \xrightarrow[\varepsilon\downarrow 0]{(d)} \Big((X_t,L_t,Z_t)\ ,\ (X_{t+s},L_{t+s},Z_{t+s})\Big).
$$

On étend aisément la démonstration à une suite finie de temps.
Ce qui achève la preuve du Théorème 1.

Références

[1] **Eisenbaum N.** Une version sans conditionnement du Théorème d'isomorphisme de Dynkin.*Sém.de Probabilités XXIX. LNM 1613 (266-289). Springer.1995*

[2] **Marcus M.B. and Rosen J.** Sample path properties of the local times of strongly symmetric Markov processes via Gaussian processes. *Annals of Proba.,20,n°4(1603-1684).1992*

[3] **Revuz D. and Yor M.** Continuous martingales and Brownian motion. *Springer Verlag.1991.Second edition 1994.*

[4] **Rosen J.** Second order limit laws for the local times of stable processes.*Sém.de Probabilités XXV.LNM 1485 (407-425).Springer.1991.*

[5] **Yor M.** Le drap brownien comme limite en loi des tempslocaux d'un mouvement brownien linéaire.*Sém de Probabilités XVII.LNM 986 (89-105).Springer.1983.*

An Itô type isometry for loops in $\mathbf{R}^d$ via the Brownian bridge

Pierre Gosselin and Tilmann Wurzbacher

Institut de Recherche Mathématique Avancée
Université Louis Pasteur et CNRS
7, rue René Descartes
67084 Strasbourg Cedex France
gosselin@math.u-strasbg.fr and wurzbach@math.u-strasbg.fr

Summary. We show that iterated stochastic integrals (in the Itô-sense) with respect to the Brownian bridge on $\mathbf{R}^d$ give an explicit unitary isomorphism between the symmetric Fock space over the $\mathbf{C}^d$-valued square-integrable functions on the unit interval having zero mean and the space of complex valued L^2-functions on based continuous loops on $\mathbf{R}^d$.

Mathematics Subject Classification (1991): 60G15, 60H05, 60J65

Introduction

The theory of Gaussian processes yields an identification of the space of square integrable functions on a given probability space provided with a Gauss process $(Z_t)_{t\in T}$ with the sum of the symmetric tensor powers of the "Gauss space", the closure of the span of the random variables Z_t (see e.g. [HT], Kap. 5.8 or [N], Ch.7). In the case of "normal martingales" the above identification is concretely realized by means of multiple stochastic integration (see [DMM], pp.199). Motivated by the important rôle recently played by loop spaces, especially with values in $\mathbf{R}^d$ or compact Lie groups, and "holomorphic L^2-sections" in appropriate line bundles over them in mathematics and physics (see e.g. [BR] or [PS]), we consider here the case of the Brownian bridge, which is only a semimartingale.

We show that iterated stochastic integrals with respect to the Brownian bridge exist and yield a unitary isomorphism between the Fock space over square-integrable, $\mathbf{C}^d$-valued functions on the unit interval having mean zero and the space of $\mathbf{C}$-valued square-integrable functions on based paths in $\mathbf{R}^d$ ("Wiener space"). Since the Brownian bridge is an adapted stochastic process that induces an isomorphism between paths and loops, transporting the Wiener measure to the conditioned Wiener measure, this result implies an orthogonal decomposition of L^2-functions on based loops in $\mathbf{R}^d$. We apply this Itô isometry in [GW] to give an analytically rigorous derivation of the Virasoro anomaly in quantization of strings.

The article is organised as follows:
In the first section we recall some notation and elementary facts about Gauss spaces and associated reproducing kernel Hilbert spaces, adapted to the case of a Brownian bridge. The second section relates the stochastic integrals of deterministic functions with respect to a Brownian motion to those with respect to the particular Brownian bridge, that we use to simulate based continuous loops.
In the final section 3 we state and prove the isometry between "mean-zero functions" on simplices and the L^2-functions on Wiener space by means of iterated stochastic integrals with respect to the Brownian bridge of the second section.

Section 1: The Gauss space of a Brownian bridge

Let $(\Omega, \mathcal{F}, \mathbf{P})$ be a probability space with a $\mathbf{R}$-valued Brownian bridge $(X_t)_{t\in[0,1]}$ from 0 to 0. On the probability space $(\tilde{\Omega}, \tilde{\mathcal{F}}, \tilde{\mathbf{P}}) = (\Omega \times \mathbf{R}, \mathcal{F} \otimes \mathcal{B}(\mathbf{R}), \mathbf{P} \otimes \nu)$, where ν denotes the probability measure $(2\pi)^{-1/2} exp(-z^2/2)dz$ on $\mathbf{R}$, we have the additional random variable Z, defined by $Z(\omega, z) = z$, who is independent of the variables $\tilde{X}_t$, given by $\tilde{X}_t(\omega, z) = X_t(\omega)$. A direct calculation shows that the process $W_t = \tilde{X}_t + tZ$ $(t \in [0,1])$ is a Brownian motion on $\tilde{\Omega}$.
Denoting the Gauss spaces of $\tilde{X}_t$ respectively W_t by $G_{\tilde{X}}$ respectively G_W, we find the following orthogonal decomposition:

$$G_W = G_{\tilde{X}} \oplus \mathbf{C} \cdot Z$$

(See e.g. [N] for the general theory of Gauss spaces). The map

$$u : G_W \to \mathbf{C}^{[0,1]},\ u(V)(t) = \mathbf{E}[V \cdot W_t]$$

is a unitary isomorphism of the closed subspace G_W of $L^2(\tilde{\Omega}, \tilde{\mathcal{F}}, \tilde{\mathbf{P}})$ onto its image $u(G_W) = \mathcal{H}_W$, the associated reproducing kernel Hilbert space (referred to as the "reproducing Hilbert space" in the sequel). Since W_t is a Brownian motion, one knows that sending h in $\mathcal{H}_W$ to $\dot{h} = f$ is a unitary isomorphism onto $L^2([0,1])$. Furthermore the inverse map u^{-1} is given as follows:

$$u^{-1}(h) = \int_0^1 \dot{h}(t) dW_t = J^W(\dot{h}) = J^W(f),$$

the notation J^W designing the stochastic integral with respect to W (see e.g. [N] for this fact).
It follows that $u(Z)(t) = \mathbf{E}[ZW_t] = t$ corresponds to the constant function 1 in $L^2([0,1])$. Thus $G_{\tilde{X}}$ is isomorphic to $L_0^2([0,1])$, the space of L^2-functions having mean zero. Since $d\tilde{X}_t = dW_t - W_1 dt$, the isomorphism between $L_0^2([0,1])$ and $G_{\tilde{X}}$ is given by a map $J^{\tilde{X}}$, which consists in integrating a deterministic function with respect to $d\tilde{X}_t$. This operator $J^{\tilde{X}}$ equals in fact the stochastic integral with respect to $\tilde{X}_t$, as can be shown by either the general theory of enlargements of filtrations (see e.g. [JY]) or by the estimates given below (see Proposition 2.2).
Remark. The above considerations extend easily to processes with values in $\mathbf{R}^d$ with $d > 1$. For the sake of clarity of the exposition, we will nevertheless work in the case $d = 1$.

Section 2: Stochastic integration with respect to the "adapted" bridge

On the Wiener space $\Omega = \{\omega \in C^0([0,1],\mathbf{R}^d) \mid \omega(0) = 0\}$ one has the natural filtration $\mathcal{F}_t = \sigma(\{B_s | 0 \le s \le t\})$ associated to the Brownian motion $B_t(\omega) = \omega(t)$, and the Wiener measure $\mathbf{P}$. We define a ("adapted") Brownian bridge by setting $X_t = (1-t)\int_0^t \frac{dB_s}{1-s}$ for $0 \le t < 1$ and $X_1 = 0$. Let us recall that X_t is a continuous semi-martingale and that $B_t = X_t + \int_0^t \frac{X_s}{1-s}ds$. It follows notably that the X_t generate the same filtration as the B_t.
Since the law of X_t equals the conditioned Wiener measure, Ω together with X can be looked upon as a substitute for the space of based, continous loops in $\mathbf{R}$.
The above inversion formula implies that the Gauss spaces G_X and G_B are equal. "Pulling-back" $G_{\tilde{X}}$ to G_X, we know that G_X is isomorphic to $L_0^2([0,1])$ by means of an integration operator J^X with respect to X. The Gauss space G_B being given by stochastic integrals with respect to B based on $L^2([0,1])$, we aim in this section for an explicit description of the relation between the two "realizations" of $G_X = G_B$. We prepare ourselves first with a lemma concerning deterministic L^2-functions :

Lemma 2.1
(i) Let f be in $L_0^2([0,1])$, then $\lim_{\xi\nearrow 1}\left(\frac{\int_0^\xi f(t)dt}{\sqrt{1-\xi}}\right) = 0.$
(ii) Let $\tilde{f} \in L^2([0,1])$ and $\gamma(\tilde{f})(t) = \int_0^t \frac{\tilde{f}(s)}{1-s}ds$, then $\|\gamma(\tilde{f})\|_{L^2} \le 2\|\tilde{f}\|_{L^2}$.
(iii) The operator $\alpha : L_0^2([0,1]) \to L^2([0,1])$, defined by

$$\alpha(f)(s) = f(s) + \frac{1}{1-s}\int_0^s f(t)dt,$$

is a unitary isomorphism, whose inverse is given by $\beta(\tilde{f}) = \tilde{f} - \gamma(\tilde{f})$.

We omit the proof, who boils down to Hardy type estimates for the parts (i) and (ii) (variants of the estimates can be found in the classical reference [HLP] and in [JY] for related probabilistic purposes as in the text) and uses partial integration plus convergence arguments on the boundary terms based on (i) and (ii) for the third part.
We can now prove the fundamental relation.

Proposition 2.2
(i) For all f in $L^2([0,1])$, the stochastic integral $J^X(f) = \int_0^1 f(t)dX_t$ exists and J^X is a unitary isometry from $L_0^2([0,1])$ to the Gauss space G_X.
(ii) A random variable Y in G_B, realized as $\int_0^1 \tilde{f}dB_t$ with $\tilde{f}$ in $L^2([0,1])$, equals $\mathbf{P}$-almost everywhere the random variable $\int_0^1 f(t)dX_t$ with $f(t) = \beta(\tilde{f})(t)$ $= \tilde{f}(t) - \int_0^t \frac{\tilde{f}(s)}{1-s}ds$.

Proof. Since $\int_0^1 1 \cdot dX_t = 0$, we can restrict ourselves to f in $L_0^2([0,1])$.
For $0 \le \xi < 1$ integration-by-parts implies then:

$$\int_0^\xi f(t)dX_t = \int_0^\xi \alpha(f)(t)dB_t + \left(\int_0^\xi f(s)ds\right)\left(\int_0^\xi \frac{dB_s}{1-s}\right).$$

The limit $\lim_{\xi\nearrow 1}$ of the first term exists in L^2 by the Itô isometry for B and the fact that $\alpha(f)$ is in $L^2([0,1])$ by Lemma 2.1. The same lemma shows, again together with the Itô isometry for B, that the limit of the second term vanishes. This shows the existence of $\int_0^1 f(t)dX_t$ as well as its $\mathbf{P}$-a.e. equality to $\int_0^1 \alpha(f)(t)dB_t$. Since α and β are mutually inverse unitary isomorphisms and J^B is isometric as well, the other assertions of the proposition follow immediately. □

Remarks.
(1) The existence of J^X as well as its unitarity can of course also be derived from the analogous properties of $J^{\tilde{X}}$ (see Section 1). We state the result nevertheless in the above proposition since it follows effortlessly as a by-product of the proof of the assertion (ii).
(2) The second assertion above can be restated by saying that the isomorphism

$$L_0^2([0,1]) \cong \mathcal{H}_X \cong G_X = G_B \cong \mathcal{H}_B \cong L^2([0,1])$$

is realized by α and the respective stochastic integrals yield the corresponding representations of elements of the Gauss space.

Section 3: Wiener chaos decomposition

The general theory of Gaussian processes gives an isomorphism between the symmetric Fock space over the Gauss space (or reproducing Hilbert space) of a Gaussian process and the space of L^2-functions of the given probability space (see again e.g. [N]). We will show in this section that, for the Brownian bridge X_t considered in Section 2, this unitary isomorphism onto $L^2(\Omega, \mathcal{F}, \mathbf{P})$ (where $\mathcal{F} = \sigma(\{B_s | 0 \le s \le 1\})$) can be realized by means of iterated stochastic integrals with respect to X and based on appropriate L^2-functions on simplices. Similar results for normal martingales are well-known (see e.g. [DMM]). We begin by assuring the existence of the iterated integrals with respect to X_t.
Let $\sum_n$ denote the n-simplex $\{(s_1, \ldots, s_n) \in \mathbf{R}^n | 0 \le s_1 \le \cdots \le s_n \le 1\}$ in the sequel.

Lemma 3.1 *Let F_n in $L^2(\sum_n)$ be the restriction to $\sum_n$ of the n-fold symmetrization $\underbrace{f \otimes \cdots \otimes f}_{n \text{ times}}$ of a function f in $L^2([0,1])$. Then the iterated integral*

$$J_t^X(F_n) = \int_0^t \left(\int_0^{s_n} \cdots \left(\int_0^{s_2} F_n(s_1, \ldots, s_n) dX_{s_1} \right) dX_{s_2} \ldots dX_{s_{n-1}} \right) dX_{s_n}$$

exists and is a continuous semimartingale.

Proof. The first chaos $Y_t = \int_0^t f(s)dX_s$ is already known to exist and is a continuous semi-martingale. Writing now $J_t^X(F_n)$ as

$$\int_0^t \left(\int_0^{s_n} \cdots \left(\int_0^{s_2} f(s_1) dX_{s_1} \right) dY_{s_2} \ldots dY_{s_{n-1}} \right) dY_{s_n}$$

and recalling that the stochastic integral of a continuous semi-martingale with respect to a continuous semi-martingale is again a continuous semi-martingale (see e.g. [HT]), the assertion follows by induction on n.

□

Let us recall that the solution of the stochastic differential equation

$$d(\mathcal{E}_t^B(h)) = \mathcal{E}_t^B(h)h(t)dB_t$$

for h in $L^2([0,1])$ is explicitely given by

$$\mathcal{E}_t^B(h) = \exp\left(\int_0^t h(s)dB_s - \frac{1}{2}\int_0^t h^2(s)ds\right)$$

(see e.g. [DMM], XXI §1.11). This process is called the "exponential process with respect to B based on h" and the random variable $\mathcal{E}^B(h) = \mathcal{E}_1^B(h)$ a "stochastic exponential vector".
The exponential process $\mathcal{E}_t^X(h) = \exp\left(\int_0^t h(s)dX_s - \frac{1}{2}\int_0^t h^2(s)ds\right)$ and the exponential vector $\mathcal{E}^X(h) = \mathcal{E}_1^X(h)$ enjoy similar properties as $\mathcal{E}_t^B(h)$ and $\mathcal{E}^B(h)$ (see [DMM], §1.9-1.11 for the case of the Brownian motion).

Lemma 3.2
Let h be in $L^2([0,1])$, then

(i) $d(\mathcal{E}_t^X(h)) = \mathcal{E}_t^X(h)h(t)dX_t$

(ii) $\mathcal{E}_t^X(h) = 1 + \int_0^t \mathcal{E}_s^X(h)h(s)dX_s$

(iii) $\mathcal{E}^X(\lambda h) = \sum_{n\geq 0}\lambda^n\left(\int_{\Sigma_n} h(s_1)\cdots h(s_n)dX_{s_1}\cdots dX_{s_n}\right)$ *for* $\lambda\in\mathbf{C}$.

Proof.
(i) This assertion follows directly from $(dX_t)^2 = dt$ and Itô's formula.
(ii) Since $\mathcal{E}_t^X(0) = 1$, the stochastic differential equation in Assertion (i) yields the result.
(iii) This assertion is given by applying $N+1$ times the Picard iteration scheme to the equality in Assertion (ii) and the fact that $\mathcal{E}_1^X(\lambda h)$ is analytic in the parameter λ. □

Lemma 3.3
For $\tilde{f}$ in $L^2([0,1])$ and $f = \beta(\tilde{f})$ in $L_0^2([0,1])$ (see Lemma 2.1 (iii) for the definition of the map β), we have equality of the corresponding exponential vectors :

$$\mathcal{E}^B(\tilde{f}) = \mathcal{E}^X(f).$$

Proof. Since $\int_0^1 \tilde{f}(s)dB_s = \int_0^1 f(s)dX_s$ by Proposition 2.2, it remains only to show that $\int_0^1 \tilde{f}^2(s)ds = \int_0^1 f^2(s)ds$. The latter equality follows now by observing that the unitary isometry α is already defined over the reals. □

Remark. Of course, we also have $\mathcal{E}^X(f) = \mathcal{E}^B(\tilde{f})$ with $\tilde{f} = \alpha(f)$.
We can now prove that the map J^X is isometric :

Lemma 3.4
Let f and g be in $L_0^2([0,1])$ and $F_n = \underbrace{f\otimes\cdots\otimes f}_{n \text{ times}}$ (resp. $G_m = \underbrace{g\otimes\cdots\otimes g}_{m \text{ times}}$) their n-fold (respectively m-fold) symmetrisation, viewed as functions on Σ_n (respectively Σ_m). Then we have
(i) $\langle J^X(F_n), J^X(G_m)\rangle_{L^2(\Omega,\mathbf{P})} = 0$ for $n \neq m$
(ii) $\langle J^X(F_n), J^X(G_m)\rangle_{L^2(\Omega,\mathbf{P})} = \frac{1}{n!}\left(\langle f,g\rangle_{L_0^2([0,1])}\right)^n = \langle F_n, G_m\rangle_{L^2(\Sigma_n)}$ for $n = m$.

Proof. Let λ, μ be in $\mathbf{R}$. We have, by Lemma 3.2 (iii)

$$J^X(F_n) = \int_{\Sigma_n} f(s_1)\cdots f(s_n)dX_{s_1}\ldots dX_{s_n} = \frac{1}{n!}\frac{d^n}{d^n\lambda}\bigg|_{\lambda=0} \mathcal{E}^X(\lambda f)$$

and analogously for $J^X(G_m)$.
By Lemma 3.3, the Itô-isometry for B, and Lemma 2.1(iii), we have

$$\begin{aligned}\langle \mathcal{E}^X(\lambda f), \mathcal{E}^X(\mu g)\rangle &= \langle \mathcal{E}^B(\lambda\tilde{f}), \mathcal{E}^B(\mu\tilde{g})\rangle \\ &= \exp(\langle \lambda\tilde{f}, \mu\tilde{g}\rangle) = \exp(\lambda\mu\langle\tilde{f},\tilde{g}\rangle) = \exp(\lambda\mu\langle f,g\rangle)\,.\end{aligned}$$

As a consequence, using again Lemma 3.2(iii), we get

$$\begin{aligned}\langle J^X(F_n), J^X(G_m) &= \frac{1}{n!}\frac{1}{m!}\frac{d^n}{d^n\lambda}\bigg|_{\lambda=0}\frac{d^m}{d^m\mu}\bigg|_{\mu=0}\langle \mathcal{E}^X(\lambda f), \mathcal{E}^X(\mu g)\rangle \\ &= \frac{1}{n!}\frac{1}{m!}\frac{d^n}{d^n\lambda}\bigg|_{\lambda=0}\frac{d^m}{d^m\mu}\bigg|_{\mu=0}\left(\sum_{k\geq 0}\frac{(\lambda\mu)^k}{k!}\langle f,g\rangle^k\right) \\ &= \delta_{n,m}\left(\frac{1}{n!}\langle f,g\rangle^n\right) = \delta_{n,m}\left(\langle F_n, G_m\rangle_{L^2(\Sigma_n)}\right)\,. \quad \square\end{aligned}$$

To complete the description of $L^2(\Omega,\mathcal{F},\mathbf{P})$ by stochastic integrals with respect to X, we define a suitable space of functions.

Definition 3.5
Let $L^2(\sum_n)$ be identified with the restrictions to $\sum_n$ of symmetric functions in $L^2(C_n)$, where $C_n = [0,1]^n$ denotes the unit cube. The space $L_0^2(\sum_n)$ is defined as the restrictions of those symmetric F_n in $L^2(C_n)$ such that

$$\int_0^1 F_n(s_1, s')ds_1 = 0$$

Lebesgue-almost-everywhere in the parameter $s' = (s_2, ..., s_n)$ in C_{n-1}.

Since finite sums of symmetric products of functions in $L_0^2([0,1])$ are dense in $L_0^2(\sum_n)$ and J^X preserves scalar products of symmetrizations by Lemma 3.4, J^X is automatically extended to an isometry defined on all of $L_0^2(\sum_n)$. One complements this by setting $L_0^2(\sum_0) = \mathbf{C}$ and $J^X(F_0) = F_0$ for a constant F_0 in $L_0^2(\sum_0)$. It follows that J^X is defined on $\overline{\bigoplus_{n\geq 0} L_0^2(\sum_n)}$, the Hilbert space completion of $\bigoplus_{n\geq 0} L_0^2(\sum_n)$.

Proposition 3.6
The map $J^X : \overline{\bigoplus_{n\geq 0} L^2_0(\sum_n)} \to L^2(\Omega, \mathbf{P})$ *, given by multiple stochastic integration with respect to the process* X*, is a surjective isometry.*

Proof. By Lemma 3.4 and the preceding discussion J^X is an isometry, thus having closed image. It suffices therefore to show that J^X has dense image.
Let us recall that the set $\{\mathcal{E}^B(\tilde{f}) | \tilde{f} \in L^2([0,1])\}$ is dense in $L^2(\Omega, \mathbf{P})$ (see e.g. [DMM]). By Lemma 3.3 each exponential vector $\mathcal{E}^B(\tilde{f})$ is in the image of J^X, indeed, by Lemma 3.2, we have

$$\mathcal{E}^B(\tilde{f}) = \mathcal{E}^X(f) = J^X\left(\sum_{n\geq 0} F_n\right),$$

where $f = \beta(\tilde{f})$ (notation of Lemma 2.1) and $F_n = f \otimes \cdots \otimes f$ is, as in Lemma 3.4, viewed as a function on Σ_n. Thus J^X is surjective. □

Remarks.
Taking care of scalar products on Fock spaces (see e.g. [M] for the right conventions) one identifies $\overline{\bigoplus_{n\geq 0} L^2_0(\sum_n)}$ unitarily with the symmetric Fock space over $L^2_0([0,1])$. Thus Proposition 3.6 gives the desired realization of the isomorphism of the Fock space over the reproducing Hilbert space $\mathcal{H}_X$ and $L^2(\Omega, \mathcal{F}, \mathbf{P})$ by means of stochastic integrals.

Acknowledgements. We would like to thank "l'équipe de probabilité de Strasbourg" and A. Sengupta for helpful discussions on several topics of stochastic calculus.

References.

[BR] M.J. Bowick and S.G. Rajeev, The holomorphic geometry of closed bosonic string theory and Diff(S^1)/S^1, Nucl. Phys. B 293 (1987) 348-384.

[DMM] C. Dellacherie, B. Maisonneuve et P.-A. Meyer, *Probabilités et Potentiel. Chapitres XVII à XXIV* (Hermann, Paris 1992).

[GW] P. Gosselin and T. Wurzbacher, A stochastic approach to the Virasoro anomaly in quantization of strings in flat space, Preprint 1996.

[HLP] G. Hardy, J.E. Littlewood and G. Pólya, *Inequalities* (Cambridge at the University Press 1934).

[HT] W. Hackenbroch und A. Thalmaier, *Stochastische Analysis* (B.G.Teubner, Stuttgart 1994).

[JY] T. Jeulin et M. Yor, Inégalité de Hardy, semi-martingales, et faux-amis, Séminaire de Probabilités XIII (1977/78), LNM 721, 332-359.

[M] P.-A. Meyer, *Quantum Probability for Probabilists* (Springer LNM 1538, Berlin Heidelberg 1993).

[N] J. Neveu, *Processus aléatoires gaussiens* (Les Presses de l'Université de Montréal 1968).

[PS] A. Pressley and G. Segal, *Loop groups* (Oxford University Press 1986).

On continuous conditional Gaussian martingales and stable convergence in law

Jean Jacod

In this paper, we start with a stochastic basis $(\Omega, \mathcal{F}, \mathbb{F} = (\mathcal{F}_t)_{t\in[0,1]}, P)$, the time interval being $[0,1]$, on which are defined a "basic" continuous local martingale M and a sequence Z^n of martingales or semimartingales, asymptotically "orthogonal to all martingales orthogonal to M". Our aim is to give some conditions under which Z^n converges "stably in law" to some limiting process which is defined on a suitable extension of $(\Omega, \mathcal{F}, \mathbb{F}, P)$.

In the first section we study systematically some, more or less known, properties of extensions of filtered spaces and of $\mathcal{F}$-conditional Gaussian martingales and so-called M-biased $\mathcal{F}$-conditional Gaussian martingales. Then we explain our limit results: in Section 2 we give a fairly general result, and in Section 3 we specialize to the case when Z^n is some "discrete-time" process adapted to the discretized filtration $\mathbb{F}^n = (\mathcal{F}^n_t)_{t\in[0,1]}$, where $\mathcal{F}^n_t = \mathcal{F}_{[nt]/n}$. Finally, Section 4 is devoted to studying the limit of a sequence of M-biased $\mathcal{F}$-conditional Gaussian martingales.

1 Extension of filtered spaces and conditionally Gaussian martingales

We begin with some general conventions. Our filtrations will always be assumed to be right-continuous. All local martingales below are supposed to be 0 at time 0, and we write $\langle M, N\rangle$ for the predictable quadratic variation between M and N if these are locally square-integrable martingales. When M and N are respectively d- and r-dimensional, then $\langle M, N^*\rangle$ is the $d\times r$ dimensional process with components $\langle M, N^*\rangle^{i,j} = \langle M^i, N^j\rangle$ (N^* stands for the transpose of N).

In all these notes, we have a basic filtered probability space $(\Omega, \mathcal{F}, \mathbb{F}, P)$.

1-1. Let us start with some definitions. We call *extension* of $(\Omega, \mathcal{F}, \mathbb{F}, P)$ another filtered probability space $(\tilde\Omega, \tilde{\mathcal{F}}, \tilde{\mathbb{F}}, \tilde P)$ constructed as follows: starting with an auxiliary filtered space $(\Omega, \mathcal{F}', \mathbb{F}' = (\mathcal{F}'_t)_{t\in[0,1]})$ such that each σ-field $\mathcal{F}'_{t-}$ is separable, and a transition probability $Q_\omega(d\omega')$ from $(\Omega, \mathcal{F})$ into $(\Omega', \mathcal{F}')$, we set

$$\tilde\Omega = \Omega\times\Omega', \quad \tilde{\mathcal{F}} = \mathcal{F}\otimes\mathcal{F}', \quad \tilde{\mathcal{F}}_t = \cap_{s>t}\mathcal{F}_s\otimes\mathcal{F}'_s, \quad \tilde P(d\omega, d\omega') = P(d\omega)Q_\omega(d\omega'). \tag{1.1}$$

According to ([3], Lemma 2.17), the extension is called *very good* if all martingales on the space $(\Omega, \mathcal{F}, \mathbb{F}, P)$ are also martingales on $(\tilde{\Omega}, \tilde{\mathcal{F}}, \tilde{\mathbb{F}}, \tilde{P})$, or equivalently, if $\omega \rightsquigarrow Q_\omega(A')$ is $\mathcal{F}_t$-measurable whenever $A' \in \mathcal{F}'_t$.

A process Z on the extension is called an *$\mathcal{F}$-conditional martingale* (resp. *$\mathcal{F}$-Gaussian process*) if for P-almost all ω the process $Z(\omega, .)$ is a martingale (resp. a centered Gaussian process) on the space $(\Omega', \mathcal{F}', (\mathcal{F}'_t)_{t\in[0,1]}, Q_\omega)$.

Let us finally denote by $\mathcal{M}_b$ the set of all bounded martingales on $(\Omega, \mathcal{F}, \mathbb{F}, P)$.

Proposition 1-1: *Let Z be a continuous adapted q-dimensional process on the very good extension $(\tilde{\Omega}, \tilde{\mathcal{F}}, \tilde{\mathbb{F}}, \tilde{P})$, with $Z_0 = 0$. The following statements are equivalent:*

(i) Z is a local martingale on the extension, orthogonal to all elements of $\mathcal{M}_b$, and the bracket $\langle Z, Z^*\rangle$ is $(\mathcal{F}_t)$-adapted.

(ii) Z is an $\mathcal{F}$-conditional Gaussian martingale.

In this case, the $\mathcal{F}$-conditional law of Z is characterized by the process $\langle Z, Z^*\rangle$ (i.e., for P-almost all ω, the law of $Z(\omega, .)$ under Q_ω depends only on the function $t \rightsquigarrow \langle Z, Z^*\rangle_t(\omega)$).

Proof. a) We first prove that, if each Z_t is $\tilde{P}$-integrable, then Z is an $\mathcal{F}$-conditional martingale iff it is an $\tilde{\mathbb{F}}$-martingale orthogonal to all bounded $\mathbb{F}$-martingales. For this, we can and will assume that Z is 1-dimensional.

Let $t \leq s$ and let U, U' be bounded measurable function on $(\Omega, \mathcal{F}_t)$ and $(\Omega', \mathcal{F}'_t)$ respectively. Let also $M \in \mathcal{M}_b$. We have

$$\tilde{E}(UU'M_sZ_s) = \int P(d\omega)U(\omega)M_s(\omega)\int Q_\omega(d\omega')U'(\omega')Z_s(\omega,\omega'), \tag{1.2}$$

$$\tilde{E}(UU'M_tZ_t) = \int P(d\omega)U(\omega)M_t(\omega)\int Q_\omega(d\omega')U'(\omega')Z_t(\omega,\omega'). \tag{1.3}$$

Assume first that Z is an $\mathcal{F}$-conditional martingale. Then for P-almost all ω we have

$$\int Q_\omega(d\omega')U'(\omega')Z_s(\omega,\omega') = \int Q_\omega(d\omega')U'(\omega')Z_t(\omega,\omega'),$$

and the latter is $\mathcal{F}_t$-measurable as a function of ω because the extension is very good. Since M is an $\mathbb{F}$-martingale, we deduce that (1.2) and (1.3) are equal: thus MZ is a martingale on the extension: then Z is a martingale (take $M \equiv 1$), orthogonal to all bounded $\mathbb{F}$-martingales.

Next we prove the sufficient condition. Take V bounded and $\mathcal{F}_s$-measurable, and consider the martingale $M_r = E(V|\mathcal{F}_r)$. With the notation above we have equality between (1.2) and (1.3), and further in (1.3) we can replace $M_t(\omega)$ by $M_s(\omega) = V(\omega)$ because the last integral is $\mathcal{F}_t$-measurable in ω. Then taking $U = 1$ we get

$$\int P(d\omega)V(\omega)\int Q_\omega(d\omega')U'(\omega')Z_s(\omega,\omega') = \int P(d\omega)V(\omega)\int Q_\omega(d\omega')U'(\omega')Z_t(\omega,\omega').$$

Hence for P-almost ω, $Q_\omega(U'Z_s(\omega, .)) = Q_\omega(U'Z_t(\omega, .))$. Using the separability of the σ-field $\mathcal{F}'_{t-}$ and the continuity of Z, we have this relation P-almost surely in

ω, simultaneously for all $t \leq s$ and all $\mathcal{F}'_{t-}$-measurable variable U': this gives the $\mathcal{F}$-conditional martingality for Z.

b) Assume that (i) holds. If $Y = \langle Z, Z^* \rangle$, a simple application of Ito's formula and the fact that Z is continuous show that, since Z is orthogonal to all $M \in \mathcal{M}_b$, the same holds for Y. Each $T_n = \inf(t : |\langle Z, Z^* \rangle_t| > n)$ is an $\mathbb{F}$-stopping time, and $T_n \uparrow \infty$ as $n \to \infty$. Then $Z(n)_t = Z_{t \wedge T_n}$ and $Y(n)_t = Y_{t \wedge T_n}$ are continuous $\tilde{\mathbb{F}}$-martingale, orthogonal to all $M \in \mathcal{M}_b$, and obviously $|Z(n)_t|$ and $|Y(n)_t|$ are integrable: by (a), and by letting $n \uparrow \infty$, we deduce that for P-almost all ω, under Q_ω the process $Z(n)(\omega, .)$ is a continous martingale with deterministic bracket $\langle Z, Z^* \rangle(\omega)$, hence it is an $\mathcal{F}$-Gaussian martingale, so we have (ii). Furthermore, it is well-known that the law of $Z(\omega)$ under Q_ω is then entirely determined by $\langle Z, Z^* \rangle(\omega)$.

c) Assume now (ii). There is a P-full set $A \in \mathcal{F}$ such that for all $\omega \in A$, under Q_ω, the process $Z(\omega, .)$ is both centered Gaussian and an $\mathbb{F}'$-martingale. Therefore if $F_t(\omega) = \int Q_\omega(d\omega') Z_t(\omega, \omega')$, the process $(ZZ^*)(\omega, .) - F(\omega)$ is an $\mathbb{F}'$-martingale under Q_ω for $\omega \in A$: that is, $ZZ^* - F$ is an $\mathcal{F}$-conditional martingale. By localizing at the $\mathbb{F}$-stopping times $T_n = \inf(t : |F_t| > n)$ and by (a), we deduce that Z and $ZZ^* - F$ are local martingales on the extension, orthogonal to all $M \in \mathcal{M}_b$. Since F is continuous, $\mathbb{F}$-adapted, and of bounded variation (since it is non-decreasing for the strong order in the set of nonnegative symmetric matrices), it follows that it is a version of $\langle Z, Z^* \rangle$, hence we have (i). □

1-2. Let now M be a continous d-dimensional local martingale, and $\mathcal{M}_b(M^\perp)$ be the class of all elements of $\mathcal{M}_b$ which are orthogonal to M (i.e., to all components of M).

A q-dimensional process Z on the extension is called an *M-biased $\mathcal{F}$-conditional Gaussian martingale* if it can be written as

$$Z_t = Z'_t + \int_0^t u_s dM_s, \tag{1.4}$$

where Z' is an $\mathcal{F}$-conditional Gaussian martingale and u is a predictable $\mathbb{R}^q \otimes \mathbb{R}^d$ on $(\Omega, \mathcal{F}, \mathbb{F}, P)$.

Proposition 1-2: *Let Z be a continuous adapted q-dimensional process on the very good extension $(\tilde{\Omega}, \tilde{\mathcal{F}}, \tilde{\mathbb{F}}, \tilde{P})$, with $Z_0 = 0$. The following statements are equivalent:*

(i) Z is a local martingale on the extension, orthogonal to all elements of $\mathcal{M}_b(M^\perp)$, and the brackets $\langle Z, Z^* \rangle$ and $\langle Z, M^* \rangle$ are $\mathbb{F}$-adapted.

(ii) Z is an M-biased $\mathcal{F}$-conditional Gaussian martingale.

In this case, the $\mathcal{F}$-conditional law of Z is characterized by the processes M, $\langle Z, Z^* \rangle$ and $\langle Z, M^* \rangle$.

Proof. Under either (i) or (ii), Z and M are continous local martingales (use the fact that the extension is very good, and use (1.4) under (ii)). We write $F = \langle Z, Z^* \rangle$, $G = \langle Z, M^* \rangle$ and $H = \langle M, M^* \rangle$.

If (ii) holds, (1.4) and Proposition 1-1 yield for all $N \in \mathcal{M}_b$:

$$G_t = \int_0^t u_s^* dH_s, \quad F_t = \langle Z', Z'^* \rangle_t + \int_0^t u_s^* dH_s u_s^*, \quad \langle Z, N \rangle_t = \int_0^t u_s^* d\langle M, N \rangle_s. \tag{1.5}$$

Then (i) readily follows. Further, (1.5) implies that u and $\langle Z', Z'^* \rangle$ are determined by F, G and H. Since $\int_0^{\cdot} u_s dM_s$ is $\mathcal{F}$-measurable, the last claim follows from (1.4) and Proposition 1-1 again.

Assume conversely (i). There are a continuous increasing process A and predictable processes f, g, h with values in $\mathbb{R}^q \otimes \mathbb{R}^q$, $\mathbb{R}^q \otimes \mathbb{R}^d$ and $\mathbb{R}^d \otimes \mathbb{R}^d$ respectively, such that $F_t = \int_0^t f_s dA_s$, $G_t = \int_0^t g_s dA_s$ and $H_t = \int_0^t h_s dA_s$.

The process (M, Z) is a continuous local martingale on the extension, with bracket $K_t = \int_0^t k_s dA_s$, where $k = \begin{pmatrix} h & g^* \\ g & f \end{pmatrix}$. By triangularization we may write $k = zz^*$, where

$$z = \begin{pmatrix} v & 0 \\ uv & w \end{pmatrix}, \tag{1.6}$$

so that $h = vv^*$, $g = uvv^*$ and $f = uvv^*u^* + ww^*$. Let us put $Y_t = \int_0^t u_s dM_s$ and $Z' = Z - Y$. Then since the extension is very good, Z' is a local martingale on the extension, and $\langle Z', Z'^* \rangle_t = \int_0^t w_s w_s^* dA_s$ is $\mathbb{F}$-adapted. Further, $\langle Z', N \rangle_t = \langle Z, N \rangle_t - \int_0^t u_s d\langle M, N \rangle_s$: first this implies that $\langle Z', N \rangle = 0$ if $N \in \mathcal{M}_b(M^\perp)$ (since then $\langle Z, N \rangle = 0$ by hypothesis), second this implies that when $N_t = \int_0^t \alpha_s dM_s$ we have $\langle Z', N \rangle_t = \int_0^t (g_s \alpha_s^* - u_s v_s v_s^* \alpha_s) dA_s = 0$. Thus Z' is orthogonal to all $N \in \mathcal{M}_b$, and it is an $\mathcal{F}$-conditional Gaussian martingale by Proposition 1-1. □

1-3. Let us denote by $\mathcal{S}_r$ the set of all symmetric nonnegative $r \times r$-matrices. In Proposition 1.1, the process $\langle Z, Z^* \rangle$ is a continuous adapted non-decreasing $\mathcal{S}_q$-valued process, null at 0. In Proposition 1-2, the bracket of (M, Z) is a continuous adapted non-decreasing $\mathcal{S}_{d+q}$-valued process, null at 0. Conversely we have:

Proposition 1-3: *a) Let F be a continuous adapted nondecreasing $\mathcal{S}_q$-valued process, with $F_0 = 0$, on the basis $(\Omega, \mathcal{F}, \mathbb{F}, P)$. There exists a continuous $\mathcal{F}$-conditional Gaussian martingale Z on a very good extension, such that $\langle Z, Z^* \rangle = F$.*

b) Let K be a continuous adapted nondecreasing $\mathcal{S}_{d+q}$-valued process, with $K_0 = 0$, and M be a continuous d-dimensional local martingale with $\langle M^i, M^j \rangle = K^{ij}$ for $1 \leq i, j \leq d$, on the basis $(\Omega, \mathcal{F}, \mathbb{F}, P)$. There exists a continuous M-biased $\mathcal{F}$-conditional Gaussian martingale Z on a very good extension, such that $\langle Z^i, M^j \rangle = K^{d+i,j}$ for $1 \leq i \leq q, 1 \leq j \leq d$, and $\langle Z^i, Z^j \rangle = K^{d+i,d+j}$ for $1 \leq i, j \leq q$.

Of course (a) is a particular case of (b) (take $M = 0$), but in the proof below (b) is obtained as a consequence of (a).

Proof. a) Take $(\Omega', \mathcal{F}', \mathbb{F}')$ to be the canonical space of all $\mathbb{R}^d$-valued continuous functions on $[0, 1]$, with the usual filtration and the canonical process $Z_t(\omega') = \omega'(t)$. For each ω, denote by Q_ω the unique probability measure on $(\Omega', \mathcal{F}')$ under which Z is a centered Gaussian process with covariance $\int Z_t Z_s^* dQ_\omega = F_{s \wedge t}(\omega)$. This structure of the covariance implies that Z has independent increments and thus is a martingale under each Q_ω: Defining $(\tilde{\Omega}, \tilde{\mathcal{F}}, \tilde{\mathbb{F}}, \tilde{P})$ by (1.1) gives the result.

b) As in the previous proof, we can write $K_t = \int_0^t k_s dA_s$ for a continuous adapted increasing process A and a predictable process $k = zz^*$ with z as in (1.6). By (a) we have a continuous $\mathcal{F}$-conditional Gaussian martingale Z' on a very good extension, with $\langle Z', Z'^* \rangle_t = \int_0^t w_s w_s^* dA_s$. We can set $Z_t = Z'_t + \int_0^t u_s dM_s$, and some computations yileds that Z satisfies our requirements. □

We even have a more "concrete" way of constructing Z above, when K is absolutely continuous w.r.t. Lebesgue measure on $[0,1]$. Let $(\Omega^W, \mathcal{F}^W, \mathbb{F}^W, P^W)$ be the q-dimensional Wiener space with the canonical Wiener process W. Then $(\tilde{\Omega}, \tilde{\mathcal{F}}, \tilde{\mathbb{F}}, \tilde{P})$ defined by

$$\tilde{\Omega} = \Omega \times \Omega^W, \quad \tilde{\mathcal{F}} = \mathcal{F} \otimes \mathcal{F}^W, \quad \tilde{\mathcal{F}}_t = \cap_{s>t} \mathcal{F}_s \otimes \mathcal{F}^W_s, \quad \tilde{P} = P \otimes P^W. \tag{1.7}$$

is a very good extension of $(\Omega, \mathcal{F}, \mathbb{F}, P)$, called the *canonical q-dimensional Wiener extension* of $(\Omega, \mathcal{F}, \mathbb{F}, P)$. Note that W is also a Wiener process on the extension.

Proposition 1-4: *Let K and M be as in Proposition 1-3(b), and assume that $K_t = \int_0^t k_s ds$ with k predictable S_{d+q}- valued. Then we can choose a version of k of the form $k = zz^*$ with $z = \begin{pmatrix} v & 0 \\ uv & w \end{pmatrix}$, and on the canonical q-dimensional Wiener extension of $(\Omega, \mathcal{F}, \mathbb{F}, P)$ the process*

$$Z_t = \int_0^t u_s dM_s + \int_0^t w_s dW_s \tag{1.8}$$

is a continuous M-biased $\mathcal{F}$-conditional Gaussian martingale, such that $\langle Z^i, M^j \rangle = K^{d+i,j}$ for $1 \le i \le q$ and $1 \le j \le d$, and $\langle Z^i, Z^j \rangle = K^{d+i,d+j}$ for $1 \le i,j \le q$.

Proof. The first claim has already been proved. (1.8) defines a continuous q-dimensional local martingale on the canonical Wiener extension and a simple computation shows that it has the required brackets. □

2 Stable convergence to conditionally Gaussian martingales

2-1. First we recall some facts about stable convergence. Let X_n be a sequence of random variables with values in a metric space E, all defined on $(\Omega, \mathcal{F}, P)$. Let $(\tilde{\Omega}, \tilde{\mathcal{F}}, \tilde{P})$ be an extension of $(\Omega, \mathcal{F}, P)$ (as in Section 1, except that there is no filtration here), and let X be an E-valued variable on the extension. Let finally $\mathcal{G}$ be a sub σ-field of $\mathcal{F}$. We say that X_n *$\mathcal{G}$-stably converges in law* to X, and write $X_n \to^{\mathcal{G}\text{-}\mathcal{L}} X$, if

$$E(Yf(X_n)) \to \tilde{E}(Yf(X)) \tag{2.1}$$

for all $f : E \to \mathbb{R}$ bounded continuous and all bounded variable Y on $(\Omega, \mathcal{G})$. This property, introduced by Renyi [6] and studied by Aldous and Eagleson [1], is (slightly) stronger than the mere convergence in law. It applies in particular when X_n, X are $\mathbb{R}^q$-valued càdlàg processes, with $E = \mathbb{D}([0,1], \mathbb{R}^q)$ the Skorokhod space.

If X'_n are some other E-valued variables, then (with δ denoting a distance on E):

$$\delta(X'_n, X_n) \to^P 0, \quad X_n \to^{\mathcal{G}-\mathcal{L}} X \quad \Rightarrow \quad X'_n \to^{\mathcal{G}-\mathcal{L}} X. \tag{2.2}$$

Also, if U_n, U are on $(\Omega, \mathcal{F})$, with values in another metric space E', then

$$U_n \to^P U, \quad X_n \to^{\mathcal{G}-\mathcal{L}} X \quad \Rightarrow \quad (U_n, X_n) \to^{\mathcal{G}-\mathcal{L}} (U, X). \tag{2.3}$$

When $\mathcal{G} = \mathcal{F}$ we simply say that X_n stably converges in law to X, and we write $X_n \to^{s-\mathcal{L}} X$.

2-2. Now we describe a rather general setting for our convergence results. We start with a continuous d-dimensional local martingale M on the basis $(\Omega, \mathcal{F}, \mathbb{F}, P)$: this will be our "reference" process. The set $\mathcal{M}_b$ is as in Section 1.

Next, for each integer n we are given a filtration $\mathbb{F}^n = (\mathcal{F}^n_t)_{t\in[0,1]}$ on $(\Omega, \mathcal{F})$ with the following property:

Property (F): We have a d-dimensional square-integrable $\mathbb{F}^n$-martingale $M(n)$ and, for each $N \in \mathcal{M}_b$, a bounded $\mathbb{F}^n$-martingale $N(n)$, such that

$$\sup_{n,t,\omega} |N(n)_t(\omega)| < \infty, \tag{2.4}$$

$$\langle M(n), M(n)^* \rangle_t \to^P \langle M, M^* \rangle_t, \quad \forall t \in [0,1], \tag{2.5}$$

(the bracket above in the predictable quadratic variation relative to $\mathbb{F}^n$) and that, for any finite family $(N^1, .., N^m)$ in $\mathcal{M}_b$,

$$(M(n), N^1(n), .., N^m(n)) \to^P (M, N^1, .., N^m) \quad \text{in } \mathbb{D}([0,1], \mathbb{R}^{d+m}).\square \tag{2.6}$$

In practice we encounter two situations: first, $\mathcal{F}^n_t = \mathcal{F}_t$, for which (F) is obvious with $M(n) = M$ and $N(n) = N$. Second, $\mathcal{F}^n_t = \mathcal{F}_{[nt]/n}$, a situation which will be examined in Section 3.

2-3. For stating our main result we need some more notation. We are interested in the behaviour of a sequence (Z^n) of q-dimensional processes, each Z^n being an $\mathbb{F}^n$-semimartingale, and we denote by (B^n, C^n, ν^n) its characteristics, relative to a given continuous truncation function h_q on $\mathbb{R}^q$ (i.e. a continuous function $h_q : \mathbb{R}^q \to \mathbb{R}^q$ with compact support and $h_q(x) = x$ for $|x|$ small enough): see [5]. If $h'_q(x) = x - h_q(x)$, we can write

$$Z^n_t = B^n_t + X^n_t + \sum_{s\leq t} h'_q(\Delta Z^n_s) \tag{2.7}$$

where X^n is an $(\mathcal{F}^n_t)$-local martingale with bounded jumps, and $\Delta Y_t = Y_t - Y_{t-}$.

Here is the main result:

Theorem 2-1: *Assume Property (F). Assume also that there are two continuous processes F and G and a continuous process B of bounded variation on $(\Omega, \mathcal{F}, \mathbb{F}, P)$ such that (the brackets below being the predictable quadratic variations relative to the filtration $\mathbb{F}^n$):*

$$\sup_t |B_t^n - B_t| \to^P 0, \tag{2.8}$$

$$F_t^n := \langle X^n, X^{n*}\rangle_t \to^P F_t, \quad \forall t \in [0,1], \tag{2.9}$$

$$G_t^n := \langle X^n, M(n)^*\rangle_t \to^P G_t, \quad \forall t \in [0,1], \tag{2.10}$$

$$U(\varepsilon)^n := \nu^n([0,1] \times \{x : |x| > \varepsilon\}) \to^P 0, \quad \forall \varepsilon > 0, \tag{2.11}$$

$$V(N)_t^n := \langle X^n, N(n)\rangle_t \to^P 0, \quad \forall t \in [0,1], \quad \forall N \in \mathcal{M}_b(M^\perp). \tag{2.12}$$

Then

(i) There is a very good extension of $(\Omega, \mathcal{F}, \mathbb{F}, P)$ and an M-biased continuous $\mathcal{F}$-conditional Gaussian martingale Z' on this extension with

$$\langle Z', Z'^*\rangle = F, \quad \langle Z', M^*\rangle = G, \tag{2.13}$$

such that $Z^n \to^{s\text{-}\mathcal{L}} Z := B + Z'$.

(ii) Assuming further that $d\langle M^i, M^i\rangle_t \ll dt$ and $dF_t^{ii} \ll dt$, there are predictable processes u, v, w with values in $\mathbb{R}^q \otimes \mathbb{R}^d$, $\mathbb{R}^d \otimes \mathbb{R}^d$ and $\mathbb{R}^q \otimes \mathbb{R}^q$ respectively, such that

$$\left.\begin{array}{l} \langle M, M^*\rangle_t = \int_0^t u_s u_s^* ds, \quad G_t = \int_0^t u_s v_s v_s^* ds, \\ F_t = \int_0^t (u_s v_s v_s^* u_s^* + w_s w_s^* ds, \end{array}\right\} \tag{2.14}$$

and the limit of Z^n can be realized on the canonical q-dimensional Wiener extension of $(\Omega, \mathcal{F}, \mathbb{F}, P)$, with the canonical Wiener process W, as

$$Z_t = B_t + \int_0^t u_s dM_s + \int_0^t w_s dW_s. \tag{2.15}$$

The proof will be divided in a number of steps.

Step 1. Let $H^n = \langle M(n), M(n)^*\rangle$ and $H = \langle M, M^*\rangle$. Consider the following processes with values in the set of symmetric $(d+q) \times (d+q)$ matrices:

$$K^n = \begin{pmatrix} H^n & G^{n*} \\ G^n & F^n \end{pmatrix}, \qquad K = \begin{pmatrix} H & G^* \\ G & F \end{pmatrix}.$$

By (2.9), (2.10) and (F), we have $K_t^n \to^P K_t$ for all t, while K^n is a nondecreasing process with values in $\mathcal{S}_{d+q}$. So there is a version of K which is also a nondecreasing $\mathcal{S}_{d+q}$-valued process. Further K is continuous in time, so by a classical result we even have

$$\sup_t |K_t^n - K_t| \to^P 0. \tag{2.16}$$

Further we can write $K_t = \int_0^t k_s dA_s$ for some continuous adapted increasing process A and some predictable $\mathcal{S}_{d+q}$-valued process k, and as seen in the proof of Proposition 1-2 we have $k = zz^*$ with z given by (1.6): under the additional assumption of (ii), we can take $A_t = t$, so we have (2.14), and the last claim of (ii) will follow from (i) and from Proposition 1-4.

Step 2. In this step we prove (2.12) can be strenghtened as such:

$$\sup_t |V(N)^n_t| \to^P 0. \tag{2.17}$$

In view of (2.12) it suffices to prove that

$$\forall \varepsilon, \eta > 0,\ \exists \theta > 0,\ \exists n_0 \in \mathbb{N}^*,\ \forall n \geq n_0 \quad \Rightarrow \quad P(w^n(\theta) > \eta) \leq \varepsilon, \tag{2.18}$$

where $w^n(\theta) = \sup_{0\leq s\leq \theta, 0\leq t\leq 1-\theta} |V(N)^n_{t+s} - V(N)^n_t|$ is the θ-modulus of continuity of $V(N)^n$. Denoting by $w'^n(\theta)$ the θ-modulus of continuity of F^n, (2.16) and the continuity of K yield

$$\forall \varepsilon, \eta > 0,\ \exists \theta > 0,\ \exists n_0 \in \mathbb{N}^*,\ \forall n \geq n_0 \quad \Rightarrow \quad P(w'^n(\theta) > \eta) \leq \varepsilon. \tag{2.19}$$

On the other hand, a classical inequality on quadratic covariations yields that for all $u > 0$ we have $2|V(N)^n_t - V(N)^n_s| \leq |F^n_t - F^n_s|/u + u(\langle N,N\rangle_t - \langle N,N\rangle_s)$ if $s < t$, so that $2w^n(\theta) \leq w'^n(\theta)/u + \langle N,N\rangle_1$, hence

$$P(w^n(\theta) > \eta) \leq P(w'^n(\theta) > u\eta) + \frac{u}{\eta} E(N(n)^2_1).$$

Then (2.18) readily follows from (2.19), $\sup_n E(N(n)^2_1) < \infty$ and from the arbitraryness of $u > 0$.

Step 3. Here we prove that, instead of proving $Z^n \to^{s-\mathcal{L}} Z$ with $Z = B + Z'$ as in (i), it is enough to prove that

$$X^n \to^{s-\mathcal{L}} Z' \tag{2.20}$$

Indeed, set $Z''^n_t = \sum_{s\leq t} h'_q(\Delta Z^n_s)$. By ([5], VI-4.22), (2.11) implies $\sup_t |\Delta Z^n_t| \to^P 0$; since $h'_q(x) = 0$ for $|x|$ small enough, we have $\sup_t |Z''^n_t| \to^P 0$. On the other hand $\Delta B^n_t = \int h_q(x)\nu^n(\{t\}, dx)$, so (2.11) again yields $\sup_t |\Delta B^n_t| \to^P 0$, hence B is continuous by (2.8). Hence the claim follows from (2.3).

Step 4. Here we prove (2.20) under the additional assumption that $\mathcal{F}$ is separable.

a) There is a sequence of bounded variables $(Y_m)_{m\in\mathbb{N}}$ which is dense in $\mathbb{L}^1(\Omega, \mathcal{F}, P)$. We set $N^m_t = E(Y_m|\mathcal{F}_t)$, so $N^m \in \mathcal{M}_b$, and we have two important properties:

(A) Every bounded martingale is the limit in $\mathbb{L}^2$, uniformly in time, of a sequence of sums of stochastic integrals w.r.t. a finite number of N^m's: see (4.15) of [2].

(B) $(\mathcal{F}_t)$ is the smallest filtration, up to P-null sets, w.r.t. which all N^m's are adapted: indeed let $(\mathcal{G}_t)$ be the above-described filtration, and $A \in \mathcal{F}_t$; there is a sequence $Y_{m(n)} \to 1_A$ in $\mathbb{L}^1$, so $N^{m(n)}_t = E(Y_{m(n)}|\mathcal{F}_t)$ is $\mathcal{G}_t$-measurable and converges in $\mathbb{L}^1$ to $E(1_A|\mathcal{F}_t) = 1_A$.

b) Introduce some more notation. First $\mathcal{N} = (N^m)_{m\in\mathbb{N}}$ and $\mathcal{N}(n) = (N^m(n))_{m\in\mathbb{N}}$ (recall Property (F)) can be considered as processes with paths in $\mathbb{D}([0,1],\mathbb{R}^{\mathbb{N}})$. Then (2.6) and (2.16) yield

$$(M(n),\mathcal{N}(n),K^n) \to^P (M,\mathcal{N},K) \quad \text{in } \mathbb{D}([0,1],\mathbb{R}^d \times \mathbb{R}^{\mathbb{N}} \times \mathbb{R}^{(d+q)^2}). \tag{2.21}$$

On the other hand, VI-4.18 and VI-4.22 in [5] and (2.11) and (2.16) imply that the sequence (X^n) is C-tight. It follows from (2.21) that the sequence $(X^n, M(n), \mathcal{N}(n))$ is tight and that any limiting process $(\hat{X},\hat{M},\hat{\mathcal{N}})$ has $\mathcal{L}(\hat{M},\hat{\mathcal{N}}) = \mathcal{L}(M,\mathcal{N})$.

c) Choose now any subsequence, indexed by n', such that $(X^{n'}, M(n'), \mathcal{N}(n'))$ converges in law. From what precedes one can realize the limit as such: consider the canonical space $(\Omega',\mathcal{F}',\mathbb{F}')$ of all continuous functions from $[0,1]$ into $\mathbb{R}^q$, with the canonical process Z', and define $(\tilde{\Omega},\tilde{\mathcal{F}},(\tilde{\mathcal{F}}_t)_{t\in[0,1]})$ by (1.1); since $\mathcal{F} = \sigma(Y_m : m \in \mathbb{N})$ up to P-null sets, there is a probability measure $\tilde{P}$ on $(\tilde{\Omega},\tilde{\mathcal{F}})$ whose Ω-marginal is P, and such that the laws of $(X^{n'}, M(n'), \mathcal{N}(n'))$ converge to the law of $(X, M, \mathcal{N})$ under $\tilde{P}$.

Therefore we have an extension $(\tilde{\Omega},\tilde{\mathcal{F}},\tilde{\mathbb{F}},\tilde{P})$ of $(\Omega,\mathcal{F},\mathbb{F},P)$ (the existence of a disintegration of $\tilde{P}$ as in (1.1) is obvious, due to the definition of $(\Omega',\mathcal{F}')$), and up to $\tilde{P}$-null sets the filtrations $\mathbb{F}$ and $\tilde{\mathbb{F}}$ are generated by $(M,\mathcal{N})$ and $(Z',M,\mathcal{N})$ respectively (use Property (B) of (a)).

Set $Y^n = (M(n), X^n)$ and $Y = (M, Z')$. By contruction, all components of Y^n, $\mathcal{N}(n)$, $Y^nY^{n*} - K^n$ are $\mathbb{F}^n$-local martingales with uniformly bounded jumps. Then IX-1.17 of [5] (applied to processes with countably many components, which does not change the proof) yields that all components of Y, $\mathcal{N}$ and $YY^* - K$ are $\tilde{\mathbb{F}}$-local martingales under $\tilde{P}$. This implies first that on our extension we have

$$F = \langle Z', Z'^* \rangle, \qquad G = \langle Z', M^* \rangle \tag{2.22}$$

(since K is continuous increasing in $\mathcal{S}_{d+q}$), and second that all N^m are $\tilde{\mathbb{F}}$-martingales. Then by (9.21) of [2] any stochastic integral $\int_0^\cdot a_s dN^m_s$ with a $\mathbb{F}$-predictable is also an $(\tilde{\mathbb{F}}$-martingale: Property (A) of (a) yields that all elements of $\mathcal{M}_b$ are $\tilde{\mathbb{F}}$-martingales, hence our extension is very good.

d) Let now $N \in \mathcal{M}_b(M^\perp)$. We could have included N in the sequence (N^m): what precedes remains valid, with the same limit, for a suitable subsequence (n'') of (n'). Moreover $X^nN(n) - V(N)^n$ is an $\mathbb{F}^n$-local martingale with bounded jumps, while by (2.17) the sequence $(X^{n''}, \mathcal{N}(n''), (n''), V(N)^{n''})$ converges in law to $(Z', \mathcal{N}, N, 0)$. The same argument as above yields that $Z'N$ is a local martingale on the extension, so Z' is othogonal to all elements of $\mathcal{M}_b(M^\perp)$.

Therefore Z' satisfies (i) of Proposition 1-2: hence Z' is an M-biased continuous $\mathcal{F}$-conditional Gaussian martingale, whose law under Q_ω, which is Q_ω itself, is determined by the processes M, F, G, and in particular it does not depend on the subsequence (n') chosen above.

In other words all convergent subsequence of $(X^n, \mathcal{N}(n))$ have the same limit $(Z', \mathcal{N})$ in law, with the same measure $\tilde{P}$, and thus the original sequence $(X^n, \mathcal{N}(n))$ converges in law to $(Z', \mathcal{N})$. In particular if f is a bounded continuous function on

$\mathbb{D}([0,1],\mathbb{R}^q)$ and since $N(n)^m$ is a component of $\mathcal{N}(n)$ bounded uniformly in n, we get

$$E(f(X^n)N(n)_1^m) \to \tilde{E}(f(Z')N_1^m).$$

Now (2.4) and (2.6) yield that $N(n)_1^m \to N_1^m$ in $\mathbb{L}^1$, hence

$$E(f(X^n)N_1^m) \to \tilde{E}(f(Z')N_1^m).$$

Since $\tilde{E}(UN_1^m) = \tilde{E}(UY_m)$ for any bounded $\tilde{\mathcal{F}}$-measurable variable U, we deduce

$$E(f(X^n)Y_m) \to \tilde{E}(f(Z')Y_m).$$

Finally any bounded $\mathcal{F}$-measurable variable Y is the $\mathbb{L}^1$-limit of a subsequence of (Y_m), hence one readily deduces that

$$E(f(X^n)Y) \to \tilde{E}(f(Z')Y), \tag{2.23}$$

which is (2.20).

Step 5. It remains to remove the separability assumption on $\mathcal{F}$. Denote by $\mathcal{H}$ the σ-field generated by the random variables $(M_t, K_t, B_t, X_t^n : t \in [0,1], n \geq 1)$, and let $\mathcal{G}$ be any separable σ-field containing $\mathcal{H}$. Let $(Y_m)_{m\in\mathbb{N}}$ be a dense sequence of bounded variables in $\mathbb{L}^1(\Omega,\mathcal{G},P)$, and $N_t^m = E(Y_m|\mathcal{F}_t)$, and set $\mathbb{G} = (\mathcal{G}_t)_{y\in[0,1]}$ for the filtration generated by the processes $(N^m)_{m\in\mathbb{N}}$.

We have $E(Y_m|\mathcal{F}_t) = E(Y_m|\mathcal{G}_t)$ for all m, so by a density argument $E(Y|\mathcal{F}_t) = E(Y|\mathcal{G}_t)$ for all $Y \in \mathbb{L}^1(\Omega,\mathcal{G},P)$: this implies that any $\mathbb{G}$-martingale is an $\mathbb{F}$-martingale, and in particular each N^m is in $\mathcal{M}_b$, and also that every $\mathbb{F}$-adapted and $\mathcal{G}$-measurable process (like K, B and M) is $\mathbb{G}$-adapted. Thus M is a $\mathbb{G}$-local martingale. Finally, any bounded $\mathbb{G}$-martingale which is orthogonal w.r.t. $\mathbb{G}$ to M is also orthogonal to M w.r.t. $\mathbb{F}$.

In other words, Property (F) is satisfied by $\mathbb{G}$ and the same filtration $\mathbb{F}^n$ and processes $M(n)$, $N(n)$, and (2.8)-(2.12) are satisfied as well with $\mathbb{G}$ instead of $\mathbb{F}$. We can thus apply Step 4 with the same space $(\Omega',\mathcal{F}',\mathbb{F}')$ and process Z', and $\tilde{\Omega} = \Omega\times\Omega'$, $\tilde{\mathcal{G}} = \mathcal{G}\otimes\mathcal{F}'$, $\tilde{\mathcal{G}}_t = \cap_{s>t}\mathcal{G}_s\otimes\mathcal{F}'_s$. We have a transition probability $Q_{\mathcal{G},\omega}(d\omega')$ from $(\Omega,\mathcal{G})$ into $(\Omega',\mathcal{F}')$, such that if $\tilde{P}_{\mathcal{G}}(d\omega,d\omega') = P_{\mathcal{G}}(d\omega)Q_{\mathcal{G},\omega}(d\omega')$ (where $P_{\mathcal{G}}$ is the restriction of P to $\mathcal{G}$), then

$$E_{\mathcal{G}}(f(X^n)Y) \to \tilde{E}_{\mathcal{G}}(f(Z')Y) \tag{2.24}$$

for all bounded continuous function f on $\mathbb{D}([0,1],\mathbb{R}^q)$ and all bounded $\mathcal{G}$-measurable variable Y.

Further, $Q_{\mathcal{G},\omega}$ only depends on M, F, G and so is indeed a transition from $(\Omega,\mathcal{H})$ into $(\Omega',\mathcal{F}')$ not depending on $\mathcal{G}$ and written Q_ω.

It remains to define $(\tilde{\Omega},\tilde{\mathcal{F}},\tilde{\mathbb{F}},\tilde{P})$ by (1.1): since $\omega \rightsquigarrow Q_\omega(A)$ is $\mathcal{F}_t$-measurable for $A \in \mathcal{F}'_t$ it is a very good extension of $(\Omega,\mathcal{F},\mathbb{F},P)$. Furthermore $E_{\mathcal{G}}(f(X^n)Y) = E(f(X^n)Y)$ and $\tilde{E}_{\mathcal{G}}(f(Z')Y) = \tilde{E}(f(Z')Y)$ for all bounded $\mathcal{G}$-measurable Y: hence (2.24) yields (2.23) for all such Y. Since any $\mathcal{F}$-measurable variable Y is also $\mathcal{G}$-measurable for some separable σ-field $\mathcal{G}$ containing $\mathcal{H}$, we deduce that (2.23) holds for all bounded $\mathcal{F}$-measurable Y, and we are finished. □

2-4. When each Z^n is $I\!\!F^n$-locally square integrable, i.e. when we can write

$$Z^n = B^n + X^n, \tag{2.25}$$

with B^n a $I\!\!F^n$-predictable with finite variation and X^n a $I\!\!F^n$-locally square-integrable martingale, we have another version, involving a Lindeberg-type condition instead of (2.11), namely:

Theorem 2-2: *Assume Property (F). Assume also that Z^n is as in (2.25), and that there are two continuous processes F and G and a continuous process B of bounded variation on $(\Omega, \mathcal{F}, I\!\!F, P)$ satisfying (2.8), (2.9), (2.10), (2.12) and*

$$W(\varepsilon)^n := \int_{|x|>\varepsilon} |x|^2 \nu^n([0,1] \times dx) \to^P 0, \quad \forall \varepsilon > 0. \tag{2.26}$$

Then all results of Theorem 2-1 hold true.

Proof. We have (2.25), and also the decomposition (2.7), i.e.:

$$Z_t^n = B_t'^n + X_t'^n + \sum_{s \le t} h_q'(\Delta Z_s^n) \tag{2.27}$$

We will denote by $F_t'^n$, $G_t'^n$ and $V'(N)_t^n$ the quantities defined in (2.9), (2.10) and (2.12) with X'^n instead of X^n. We will prove that the assumptions of Theorem 2-1 are met, i.e. we have (2.11) and

$$\sup_t |B_t'^n - B_t| \to^P 0, \tag{2.28}$$

$$F_t'^n \to^P F_t, \quad \forall t \in [0,1], \tag{2.29}$$

$$G_t'^n \to^P G_t, \quad \forall t \in [0,1], \tag{2.30}$$

$$V'(N)_t^n \to^P 0, \quad \forall t \in [0,1], \quad \forall N \in \mathcal{M}_b \text{ orthogonal to } M. \tag{2.31}$$

First (2.11) readily follows from (2.26). Next, comparing (2.25) and (2.27), and if μ^n denotes the jump measure of Z^n, we get

$$B_t'^n = B_t^n + \int h_q'(x)\nu^n([0,t] \times dx), \quad X''^n := X^n - X'^n = h_q' \star (\mu^n - \nu^n).$$

We have $|h_q'(x)| \le C|x| 1_{\{|x|>\theta\}}$ for some constants $\theta > 0$ and C. This implies first that (2.28) follows from (2.8) and (2.26). It also implies

$$\sum_{i=1}^q \langle X''^{i,n}, X''^{i,n}\rangle_t \le \int |h_q'(x)|^2 \nu^n((0,t] \times dx) \le C^2 W^n(\theta). \tag{2.32}$$

We have

$$|F_t^n - F_t'^n| \le |\langle X''^n, X''^{n\star}\rangle_t| + \sqrt{|\langle X^n, X^{n\star}\rangle_t||\langle X''^n, X''^{n\star}\rangle_t|},$$

so (2.9), (2.26) and (2.32) yield (2.29). Similarly, (2.30) follows from (2.5), (2.10), (2.26), (2.32) and from the following inequality:

$$|G_t^n - G_t'^n| \le \sqrt{|\langle M(n), M(n)^\star\rangle_t||\langle X''^n, X''^{n\star}\rangle_t|}.$$

Finally we have

$$|V(N)_t^n - V'(N)_t^n| \ \le \ \sqrt{\langle N(n), N(n)\rangle_t |\langle X''^n, X''^{n*}\rangle_t|},$$

while $E(\langle N(n), N(n)\rangle_t^2) \le E(N(n)_1^2)$, which is bounded by a constant by (2.4): hence (2.31) follows as above. □

3 Convergence of discretized processes

In this section we specialize the previous results to the case when the filtration $\mathbb{F}^n$ is the "discretized" filtration defined by $\mathcal{F}_t^n = \mathcal{F}_{[nt]/n}$. For every càdlàg process Y write

$$Y_t^n \ = Y_{[nt]/n}, \qquad \Delta_i^n Y \ = \ Y_{i/n} - Y_{(i-1)/n}. \tag{3.1}$$

Here again we have a continuous d-dimensional local martingale M on the stochastic basis $(\Omega, \mathcal{F}, \mathbb{F}, P)$. We denote by h_d a continuous truncation function on $\mathbb{R}^d$. We also consider for each n an $\mathbb{F}^n$-semimartingale, i.e. a process of the form

$$Z_t^n \ = \ \sum_{i=1}^{[nt]} \chi_i^n \tag{3.2}$$

where each χ_i^n is $\mathcal{F}_{i/n}$-measurable. We then have:

Theorem 3-1: *Assume that there are two continuous processes F and G and a continuous process B of bounded variation on $(\Omega, \mathcal{F}, \mathbb{F}, P)$ such that*

$$\sup_t |\sum_{i=1}^{[nt]} E(h_q(\chi_i^n)|\mathcal{F}_{\frac{i-1}{n}}) - B_t| \ \to^P \ 0, \tag{3.3}$$

$$\sum_{i=1}^{[nt]} (E(h_q(\chi_i^n)h_q(\chi_i^n)^*|\mathcal{F}_{\frac{i-1}{n}}) - E(h_q(\chi_i^n)|\mathcal{F}_{\frac{i-1}{n}})E(h_q(\chi_i^n)^*|\mathcal{F}_{\frac{i-1}{n}})) \ \to^P \ F_t, \ \forall t \in [0,1], \tag{3.4}$$

$$\sum_{i=1}^{[nt]} (E(h_q(\chi_i^n)h_d(\Delta_i^n M)^*|\mathcal{F}_{\frac{i-1}{n}}) - E(h_q(\chi_i^n)|\mathcal{F}_{\frac{i-1}{n}})E(h_d(\Delta_i^n M)^*|\mathcal{F}_{\frac{i-1}{n}})) \to^P \ G_t, \quad \forall t \in [0,1], \tag{3.5}$$

$$\sum_{i=1}^{n} P(|\chi_i^n| > \varepsilon|\mathcal{F}_{\frac{i-1}{n}}) \ \to^P \ 0, \quad \forall \varepsilon > 0, \tag{3.6}$$

$$\sum_{i=1}^{[nt]} E(h_q(\chi_i^n)\Delta_i^n N|\mathcal{F}_{\frac{i-1}{n}}) \ \to^P \ 0, \quad \forall t \in [0,1], \quad \forall N \in \mathcal{M}_b(M^\perp). \tag{3.7}$$

Then all results of Theorem 2-1 hold true.

Proof. We will prove that the assumptions of Theorem 2-1 are in force.

a) First we check Property (F). We will take $N(n)=N^n$, as defined in (3.1), for all $N\in\mathcal{M}_b$, so (2.4) is obvious. Note also that that if $N^1,..,N^m$ are in $\mathcal{M}_b$, then

$$(M^n,N(n)^1,..,N(n)^m)\ \to^P\ (M,N^1,..,N^m)\quad\text{in } \mathbb{D}([0,1],\mathbb{R}^{d+m}). \tag{3.8}$$

Next, $M(n)$ is:

$$M(n)_t=\sum_{i=1}^{[nt]}\left(h_d(\Delta_i^nM)-E(h_d(\Delta_i^nM)|\mathcal{F}_{\frac{i-1}{n}})\right), \tag{3.9}$$

so $M^n-M(n)=A^n+A'^n$, where we have put $A_t^n=\sum_{i=1}^{[nt]}E(h_d(\Delta_i^nM)|\mathcal{F}_{\frac{i-1}{n}})$ and $A_t'^n=\sum_{i=1}^{[nt]}h_d'(\Delta_i^nM)$ (with $h_d'(x)=x-h_d(x)$). Then (2.5) follows from combining the results (1.15) and (2.12) in [4] (since M is continuous). These results also yield $\sup_t|A_t^n|\ \to^P\ 0$, and for all $\varepsilon>0$:

$$\sum_{i=1}^{n}P(|\Delta_i^nM|>\varepsilon|\mathcal{F}_{\frac{i-1}{n}})\ \to^P\ 0.$$

This and VI-4.22 of [5], together with the fact that $h_d'(x)=0$ for $|x|$ small enough, imply that $\sup_t|A_t'^n|\ \to^P\ 0$, so finally $\sup_t|M_t^n-M(n)_t|\ \to^P\ 0$ and (2.6) follows from (3.9): we thus have (F).

b) The decomposition (2.7) of Z^n has $B_t^n=\sum_{i=1}^{[nt]}E(h_q(\chi_i^n)|\mathcal{F}_{\frac{i-1}{n}})$ and $X_t^n=\sum_{i=1}^{[nt]}(h_q(\chi_i^n)-E(h_q(\chi_i^n)|\mathcal{F}_{\frac{i-1}{n}}))$. Hence (3.3) is (2.8), and the left-hand sides of (3.4), (3.5) and (3.7) are those of (2.9), (2.10) and (2.12). Finally the left-hand sides of (3.6) and of (2.11) are also the same, so we are finished. □

Finally, we could state the "discrete" version of Theorem 2-2. We will rather specialize a little bit more, by supposing that M is square-integrable and that each χ_i^n is square-integrable. This reads as:

Theorem 3-2: *Assume that M is a square-integrable continuous martingale, and that each χ_i^n is square-integrable. Assume also that there are two continuous processes F and G and a continuous process B of bounded variation on $(\Omega,\mathcal{F},\mathbb{F},P)$ such that*

$$\sup_t|\sum_{i=1}^{[nt]}E(\chi_i^n|\mathcal{F}_{\frac{i-1}{n}})-B_t|\ \to^P\ 0, \tag{3.10}$$

$$\sum_{i=1}^{[nt]}\left(E(\chi_i^n\chi_i^{n*}|\mathcal{F}_{\frac{i-1}{n}})-E(\chi_i^n|\mathcal{F}_{\frac{i-1}{n}})E(\chi_i^{n*}|\mathcal{F}_{\frac{i-1}{n}})\right)\ \to^P\ F_t,\quad\forall t\in[0,1]; \tag{3.11}$$

$$\sum_{i=1}^{[nt]}E(\chi_i^n\Delta_i^nM^*|\mathcal{F}_{\frac{i-1}{n}})\ \to^P\ G_t,\quad\forall t\in[0,1]; \tag{3.12}$$

$$\sum_{i=1}^{n}E(|\chi_i^n|^2 1_{\{|\chi_i^n|>\varepsilon\}}|\mathcal{F}_{\frac{i-1}{n}})\ \to^P\ 0,\quad\forall\varepsilon>0, \tag{3.13}$$

$$\sum_{i=1}^{[nt]} E(\chi_i^n \Delta_i^n N | \mathcal{F}_{\frac{i-1}{n}}) \to^P 0, \quad \forall t \in [0,1], \quad \forall N \in \mathcal{M}_b(M^\perp). \tag{3.14}$$

Then all results of Theorem 2-1 hold true.

Proof. If we write the decomposition (2.26) for Z^n, the left-hand sides of (3.10), (3.11), (3.12), (3.13) and (3.14) are the left-hand sides of (2.8), (2.9), (2.10) with M^n instead of $M(n)$, (2.26) and (2.12). By Theorem 2-2 it thus suffices to prove that (F) is satisfied if $N(n) = N^n$ and $M(n) = M^n$. We have seen (2.4) and (2.6) in the proof of Theorem 3-1, so it remains to prove that $\langle M^n, M^{n*}\rangle_t \to^P \langle M, M^*\rangle_t$ for all t.

Let us consider $M(n)$ as in (3.9): we have seen that it has (2.5), so it is enough to prove that if $Y^n = M^n - M(n)$, then

$$\langle Y^n, Y^{n*}\rangle_1 \to^P 0. \tag{3.15}$$

The process $\langle Y^n, Y^{n*}\rangle_t$ is L-dominated by $D_t^n = \sup_{s\leq t} |Y_s^n|$, and $W = \sup_{n,t} |\Delta D_t^n|$ satisfies $W \leq 2C + 2\sup_t |M_t|$ where $C = \sup |h_d|$: hence $E(W) < \infty$. We have seen in the proof of Theorem 3-1 that $D_1^n \to^P 0$, so the "optional" Lenglart inequality I-3.32 of [5] yields (3.15), and the proof is finished. □

4 Convergence of conditionally Gaussian martingales

Here we still have our basic continuous d-dimensional local martingale M on the basis $(\Omega, \mathcal{F}, \mathbb{F}, P)$, and a sequence Z^n of M-biased continuous $\mathcal{F}$-conditional Gaussian martingales: each one is defined on its own very good extension $(\tilde{\Omega}^n, \tilde{\mathcal{F}}^n, \tilde{\mathbb{F}}^n, \tilde{P}^n)$. Note that $\mathcal{F}$ can be considered as a sub σ-field of $\tilde{\mathcal{F}}^n$ for each n.

Theorem 4-1: *Assume that there are two continuous processes F and G on $(\Omega, \mathcal{F}, \mathbb{F}, P)$ such that*

$$F_t^n := \langle Z^n, Z^{n*}\rangle_t \to^P F_t, \quad \forall t \in [0,1], \tag{4.1}$$

$$G_t^n := \langle Z^n, M(n)^*\rangle_t \to^P G_t, \quad \forall t \in [0,1], \tag{4.2}$$

Then there is a very good extension of $(\Omega, \mathcal{F}, \mathbb{F}, P)$ and an M-biased $\mathcal{F}$-conditional Gaussian martingale Z on this extension with

$$\langle Z, Z^*\rangle = F, \quad \langle Z, M^*\rangle = G, \tag{4.3}$$

such that $Z^n \to^{\mathcal{F}\text{-}\mathcal{L}} Z$.

Proof. Set $H^n = H = \langle M, M^*\rangle$, and define K^n and K as in Step 1 of the proof of Theorem 2-1. (4.1) and (4.2) imply that $K_t^n \to^P K_t$ for all t, and since K^n is continuous in time the same holds for K, and we have (2.16). Further, if $V(N)^n = \langle Z^n, N\rangle$, by assumption on Z^n we know that $V(N)^n = 0$ for all $N \in \mathcal{M}_b(M^\perp)$.

We can then reproduce Step 4 of the proof of Theorem 2-1, with $M(n) = M$ and $N^m(n) = N^m$ and Z^n and Z instead of X^n and Z'. In place of (2.23), we get

$$\tilde{E}^n(f(Z^n)Y) \to \tilde{E}(f(Z)Y)$$

for all bounded $\mathcal{F}$-measurable variables Y and all bounded continuous functions f on $\mathbb{D}([0,1],\mathbb{R}^q)$: this is the desired convergence result when $\mathcal{F}$ is separable. Finally, Step 5 of the same proof may be reproduced here, to relax the separability assumption on $\mathcal{F}$, and the proof is complete. □

References

[1] Aldous, D.J. and Eagleson, G.K. (1978): On mixing and stability of limit theorems. *Ann. Probab.* **6** 325-331.

[2] Jacod, J. (1979): *Calcul stochastique et problèmes des martingales.* Lect. Notes in Math. **714**, Springer Verlag: Berlin.

[3] Jacod, J. and Mémin, J. (1981): Weak and strong solutions of stochastic differential equations; existence and stability. In *Stochastic Integrals*, D. Williams ed., Proc. LMS Symp., Lect. Notes in Math. **851**, 169-212, Springer Verlag: Berlin.

[4] Jacod, J. (1984): Une généralisation des semimartingales: les processus admettant un processus à accroissements indépendants tangent. §éminaire Proba. XVIII, Lect. Notes in Math. **1059**, 91-118, Springer Verlag: Berlin.

[5] Jacod, J. and Shiryaev, A. (1987): *Limit Theorems for Stochastic Processes.* Springer-Verlag: Berlin.

[6] Renyi, A. (1963): On stable sequences of events. *Sankya* Ser. A, **25**, 293-302.

Laboratoire de Probabilités (CNRS, URA 224), Université Paris VI, Tour 56, 4, Place Jussieu, 75252 Paris Cedex 05, France.

Simple examples of non-generating Girsanov processes

J.Feldman*and M.Smorodinsky
University of California, Berkeley and Tel-Aviv University

Let $B(t), 0 \leq t < \infty$ be a Brownian motion on $(\Omega, \mathcal{F}, \mathcal{P})$ with $B_0 = 0$. Let $\mathcal{F}(t), 0 \leq t \leq \infty$ be its filtration, with $\mathcal{F}(\infty) = \mathcal{F}$. We construct simple examples of probability measures $P' \sim P$ for which this filtration is not generated by the corresponding Girsanov process, but is nevertheless generated by *some* process which is a Brownian motion for the measure P'.

1. Introduction. Given a Brownian motion and $P' \sim P$ as above, the corresponding Radon-Nikodym derivative may be written in the form

$$dP'/dP = exp\{\int_0^\infty \Phi(t)dB(t) - (1/2)\int_0^\infty |\Phi(t)|^2 dt\},$$

where Φ is a process on $[0, \infty)$ adapted to the filtration of B and satisfying certain other conditions which, in particular, cause the expression to make sense and have expectation 1. This is the *Cameron-Martin-Girsanov formula.* $\Phi(t)$ is uniquely determined *a.e.* in t. While it is not easy to characterize exactly those processes Φ which arise in this manner(See Kazamaki's recent monograph [K]), we note that if Φ is adapted to the filtration of B, and $\int_0^\infty |\Phi|^2(t)dt$ is bounded by a fixed constant, then Φ arises in such a way. The associated *Girsanov Process G* defined by

$$G(t) = B(t) - \int_0^t \Phi(s)ds, 0 \leq t < \infty$$

is a Brownian motion with respect to the measure P' (*"Girsanov's Theorem"*). Let $\mathcal{G}(t), 0 \leq t \leq \infty$ be its filtration. Because Φ is adapted to the filtration of the original Brownian motion, we have always $\mathcal{G}(t) \subset \mathcal{F}(t)$ for all t. A good reference for these matters is [RY].

The question then arises, whether the Girsanov Process always generates the filtration of B, i.e. whether $\mathcal{G}(t) = \mathcal{F}(t)$ for all t in $[0, \infty)$; note that this will follow for all such t if it holds for $t = \infty$. This question is of relevance for Stochastic Differential Equations. In 1975 B.Tsirelson showed that the answer is *no*: in [T] he constructed a P' for which the Girsanov Process does *not* generate this filtration. For further discussion of this important example see [Y].

An obvious next question, explicitly asked in [RY], is whether at least there is for

*Supported in part by NSF Grant #DMS 9113642.

every $P' \sim P$ *some* process which is a Brownian motion for P' and whose filtration coincides with that of the original Brownian motion. The answer again is *no*: in [DFST] there are constructed measures $P' \sim P$ for which no such Brownian motion for P' exists, i.e for which the filtration $\mathcal{F}(t), 0 \leq t \leq \infty$ is not Brownian with respect to P'.

This leaves open the following question: can it happen that $\mathcal{F}(t), 0 \leq t \leq \infty$ is Brownian for P' but the corresponding Girsanov process does not generate the filtration of B? One would expect so, and this is what we show here:

Theorem: *Given a Brownian motion $B(t), 0 \leq t < \infty$ on $(\Omega, \mathcal{F}, \mathcal{P})$, there exist probability measures $P' \sim P$ for which the filtration of B is Brownian with respect to P' as well, but for which the corresponding Girsanov Process does not generate.*

Our examples are simple, and the described properties are easy to demonstrate. What about the P' of Tsirelson's 1975 example? We have not determined whether the filtration of B is Brownian for this P'. However, A.M.Vershik tells us he can show that it is.

It should be remarked that although this paper is in no way dependent on it, our examples were motivated by considerations coming from the study of decreasing sequences of sigma fields, for which see [V]. We also note that questions of this type for discrete time processes were studied by M.Rosenblatt [R], and one of his constructions there may be viewed as analogous to ours.

We thank Marc Yor for his careful reading and helpful suggestions.

2. A class of examples. Choose $\infty > t_1 > t_2 > \cdots \to 0$, and a sequence $a_1, a_2, \cdots$ of positive numbers with $\sum_{n=0}^{\infty} a_n^2(t_n - t_{n+1}) < \infty$; for example, any *bounded* sequence will do. For a subset S of the reals, let $\chi_S = -1$ on S and 1 elsewhere. Let $S_1, S_2, \cdots$ be measurable subsets of the reals, and $\sigma_n = \chi_{S_n}(B(t_n) - B(t_{n+1}))$, i.e. -1 if $B(t_n) - B(t_{n+1})$ lies in S_n and 1 otherwise. Let $\Phi(t) = \sigma_{n+1} a_n$ if $t_{n+1} < t \leq t_n$ and zero if $t_1 < t$. The following lemma is just a small calculation:

Lemma 1: *$\int_0^{\infty} |\Phi|^2(t)dt$ is bounded by a constant, in fact equals $\sum_{n=1}^{\infty} a_n^2(t_n - t_{n+1})$, so it defines a probability measure $P' \sim P$ with Radon-Nikodym derivative*

$$dP'/dP = exp\{\sum_{n=1}^{\infty} \sigma_{n+} a_n(B(t_n) - B(t_{n+1})) - (1/2) \sum_{n=1}^{\infty} a_n^2(t_n - t_{n+1})\}.$$

Denote by $N(m, v)$ the normal distribution with mean m and variance v. The following remarks follow from Lemma 1 and Girsanov's Theorem.

Remark 2: The random variables $y_1, y_2, \cdots$ defined by

$$y_n = G(t_n) - G(t_{n+1}) = B(t_n) - B(t_{n+1}) - \sigma_{n+1} a_n(t_n - t_{n+1}).$$

form an independent sequence of random variables with respect to P', y_n having distribution $N(0, t_n - t_{n+1})$.

Remark 3: The sequence of random variables $x_1, x_2, \cdots$ defined by

$$x_n = B(t_n) - B(t_{n+1})$$

forms, with respect to P', a Markov stochastic process; the conditional measure $P'(\cdot|x_{n+1})$ is either $N(b_n, t_n - t_{n+1})$ or $N(-b_n, t_n - t_{n+1})$, depending on whether x_{n+1} is in S_{n+1} or not, where $b_n = a_n(t_n - t_{n+1})$. The total distribution of x_n, call it μ_n, is just their average, weighted by $(\mu_{n+1}(S_{n+1}), 1 - \mu_{n+1}(S_{n+1}))$. Let us further define

$$y(t) = G(t) - G(t_{n(t)}), 0 \leq t < \infty,$$

where $n(t) = min\{n : t_n \leq t\}$, and let $\mathcal{G}_{n-1}$ be the sigma-field generated by $y(t), t_n < t \leq t_{n-1}$. Then the following stronger Markov relation holds: x_n is P' conditionally independent of $\mathcal{F}(t_{n+1}) \vee \mathcal{G}_n$ given y_n.

3. Proof of Theorem.

Lemma 4: *For each nonatomic probability measure μ on the real line and each $b > 0$ there exists a measurable subset S of the real line, in fact a countable union of intervals, with $\mu(S) = 1/2$, and a.e. symmetric about zero, so that for each real number η exactly one of the two numbers $\eta + b, \eta - b$ lies in S.*

Proof: Let $I(n)$ be the half-open interval $[(n-1)b, (n+1)b)$,and for $0 \leq r \leq b$ let $J(n, r)$ be the half-open interval $[nb - r, nb + r)$. Let $S(r)$ be the union of the sets $J(4n+2, r)$ and $I(4n) \cap J(4n, r)^c$ over all integers n. This set is *a.e.* symmetric: the only differences between $S(r)$ and $-S(r)$ occur at end points of the constituent intervals. The sets $S(0)$ and $S(b)$ form a partition of the real line, so $\mu(S(0)) + \mu(S(b)) = 1$. The map $r \mapsto \mu(S(r))$ is continuous, so there is some r in $(0, b)$ with $\mu(S(r)) = 1/2$. Setting $S = S(r)$, we are done.

Now choose the sets S_n by setting $b = b_n$ in Lemma 4 and μ equal to the $(1/2, 1/2)$ average of $N(b_n, t_n - t_{n+1})$ and $N(-b_n, t_n - t_{n+1})$. Then $P'[x_n \in S_n]$ will be $1/2$ for each n. Each value η of y_n could have come from either of the two values $\eta + b_n$ or $\eta - b_n$ for x_n; and S_n has been so chosen that for each η exactly one of these lies in S_n. Additionally, each S_n is *a.e.* symmetric about zero. We proceed to prove that the Girsanov process constructed by means of this sequence of sets does not generate the filtration of B.

Denote by $E'(\cdot|\cdot)$ conditional expectation with respect to P'. Then $E'(\sigma_1|y_1; x_2) = E'(\sigma_1|y_1; \sigma_2) = 1$ or -1 with equal P' probability. Integrating out x_2 gives $E'(\sigma_1|y_1) = 0$ *a.e.* Repeating this argument inductively with $E'(\sigma_1|y_1, y_2, ..., y_n; \sigma_{n+1})$ gives with probability one:

$$E'(\sigma_1|y_1, y_2, ..., y_n) = 0$$

for all integers $n > 0$. It follows from the Markov relation in Remark 3 that $E'(\sigma_1|\mathcal{G}(t_1)) = E'(\sigma_1|y_1, y_2, ...)$, so σ_1 is P' independent of $\mathcal{G}(t_1)$. But σ_1 is measurable with respect to $\mathcal{F}(t_1)$. So $\mathcal{G}(t_1)$ is not all of $\mathcal{F}(t_1)$.

Next we show that there is a process $B'(t), 0 \leq t < \infty$ which is a Brownian motion under P' and whose filtration is precisely that of B. First introduce the random variables $y'_n = \sigma_{n+1} y_n$ and $y'(t) = \sigma_{n(t)+1} y(t)$. Then put

$$B'(t) = y'(t) + \sum_{i=n(t)}^{\infty} y'_i = \int_0^t \sigma_{n(s)+1} dy(s).$$

It is clear that B' is a Brownian motion with respect to P'.
Let $x(t) = B(t) - B(t_{n(t)})$. To prove our claim it suffices to show that $x(t)$ is $\mathcal{F}'(t)$ measurable for all $t > 0$, where $\mathcal{F}'(t), 0 \leq t \leq \infty$ is the filtration generated by B'.

We claim that for any $t > 0$, $x(t)$ is *a.e* a function of the variables $y'(t), y'_{n(t)}, y'_{n(t)+1}$. For if $y'_{n(t)+1}$ takes on the value η then either:

(1) $\sigma_{n(t)+2} = -1$, so $y_{n(t)+1} = -\eta$, and $x_{n(t)+1} = -\eta + b_n$, or

(2) $\sigma_{n(t)+2} = 1$, so $y_{n(t)+1} = \eta$, and $x_{n(t)+1} = \eta - b_n$.

Thus $y'_{n(t)+1}$ completely determines $|x_{n(t)+1}|$. Since $S_{n(t)+1}$ is *a.e.* symmetric, $y'_{n(t)+1}$ determines with probability one the distribution of $x_{n(t)}$, so $y'_{n(t)}$ and $y'_{n(t)+1}$ together determine $x_{n(t)}$ with probability one. But clearly the pair $(y'(t), x_{n(t)})$ determines $x(t)$ with probability one. This completes the proof of the theorem.

Remark 5: Even without the symmetry assumption on the sets $S_1, S_2, \cdots$ the process B' generates the filtration of B, but the argument is a little more involved.

References:

[DFST] L. Dubins, J.Feldman, M.Smorodinsky, and B.S.Tsirelson(1995): *Decreasing sequences of σ-fields and a measure change for Brownian motion*, to appear, Ann.Prob.

[K] N.Kazamaki (1994): *Continuous exponential martingales and BMO*, Lec.Notes Math.1579, Springer-Verlag.

[R] M.Rosenblatt (1959): *Stationary processes as shifts of functions of independent random variables*, Jour.Math.Mech. 8, pp.665-681.

[RY] D.Revuz, and M.Yor (1991): *Continuous Martingales and Brownian Motion*, Springer, Berlin.

[T] B.S.Tsirelson, (1975): *An example of a stochastic differential equation having no strong solution*, Theor.Prob.Appl. 20, pp.416-418.

[V] A.M.Vershik, (1994): *Theory of decreasing sequences of measurable partitions*, Alg. Anal. 6, pp.1-68,in Russian; English version to appear in St.Petersburg Math. Jour. 6.

[Y] M.Yor, (1992): *Tsirelson's equation in discrete time*, Prob.Th. Related Fields, 91, pp.135-152.

Formule d'Ito généralisée pour le mouvement brownien linéaire

D'après Föllmer, Protter et Shiryaev [1], par P.A. Meyer*.

On a du mal à croire que l'on puisse encore découvrir des résultats simples sur le mouvement brownien. C'est pourtant le cas avec ce remarquable article.

Soit (X_t) le mouvement brownien linéaire issu de 0. Considérons une fonction $F(x)$ de classe C^2, et posons $f(x) = F'(x)$. Alors

$$F(X_t) = F(0) + \int_0^t f(X_s)\,dX_s + A_t \ ,$$

où A_t est le processus à variation finie $\frac{1}{2}\int_0^t f'(X_s)\,ds$. De plus, en utilisant comme d'habitude des subdivisions dyadiques de $[0,t]$,

$$\text{(2)} \qquad \int_0^t f(X_s)\,dX_s = \lim \sum_i f(X_{t_i})(X_{t_{i+1}} - X_{t_i})$$

$$\text{(3)} \qquad A_t = \lim \sum_i (f(X_{t_{i+1}}) - f(X_{t_i}))(X_{t_{i+1}} - X_{t_i}) \ ;$$

(A_t) est la covariation $[f(X), X]_t$. On sait depuis longtemps que l'hypothèse de classe C^2 est trop forte : si F est de classe C^1 (*i.e.* si f est continue), les résultats ci-dessus restent vrais, avec la seule différence que (A_t) n'est plus l'intégrale (maintenant dépourvue de sens) de $f'(X_s)$, et qu'au lieu d'être à variation finie il est seulement continu à variation quadratique nulle.

Le problème traité par les trois auteurs consiste à étendre ces derniers résultats au cas où f est seulement *localement de carré intégrable*. La démonstration exige que l'on travaille, non seulement sur le mouvement brownien (X_t), mais sur le pont brownien (Y_t), nul aux deux extrémités de l'intervalle $[0,1]$. Rappelons que $\mathbb{E}[Y_s^2] = s(1-s)$, et que

$$\text{(4)} \qquad Y_t = W_t - \int_0^t \frac{Y_s}{1-s}\,ds \ ,$$

où (W_t) est un mouvement brownien.

1. Approximations d'intégrales stochastiques. Nous allons établir ici le principal résultat technique de l'article — d'ailleurs simple et intéressant. Soit $f(t,x)$ une fonction borélienne telle que :

i) $f(t,\cdot)$ appartienne à $L^2(\mathbb{R})$ pour tout $t > 0$;

ii) l'application $t \longmapsto f(t,\cdot)$ à valeurs dans $L^2(\mathbb{R})$ soit fortement continue sur $]0,\infty[$, et bornée au voisinage de 0.

Alors l'intégrale stochastique $I_t = \int_0^t f(s,X_s)\,dX_s$ existe pour t fini, appartient à L^2, et de plus elle est approchée en norme L^2 par les sommes de Riemann dyadiques usuelles sur $[0,t]$,

$$I_t = \lim \sum_i f(t_i, X_{t_i})(X_{t_{i+1}} - X_{t_i}) \ .$$

---

* Exposé de décembre 1994

Avant toute chose, remarquons que ces notations ont un sens : la loi de (X_t) étant absolument continue, la v.a. $f(t,X_t)$ ne dépend que de la classe de $f(t,\cdot)$.

On a le même résultat en remplaçant X_t par le pont brownien Y_t (et en restreignant le temps à $[0,1]$). Cependant, dans ce cas l'intégrale stochastique et les sommes de Riemann n'appartiennent qu'à L^1, et l'approximation a lieu en norme L^1. Cette extension sera essentielle pour la suite.

Enfin, nous aurons besoin du même résultat pour un pont brownien prenant des valeurs quelconques aux extrémités de l'intervalle $[0,1]$; l'extension est facile, et nous en dirons un mot à la fin.

DÉMONSTRATION. Commençons par l'appartenance à L^2 de l'intégrale relative à X. Nous partons de l'inégalité

$$\int_a^b \mathbb{E}[f^2(s,X_s)]\,ds = \int_a^b f^2(s,x)e^{-x^2/2s}\frac{dxds}{\sqrt{2\pi s}} \le \int_a^b \|f(s,\cdot)\|^2\frac{ds}{\sqrt{2\pi s}}$$

intégrale finie puisque $\|f(s,\cdot)\|$ est localement borné. Noter que l'on peut prendre $a=0$, et que pour b petit les intégrales sont alors petites aussi.

Passons au cas du pont brownien (4). Il y a deux termes à considérer. D'abord un terme qui appartient à L^2, l'intégrale de $f(s,Y_s)$ par rapport à W ; il se traite comme ci-dessus, à cela près que la variance de Y_s n'est pas s mais $s(1-s)$. D'autre part, nous avons à évaluer une norme L^1 :

$$\mathbb{E}\Big[\int_a^b \frac{|f(s,Y_s)|\,|Y_s|}{1-s}\,ds\Big] = \int_a^b \frac{ds}{1-s}\frac{1}{\sqrt{2\pi s(1-s)}}\int |f(s,x)x|e^{-x^2/s(1-s)}dx\ .$$

Dans l'intégrale de droite, on applique l'inégalité de Schwarz, en faisant apparaître d'une part $\|f(s,\cdot)\|_2$, et d'autre part $(\int x^2e^{-x^2/s(1-s)}dx)^{1/2}$. L'intégrale gaussienne est en $(s(1-s))^{3/2}$, d'où

$$\mathbb{E}\Big[\int_a^b \frac{|f(s,Y_s)|\,|Y_s|}{1-s}\,ds\Big] \le C\int_a^b \|f(s,\cdot)\|_2\frac{s^{1/4}ds}{(1-s)^{3/4}} < \infty\ .$$

Nous passons à la seconde étape : sachant que l'intégrale existe, montrer qu'elle est approchée par ses sommes de Riemann. Revenons à X qui est plus simple. L'idée essentielle est que, bien que la fonction $f(s,\cdot)$ ne soit pas continue, *le processus* $f(s,X_s)$ *est continu dans* L^2 *pour* $s>0$. La situation près de 0 est d'ailleurs bien contrôlée en norme, et on n'a pas besoin de s'en occuper.

Nous n'aurons besoin en fait que de la continuité à droite. Nous fixons $s>0$, et prenons $t\in[s,s+h]$, h petit. En posant $t-s=u$, et en désignant par (μ_r) le semi-groupe de convolution brownien, on a

$$\|f(s,X_s)-f(s,X_t)\|^2 = \int \mu_s(dx)\,(f(s,x)-f(s,x+y))^2\mu_u(dy)$$

Mais il est bien connu que les translations opèrent continûment dans $L^2(\mathbb{R})$, donc (μ_s ayant une densité bornée) $\int\mu_s(dx)(f(s,x)-f(s,x+y))^2$ est uniformément borné et petit pour y petit; comme $\mu_u(dy)$ est concentrée près de 0 pour u petit, la norme du côté gauche est petite aussi.

Reste à considérer $f(t,X_t) - f(s,X_t) = g(X_t)$ en posant $g = f(t,\cdot) - f(s,\cdot)$: la norme de g dans $L^2(\mathbb{R})$ est petite par hypothèse, et l'on utilise la majoration grossière $\| g(X_t) \|_2 \leq \| g \|_2/\sqrt{2\pi t}$. □

La norme L^2 de la différence entre l'intégrale stochastique $\int_a^b f(t,X_t)\,dX_t$ et la somme de Riemann correspondante est alors

$$\sum_i \int_{s_i}^{s_{i+1}} \mathbb{E}\left[(f(t,X_t) - f(s_i,X_{s_i}))^2\right] ds$$

et ceci est petit.

Passons au pont brownien, pour lequel on ne peut utiliser la convolution. Nous utiliserons la remarque suivante (Revuz-Yor, p. 37) : on peut construire le pont brownien Y_t sur $[0,1]$ comme $(1-t)B_{t/1-t}$ où B est un mouvement brownien. Donc la continuité L^2 de $f(s,Y_s)$ se ramène à celle de $g(s,X_s)$ pour la fonction

$$g(s,x) = f\left(\frac{t}{1+t}, \frac{x}{1+t}\right)$$

tant que l'on va pas trop près de 1 (et de 0, où elle n'a pas lieu pour le brownien). Cela permet — compte tenu des majorations du début près des bornes — d'approcher par des sommes de Riemann dans L^2 l'intégrale stochastique $\int_0^1 f(s,Y_s)\,dW_s$. Quant au produit $f(s,Y_s)Y_s/(1-s)$, il est continu dans L^1 sur l'intervalle $]0,1[$ ouvert, et l'intégrale $\int_0^1 f(s,Y_s)Y_s ds/1-s$ peut donc être approchée par ses sommes de Riemann.

Dernier point, on peut obtenir le même résultat pour un pont brownien sur $[0,1]$ entre deux valeurs quelconques. En effet, un tel pont s'écrit $Y_t + a + bt$, et cela revient encore à changer de fonction f.

2. Passage au résultat principal. Puisque les retournés des ponts browniens sont des ponts browniens, on voit que les sommes de Riemann en avant et en arrière, relatives à un intervalle $[u,v]$ avec $0 \leq u < v \leq 1$

$$\sum_i f(s_i, Y_{s_i})(Y_{s_{i+1}} - Y_{s_i}) \quad \text{et} \quad \sum_i f(s_{i+1}, Y_{s_{i+1}})(Y_{s_{i+1}} - Y_{s_i}),$$

convergent en probabilité vers l'intégrale stochastique correspondante. Prenant la différence, on obtient l'existence de la covariation

$$[f(\cdot,Y),Y] = \lim \sum_i (f(s_{i+1},Y_{s_{i+1}}) - f(s_i,Y_{s_i}))(Y_{s_{i+1}} - Y_{s_i})$$

et du même coup, de l'intégrale stochastique de Stratonovich.

Ce résultat s'applique alors aussi au mouvement brownien sur $[0,1]$, en le conditionnant par ses valeurs initiale et finale.

Résumons brièvement les autres résultats de l'article, qui sont des conséquences assez simples du résultat principal.

Puisque la covariation $[f(X),X]$ est la différence de deux intégrales stochastiques browniennes, elle possède le même type de continuité en f que les intégrales stochastiques. Mais alors il n'est pas difficile d'étendre la formule d'Ito classique

$$F(X_t) - F(X_0) = \int_0^t F'(X_s)\,dX_s + \frac{1}{2}[F'(X),X]_t$$

du cas où F' est de classe C^1 au cas où elle appartient seulement à L^2, puis à L^2_{loc}, ce qui a été annoncé au début.

Une autre application très intéressante est l'extension de la notion de temps local. Le temps local de a s'obtient en appliquant la formule d'Ito à $F(x) = (x-a)^+$, autrement dit, c'est la covariation de X avec $f = I_{[a,\infty[}$. On a maintenant le moyen de définir le temps local de X sur une courbe continue $a(t)$, en prenant comme fonction $f(s,x) = I_{x>a(s)}$. Les auteurs montrent que c'est un processus croissant continu, qui ne croît que sur l'ensemble où le brownien rencontre la courbe.

En tant que fonction de t, la covariation $V_t(f) = [f(\cdot,X),X]_t$, différence de deux intégrales stochastiques, est un processus adapté à trajectoires continues. Contrairement au cas où f est de classe C^1, ce n'est pas en général un processus à variation finie. Mais on s'attend à ce qu'il ait une variation quadratique nulle : les auteurs établissent ce résultat, qui n'exige pas de nouvelles techniques.

En effet, comme la covariation quadratique $V_t(f)$ est la différence de deux intégrales stochastiques, l'une en avant et l'autre en arrière, on a une majoration a priori pour les sommes de carrés d'accroissements sur un intervalle $[a,b]$

$$\mathbb{E}\,[\sum\nolimits_i (V_{t_{i+1}}(f) - V_{t_i}(f))^2] \le 2\mathbb{E}\,[\int_a^b f^2(s,X_s)\,ds]\ .$$

Nous écrivons ensuite que $V_t(f) = V_t(h) + V_t(f-h)$, où h est une approximation régulière de f ; d'où pour les sommes de carrés d'accroissements une inégalité (avec un facteur 2). Comme $V_t(h)$ est à variation finie d'après la formule d'Ito classique, les sommes de carrés correspondantes tendent vers 0 dans L^1 lorsque le pas de la subdivision tend vers 0 ; d'autre part, la majoration précédente s'applique à $V_t(f-h)$, et les sommes correspondantes sont donc petites dans L^1.

Enfin, l'existence de la covariation permet de définir une intégrale de Stratonovich, et d'établir une version générale de la formule d'Ito.

RÉFÉRENCE

[1] FÖLLMER (H.), PROTTER (P.) et SHIRYAYEV (A.N.). Quadratic covariation and an extension of Ito's formula, *Bernoulli* **1/2**, 1995, p. 149–169.

On the martingales obtained by an extension due to Saisho, Tanemura and Yor of Pitman's theorem

Koichiro TAKAOKA
Dept. of Applied Physics, Tokyo Institute of Technology*

Abstract

M. Yor constructed a family of one-dimensional continuous martingales in connection with Saisho and Tanemura's extension of Pitman's theorem. This paper reveals some properties of these martingales and the corresponding stochastic differential equations. In particular, this implies that the pathwise uniqueness theorem by Yamada and Watanabe cannot be generalized to a non-diffusion case.

1 Introduction

M. Yor has recently showed the following property based on Saisho and Tanemura's generalization [5] of Pitman's theorem [2].

Theorem 1.1 (Yor [9], Corollary 12.5.1) *Let $\left(R_\alpha(t)\right)_{t\in[0,\infty)}$ be an α-dimensional Bessel process starting from the origin on a certain probability space $(\Omega, \mathcal{F}, P)$. Define*

$$X_\alpha(t) \overset{\text{def}}{=} 2 \min_{s\in[t,\infty)} R^\alpha_{\alpha+2}(s) - R^\alpha_{\alpha+2}(t) \qquad \text{for } t \in [0,\infty).$$

Then, for each $\alpha > 0$, $\left(X_\alpha(t)\right)_{t\in[0,\infty)}$ is an $\mathcal{F}^{X_\alpha}$-martingale, where $\mathcal{F}^{X_\alpha} = \left(\mathcal{F}^{X_\alpha}_t\right)$ denotes the filtration generated by $\left(X_\alpha(t)\right)$.

Remarks. (i) As shown in Revuz-Yor [4] Theorem VI.3.5, we see from Theorem 1.1 with $\alpha = 1$ and from Lévy's characterization theorem that $\left(X_1(t)\right)$ is a one-dimensional Brownian motion. Therefore, $\{X_\alpha;\ \alpha > 0\}$ is a family of $\boldsymbol{R}$-valued continuous martingales that includes one-dimensional Brownian motion.

*Oh-okayama Meguro-ku Tokyo 152, Japan. E-mail:takaoka@neptune.ap.titech.ac.jp

(ii) As known from the literature (e.g. the above cited book of Revuz-Yor), it holds that

$$\max_{s\in[0,t]} X_\alpha(s) = \min_{s\in[t,\infty)} R^\alpha_{\alpha+2}(s), \qquad t \in [0,\infty), \quad \text{a.s.},$$

and hence Pitman's theorem is equivalent to the above mentioned fact that $(X_1(t))$ is a one-dimensional Brownian motion. Thus, Theorem 1.1 can be viewed as an extension of Pitman's theorem. Yor [9] has actually proved that Theorem 1.1 holds for a larger class of diffusions; an even further extension is done recently by Rauscher [3]. It should also be mentioned that in several works, different generalizations of Pitman's theorem were studied; e.g. Bertoin [1] and Tanaka [6] [7].

(iii) Theorem 1.1 is proved in Yor's book by using the "enlargement of filtration" technique first introduced by T. Jeulin. Note here that the filtration $\mathcal{F}^{X_\alpha}$ is strictly larger than $\mathcal{F}^{R_{\alpha+2}}$:

$$\forall t \geq 0, \qquad \mathcal{F}^{X_\alpha}_t = \mathcal{F}^{R_{\alpha+2}}_t \vee \sigma\left(\min_{s\in[t,\infty)} R_{\alpha+2}(s)\right).$$

The aim of the present paper is to investigate which properties of one-dimensional Brownian motion hold for other members of our martingale family $\{X_\alpha;\ \alpha > 0\}$ and which do not. Among others, the following two properties will be shown:

1) The stochastic differential equations (henceforth SDEs) satisfied by $(X_\alpha(t))$, $\alpha \neq 1$, are of non-diffusion type and do not fulfill the Lipschitz condition. If $\alpha \leq 1$, then pathwise uniqueness holds for our SDE. On the other hand, if $\alpha > 1$, even uniqueness in law fails; in particular, for $\alpha \geq 2$, our SDEs are counterexamples showing that the famous Yamada-Watanabe pathwise uniqueness theorem for one-dimensional diffusion-type SDEs cannot be extended to non-diffusion cases (Theorem 2.4).

2) For each fixed $t \geq 0$, the random variable $X_\alpha(t)$ is symmetrically distributed with respect to the origin, while the processes $(X_\alpha(t))$ and $(-X_\alpha(t))$ do not have the same law if $\alpha \neq 1$ (Proposition 2.2 and Theorem 2.3).

This paper is organized as follows. In Section 2 we state our results. The proofs of these properties will be given in Section 3. Throughout this paper, we frequently cite the book of Revuz-Yor [4] as the basic reference.

Acknowledgements. A stimulating conversation with Professor T. Shiga has improved Theorem 2.4(iii); sincere thanks are due to him. The author also wishes to thank Prof. V. Vinogradov and Dr. J. Akahori for their helpful comments.

2 Statement of the results: some properties of the martingales $(X_\alpha(t))$

As mentioned above in the Introduction, the proofs of all the properties listed in this section will be given in Section 3.

Proposition 2.1 *(i) If $\alpha \neq 1$, then $(X_\alpha(t))$ is not a Markov process, while the $\boldsymbol{R}^2$-valued process*

$$\left(X_\alpha(t),\ \max_{s\in[0,t]} X_\alpha(s) \right)_{t\in[0,\infty)}$$

is Markov for any $\alpha > 0$.
(ii) For each $\alpha > 0$, $(X_\alpha(t))$ is self-similar in the sense that

$$\forall c > 0, \qquad \left(c^{-\alpha/2} X_\alpha(ct)\right)_{t\in[0,\infty)} \overset{(d)}{=} \left(X_\alpha(t)\right)_{t\in[0,\infty)}.$$

(iii) For each $\alpha > 0$, $(X_\alpha(t))$ is a divergent martingale:

$$\lim_{t\uparrow\infty} [X_\alpha]_t = \infty \qquad a.s.,$$

where $\left([X_\alpha]_t\right)_{t\in[0,\infty)}$ denotes the quadratic variation process of $(X_\alpha(t))$.

The next proposition generalizes well-known results for one-dimensional Brownian motion.

Proposition 2.2 *Fix $\alpha > 0$ and $t > 0$.*
(i) The distribution of $X_\alpha(t)$ is symmetric with respect to the origin:

$$X_\alpha(t) \overset{(d)}{=} -X_\alpha(t).$$

In more detail, we have

$$P[\,X_\alpha(t) \in dx\,] = \frac{1}{\alpha\,(2t)^{\alpha/2}\,\Gamma\left(\frac{\alpha}{2}\right)} \exp\left(-\frac{|x|^{2/\alpha}}{2t}\right) dx, \qquad x \in \boldsymbol{R}.$$

(ii) The following four random variables are all identically distributed:

$$\begin{aligned}
&(a) \qquad \max_{s\in[0,t]} X_\alpha(s) && \left(= \min_{s\in[t,\infty)} R^\alpha_{\alpha+2}(s)\right); \\
&(b) \qquad \max_{s\in[0,t]} X_\alpha(s) - X_\alpha(t) && \left(= R^\alpha_{\alpha+2}(t) - \min_{s\in[t,\infty)} R^\alpha_{\alpha+2}(s)\right); \\
&(c) \qquad |X_\alpha(t)| && \left(= \left|2 \min_{s\in[t,\infty)} R^\alpha_{\alpha+2}(s) - R^\alpha_{\alpha+2}(t)\right|\right); \\
&(d) \qquad R^\alpha_\alpha(t).
\end{aligned}$$

The two questions which arise naturally from Proposition 2.2 are as follows:

- *Is $(X_\alpha(t))$, as a process, symmetric with respect to the origin?*
- *Do (b), (c) and (d) of Proposition 2.2(ii) have, as processes, the same law?*

since it is well known that the answer is "yes" for both of them if $\alpha = 1$. The next theorem, however, answers these questions in the negative for $\alpha \neq 1$.

Theorem 2.3 *Suppose $\alpha \neq 1$.*
(i) The following three martingales have different laws from one another:

$$\big(X_\alpha(t)\big)_{t\in[0,\infty)};$$
$$\big(-X_\alpha(t)\big)_{t\in[0,\infty)};$$
$$\left(\int_0^t \operatorname{sgn}(X_\alpha(s))\,dX_\alpha(s)\right)_{t\in[0,\infty)}.$$

(ii) It also holds that (b), (c) and (d) of Proposition 2.2(ii) have, as processes, different laws from one another.

We now turn our attention to the SDEs satisfied by our martingales.

Theorem 2.4 *(i) For each $\alpha > 0$, $\big(X_\alpha(t)\big)$ is a weak solution to the one-dimensional SDE*

$$(2.1)\qquad \begin{cases} dX_t &= \alpha\left(2\max\limits_{s\in[0,t]} X_s - X_t\right)^{\frac{\alpha-1}{\alpha}} dW_t; \\ X_0 &= 0; \end{cases}$$

where (W_t) is one-dimensional standard Brownian motion.
(ii) If $\alpha > 1$, then the above SDE also has the trivial solution $X \equiv 0$, and so uniqueness in law fails. Among the solutions of the SDE, the law of $\big(X_\alpha(t)\big)$ is characterized as follows: if a weak solution (X_t) satisfies

$$(2.2)\qquad \inf\{\, t > 0 \,|\, X_t \neq 0 \,\} = 0 \qquad a.s.,$$

then it is identical in law to $\big(X_\alpha(t)\big)$.
(iii) If $\alpha \leq 1$, then pathwise uniqueness holds; $\big(X_\alpha(t)\big)$ is the unique strong solution.

Remarks. If $\alpha \geq 2$, then $\frac{1}{2} \leq \frac{\alpha-1}{\alpha} < 1$. The first assertion of Theorem 2.4(ii) thus implies that *uniqueness in law does not, in general, hold for the one-dimensional SDE*

$$dX_t = \sigma(t, X.)\,dW_t,$$

where $\sigma(t, X.)$ is a predictable functional and (W_t) is one-dimensional Brownian motion, even if
(2.3)

$$\frac{1}{2} \leq \exists\eta < 1, \quad \exists K > 0 \quad \textit{such that} \quad \big|\sigma(t,x.) - \sigma(t,y.)\big| \leq K \max_{s\in[0,t]} |x_s - y_s|^\eta.$$

In contrast, note that if $\sigma(t, X.)$ depends only on t and X_t, *i.e.* if $\sigma(t, X.) \equiv \sigma(t, X_t)$, and if

$$\exists\eta \geq \frac{1}{2}, \quad \exists K > 0 \quad \textit{such that} \quad \big|\sigma(t,x) - \sigma(t,y)\big| \leq K|x-y|^\eta,$$

then pathwise uniqueness follows from the Yamada-Watanabe theorem [8]. It should also be mentioned that if $\eta \geq 1$ instead of $\frac{1}{2} \leq \eta < 1$ in (2.3), then the Lipschitz condition is satisfied and pathwise uniqueness holds.

Finally, we deduce the following property from the proof of Theorem 2.4.

Corollary 2.5 *$(X_\alpha(t))$ is a pure martingale, i.e., $\mathcal{F}_\infty^{X_\alpha} = \mathcal{F}_\infty^\beta$ with β being the time-changed Brownian motion. Consequently, $(X_\alpha(t))$ has the martingale representation property.*

3 Proofs

Proof of Proposition 2.1 (i) If $(X_\alpha(t))$ were Markov for $\alpha \neq 1$, then $d[X_\alpha]_t$ would depend only on the value of $X_\alpha(t)$ and not on the past history. It holds, however, that

$$\begin{aligned} d[X_\alpha]_t &= d[R_{\alpha+2}^\alpha]_t \\ &= \left(\alpha R_{\alpha+2}^{\alpha-1}(t)\right)^2 dt \\ &= \alpha^2 \Big(2 \max_{s\in[0,t]} X_\alpha(s) - X_\alpha(t) \Big)^{\frac{2(\alpha-1)}{\alpha}} dt. \end{aligned}$$

The second assertion follows from the Markov property of the $\boldsymbol{R}^2$-valued process

$$\left(R_{\alpha+2}(t), \min_{s\in[t,\infty)} R_{\alpha+2}(s) \right)_{t\in[0,\infty)}.$$

(ii) The scaling property of $(X_\alpha(t))$ follows from that of $(R_{\alpha+2}^\alpha(t))$.
(iii) It follows from the scaling property of $(R_{\alpha+2}(t))$ that

$$\begin{aligned} \forall t \geq 0, \qquad [X_\alpha]_t &= \int_0^t \alpha^2 R_{\alpha+2}^{2(\alpha-1)}(s)\, ds \\ &\overset{(d)}{=} t^\alpha \int_0^1 \alpha^2 R_{\alpha+2}^{2(\alpha-1)}(s)\, ds, \end{aligned}$$

hence

$$\begin{aligned} \forall M > 0, \qquad P\Big[\lim_{t\uparrow\infty} [X_\alpha]_t \leq M\Big] &= \lim_{t\uparrow\infty} P\Big[\int_0^t \alpha^2 R_{\alpha+2}^{2(\alpha-1)}(s)\, ds \leq M\Big] \\ &= \lim_{t\uparrow\infty} P\Big[\int_0^1 \alpha^2 R_{\alpha+2}^{2(\alpha-1)}(s)\, ds \leq \frac{M}{t^\alpha}\Big] \\ &= 0. \qquad \square \end{aligned}$$

The next trivial fact will be used in the proof of Proposition 2.2.

Lemma 3.1 *Let $r > 0$. Suppose Y is a random variable uniformly distributed on the interval $[0, r]$. Then $2Y - r$ is uniformly distributed on the interval $[-r, r]$; in particular,*

$$2Y - r \stackrel{(d)}{=} -(2Y - r).$$

Furthermore, the three random variables Y, $r - Y$ and $|2Y - r|$ are all identically distributed.

Proof of Proposition 2.2 (i) Conditioned by $\mathcal{F}_t^{R_{\alpha+2}}$, $\min_{s\in[t,\infty)} R_{\alpha+2}^{\alpha}(s)$ is a random variable uniformly distributed on the interval $\left[0, R_{\alpha+2}^{\alpha}(t)\right]$, since the scale function of the diffusion $\left(R_{\alpha+2}^{\alpha}(t)\right)$ is $s(x) = -\frac{1}{x}$. This and Lemma 3.1 imply that for each fixed $t \geq 0$, $X_\alpha(t)$ and $-X_\alpha(t)$ have the same distribution conditioned by $\mathcal{F}_t^{R_{\alpha+2}}$, which yields the desired result. The calculation of the density function follows along the same lines.
(ii) The same reasoning as in (i) leads to the equi-distribution property of (a), (b) and (c). Revuz-Yor [4] Exercise XI.1.18 shows that

$$\forall \alpha > 0, \quad \forall t \geq 0, \quad \min_{s\in[t,\infty)} R_{\alpha+2}(s) \stackrel{(d)}{=} R_\alpha(t),$$

so (a) and (d) have the same distribution. □

We also need the following lemma to prove Theorem 2.3.

Lemma 3.2 (c.f. Revuz-Yor [4] Exercise VI.2.32) *Suppose M and N are divergent continuous local martingales starting from the origin. Let β and γ denote the time-changed Brownian motions of M and N, respectively. Then*

$$(M_t) \stackrel{(d)}{=} (N_t) \iff \left(\beta_t, [M]_t\right) \stackrel{(d)}{=} \left(\gamma_t, [N]_t\right).$$

Proof of Theorem 2.3 (i) We have already shown in Proposition 2.1(iii) that $\left(X_\alpha(t)\right)$ is divergent. Let β be its time-changed Brownian motion. Then by Lemma 3.2 we have

$$\begin{aligned}
\left(X_\alpha(t)\right) \stackrel{(d)}{=} \left(-X_\alpha(t)\right) \quad &\iff \quad \left(\beta_t, [X_\alpha]_t\right) \stackrel{(d)}{=} \left(-\beta_t, [X_\alpha]_t\right) \\
&\iff \quad (\beta_t) \stackrel{(d)}{=} (-\beta_t) \text{ conditioned by } \mathcal{F}_\infty^{[X_\alpha]}.
\end{aligned}$$

Thus, to prove $\left(X_\alpha(t)\right) \stackrel{(d)}{\neq} \left(-X_\alpha(t)\right)$ it is sufficient to show that (β_t) and $(-\beta_t)$ have different laws conditioned by $\mathcal{F}_\infty^{[X_\alpha]}$. This is a consequence of the following fact:

$$\begin{aligned}
\mathcal{F}_\infty^{[X_\alpha]} &= \mathcal{F}_\infty^{R_{\alpha+2}^{\alpha-1}} \\
&= \mathcal{F}_\infty^{R_{\alpha+2}} \qquad \text{since} \quad \alpha \neq 1 \\
&= \mathcal{F}_\infty^{X_\alpha} \\
&\supset \mathcal{F}_\infty^{\beta}.
\end{aligned}$$

Similarly, we can show that the third martingale in the statement of the theorem is not identical in law to the other two.

(ii) First, it is easy to see that the law of (d) is different from those of the other two processes, since only (d) is Markov among the three. Furthermore, as stated in Revuz-Yor [4] Exercise VI.2.32,

$$\big(|X_\alpha(t)|\big) \stackrel{(d)}{=} \Big(\max_{s\in[0,t]} X_\alpha(s) - X_\alpha(t)\Big)$$

$$\iff \Big(\int_0^t \operatorname{sgn}(X_\alpha(s))\, dX_\alpha(s)\Big) \stackrel{(d)}{=} \big(-X_\alpha(t)\big),$$

so by (i) we see that (b) and (c), as processes, do not have the same law. □

Proof of Theorem 2.4 (i) Straightforward.

(ii) We only have to prove the second assertion. First observe that any weak solution of the SDE satisfying the additional condition (2.2) is a divergent continuous local martingale. Indeed, for almost all $\omega \in \Omega$, there exists some $t_0 = t_0(\omega) > 0$ such that $X_{t_0}(\omega) > 0$, and hence

$$\begin{aligned}
\lim_{t\uparrow\infty}[X]_t(\omega) &= \int_0^\infty \alpha^2 \Big(2\max_{s\in[0,t]} X_s(\omega) - X_t(\omega)\Big)^{\frac{2(\alpha-1)}{\alpha}} dt \\
&\geq \int_0^\infty \alpha^2 \Big(\max_{s\in[0,t]} X_s(\omega)\Big)^{\frac{2(\alpha-1)}{\alpha}} dt \\
&\geq \int_{t_0(\omega)}^\infty \alpha^2 \big(X_{t_0}(\omega)\big)^{\frac{2(\alpha-1)}{\alpha}} dt \\
&= \infty.
\end{aligned}$$

Next, define

$$\begin{aligned}
\tau_t &\stackrel{\text{def}}{=} \inf\{s > 0 \,|\, [X]_s > t\}; \\
\beta_t &\stackrel{\text{def}}{=} X(\tau_t).
\end{aligned}$$

Clearly, (β_t) is a one-dimensional Brownian motion starting from the origin. By the inverse function theorem we have

$$\begin{aligned}
\forall t > 0, \qquad \frac{d\tau_t}{dt} &= \alpha^{-2}\Big(2\max_{u\in[0,\tau_t]} X_u - X_{\tau_t}\Big)^{-\frac{2(\alpha-1)}{\alpha}} \\
&= \alpha^{-2}\Big(2\max_{u\in[0,t]} \beta_u - \beta_t\Big)^{-\frac{2(\alpha-1)}{\alpha}},
\end{aligned}$$

thus

$$\forall t \geq 0, \qquad \tau_t = \int_0^t \alpha^{-2}\Big(2\max_{u\in[0,s]} \beta_u - \beta_s\Big)^{-\frac{2(\alpha-1)}{\alpha}} ds.$$

(Note that $\tau_0 = 0$ a.s. by the condition (2.2).) This implies that

$$X_t = \beta\left(\inf\left\{s > 0 \,\Big|\, \int_0^s \alpha^{-2}\Big(2\max_{v\in[0,u]} \beta_v - \beta_u\Big)^{-\frac{2(\alpha-1)}{\alpha}} du > t\right\}\right).$$

The law of a weak solution of (2.1) satisfying the condition (2.2) is therefore uniquely determined.

(iii) The assertion is trivial if $\alpha = 1$, so we assume $\alpha < 1$ in the sequel. We divide the proof into three steps.

Step 1. We first show uniqueness in law. Since $\alpha < 1$, it is easy to see that any weak solution of this SDE must satisfy the condition (2.2). Also, any weak solution is a divergent continuous local martingale. Indeed, if a solution (X_t) were not divergent, then, with positive probability, $X_t(\omega)$ would converge to a real number as $t \uparrow \infty$. Then $2 \max_{u\in[0,t]} X_u(\omega) - X_t(\omega)$ would also converge to a real number for such an ω as $t \uparrow \infty$ and hence

$$\lim_{t\uparrow\infty} [X]_t(\omega) = \lim_{t\uparrow\infty} \int_0^t \alpha^2 \left(2 \max_{u\in[0,s]} X_u(\omega) - X_s(\omega)\right)^{\frac{2(\alpha-1)}{\alpha}} ds = \infty,$$

a contradiction. The rest of the proof of uniqueness in law is exactly the same as in (ii).

Step 2. Suppose $\left(\Omega, \mathcal{F}, P, (\mathcal{F}_t), (W_t), (X_t)\right)$ is a weak solution to the SDE (2.1), *i.e.*,
$\left(\Omega, \mathcal{F}, P, (\mathcal{F}_t)\right)$ is a filtered probability space, (X_t) is a semimartingale on it, (W_t) is an $(\mathcal{F}_t)$-Brownian motion starting from the origin, and they satisfy (2.1). Define

$$R_t \stackrel{\text{def}}{=} \left(2 \max_{s\in[0,t]} X_s - X_t\right)^{1/\alpha};$$
$$Z_t \stackrel{\text{def}}{=} 2 \min_{s\in[t,\infty)} R_s - R_t.$$

It then follows from the Itô formula that $\left(\Omega, \mathcal{F}, P, (\mathcal{F}_t), (W_t), (Z_t)\right)$ satisfies

$$Z_t = W_t + \frac{\alpha-1}{2} \int_0^t \frac{ds}{2 \max_{u\in[0,s]} Z_u - Z_s}. \tag{3.1}$$

(For this equation, see also Revuz-Yor [4] Exercise XI.1.29 and Yor [9] Corollary 12.5.1.) Conversely, if $\left(\Omega, \mathcal{F}, P, (\mathcal{F}_t), (W_t), (Z_t)\right)$ satisfies (3.1) and

$$\inf\left\{ t > 0 \,\middle|\, \max_{u\in[0,t]} Z_u > 0 \right\} = 0 \quad \text{a.s.}, \tag{3.2}$$

then we define

$$R_t \stackrel{\text{def}}{=} 2 \max_{s\in[0,t]} Z_s - Z_t;$$
$$X_t \stackrel{\text{def}}{=} 2 \min_{s\in[t,\infty)} R_s^\alpha - R_t^\alpha;$$

and it is not hard to see that (X_t) satisfies (2.1). Therefore, pathwise uniqueness holds for (2.1) if and only if a solution of the SDE (3.1) satisfying the condition (3.2) is pathwisely unique. We have already shown in Step 1 the uniqueness in law, so it suffices to show that if both $Z_t^{(1)}$ and $Z_t^{(2)}$ satisfy (3.1) in the same set-up, then so does $Y_t \stackrel{\text{def}}{=} Z_t^{(1)} \vee Z_t^{(2)}$.

Since $\left(Z_t^{(1)} - Z_t^{(2)}\right)$ is a process with continuously differentiable trajectories, it is easy to verify that

$$\begin{aligned} dY_t &= dZ_t^{(2)} + d\left(Z_t^{(1)} - Z_t^{(2)}\right)^+ \\ &= 1_{\{Z_t^{(1)} > Z_t^{(2)}\}}\, dZ_t^{(1)} + 1_{\{Z_t^{(1)} \le Z_t^{(2)}\}}\, dZ_t^{(2)}, \end{aligned}$$

and thus

$$(3.3)\; Y_t = W_t + \frac{\alpha - 1}{2} \int_0^t ds \left\{ \frac{1_{\{Z_s^{(1)} > Z_s^{(2)}\}}}{2 \max\limits_{u \in [0,s]} Z_u^{(1)} - Z_s^{(1)}} + \frac{1_{\{Z_s^{(1)} \le Z_s^{(2)}\}}}{2 \max\limits_{u \in [0,s]} Z_u^{(2)} - Z_s^{(2)}} \right\}.$$

If $Z_s^{(1)}(\omega) > Z_s^{(2)}(\omega)$ then $\max\limits_{u \in [0,s]} Z_u^{(1)}(\omega) \ge \max\limits_{u \in [0,s]} Z_u^{(2)}(\omega)$, as we will see in Step 3, so we can rewrite (3.3) as

$$(3.4)\; Y_t = W_t + \frac{\alpha - 1}{2} \int_0^t ds \left\{ \frac{1_{\{Z_s^{(1)} \ne Z_s^{(2)}\}}}{2 \max\limits_{u \in [0,s]} Y_u - Y_s} + \frac{1_{\{Z_s^{(1)} = Z_s^{(2)}\}}}{2 \max\limits_{u \in [0,s]} Z_u^{(2)} - Z_s^{(2)}} \right\}.$$

(Note that $\max\limits_{u \in [0,s]} Y_u = \max\limits_{u \in [0,s]} Z_u^{(1)} \vee \max\limits_{u \in [0,s]} Z_u^{(2)}$.) Similarly, we have

$$(3.5)\; Y_t = W_t + \frac{\alpha - 1}{2} \int_0^t ds \left\{ \frac{1_{\{Z_s^{(1)} \ne Z_s^{(2)}\}}}{2 \max\limits_{u \in [0,s]} Y_u - Y_s} + \frac{1_{\{Z_s^{(1)} = Z_s^{(2)}\}}}{2 \max\limits_{u \in [0,s]} Z_u^{(1)} - Z_s^{(1)}} \right\}.$$

Comparing (3.4) and (3.5), we see that for almost all $\omega \in \Omega$:

$$\mu \left\{ t \in [0,\infty) \,\middle|\, Z_t^{(1)}(\omega) = Z_t^{(2)}(\omega) \text{ and } \max_{u \in [0,t]} Z_u^{(1)}(\omega) \ne \max_{u \in [0,t]} Z_u^{(2)}(\omega) \right\} = 0,$$

where μ denotes the Lebesgue measure, and hence

$$\int_0^t \frac{1_{\{Z_s^{(1)} = Z_s^{(2)}\}}}{2 \max\limits_{u \in [0,s]} Z_u^{(2)} - Z_s^{(2)}}\, ds = \int_0^t \frac{1_{\{Z_s^{(1)} = Z_s^{(2)}\}}}{2 \max\limits_{u \in [0,s]} Y_u - Y_s}\, ds.$$

This together with (3.4) implies that

$$Y_t = W_t + \frac{\alpha - 1}{2} \int_0^t \frac{ds}{2 \max\limits_{u \in [0,s]} Y_u - Y_s}.$$

Step 3. It remains to show that if $Z_s^{(1)}(\omega) > Z_s^{(2)}(\omega)$ then $\max_{u\in[0,s]} Z_u^{(1)}(\omega) \geq \max_{u\in[0,s]} Z_u^{(2)}(\omega)$. Let

$$s_0 \stackrel{\text{def}}{=} \sup\{u \in [0,s) \mid Z_u^{(1)}(\omega) = Z_u^{(2)}(\omega)\}.$$

Then

$$\frac{d}{du}\left(Z_u^{(1)}(\omega) - Z_u^{(2)}(\omega)\right)\Bigg|_{u=s_0} \geq 0,$$

which implies that

$$\frac{\alpha-1}{2}\,\frac{1}{2\max_{u\in[0,s_0]} Z_u^{(1)}(\omega) - Z_{s_0}^{(1)}(\omega)} \geq \frac{\alpha-1}{2}\,\frac{1}{2\max_{u\in[0,s_0]} Z_u^{(2)}(\omega) - Z_{s_0}^{(2)}(\omega)},$$

$$\max_{u\in[0,s_0]} Z_u^{(1)}(\omega) \geq \max_{u\in[0,s_0]} Z_u^{(2)}(\omega) \qquad \text{since } \alpha < 1.$$

Also $Z_u^{(1)}(\omega) > Z_u^{(2)}(\omega)$ for $u \in (s_0, s]$, thus we obtain the desired property. □

Proof of Corollary 2.5 We have already shown in the proof of Theorem 2.4 that $(X_\alpha(t))$ is pure. Note that every pure local martingale has the martingale representation property; see, for instance, Revuz-Yor [4] §V.4. □

References

[1] Bertoin, J., *An extension of Pitman's theorem for spectrally positive Lévy processes*, Ann. Prob. **20** (1993), 1463–1483.

[2] Pitman, J., *One-dimensional Brownian motion and the three-dimensional Bessel process*, Adv. Appl. Prob. **7** (1975), 511–526.

[3] Rauscher, B., *Some remarks on Pitman's theorem*, in this volume of the *Séminaire de Probabilités.*

[4] Revuz, D. & Yor, M., *Continuous martingales and Brownian motion*, Second edition, Springer (1994).

[5] Saisho, Y. & Tanemura, H., *Pitman type theorem for one-dimensional diffusion processes*, Tokyo J. Math. **13** (1990), 429–440.

[6] Tanaka, H., *Time reversal of random walks in dimension one*, Tokyo J. Math. **12** (1989), 159–174.

[7] ———, *Time reversal of random walks in* $\boldsymbol{R}^d$, Tokyo J. Math. **13** (1990), 375–389.

[8] Yamada, T. & Watanabe, S., *On the uniqueness of solutions of stochastic differential equations*, J. Math. Kyoto Univ. **11** (1971), 155–167.

[9] Yor, M., *Some Aspects of Brownian Motion Part II: Some Recent Martingale Problems*, ETH Lecture Notes in Mathematics, Birkhäuser (to appear).

Some Remarks on Pitman's Theorem

BERNHARD RAUSCHER
NWF I - Mathematik, Universität Regensburg
D-93040 Regensburg, Germany

Abstract — Pitman constructs $\mathrm{BES}^3(0)$ as $2M-X$ where X is $\mathrm{BM}^1(0)$ and $M_t = \sup_{r\le t} X_r$. Equivalently, $X - 2J$ is $\mathrm{BM}^1(0)$ when X is $\mathrm{BES}^3(0)$ and $J_t = \inf_{r\ge t} X_r$. Now the fact that $X - 2J$ gives a local martingale may be extended to a general result for linear diffusions. In particular, if X is a linear diffusion, we introduce a general class of nontrivial transformations φ such that $Z = \varphi(X, J)$ is a local martingale.

In his fundamental paper [6] Pitman constructed $\mathrm{BES}^3(0)$ as $2M - X$ where X is $\mathrm{BM}^1(0)$ and $M_t = \sup_{r\le t} X_r$ using random walk approximations. This result immediately leads to the path decompositions for BES^3 obtained by Williams [16], so it was of natural interest whether the path transformation $2M - X$ gives again a diffusion when X is a linear diffusion and $M_t = \sup_{r\le t} X_r$ is defined as above. Yet, this $2M - X$ property just holds for a few more types of diffusions including BM with constant drift and may not be extended to a wider class of diffusions as was proved in Rogers-Pitman [9], Rogers [8], see also Salminen [10].

Apart from diffusions on the real line, further generalizations of Pitman's Theorem in different directions were obtained by Tanaka [13], [14] for random walks, Bertoin [1], [2] for certain Lévy processes, and Biane [3] for the planar Brownian motion in a cone. Recently, the case of a general class of diffusions was taken up again by Saisho and Tanemura [11] using solutions of Skorokhod type SDEs with globally Lipschitz continuous coefficients.

Now Pitman's result amounts to the same as saying that $X - 2J$ is a $\mathrm{BM}^1(0)$ when X is $\mathrm{BES}^3(0)$ and $J_t = \inf_{r\ge t} X_r$. In particular, the process $X - 2J$ is observed to be a local martingale, and this fact was recently extended by Yor [17] to a general result about diffusions on the real line, hereby using techniques from the theory of enlargement of filtrations initiated by Jeulin [4], [5]. For instance, let $(X_t)_{t\ge 0}$ be a transient diffusion in $]0,\infty[$ which satisfies the Itô type SDE

$$dX = d\beta + b(X)\,dt$$

where $b\colon]0,\infty[\to \mathbb{R}$ allows uniqueness in law for this equation. Further assume that a scale function s can be chosen to satisfy $s(]0,\infty[) =]-\infty, 0[$. Then, by Theorem 12.7 in Yor's book [17], the process

$$Z = -\frac{1}{s(X)} + \frac{2}{s(J)}$$

is a local martingale where $J_t = \inf_{r\ge t} X_r$. In the particular case of BES^δ, these martingales were studied in depth by Takaoka [15] thus revealing some interesting relations and differences to Brownian motion. In this paper, using elementary methods, we construct local martingales of the type $Z = \varphi(X, J)$ for linear diffusions and a whole class of transformations φ.

Consider an open interval $E =]c_1, c_2[\subseteq \mathbb{R}$ with $-\infty \le c_1 < c_2 \le \infty$ and let $\sigma, b\colon E \to \mathbb{R}$ be continuous functions with $\sigma > 0$. By $\bar{E} = E \cup \{c_1, c_2\}$ we denote the two-point-compactification of E. Let (X, β) denote a weak solution of the Itô type SDE

$$(*) \qquad dX = \sigma(X)\, d\beta + b(X)\, dt$$

in the following sense: Both X and β are adapted processes defined on a filtered probability space $(\Omega, \mathscr{F}, P; (\mathscr{F}_t))$ satisfying the usual conditions such that β is a linear $(\mathscr{F}_t)$-BM, X is an $\bar{E}$-valued continuous adapted process, and such that for any $f \in C_c^2(E)$ (compactly supported C^2 functions on E)

$$d\, f(X) = (f'\sigma)(X)\, d\beta + (Lf)(X)\, dt$$

holds. Hereby, $L = \frac{1}{2}\sigma^2 D^2 + bD$ denotes the generator of X. Note that X is allowed to launch from the boundary and hit the boundary in possibly finite life time

$$\zeta = \inf\{t > 0 \colon X_t \in \{c_1, c_2\}\}$$

which gives a predictable stopping time. We suppose $\zeta > 0$. Clearly, the solutions of the equation $(*)$ become uncontrolled when reaching the boundary, yet there is uniqueness in law up to the life time as may be proved by changing scale and speed. Using the same method we obtain general existence of solutions when starting from a point within E whereas coming in from the boundary requires the drift to be enough singular there.

Further we pick a function $s \in C^2(E)$ with $Ls = 0$ and $s' > 0$, called *scale function*. We suppose $s(E) =]-\infty, 0[$ which enforces $X_0 < c_2$ a.s. and the process X to be transient, namely $X_t \to c_2$ as $t \to \zeta$, a.s.

Associated with X we consider the continuous increasing process $J_t = \inf_{r \ge t} X_r^\zeta$ valued in $\bar{E}$, representing the absolute minimum of X past time t. Here, $X_r^\zeta = X_{\zeta \wedge r}$. In order to get both X and J adapted, we introduce the filtration $(\mathscr{G}_t)$ defined as the usual augmentation of $\mathscr{G}_t^0 = \sigma(X_r, J_r : r \le t)$, $t \ge 0$. Then the main result is as follows.

THEOREM. *Let X denote a solution of the equation $(*)$ with positive life time ζ and scale function s as described above. Pick $\mu \in C^1(E)$ and consider φ: $E \times E \to \mathbb{R}$,*

$$\varphi(x, y) = \left(\frac{s^2 \mu'}{s'}\right)(y) - \frac{1}{s(x)} \left(\frac{s^2 (\mu s)'}{s'}\right)(y) .$$

Then the process $Z = \varphi(X, J)$ is a local martingale with respect to $(\mathscr{G}_t)$ on $]0, \zeta[$.

We postpone the proof. Note that on $]0, \zeta[$ both the processes X and J take their values in E, hence Z is defined. The assertion of the theorem is to be understood in the sense of Sharpe [12] as follows. Let Ω_0 denote the set where $Z_0 = \lim_{t \searrow 0} Z_t$ exists in $\mathbb{R}$. Then there is a sequence (R_n) of $(\mathscr{G}_t)$-stopping times with $0 \le R_n < \zeta$, $R_n \searrow 0$, and $\{R_n = 0\} \subseteq \Omega_0$ such that $(Z_{R_n + t})_{t < \zeta - R_n}$ is a local martingale over $(\mathscr{G}_{R_n + t})$, for each n. The reason for this technical statement is that X may start from the boundary c_1 and then Z_0 is not defined. Nevertheless, conditionally on Ω_0, the process $(Z_t)_{0 \le t < \zeta}$ gives a local martingale over $(\mathscr{G}_t)$ in the usual sense, cf. [12].

Before entering the proof of our theorem we present some corollaries and discuss the connections with the previous results mentioned above.

First of all, taking $\mu = -1/s^2$ and $\mu = 1$ in the theorem, we obtain Yor's result as well as a multiplicative version thereof.

COROLLARY 1. *Both the processes*

$$-\frac{1}{s(X)} + \frac{2}{s(J)}, \quad -\frac{s^2(J)}{s(X)}$$

are local martingales with respect to $(\mathscr{G}_t)$ *on* $]0,\zeta[$.

Considering $Z = -1/s(X) + 2/s(J)$ it is easily seen that $\lim_{t\searrow 0} Z_t$ exists a.s. in $\mathbb{R}$, hence the process Z is a local martingale even on the left-closed interval $[0,\zeta[$ with quadratic variation

$$\langle Z, Z\rangle = \int (s'\sigma/s^2)^2(X)\,dt.$$

Thus, we may integrate $(s^2/s'\sigma)(X)\,1_{]0,\zeta[}$ with respect to Z and the process

$$B = \int (s^2/s'\sigma)(X)\,dZ$$

is a BM on $[0,\zeta[$ which will be referred to in the next corollary. We further introduce the 'dual' coefficients $\sigma^* = \sigma$ and $b^* = b + \sigma^2 s'/s$ so that $L^* = \frac{1}{2}(\sigma^*)^2 D^2 + b^* D = s^{-1}Ls$. Then $s^* = -1/s$ serves as a scale function for L^* and $b = b^* + (\sigma^*)^2(s^*)'/s^*$ which shows the dual character of this construction.

COROLLARY 2. *On* $]0,\zeta[$ *we have* $d(X-2J) = \sigma^*(X)\,dB + b^*(X)\,dt$.

Proof. This is an application of Itô's formula to $\varphi(x) = s^{-1}(-1/x)$ and the $]0,\infty[$-valued semimartingale $-1/s(X) = Z - 2/s(J)$ on $]0,\zeta[$ where $Z = -1/s(X) + 2/s(J)$ as in Corollary 1. Hereby, the function s^{-1} denotes the inverse of s. Note that the measure dJ is carried by the set $\{t\colon X_t = J_t\}$. □

Now, if X is a $\mathrm{BES}^3(0)$, we obtain $X - 2J = B$ which is Pitman's Theorem. More generally, let X be a Bessel process of index $\nu > 0$ on $]0,\infty[$ starting from 0, then the corollary says

$$X_t = B_t + 2J_t + (-\nu + \tfrac{1}{2})\int_0^t \frac{dr}{X_r}$$

which is also stated in Corollary 12.7.1 in Yor's book [17].

Further, if both σ^* and b^* are constant functions, the process $X - 2J$ is a diffusion again which goes with [8]. But due to [9], in case of σ^* or b^* being nonconstant, $X - 2J$ is not a diffusion any longer which may be at least heuristically apparent by the corollary.

The contents of the corollary are also closely related to the results of Saisho and Tanemura [11]. Namely, putting $Y = X - J$ and $K = J$, the corollary may be stated as

$$dY = \sigma^*(Y+K)\,dB + b^*(Y+K)\,dt + dK$$

where Y is adapted, continuous, nonnegative and K is adapted, continuous, and nondecreasing with $dK = 1_{\{0\}}(Y)\,dK$. We further know from the beginning that the process $X = Y + K$ is a diffusion with generator L. This is in accord with the main part of [11] where it is proved, in the particular case of both σ^* and b^* being globally Lipschitz continuous functions on $\mathbb{R}$, that the above SDE of Skorokhod type can be solved uniquely by a pair (Y, K) and that the process $Y + K$ is a diffusion with generator L.

Further, we recall the first part of Williams' result on path decomposition of diffusions [16] which fits into the result of the corollary. Suppose $X_0 = e \in E$, then there is a.s. a unique time $T < \zeta$ with $X_T = J_0$ and, conditionally on $J_0 = c$, the process $(X_t)_{t<T}$ is a diffusion with generator L^* on $]c, c_2[$.

We finally remark that the theorem may be related to a result of Azema and Yor, cf. Revuz-Yor [7] Chap. VI, §4. Indeed, considering the particular case of X a $\mathrm{BES}^3(0)$ on $]0, \infty[$ with $s(x) = -1/x$ reveals $\mu'(J) + X\,(\mu s)'(J)$ to be a local martingale. Consequently, by Pitman's Theorem, the process $\mu'(S) + (2S - B)\,(\mu s)'(S) = (-\mu s)(S) - (S - B)\,(-\mu s)'(S)$ is a local martingale where B denotes a $\mathrm{BM}^1(0)$ and $S_t = \sup_{r \le t} B_r$.

Now we turn to the proof of the theorem which is prepared by three lemmas.

LEMMA 1. *Consider the kernel K in $\bar{E}$ defined by*

$$K(e, \cdot\,) = \begin{Bmatrix} -s(e)\, s'(t)\, s(t)^{-2}\, 1_{]c_1, e[}(t)\, dt & \textit{for } e \in E \\ \delta_e & \textit{for } e = c_i \end{Bmatrix}.$$

Then $K\big(X_0(\cdot\,), \cdot\,\big)$ is a conditional distribution for $J_0 = \inf_{t \le \zeta} X_t$ given $\mathscr{F}_0$.

Proof. For every $x \in \bar{E}$ we need to show $P\{J_0 \le x \,|\, A\} = E\big[K\big(X_0, [c_1, x]\big) \big| A\big]$ for all $A \in \mathscr{F}_0$ with $P(A) > 0$. This may further be reduced to proving $P\{J_0 \le x\} = E\big[K\big(X_0, [c_1, x]\big)\big]$ where $X_0 > x$ a.s. But, using the optional stopping theorem, we easily compute

$$P\{J_0 \le x\} = \frac{E[s(X_0)]}{s(x)} = E\big[K\big(X_0, [c_1, x]\big)\big],$$

where the last equation stems from $K\big(e, [c_1, x]\big) = s(e)/s(x)$ for all $e \in\,]x, c_2]$. □

LEMMA 2. *Let $\tau < \zeta$ be a $(\mathscr{F}_t)$-stopping time with $X_\tau \le y$ a.s. for some $y \in E$. Then, for any $z \in E$, the random variable $Z_\tau 1_{\{J_\tau > z\}}$ is integrable and the equation*

$$\mu(z)\big[s(X_\tau) - s(z)\big]\, 1_{\{X_\tau > z\}} = E\big[Z_\tau 1_{\{J_\tau > z\}} \big| \mathscr{F}_\tau\big] \quad \text{a.s.}$$

holds.

Proof. Since the pair $(X_{\tau+t}, \beta_{\tau+t})$ over $(\mathscr{F}_{\tau+t})$ is a solution to $(*)$, too, we may assume $\tau = 0$ without loss of generality. On $\{J_0 > z\}$ we clearly have $z < J_0 \le X_0 \le y$ a.s., hence all integrands in the sequel are integrable and we get

$$\begin{aligned} E\big[Z_0 1_{\{J_0 > z\}} \big| \mathscr{F}_0\big] &= E\left[\left(\frac{s^2 \mu'}{s'}\right)(J_0)\, 1_{\{J_0 > z\}} \middle| \mathscr{F}_0\right] \\ &\quad - \frac{1}{s(X_0)}\, E\left[\left(\frac{s^2 (\mu s)'}{s'}\right)(J_0)\, 1_{\{J_0 > z\}} \middle| \mathscr{F}_0\right]. \end{aligned}$$

By virtue of Lemma 1 these conditional expectations may be explicitely computed:

$$\begin{aligned} E\left[\left(\frac{s^2 \mu'}{s'}\right)(J_0)\, 1_{\{J_0 > z\}} \middle| \mathscr{F}_0\right] &= -s(X_0)\, 1_{\{X_0 > z\}} \int_{]z, X_0[} \left(\frac{s^2 \mu'}{s'} \cdot \frac{s'}{s^2}\right)(t)\, dt \\ &= -s(X_0)\big[\mu(X_0) - \mu(z)\big]\, 1_{\{X_0 > z\}}. \end{aligned}$$

So we finally get to

$$\begin{aligned} E\big[Z_0 1_{\{J_0>z\}}\big|\mathscr{F}_0\big] &= -s(X_0)\big[\mu(X_0)-\mu(z)\big]\,1_{\{X_0>z\}} \\ &\quad + \frac{1}{s(X_0)}s(X_0)\big[(\mu s)(X_0)-(\mu s)(z)\big]\,1_{\{X_0>z\}} \\ &= \mu(z)\big[s(X_0)-s(z)\big]\,1_{\{X_0>z\}} \end{aligned}$$

which is the desired result. □

LEMMA 3. *Let $R \le \rho \le \tau \le S < \zeta$ be stopping times with respect to $(\mathscr{F}_t)$ such that $X^S_{R+t} \le y$ a.s. on $\{S > R\}$ for some $y \in E$. Then, for any $z \in E$, both $Z_\rho 1_{\{S>R,J_\rho>z\}}$ and $Z_\tau 1_{\{S>R,J_\rho>z\}}$ are integrable and*

$$E\big[Z_\rho 1_{\{S>R,J_\rho>z\}}\big|\mathscr{F}_\rho\big] = E\big[Z_\tau 1_{\{S>R,J_\rho>z\}}\big|\mathscr{F}_\rho\big] \quad \text{a.s.}$$

holds.

Proof. We may assume $\rho = 0$ and $S > R$ a.s. without loss of generality. On $\{J_0 > z\}$ we have $z < J_0 \le J_\tau \le X_\tau \le y$ a.s., hence $Z_0 1_{\{J_0>z\}}$ and $Z_\tau 1_{\{J_0>z\}}$ are integrable. Next we put $\tau_z = \inf\{t \ge 0 : X_t \le z\}$ and compute step by step

$$\begin{aligned} E\big[Z_\tau 1_{\{J_0>z\}}\big|\mathscr{F}_0\big] &= E\big[Z_\tau 1_{\{\inf_{t\le\tau} X_t>z\}} 1_{\{J_\tau>z\}}\big|\mathscr{F}_0\big] \\ &= E\big[1_{\{\inf_{t\le\tau} X_t>z\}} E\big[Z_\tau 1_{\{J_\tau>z\}}\big|\mathscr{F}_\tau\big]\big|\mathscr{F}_0\big] \\ &= E\big[1_{\{\inf_{t\le\tau} X_t>z\}} \mu(z)\big[s(X_\tau)-s(z)\big]\big|\mathscr{F}_0\big] \\ &= E\big[1_{\{X_0>z\}} \mu(z)\big[s(X_{\tau\wedge\tau_z})-s(z)\big]\big|\mathscr{F}_0\big] \\ &= 1_{\{X_0>z\}} \mu(z)\big[s(X_0)-s(z)\big] \\ &= E\big[Z_0 1_{\{J_0>z\}}\big|\mathscr{F}_0\big], \end{aligned}$$

where in the last but one equation we made use of the fact that $s(X^{\tau\wedge\tau_z})\,1_{\{X_0>z\}}$ is a bounded martingale with respect to $(\mathscr{F}_t)$. □

Proof (of the theorem). To begin with, we fix points $x < e < y$ in E and consider the first entry times $R = \inf\{t : X_t \ge e\}$ and $S = \inf\{t \ge R : X_t \ge y\}$. Then the map $T = S\,1_{\{J_R>x\}}$ clearly is a stopping time with respect to $(\mathscr{G}^0_{R+t})$, and we shall prove that the process $(Z^T_{R+t} 1_{\{T>R\}})$ is a martingale over $(\mathscr{G}^0_{R+t})$. Namely, we have $X^S_{R+t} \le y$ a.s. on $\{S > R\}$ and for any $t \ge 0$, putting $\tau = (R+t) \wedge S$, the random variable

$$Z^T_{R+t} 1_{\{T>R\}} = Z_\tau 1_{\{S>R,J_R>x\}}$$

is integrable by Lemma 3. The same lemma, for any $z \in E$, $r \le t$ and considering the times $\rho = (R+r)\wedge S$, $\tau = (R+t) \wedge S$, yields

$$\begin{aligned} E\big[Z_\tau 1_{\{S>R,J_R>x\}}\,1_{\{J_\rho>z\}}\big|\mathscr{F}_\rho\big] &= 1_{\{\inf_{R\le t\le\rho} X_t>x\}}\,E\big[Z_\tau 1_{\{S>R,J_\rho>x\vee z\}}\big|\mathscr{F}_\rho\big] \\ &= 1_{\{\inf_{R\le t\le\rho} X_t>x\}}\,E\big[Z_\rho 1_{\{S>R,J_\rho>x\vee z\}}\big|\mathscr{F}_\rho\big] \\ &= E\big[Z_\rho 1_{\{S>R,J_R>x\}}\,1_{\{J_\rho>z\}}\big|\mathscr{F}_\rho\big] \quad \text{a.s.} \end{aligned}$$

But, using $J^\rho_t = \inf_{r\ge t} X^\rho_r \wedge J_\rho$, we have $\mathscr{G}^0_\rho = \mathscr{F}^X_\rho \vee \sigma(J_\rho) \subseteq \mathscr{F}_\rho \vee \sigma(J_\rho)$, hence

$$\begin{aligned} Z^T_{R+r} 1_{\{T>R\}} &= E\big[Z^T_{R+t} 1_{\{T>R\}}\big|\mathscr{G}^0_\rho\big] \\ &= E\big[Z^T_{R+t} 1_{\{T>R\}}\big|\mathscr{G}^0_{R+r}\big] \quad \text{a.s.} \end{aligned}$$

which is the assertion on the martingale property.

Now letting $x \searrow c_1$ and $y \nearrow c_2$ the process (Z_{R+t}) is seen to be a local martingale over $(\mathcal{G}_{R+t})$, and the proof is completed by letting $e \searrow c_1$. □

References

[1] J. BERTOIN. Sur la décomposition de la trajectoire d'un processus de Lévy spectralement positif en son infimum. *Ann. Inst. Henri Poincaré, Probab. Stat.* **27** 4 (1991) 537–547.

[2] J. BERTOIN. An extension of Pitman's theorem for spectrally positive Lévy processes. *Ann. Probab.* **20** 3 (1992) 1464–1483.

[3] P. BIANE. Quelques proprietes du mouvement Brownien dans un cone. *Stochastic Processes Appl.* **53** 2 (1994) 233-240.

[4] T. JEULIN. Un théorème de J. W. Pitman. *Séminaire de Probabilités* XIII. Lect. Notes in Math. **721**. Springer, Berlin Heidelberg New York (1979) 521–531.

[5] T. JEULIN. Semi-martingales et grossissement d'une filtration. Lect. Notes in Math. **833**. Springer, Berlin Heidelberg New York (1980).

[6] J. W. PITMAN. One-dimensional Brownian motion and the three-dimensional Bessel process. *Adv. Appl. Prob.* **7** (1975) 511–526.

[7] D. REVUZ and M. YOR. *Continuous Martingales and Brownian Motion.* 2nd edition. Berlin: Springer, 1994.

[8] L. C. G. ROGERS. Characterizing all diffusions with the $2M - X$ property. *Ann. Prob.* **9** (1981) 561–572.

[9] L. C. G. ROGERS and J. W. PITMAN. Markov functions. *Ann. Prob.* **9** (1981) 573–582.

[10] P. SALMINEN. Mixing Markovian Laws; With an Application to Path Decompositions. *Stochastics* **9** (1983) 223–231.

[11] Y. SAISHO and H. TANEMURA. Pitman type theorem for one-dimensional diffusion processes. *Tokyo J. Math.* **13** (2) (1990) 429–440.

[12] M. J. SHARPE. Local times and singularities of continuous local martingales. In: J. Azéma, P. A. Meyer and M. Yor (Eds.) *Séminaire de Probabilités* XIV. Lect. Notes in Math. **784**. Springer, Berlin Heidelberg New York (1980) 76–101.

[13] H. TANAKA. Time reversal of random walks in one dimension. *Tokyo J. Math.* **12** (1989) 159–174.

[14] H. TANAKA. Time reversal of random walks in $\mathbb{R}^d$. *Tokyo J. Math.* **13** (1989) 375–389.

[15] K. TAKAOKA. On the martingales obtained by an extension due to Saisho, Tanemura and Yor of Pitman's theorem. To appear in: *Séminaire de Probabilités* XXXI. Lect. Notes in Math. Springer, Berlin Heidelberg New York (1997).

[16] D. WILLIAMS. Path Decomposition and Continuity of Local Time for One-dimensional Diffusions, I. *Proc. London Math. Soc.* (3) **61** (1974) 738–768.

[17] M. YOR. *Some Aspects of Brownian Motion. Part II: Some recent martingale problems.* To appear: Lectures in Mathematics, ETH Zürich, Basel: Birkhäuser.

E-mail: `bernhard.rauscher@mathematik.uni-regensburg.de`

On the lengths of excursions of some Markov processes*

Jim Pitman[(1)] and Marc Yor[(2)]

(1) Department of Statistics, University of California, 367 Evans Hall # 3860, Berkeley, CA 94720-3860, USA
(2) Laboratoire de Probabilités, Université Paris VI, 4 Place Jussieu, 75252 Paris, France

Abstract. Results are obtained regarding the distribution of the ranked lengths of component intervals in the complement of the random set of times when a recurrent Markov process returns to its starting point. Various martingales are described in terms of the Lévy measure of the Poisson point process of interval lengths on the local time scale. The martingales derived from the zero set of a one-dimensional diffusion are related to martingales studied by Azéma and Rainer. Formulae are obtained which show how the distribution of interval lengths is affected when the underlying process is subjected to a Girsanov transformation. In particular, results for the zero set of an Ornstein-Uhlenbeck process or a Cox-Ingersoll-Ross process are derived from results for a Brownian motion or recurrent Bessel process, when the zero set is the range of a stable subordinator.

1 Introduction

Let Z be the random set of times that a recurrent diffusion process X returns to its starting state 0. For a fixed or random time T, let $\mathbf{V}(T) = (V_1(T), V_2(T), \cdots)$ where

$$V_1(T) \geq V_2(T) \geq \cdots \tag{1}$$

are the ranked lengths of component intervals of the random open set $(0,T)\backslash Z$. Features of the distribution of the random sequence $\mathbf{V}(T)$ have been studied by a number of authors [17, 32, 11, 15, 18, 19, 24, 25, 26]. It is well known that Z is the closure of the range of the *subordinator* $(\tau_s, s \geq 0)$ which is the inverse of the local time process of X at zero. If (τ_s) is a stable(α) subordinator for some $0 < \alpha < 1$, as is the case if X is a Brownian motion without drift ($\alpha = 1/2$) or a Bessel process of dimension $2 - 2\alpha$, it is obvious that the law of $\mathbf{V}(t)/t$ is the same for all t, and that the law of $\mathbf{V}(\tau_s)/\tau_s$ is the same for all s. It is less obvious, but nonetheless true [24], that the common law of $\mathbf{V}(t)/t$ for $t > 0$ is identical to the common law of $\mathbf{V}(\tau_s)/\tau_s$ for all

*Research supported in part by N.S.F. Grant DMS-9404345

$s > 0$. See [25] for a detailed study of this probability law on decreasing sequences of positive reals with sum 1, and relations between this distribution and Kingman's [11] Poisson-Dirichlet distribution on the same set of sequences.

If Z is the zero set of a real valued diffusion, the law of which is locally equivalent either to Wiener measure, or to the distribution of a Bessel process of dimension $2-2\alpha$ started at 0, it follows from the identities in distribution mentioned above that for each $t > 0$ and $s > 0$ the laws of $\mathbf{V}(t)/t$ and $\mathbf{V}(\tau_s)/\tau_s$ are *equivalent*, that is to say mutually absolutely continuous. Our interest here is in describing explicitly the Radon-Nikodym densities relating these various laws, and thereby extending various aspects of our previous studies of zero sets derived from a stable(α) subordinator to this more general case. We start in Section 2 by treating the example of Ornstein-Uhlenbeck processes. In particular, we obtain various generalizations of results of Truman-Williams [30, 31] and Hawkes-Truman [5] regarding the zero set of the simplest Gaussian-Ornstein-Uhlenbeck process derived from Brownian motion. The results of Section 2 lead to the study in Section 3 of various martingales associated with the range of a subordinator which arise from a change in the Lévy measure of the subordinator. Finally, in Section 4 we compare the results of Sections 2 and 3 to some relations between the stationary distribution of a recurrent Markov process and the Lévy measure of the inverse local time process at a point in the state space. While the basic relations are known to hold in great generality [20], the application of these relations to the zero sets of diffusion processes has been rather neglected in the literature.

2 Lengths of excursions of Ornstein-Uhlenbeck processes

The Ornstein-Uhlenbeck process $(U_t, t \geq 0)$ with parameter $\mu > 0$ is the solution of Langevin's equation

$$dU_t = dB_t - \mu U_t\, dt \tag{2}$$

where B is a Brownian motion. So far as the zero set of U is concerned, we may as well consider the process $X := U^2$. More generally, we consider for $0 < \alpha < 1$ and $\mu > 0$ the *squared OU process with dimension* $\delta = 2 - 2\alpha$ *and drift parameter* μ, that is the non-negative solution X of

$$dX_t = 2\sqrt{X_t}dB_t + (\delta - 2\mu X_t)\, dt \tag{3}$$

where we assume $X_0 = 0$. Denote by $Q^{\delta,\mu}$ the law of this process X on the usual path space $C[0,\infty)$. See [22, 23, 6] for further background and motivation for the study of these processes, known in mathematical finance as Cox-Ingersoll-Ross processes. Note that for a positive integer δ, if U solves (2), where we now suppose that the equation concerns $\mathbb{R}^\delta$-valued processes, then $X = |U|^2$ solves (3). Let Z denote the zero set of X, now taken to be the coordinate process on $C[0,\infty)$, and define $V_n(T)$ in terms of Z as in (1). Let $Q^\delta = Q^{\delta,0}$, so Q^δ is the law of the square of a δ-dimensional Bessel process [29, 22]. Let $(S_t, t \geq 0)$ denote a local time process for X at zero, and let (τ_s) be the right continuous inverse of this local time process. Then (τ_s) is a stable (α) subordinator, and $Q^{\delta,0}$ almost surely the zero set Z of X is the closure of the range of (τ_s). Note that while the definition of both (S_t) and (τ_s) depends on the value of δ, this dependence is hidden in the notation.

We recall the Cameron-Martin-Girsanov relationship between $Q^{\delta,\mu}$ and Q^{δ}: for every $t > 0$

$$\left.\frac{dQ^{\delta,\mu}}{dQ^{\delta}}\right|_{\mathcal{F}_t} = \exp\left(-\frac{\mu}{2}(X_t - \delta t) - \frac{\mu^2}{2}\int_0^t du X_u\right) \tag{4}$$

As a consequence of (4) and the recurrence of X under $Q^{\delta,\mu}$ for every $\mu > 0$, we have also for every $s > 0$ that

$$\left.\frac{dQ^{\delta,\mu}}{dQ^{\delta}}\right|_{\mathcal{F}_{\tau_s}} = \exp\left(\frac{\mu\delta\tau_s}{2} - \frac{\mu^2}{2}\int_0^{\tau_s} du X_u\right) \tag{5}$$

From this absolute continuity relation, it is immediate the zero set Z of X is represented $Q^{\delta,\mu}$ almost surely for all $\mu \geq 0$ as the closed range of the process (τ_s), which is a subordinator under $Q^{\delta,\mu}$ for each $\mu \geq 0$, a subordinator that is stable for $\mu = 0$ but not for $\mu > 0$. The Lévy measure of (τ_s) under $Q^{\delta,\mu}$ can be computed from (5) as indicated below.

Theorem 1 *For a random time T let $\mathcal{V}_T = \sigma(V_n(T), n = 1, 2, \cdots)$. Then for each $t > 0$*

$$\left.\frac{dQ^{\delta,\mu}}{dQ^{\delta}}\right|_{\mathcal{V}_t} = \exp\left(\frac{\mu\delta t}{2}\right)\Sigma(\mu,t)\Pi(\mu,t)^{2-\frac{\delta}{2}} \tag{6}$$

and for each $s > 0$

$$\left.\frac{dQ^{\delta,\mu}}{dQ^{\delta}}\right|_{\mathcal{V}_{\tau_s}} = \exp\left(\frac{\mu\delta\tau_s}{2}\right)\Pi(\mu,\tau_s)^{2-\frac{\delta}{2}} \tag{7}$$

where

$$\Sigma(\mu,t) = \sum_n \frac{1 - e^{-2\mu V_n(t)}}{2\mu t} \qquad \textit{and} \qquad \Pi(\mu,t) = \prod_n \frac{\mu V_n(t)}{\sinh(\mu V_n(t))}$$

Proof. Let $G_t = \sup(Z \cap [0,t))$. Note first that for fixed t,

$$\mathcal{V}_t \subset \mathcal{H}_t \subset \mathcal{F}_{G_t}$$

where $\mathcal{H}_t = \sigma(G_s, 0 \leq s \leq t)$ and $\mathcal{F}_{G_t} = \sigma(X_s 1_{(s \leq G_t)}, 0 \leq s \leq t)$. Moreover, for each $s > 0$, the random time τ_s is an $(\mathcal{H}_t)$ stopping time with $\tau_s = G_{\tau_s}$ a.s., and

$$\mathcal{V}_{\tau_s} \subset \mathcal{H}_{\tau_s} \subset \mathcal{F}_{\tau_s}$$

modulo Q^{δ} null sets. Consequently, we will be able to prove the formulae of the theorem by projecting the Q^{δ} martingale which appears in (4), first on $(\mathcal{F}_{G_t})$, then on $(\mathcal{H}_t)$, and finally on the σ-field $\mathcal{V}_t$. (Note that $\mathcal{V}_s$ is not contained in $\mathcal{V}_t$ for $s < t$. So unlike the other families considered above, the family $(\mathcal{V}_t, t > 0)$ does not constitute a filtration.)

Projection on $(\mathcal{F}_{G_t})$. Here we will use the fact that under Q^{δ} the squared meander

$$\left(m_u^2 := \frac{1}{t - G_t} X_{G_t + u(t - G_t)}, 0 \leq u \leq 1\right)$$

is independent of $\mathcal{F}_{G_t}$, and satisfies

$$(m_u^2, 0 \leq u \leq 1) \stackrel{d}{=} (\rho_u^2 + R_u^2, 0 \leq u \leq 1)$$

where $(\rho_u, 0 \le u \le 1)$ is a standard Bessel bridge of dimension $2-\delta$, and R is an independent 2-dimensional Bessel process. See [33, Corollary 3.9.1, page 44]. From the above description of $(m_u^2, 0 \le u \le 1)$, as a special case of the extended Lévy area formulae given in [33, (2.1) and (2.5)], and in [22, (2.k)], we easily deduce the following formula: for all $\gamma, \nu \ge 0$

$$Q^\delta\left[\exp\left(-\gamma m_1^2 - \frac{\nu^2}{2}\int_0^1 ds\, m_s^2\right)\right] = \left(\frac{\nu}{\sinh\nu}\right)^{1-\frac{\delta}{2}}\left(\cosh\nu + \frac{2\gamma}{\nu}\sinh\nu\right)^{-1} \tag{8}$$

In particular, for $\gamma = \nu/2$,

$$Q^\delta\left[\exp\left(-\frac{\nu}{2}m_1^2 - \frac{\nu^2}{2}\int_0^1 ds\, m_s^2\right)\right] = \Phi_\delta(\nu) := \left(\frac{\nu}{\sinh\nu}\right)^{1-\frac{\delta}{2}} e^{-\nu} \tag{9}$$

We deduce from (4) and (9) that

$$\left.\frac{dQ^{\delta,\mu}}{dQ^\delta}\right|_{\mathcal{F}_{G_t}} = \exp\left(\frac{\mu\delta t}{2}\right)\Phi_\delta(\mu(t-G_t))\exp\left(-\frac{\mu^2}{2}\int_0^{G_t} du\, X_u\right) \tag{10}$$

Projection on $(\mathcal{H}_t)$. From the previous formula we obtain

$$\left.\frac{dQ^{\delta,\mu}}{dQ^\delta}\right|_{\mathcal{H}_t} = \exp\left(\frac{\mu\delta t}{2}\right)\Phi_\delta(\mu(t-G_t))\Pi(\mu, G_t)^{2-\frac{\delta}{2}} \tag{11}$$

We derive (11) from (10) using the excursion theory under Q^δ, in particular, the fact that under n_δ, the corresponding Itô law of excursions, given that the lifetime equals v, the excursion process $(\epsilon_u, u \le v)$ is a Bessel bridge of dimension $4-\delta$, and we have used the Lévy-type formula [22, 33]

$$Q^{4-\delta}\left(\exp -\frac{\mu^2}{2}\int_0^v ds\, X_s \,\middle|\, X_v = 0\right) = \left(\frac{\mu v}{\sinh(\mu v)}\right)^{2-\frac{\delta}{2}}$$

Since

$$\Pi(\mu, G_t) = \Pi(\mu, t)\left(\frac{\sinh(\mu(t-G_t))}{\mu(t-G_t)}\right)$$

and

$$\Phi_\delta(x)\left(\frac{\sinh x}{x}\right)^{2-\frac{\delta}{2}} = \left(\frac{1-e^{-2x}}{2x}\right)$$

we can write (11) as

$$\left.\frac{dQ^{\delta,\mu}}{dQ^\delta}\right|_{\mathcal{H}_t} = \exp\left(\frac{\mu\delta t}{2}\right)\left(\frac{1-e^{-2\mu(t-G_t)}}{2\mu(t-G_t)}\right)\Pi(\mu, t)^{2-\frac{\delta}{2}} \tag{12}$$

Projection on $(\mathcal{V}_t)$. Formula (6) follows from the previous formula (12) and the result of [24] that

$$Q^\delta(t - G_t = V_n(t) \mid \mathcal{V}_t) = \frac{V_n(t)}{t} \tag{13}$$

□

Let $\Lambda^{\delta,\mu}$ denote the Lévy measure of (τ_s) under $Q^{\delta,\mu}$. So by definition

$$Q^{\delta,\mu}(\exp(-\theta\tau_s)) = \exp\left(-s\int_0^\infty (1-e^{-\theta x})\Lambda^{\delta,\mu}(dx)\right) \tag{14}$$

Write simply Λ^δ for $\Lambda^{\delta,0}$. From Theorem 1 and the basic formula

$$\Lambda^\delta(dy) = C dy/y^{2-\frac{\delta}{2}} = C dy/y^{1+\alpha} \tag{15}$$

where C is a constant depending on the choice of normalization of local time, we obtain for $\mu > 0$ the formula

$$\Lambda^{\delta,\mu}(dy) = C\left(\frac{\mu}{\sinh(\mu y)}\right)^2\left(\frac{e^{2\mu y}-1}{2\mu}\right)^{\delta/2} dy \tag{16}$$

To check, we recover (15) from (16) in the limit as $\mu \downarrow 0$. And for $\delta = 1$ we recover the result of Hawkes-Truman [5] for the zero set of the Gaussian-Ornstein-Uhlenbeck process. See also Section 4 for another confirmation of the formula (16) which involves almost no calculation. By combination of (14) and (12) we obtain for $\mu > 0$ the formula

$$Q^{\delta,\mu}(t - G_t = V_n(t) \mid \mathcal{V}_t) = \frac{1-e^{-2\mu V_n(t)}}{\sum_m (1-e^{-2\mu V_m(t)})} \tag{17}$$

which is a particular case of formula (7.d) of [24]. From the proof of Theorem 1, we extract also the following corollary, which is a particular case of more general results presented in the next section.

Corollary 2 *Let $G_t = \sup(Z \cap [0,t))$ where Z is the range of a stable(α) subordinator and let $\mathcal{H}_t = \sigma(G_s, 0 \le s \le t)$. Then for every $\mu > 0$*

$$\psi(\mu(t-G_t))\exp(\mu(1-\alpha)t)\prod_{n=1}^\infty\left(\frac{\mu V_n(t)}{\sinh(\mu V_n(t))}\right)^{1+\alpha} \tag{18}$$

is an $(\mathcal{H}_t)$-martingale, where $\psi(x) := (1-e^{-2x})/(2x)$, and

$$\exp(\mu(1-\alpha)\tau_s)\prod_{n=1}^\infty\left(\frac{\mu V_n(\tau_s)}{\sinh(\mu V_n(\tau_s))}\right)^{1+\alpha} \tag{19}$$

is an $(\mathcal{H}_{\tau_s})$-martingale.

Remark 3 The formula (12) and the more general formula (30) presented in the next section are closely related to the studies by Azéma [1] and Rainer [27] of martingales relative to the filtration $(\mathcal{H}_t)$ generated by the zero set of a real valued diffusion. In particular, if $(X_t, t \ge 0)$ is a recurrent diffusion on natural scale on a subinterval of the line containing 0, and $\Lambda = \Lambda_+ + \Lambda_-$ is the decomposition of the Lévy measure Λ induced by positive and negative excursions, as discussed further in Section 4, then the process

$$\mu_t := \frac{1(X_t > 0)}{\Lambda_+(t-G_t,\infty)} - \frac{1(X_t < 0)}{\Lambda_-(t-G_t,\infty)} \tag{20}$$

is an $(\mathcal{H}_t)$ local martingale. (This is, up to a factor of 1/2, the formula at the end of the introduction of [27], after correction of a misprint in that formula as indicated at the end of the present volume.) Our martingales (18) and (30) can be recovered by application of Itô's formula. If (X_t) is Brownian motion, then μ_t is a constant multiple of Azéma's martingale $\operatorname{sgn}(X_t)\sqrt{t-G_t}$.

3 Change of measure formulae for subordinators

Let probability distributions P and Q on the same basic measurable space $(\Omega, \mathcal{F})$ govern a process $(\tau_s, s \geq 0)$ as a subordinator, with Lévy measures Λ_P and Λ_Q respectively. We assume that

$$\Lambda_Q(dy) = \Phi(y)\Lambda_P(dy) \text{ for a } \Phi \text{ such that} \tag{21}$$

$$\int_0^\infty |1 - \Phi(y)|\Lambda_P(dy) < \infty \tag{22}$$

and use the notation

$$\bar{\Lambda}_P(x) = \Lambda_P(x, \infty); \quad \bar{\Lambda}_Q(x) = \Lambda_Q(x, \infty) = \int_{(x,\infty)} \Phi(y)\Lambda_P(dy) \tag{23}$$

Let Z be the range of (τ_s), $V_n(T)$ as in (1). Let $(S_t, t \geq 0)$ be the continuous local time inverse of $(\tau_s, s \geq 0)$.

Theorem 4 *Under the hypothesis* (22) *on the function* $\Phi = d\Lambda_Q/d\Lambda_P$, *define a function* Ψ *and a real number* γ *by*

$$\Psi(0) = 1; \quad \Psi(x) = \frac{\bar{\Lambda}_Q(x)}{\bar{\Lambda}_P(x)} \qquad (x > 0 : \bar{\Lambda}_P(x) > 0), \tag{24}$$

$$\gamma = \int_0^\infty (\Phi(x) - 1)\Lambda_P(dx) = \int_0^\infty (\Lambda_Q - \Lambda_P)(dx) \tag{25}$$

and define processes $(\prod_\Phi(t), t \geq 0)$ *and* $(M_\Phi(t), t \geq 0)$ *by*

$$\textstyle\prod_\Phi(0) = 1; \quad \prod_\Phi(t) = \prod_n \Phi(V_n(t)) \qquad (t > 0) \tag{26}$$

$$M_\Phi(t) = \Psi(t - G_t)\textstyle\prod_\Phi(G_t)\exp(-\gamma S_t) \tag{27}$$

Then for each $(\mathcal{H}_t)$*-stopping time* T *such that* $P(T < \infty) = Q(T < \infty) = 1$, *the law* Q *is absolutely continuous with respect to* P *on* $\mathcal{H}_T$, *with density*

$$\left.\frac{dQ}{dP}\right|_{\mathcal{H}_T} = M_\Phi(T) \tag{28}$$

In particular this formula holds for every fixed time T, *and for* $T = \tau_s$ *for every* $s > 0$, *in which case the right side of* (28) *is*

$$M_\Phi(\tau_s) = \textstyle\prod_\Phi(\tau_s)\exp(-\gamma s) \qquad (s \geq 0) \tag{29}$$

Consequently,

$$(M_\Phi(t), t \geq 0) \text{ is an } ((\mathcal{H}_t), P)\text{-martingale} \tag{30}$$

and

$$(M_\Phi(\tau_s), s \geq 0) \text{ is an } ((\mathcal{H}_{\tau_s}), P)\text{-martingale.} \tag{31}$$

By combination of Theorem 4 with Theorem 7.1 of [24] we obtain the following:

Corollary 5 *Suppose further that* $\Lambda_P(dy) = \rho_P(y)dy$ *and* $\Lambda_Q(dy) = \rho_Q(y)dy$ *for some densities* ρ_P *and* ρ_Q *which are strictly positive on* $(0,\infty)$. *For* $y > 0$ *and* $t > 0$ *let*

$$h_P(y) = \frac{\bar{\Lambda}_P(y)}{\rho_P(y)}; \qquad H_P(t) = \sum_n h_P(V_n(t)); \tag{32}$$

and define $h_Q(y)$ *and* $H_Q(t)$ *similarly with* Q *instead of* P. *Then*

$$\Phi(x) = \frac{\rho_Q(x)}{\rho_P(x)} \qquad (x > 0);\ \gamma = \int_0^\infty (\rho_Q(y) - \rho_P(y))\, dy \tag{33}$$

For fixed $t > 0$, *let* $\mathcal{V}_t = \sigma(V_n(t), n = 1, 2, \cdots)$. *Then*

$$\left.\frac{dQ}{dP}\right|_{\mathcal{V}_t} = \frac{H_Q(t)}{H_P(t)} \prod_\Phi(t) \exp(-\gamma S_t) \tag{34}$$

Proof of Corollary 5. The formulae (33) are immediate. To deduce (34) from (28), it suffices to take $T = t$ and project the density in (28) onto the σ-field $\mathcal{V}_t$, using the fact that $\prod_\Phi(t)$ and S_t are $\mathcal{V}_t$-measurable, the fact that $\prod_\Phi(G_t) = \prod_\Phi(t)/\Phi(t - G_t)$ and the formula

$$E_P\left[\left(\frac{h_Q}{h_P}\right)(t - G_t) \,\middle|\, \mathcal{V}_t\right] = \frac{H_Q(t)}{H_P(t)} \tag{35}$$

which is obtained by evaluation of the left side of (35) using the *sampling formula*

$$P(t - G_t = V_n(t) \,|\, \mathcal{V}_t) = \frac{h_P(V_n(t))}{H_P(t)} \tag{36}$$

established in Theorem 7.1 of [24]. This shows that the right side of (35) equals

$$\sum_n \frac{h_Q(V_n(t))}{h_P(V_n(t))} \frac{h_P(V_n(t))}{H_P(t)} = \frac{H_Q(t)}{H_P(t)} \tag{37}$$

Proof of Theorem 4.

Step 1. Proof for $T = \tau_s$ *for fixed* $s > 0$. In this case we have $\tau_s - G_{\tau_s} = 0$ a.s. so $\Psi(\tau_s - G_{\tau_s}) = 1$, and the task is to show that for every non-negative $\mathcal{H}_{\tau_s}$-measurable random variable X,

$$E_Q(X) = E_P[X M_\Phi(\tau_s)] \text{ where } M_\Phi(\tau_s) = \prod_\Phi(\tau_s) \exp(-\gamma s) \tag{38}$$

This is a consequence of the following variation of Campbell's formula [12, (3.35)]: for Φ satisfying (22),

$$E_P\left[\prod_n \Phi(V_n(\tau_s))\right] = \exp\left(s \int_0^\infty (\Phi(x) - 1)\Lambda_P(dx)\right) = \exp(\gamma s) \tag{39}$$

Apply (39) with Q instead of P and g instead of Φ, for non-negative g with $\int |g - 1| d\Lambda_Q < \infty$. Then write $(g-1)\Phi = (g\Phi - 1) - (\Phi - 1)$ and use (39) again twice under P to see that (38) holds for $X = \prod_n g(V_n(\tau_s))$. Varying g provides enough X's to deduce that (38) holds for all non-negative $\mathcal{V}_{\tau_s}$-measurable X's. But the σ-field $\mathcal{V}_{\tau_s}$ is contained in $\mathcal{H}_{\tau_s} = \sigma(\tau_u, 0 \le u \le s)$, and the identity (38) extends to all non-negative

$\mathcal{H}_{\tau_s}$-measurable X because P and Q share a common conditional distribution for $(\tau_u, 0 \leq u \leq s)$ given $\mathcal{V}_{\tau_s}$, that is the unique law of an increasing process parameterized by $[0, s]$ with exchangeable increments and jumps of the prescribed sizes $V_n(\tau_s)$, $n = 1, 2, \cdots$. (Assuming for simplicity that Λ_P is continuous, to avoid ties among the $V_n(\tau_s)$, we can write

$$\tau_u = \sum_n V_n(\tau_s) 1(\sigma_n \leq u) \quad (0 \leq u \leq s) \tag{40}$$

where σ_n is the a.s. unique local time u such that $\tau_u - \tau_{u-} = V_n(\tau_s)$. The common conditional law of $(\tau_u, 0 \leq u \leq s)$ given $\mathcal{V}_{\tau_s}$ is then specified by the fact that under both P and Q the σ_n are i.i.d random variables with uniform $[0, s]$ distribution, independent of $\mathcal{V}_{\tau_s}$. See [10] regarding this decomposition of τ_u and the corresponding result allowing Λ_P to have a discrete component).

Step 2. Proof for $T = t$ for a fixed $t > 0$. From the previous result for $T = \tau_s$, for all $s, t > 0$ we can compute

$$\left.\frac{dQ}{dP}\right|_{\mathcal{H}_t} = E_P\left(M_\Phi(\tau_s) \,|\, \mathcal{H}_{t \wedge \tau_s}\right) \qquad \text{on } (t < \tau_s) \tag{41}$$

But on $(t < \tau_s)$ we find that

$$M_\Phi(\tau_s) = \exp(-\gamma s) \Pi_\Phi(G_t) \Phi(D_t - G_t) \Pi^* \tag{42}$$

where Π^* is the product of $\Phi(V_n(\tau_s))$ over n corresponding to those component intervals of $[0, \tau_s] \setminus Z$ that are contained in $[D_t, \tau_s]$. Let (S_t) denote the continuous local time inverse of (τ_s). By the strong Markov property of $(\tau_u, u \geq 0)$ at the stopping time S_t, when $\tau_{S_t} = D_t$, and (39),

$$E_P\left(\Pi^* \,\middle|\, \mathcal{H}_{t \wedge \tau_s}\right) = \exp(\gamma(s - S_t)) \tag{43}$$

Also, by the last exit decomposition at G_t,on the event $(\tau_s > t)$, which is identical to $(S_t < s)$,

$$P(D_t - G_t \in dy \,|\, \mathcal{H}_{t \wedge \tau_s}, \tau_s > t, t - G_t = x) = \Lambda_P(dy) / \bar{\Lambda}_P(x) \quad (y > x) \tag{44}$$

Combining these observations shows that

$$E_P\left(M_\Phi(\tau_s) \,\middle|\, \mathcal{H}_{t \wedge \tau_s}\right) = M_\Phi(t) \qquad \text{on } (t < \tau_s) \tag{45}$$

That is to say, for every non-negative $\mathcal{H}_t$-measurable random variable H_t

$$E_Q[H_t 1_{(\tau_s > t)}] = E_P[H_t M_\Phi(t) 1_{(\tau_s > t)}] \tag{46}$$

Now for each $t > 0$ we can let $s \to \infty$, and use the fact that $1_{(\tau_s > t))} \uparrow 1$ both P and Q a.s. to deduce $E_Q(H_t) = E_P(H_t Z_t)$, which is the desired result.

Step 3. Proof for a general $(\mathcal{H}_t)$-stopping time T with $P(T < \infty) = Q(T < \infty) = 1$
This is a reprise of the previous argument, first using the optional sampling theorem for $T \wedge t$, then letting $t \to \infty$.

Example 6 As an example of the situation described in Theorem 4 where $\gamma \neq 0$, following Kinkladze [13] we now consider the pair of diffusions B and $X^{(\mu)}$, where B is a Brownian motion, and $X^{(\mu)}$ with law $P^{(\mu)}$ is the solution of

$$dX_t = dB_t - \mu \operatorname{sgn}(X_t)\, dt$$

We have

$$P^{(\mu)}\Big|_{\mathcal{F}_t} = \exp\left(-\mu\{|X_t| - S_t\} - \tfrac{1}{2}\mu^2 t\right) \cdot P\Big|_{\mathcal{F}_t} \tag{47}$$

where $(S_t, t \geq 0)$ denotes the local time of X at 0. From (47) we deduce

$$P^{(\mu)}\Big|_{\mathcal{H}_t} = f(-\mu\sqrt{t - G_t}) \exp\left(\mu S_t - \tfrac{1}{2}\mu^2 t\right) \cdot P\Big|_{\mathcal{H}_t} \tag{48}$$

where $f(\lambda) := E[\exp(\lambda m_1)]$ for m_1 the value at time 1 of a Brownian meander, that is

$$f(\lambda) = \int_0^\infty r \exp(-r^2/2) \exp(\lambda r)\, dr$$

It follows that

$$\frac{dP^{(\mu)}}{dP}\Big|_{\mathcal{H}_t} = \Psi(t - G_t)\Pi_\Phi(G_t) \exp(-\gamma S_t)$$

where

$$\Psi(x) = f(-\mu\sqrt{x}) \exp(-\tfrac{1}{2}\mu^2 x); \quad \Phi(x) = \exp(-\tfrac{1}{2}\mu^2 x); \quad \gamma = -\mu$$

4 The Lévy measure of the inverse local time process

Let 0 be a recurrent point in the state-space E of a nice recurrent strong Markov process X. Let $T_0 = \inf\{t : t > 0, X_t = 0\}$. Assuming that 0 is regular for itself, that is $P_0(T_0 = 0) = 1$, it is well known that there exists a continuous increasing local time process for X at 0, say $(L_t, t \geq 0)$, whose right-continuous inverse, say $(\tau_\ell, \ell \geq 0)$ is a subordinator under P_0. Let Λ denote the Lévy measure of this subordinator. Due to different conventions about the normalization of local time processes in different settings, let us allow an arbitrary normalization of (L_t) in this generality. So Λ is unique up to constant factors: multiplying L by c divides Λ by c. It is known [4] that such a Markov process X admits a σ-finite invariant measure m such that $P^m(T_0 = \infty) = 0$. As a consequence of a general Palm formula for excursions of stationary (not necessarily Markovian) processes established in [20], this m is unique up to constant multiples and there is the identity

$$P^m(T_0 \in da) = m\{0\}\delta_0(da) + c\Lambda(a, \infty)da \quad (a \geq 0) \tag{49}$$

for some $c > 0$ depending on the choice of m and the choice of normalization of local time. That is to say, the P^m distribution of T_0 has an atom at 0 of magnitude $m\{0\}$, and has a density on $(0, \infty)$ given by $c\Lambda(a, \infty)$ for $0 < a < \infty$.

The connection between the invariant measure m on the state-space of X and the Lévy measure Λ on $(0, \infty)$ is made via Itô's law n for excursions ε of X away from 0. Assume that an excursion $\varepsilon = (\varepsilon_t, t \geq 0)$ is absorbed at 0 at time $T_0 = T_0(\varepsilon) = \inf\{t :$

$t > 0, \varepsilon_t = 0\}$. And assume for simplicity that $m\{0\} = 0$, which is to say that the Lebesgue measure of the zero set of X is 0 a.s. P^x for all $x \in E$. By definition of n [7, 28]), the Lévy measure Λ of the inverse local time process at 0 is the n distribution of T_0:

$$\Lambda(a, \infty) = n(T_0 > a) \qquad (a > 0) \tag{50}$$

Also, the formula

$$mf = \int n(d\varepsilon) \int_0^{T_0} f(\varepsilon_t)dt \quad (f \geq 0) \tag{51}$$

for non-negative measurable functions f on E defines an invariant measure m for X [4, 20], and if we take this m in (49) the constant c is forced equal to 1. That is to say, for m defined by (51)

$$P^m(T_0 \in da) = \Lambda(a, \infty)da \quad (a \geq 0) \tag{52}$$

As shown in [20], this identity is a consequence of the following more general identity. Let n^* denote Maisonneuve's exit law for state 0, that is the distribution on path-space under which $(X_t, 0 \leq t \leq T_0)$ and $(X_{T_0+u}, 0 \leq u \leq \infty)$ are independent with laws n and P_0 respectively. Then for an arbitrary non-negative measurable Y defined on path-space

$$P^m(Y) = n^*\left(\int_0^{T_0} Y(\theta_t)dt\right) \tag{53}$$

where θ_t is the usual shift operator on path space, so

$$Y(\theta_t) = Y(X_{t+u}, 0 \leq u < \infty)$$

Taking $Y = f(X_0)$ yields (51), while taking $Y = h(T_0)$ for a non-negative measurable h on $(0, \infty)$ and using (50) yields (49).

Suppose now that X is a recurrent diffusion process on a subinterval of the line containing 0. Let m_+ and m_- denote the restrictions of m to $(0, \infty)$ and $(-\infty, 0)$ respectively, so $m = m_+ + m_-$. By path continuity of X, each excursion is either positive or negative, and there are corresponding decompositions $n = n_+ + n_-$ and $\Lambda = \Lambda_+ + \Lambda_-$ which imply via (53) that (49) holds just as well with m replaced by $m_\pm$ and Λ replaced by $\Lambda_\pm$, where $\pm$ is either $+$ or $-$.

The decomposition $\Lambda = \Lambda_+ + \Lambda_-$ and reflection through 0 reduces computation of Λ to computation of Λ_+.

Put another way, there no loss of generality in assuming, as we shall from now on, that the statespace E of the diffusion is either $[0, \infty)$ or $[0, b]$ for some $b > 0$. To be definite, assume $E = [0, \infty)$.

Example 7 It is known [22] and easily checked that if X has distribution $Q^{\delta,0}$, then the process $X^{\delta,\mu}$ defined by

$$X^{\delta,\mu}(t) = e^{-2\mu t} X(e^{2\mu t}/2\mu) \quad (-\infty < t < \infty) \tag{54}$$

is a two sided stationary process governed by the stochastic differential equation (3) for $t > 0$. Let $\eta = \eta^{\delta,\mu}$ denote the $Q^{\delta,0}$ distribution of

$$X^{\delta,\mu}(0) = X(1/2\mu) \stackrel{d}{=} (2\mu)^{-1} X(1) \stackrel{d}{=} \mu^{-1} Z_{\delta/2} \tag{55}$$

where Z_a denotes a gamma(a) variable. Then the P^{η} distribution of T_0 considered in (49) is immediately identified in this example with the $Q^{\delta,0}$ distribution of

$$\inf\{t > 0 : X^{\delta,\mu}(t) = 0\} = \inf\{t > 0 : X(e^{2\mu t}/2\mu) = 0\} \tag{56}$$

$$= \frac{1}{2\mu}\log(2\mu D_{1/2\mu}) \stackrel{d}{=} \frac{1}{2\mu}\log(D_1) \tag{57}$$

where $D_t = \inf\{u > t : X(u) = 0\}$. Since the distribution of D_t for a stable(α) zero set is given by

$$D_t \stackrel{d}{=} tD_1 \stackrel{d}{=} \frac{t}{G_1} \stackrel{d}{=} \frac{t}{Z_{\alpha,1-\alpha}} \tag{58}$$

where $Z_{a,b}$ denotes a beta(a,b) variable [3, 17], a simple change of variables yields the following formula for the density of $(2\mu)^{-1}\log(D_1)$ in (57), hence for $\Lambda(a,\infty)$ in (52):

$$\Lambda^{\mu,\delta}(a,\infty) = \frac{P[(2\mu)^{-1}(\log D_1) \in da]}{da} = \frac{2\mu}{\Gamma(\alpha)\Gamma(1-\alpha)}\frac{e^{-2\mu\alpha a}}{(1-e^{-2\mu a})^{\alpha}} \tag{59}$$

where $\alpha = 1-\delta/2$. It is easily verified that this formula is consistent with the previous formula (16).

Some general formulae for diffusions. In the case of one-dimensional diffusion processes, there is an alternative *local formula* for Λ which has been known for much longer than the *global formula* (52). Assuming for simplicity that the statespace is $[0,\infty)$, the local formula for Λ is

$$\Lambda(a,\infty) = c\lim_{x\downarrow 0}\frac{P^x(T_0 > a)}{s(x) - s(0)} \tag{60}$$

where s is the scale function of the diffusion and c is a constant depending on normalization conventions for the scale function and the local time process. This formula appears in Section 6.2 of Itô-McKean[8], along with various Laplace transformed expressions of this formula now discussed. There are also corresponding local formulae for Itô's excursion law n and for Maisonneuve's exit law n^* in this setting, for instance $n^*(Y) = c\lim_{x\downarrow 0}\frac{P^x(Y)}{s(x)-s(0)}$ for appropriately regular Y. See e.g. Section 3 of [22] for further discussion and other descriptions of n.

So far as the zero set of X is concerned, there is no loss of generality in replacing X by $s(X)$ where s is a scale function for X chosen so that $s(0) = 0$, such a choice being possible due to the assumed recurrence of the boundary state 0. So let us assume that X is already on natural scale, i.e. that $s(x) = x$, so the generator G of X, acting on smooth functions vanishing in a neighbourhood of 0 is

$$G = \frac{1}{2}\frac{d}{dm}\frac{d}{dx} \tag{61}$$

where m is the *speed measure* of X on $[0,\infty)$ and we assume for simplicity that $m\{0\} = 0$. Now in (60) we obtain

$$\Lambda(a,\infty) = \frac{1}{2}\frac{d}{dx}P^x(T_0 > a)\bigg|_{x=0+} \tag{62}$$

provided the local time process (L_t) at 0 is defined as $L_t = L_t^0$ where

$$(L_t^x; t \geq 0, x \geq 0)$$

is a jointly continuous version of the local times normalized as occupation densities relative to the speed measure m of X. See e.g. [8].

In terms of the Laplace exponent

$$\Theta(\lambda) := \int_0^\infty (1 - e^{-\lambda x})\Lambda(dx) = \lambda \int_0^\infty e^{-\lambda a}\Lambda(a, \infty)\, da \tag{63}$$

taking a Laplace transform converts (62) into

$$\Theta(\lambda) = -\frac{1}{2}\frac{d}{dx}\phi_\lambda(x)\bigg|_{x=0+} \tag{64}$$

where

$$\phi_\lambda(x) = P^x(e^{-\lambda T_0}) \tag{65}$$

is well known to be the unique solution ϕ of the Sturm-Liouville equation

$$G\phi = \lambda\phi \text{ on } (0, \infty) \text{ with } \phi(0) = 1,\ 0 \leq \phi \leq 1, \tag{66}$$

which can be written alternatively as

$$\tfrac{1}{2}\phi'' = \lambda m \cdot \phi \text{ on } (0, \infty) \text{ with } \phi(0) = 1,\ 0 \leq \phi \leq 1. \tag{67}$$

Another well known formula in this setting is

$$\Theta(\lambda)^{-1} = \int_0^\infty e^{-\lambda t} p(t, 0, 0) dt \tag{68}$$

where $p(t, x, y)$ is a smooth transition density for X relative to m, that is $P^x(X_t \in dy) = p(t, x, y)m(dy)$. See our papers [22, 23, 21] regarding the relation between the above formulae, the Ray-Knight theorems for Brownian local times, and the distribution of quadratic functionals of Bessel processes, and see the work of Knight [14] and Kotani-Watanabe [16] regarding the relation of these formulae to Krein's spectral theory for vibrating strings [9, 2]. Since the speed measure m is an invariant measure for X, in this setting the global formula (52) gives

$$\Lambda(a, \infty) = P^m(T_0 \in da)/da \tag{69}$$

which when Laplace transformed amounts via (63) to

$$\Theta(\lambda) = \lambda P^m(e^{-\lambda T_0}) = \lambda \int_0^\infty \phi_\lambda(x) m(dx) \tag{70}$$

Note that this formula holds just as well in the general Markov setting discussed earlier. Comparison of (64) and (70) shows that the agreement of the local and global formulae for Λ amounts to the following about the unique solution ϕ_λ of the Sturm-Liouville equation (67):

$$-\frac{1}{2}\frac{d}{dx}\phi_\lambda(x)\bigg|_{x=0+} = \lambda \int_0^\infty \phi_\lambda(x) m(dx) \tag{71}$$

This is easily checked from (67), since from that equation the right side of (71) is

$$\frac{1}{2}\int_0^\infty dx\,\phi''_\lambda(x) = \frac{1}{2}(\phi'_\lambda(\infty) - \phi'_\lambda(0+)) \tag{72}$$

and since ϕ'_λ is an increasing function of x the constraint that ϕ_λ is bounded forces $\phi'_\lambda(\infty) = 0$. The formula (71) is a generalization of an identity of Truman-Williams [30, (77) and (92)].

Example 8 *Reflecting BM.* Let X be RBM on $[0,\infty)$. We take $m(dx) = dx$, local time at zero is occupation density at 0+ relative to dx. The Laplace exponent is $\Theta(\lambda) = \sqrt{2\lambda}/2$, and we find $\phi_\lambda(x) = \sqrt{2\lambda}x$, $\phi'_\lambda(0+) = \sqrt{2\lambda}$ and $\lambda\int_0^\infty \phi_\lambda(x)dx = \lambda/\sqrt{2\lambda}$.

References

[1] J. Azéma. Sur les fermés aléatoires. In *Séminaire de Probabilités XIX*, pages 397–495. Springer, 1985. Lecture Notes in Math. 1123.

[2] H. Dym and H.P. McKean. *Gaussian Processes, Function Theory and the Inverse Spectral Problem.* Academic Press, 1976.

[3] E. B. Dynkin. Some limit theorems for sums of independent random variables with infinite mathematical expectations. *IMS-AMS Selected Translations in Math. Stat. and Prob.*, 1:171–189, 1961.

[4] R. K. Getoor. Excursions of a Markov process. *Annals of Probability*, 7:244 – 266, 1979.

[5] J. Hawkes and A. Truman. Statistics of local time and excursions for the Ornstein-Uhlenbeck process. In M. Barlow and N. Bingham, editors, *Stochastic Analysis*, volume 167 of *Lect. Note Series*, pages 91–101. London Math. Soc., 1991.

[6] Y. Hu, Z. Shi, and M. Yor. Some applications of Lévy's area formula to pseudo-Brownian and pseudo-Bessel bridges. In *Exponential functionals and principal values of Brownian motion.* Biblioteca de la Revista Matematica Ibero-Americana, Madrid, 1996/1997.

[7] K. Itô. Poisson point processes attached to Markov processes. In *Proc. 6th Berk. Symp. Math. Stat. Prob.*, volume 3, pages 225–240, 1971.

[8] K. Itô and H.P. McKean. *Diffusion Processes and their Sample Paths.* Springer, 1965.

[9] I.S. Kac and M. G. Krein. On the spectral function of the string. *Amer. Math. Society Translations*, 103:19–102, 1974.

[10] O. Kallenberg. Canonical representations and convergence criteria for processes with interchangeable increments. *Z. Wahrsch. Verw. Gebiete*, 27:23–36, 1973.

[11] J. F. C. Kingman. Random discrete distributions. *J. Roy. Statist. Soc. B*, 37:1–22, 1975.

[12] J.F.C. Kingman. *Poisson Processes.* Clarendon Press, Oxford, 1993.

[13] G. N. Kinkladze. A note on the structure of processes the measure of which is absolutely continuous with respect to the Wiener process modulus measure. *Stochastics*, 8:39 - 44, 1982.

[14] F.B. Knight. Characterization of the Lévy measure of inverse local times of gap diffusions. In *Seminar on Stochastic Processes, 1981*, pages 53–78. Birkhäuser, Boston, 1981.

[15] F.B. Knight. On the duration of the longest excursion. In E. Cinlar, K.L. Chung, and R.K. Getoor, editors, *Seminar on Stochastic Processes*, pages 117–148. Birkhäuser, 1985.

[16] S. Kotani and S. Watanabe. Krein's spectral theory of strings and generalized diffusion processes. In M. Fukushima, editor, *Functional Analysis in Markov Processes*, pages 235–249. Springer, 1982. Lecture Notes in Math. 923.

[17] J. Lamperti. An invariance principle in renewal theory. *Ann. Math. Stat.*, 33:685 - 696, 1962.

[18] M. Perman. Order statistics for jumps of normalized subordinators. *Stoch. Proc. Appl.*, 46:267–281, 1993.

[19] M. Perman, J. Pitman, and M. Yor. Size-biased sampling of Poisson point processes and excursions. *Probability Theory and Related Fields*, 92:21–39, 1992.

[20] J. Pitman. Stationary excursions. In *Séminaire de Probabilités XXI*, pages 289–302. Springer, 1986. Lecture Notes in Math. 1247.

[21] J. Pitman. Cyclically stationary Brownian local time processes. To appear in *Probability Theory and Related Fields*, 1996.

[22] J. Pitman and M. Yor. A decomposition of Bessel bridges. *Zeitschrift für Wahrscheinlichkeitstheorie und Verwandte Gebiete*, 59:425–457, 1982.

[23] J. Pitman and M. Yor. Sur une décomposition des ponts de Bessel. In M. Fukushima, editor, *Functional Analysis in Markov Processes*, pages 276–285. Springer, 1982. Lecture Notes in Math. 923.

[24] J. Pitman and M. Yor. Arcsine laws and interval partitions derived from a stable subordinator. *Proc. London Math. Soc. (3)*, 65:326–356, 1992.

[25] J. Pitman and M. Yor. The two-parameter Poisson-Dirichlet distribution derived from a stable subordinator. Technical Report 433, Dept. Statistics, U.C. Berkeley, 1995. To appear in *The Annals of Probability.*

[26] J. Pitman and M. Yor. On the relative lengths of excursions derived from a stable subordinator. Technical Report 469, Dept. Statistics, U.C. Berkeley, 1996. To appear in *Séminaire de Probabilités XXXI.*

[27] C. Rainer. Projection d'une diffusion sur sa filtration lente. In J. Azéma, M. Emery, and M. Yor, editors, *Séminaire de Probabilités XXX*, pages 228–242. Springer, 1996. Lecture Notes in Math. 1626.

[28] D. Revuz and M. Yor. *Continuous martingales and Brownian motion.* Springer, Berlin-Heidelberg, 1994. 2nd edition.

[29] T. Shiga and S. Watanabe. Bessel diffusions as a one-parameter family of diffusion processes. *Z. Wahrsch. Verw. Gebiete*, 27:37–46, 1973.

[30] A. Truman and D. Williams. A generalised arc sine law and Nelson's stochastic mechanics of one-dimensional time homogeneous diffusions. In M. A. Pinsky, editor, *Diffusion Processes and Related Problems in Analysis, Vol.1*, volume 22 of *Progress in Probability*, pages 117–135. Birkhäuser, 1990.

[31] A. Truman and D. Williams. Excursions and Itô calculus in Nelson's stochastic mechanics. In A. Boutet de Monvel et. al., editor, *Recent Developments in Quantum Mechanics: Proceedings of the Brasov Conference, Poiana Brasov 1989*, volume 12 of *Mathematical Physics Studies.* Kluwer Academic, 1991.

[32] J.G. Wendel. Zero-free intervals of semi-stable Markov processes. *Math. Scand.*, 14:21 - 34, 1964.

[33] M. Yor. *Some Aspects of Brownian Motion.* Lectures in Math., ETH Zürich. Birkhaüser, 1992. Part I: Some Special Functionals.

On the relative lengths of excursions derived from a stable subordinator*

Jim Pitman[(1)] and Marc Yor[(2)]

(1) Department of Statistics, University of California, 367 Evans Hall # 3860, Berkeley, CA 94720-3860, USA
(2) Laboratoire de Probabilités, Université Paris VI, 4 Place Jussieu, 75252 Paris, France

Abstract. Results are obtained concerning the distribution of ranked relative lengths of excursions of a recurrent Markov process from a point in its state space whose inverse local time process is a stable subordinator. It is shown that for a large class of random times T the distribution of relative excursion lengths prior to T is the same as if T were a fixed time. It follows that the generalized arc-sine laws of Lamperti extend to such random times T. For some other random times T, absolute continuity relations are obtained which relate the law of the relative lengths at time T to the law at a fixed time.

1 Introduction

Following Lamperti [10], Wendel [24], Kingman [7], Knight [8], Perman-Pitman-Yor [12, 13, 15], consider the sequence

$$V_1(T) \geq V_2(T) \geq \cdots \tag{1}$$

of ranked lengths of component intervals of the set $[0,T]\backslash Z$, where T is a strictly positive random time, and Z is the zero set of a Markov process X started at zero, such as a Brownian motion or Bessel process, for which the inverse $(\tau_s, s \geq 0)$ of the local time process of X at zero is a *stable*(α) *subordinator*, that is an increasing process with stationary independent increments and Lévy measure Λ_α where

$$\Lambda_\alpha(x,\infty) = Cx^{-\alpha} \qquad (x>0) \tag{2}$$

for some constant $C>0$, and $0<\alpha<1$. That is, for $\lambda > 0$

$$E[\exp(-\lambda\tau_s)] = \exp(-sK\lambda^\alpha) \text{ where } K = C\Gamma(1-\alpha). \tag{3}$$

*Research supported in part by N.S.F. Grant DMS-9404345

It was shown in [15] that for all $t > 0$ and $s > 0$

$$\left(\frac{V_1(t)}{t}, \frac{V_2(t)}{t}, \ldots\right) \stackrel{d}{=} \left(\frac{V_1(\tau_s)}{\tau_s}, \frac{V_2(\tau_s)}{\tau_s}, \ldots\right) \tag{4}$$

where $\stackrel{d}{=}$ denotes equality in distribution. Write simply V_n instead of $V_n(1)$, so $(V_1, V_2, \ldots)$ is a convenient notation for a sequence of random variables with the common joint distribution of the sequences displayed in (4) for all $s > 0$ and $t > 0$. The distribution of (V_n) of course depends on α, but we suppress α in the notation. Note that

$$V_1 > V_2 > \cdots > 0 \text{ a.s. and } \textstyle\sum_n V_n = 1 \text{ a.s.} \tag{5}$$

For a detailed account of features of the distribution of (V_n) with a parameter $0 < \alpha < 1$, references to earlier work, and connections with Kingman's [7] Poisson-Dirichlet distribution, see [16]. Our main purpose in this paper is to point out that beyond the fixed times t and inverse local times τ_s featured in (4), there are many more random times T such that

$$\left(\frac{V_1(T)}{T}, \frac{V_2(T)}{T}, \ldots\right) \stackrel{d}{=} (V_1, V_2, \ldots) \tag{6}$$

Definition 1 Call T *admissible*, or to be more precise *admissible for* Z, if (6) holds. Call T *inadmissible* otherwise.

Note that Definition 1 makes sense for any random closed subset Z of $\mathbb{R}^+$, and any $\mathbb{R}^+$-valued random variable T, with $V_n(T)$ defined as the nth longest component interval of $[0, T]\backslash Z$ and $V_n := V_n(1)$. In this paper we obtain some general results which clarify the relation between stability properties of Z and admissibility of various random times T for Z. But for the rest of the introduction we continue to assume that Z is the closure of the range of a stable(α) subordinator.

We showed in [16] by direct calculation that

$$H_m := \inf\{t : V_m(t) \geq 1\} \textit{ is admissible for each } m = 1, 2, \ldots \tag{7}$$

Here we provide a criterion for a random time T to be admissible, which yields a large family of random times, including the times t, τ_s and H_m mentioned above, which are admissible for Z derived from a stable(α) subordinator. Let

$$G_t = \sup(Z \cap [0, t)); \qquad D_t = \inf(Z \cap [t, \infty)) \tag{8}$$

The admissibility of H_m turns out to be intimately connected with the following *sampling property* of Z, established in [15], which finds several applications in this paper:

$$P(1 - G_1 = V_n | V_1, V_2, \ldots) = V_n \qquad (n = 1, 2, \ldots) \tag{9}$$

See [14, 17] for further discussion of this property and related results.

The rest of this paper is organized as follows. The main results for the range of a stable subordinator are presented in Section 2 and proved in Section 3. Besides finding times that are admissible, we show for some inadmissible random times T, in particular for $T = G_t$ and $T = D_t$ for a fixed time t, that the distribution of the sequence on the left side of (6) has a simple density relative to that of $(V_1, V_2, \ldots)$. In Section 4 we relate our study of admissible times to the generalized arc-sine laws of Lamperti [9, 10], studied also in [2, 15, 23]. In particular, we describe the distribution of time spent positive by a skew Bessel process or skew Bessel bridge.

2 Results for a Stable Subordinator

Throughout this section, let $0 < \alpha < 1$, and let E_α denote expectation with respect to a probability distribution P_α which governs $(\tau_s, s \geq 0)$ as a stable(α) subordinator, and let Z be the closure of the range of (τ_s). Let $(S_t, t \geq 0)$ denote the continuous *local time process* defined by $S_t = \inf\{s : \tau_s > t\}$. While many approximations of local time are known [4], a useful one in the present setting is the following:

Proposition 2 *For each* $t > 0$,

$$n^{1/\alpha} V_n(t) \rightarrow (CS_t)^{1/\alpha} \textit{almost surely } (P_\alpha) \textit{ as } n \rightarrow \infty. \tag{10}$$

where the limit holds uniformly in $0 \leq t \leq t_0$ *almost surely* (P_α) *for every* $t_0 > 0$, *and also in pth mean for every* $p > 0$.

Proof. The convergence both a.s. and in pth mean for a fixed $t > 0$ is established in Proposition 10 of [16]. As observed by Kingman [7], (10) holds almost surely with the random time τ_s substituted instead of the fixed time t, and $S_{\tau_s} = s$ instead of S_t. Since $(V_n(t), t \geq 0)$ is an increasing process in t for each n, and $(S_t, t \geq 0)$ is a continuous increasing process, the claimed almost sure convergence can be deduced by a standard argument. See for instance Lemma 2.5 of [5]. □

2.1 Admissible Times

Proposition 3 *Given* $c_n \geq 0$ *with* $\sup_n c_n < \infty$ *and* $c \geq 0$, *let*

$$A_t := \sum_n c_n V_n(t) + c S_t^{1/\alpha} \tag{11}$$

and for $u > 0$ *let*

$$\alpha_u := \inf\{t : A_t > u\} \tag{12}$$

Then α_u *is an admissible time.*

Proposition 2 has the following immediate corollary:

Corollary 4 *If* T *is admissible then*

$$\left(\frac{S_T}{T^\alpha}, \frac{V_1(T)}{T}, \frac{V_2(T)}{T}, \ldots\right) \stackrel{d}{=} (S_1, V_1, V_2, \ldots) \tag{13}$$

where

$$S_1 := C^{-1} \lim_n n V_n^\alpha \textit{ almost surely } (P_\alpha) \textit{ and in pth mean for all } p > 0 \tag{14}$$

2.2 Inadmissible Times

Corollary 4 implies that if T is an admissible time such that $P_\alpha(G_T < T) > 0$, then G_T is not admissible. Indeed

$$\frac{S_{G_T}}{G_T^\alpha} = \frac{S_T}{G_T^\alpha} \geq \frac{S_T}{T^\alpha}$$

and the inequality is strict on the event $(G_T < T)$. So S_{G_T}/G_T^α cannot have the same distribution as S_T/T^α if $P_\alpha(G_T < T) > 0$. Similar remarks apply to D_T. For a constant time t, the sequence $\left(\frac{V_1(G_t)}{G_t}, \frac{V_2(G_t)}{G_t}, \ldots\right)$ is independent of G_t with the same distribution as the sequence of ranked lengths of excursion intervals of the corresponding bridge of length 1. This follows from the fact (easily verified using the invariance of Bessel processes under time inversion [22]) that if $(R_t, t \geq 0)$ is a Bessel process of dimension $2 - 2\alpha$ starting at 0, then $(G_t^{-1/2} R_{uG_t}, 0 \leq u \leq 1)$ is a standard Bessel bridge of the same dimension independent of G_t. From Theorem 5.3 of [15], there is the following density formula relative to the distribution of $(V_1, V_2, \ldots)$: for all non-negative product measurable f

$$E_\alpha\left[f\left(\frac{V_1(G_t)}{G_t}, \frac{V_2(G_t)}{G_t}, \ldots\right)\right] = \frac{E_\alpha\left[S_1 f(V_1, V_2, \ldots)\right]}{E_\alpha(S_1)} \tag{15}$$

Let N_t be the rank of the meander length $t - G_t$ in the sequence of excursion lengths $V_1(t) > V_2(t) > \cdots$, so $t - G_t = V_{N_t}(t)$. Formula (9) amounts to the formula

$$E_\alpha\left[1(N_t = n) f\left(\frac{V_1(t)}{t}, \frac{V_2(t)}{t}, \ldots\right)\right] = E_\alpha[V_n f(V_1, V_2, \ldots)] \tag{16}$$

for all $n = 1, 2, \ldots$ and all non-negative product measurable functions f. Consider now N_{D_t}, the rank of the excursion length $D_t - G_t$ straddling t in the sequence of complete excursion lengths $V_1(D_t) > V_2(D_t) > \cdots$. So $N_t - 1$ is the number of excursions completed by time t whose lengths exceed $t - G_t$, and $N_{D_t} - 1$ is the smaller number of such excursions whose lengths exceed $D_t - G_t$.

Proposition 5 *For each $t > 0$ and $n = 1, 2, \ldots$,*

$$E_\alpha\left[1(N_{D_t} = n) f\left(\frac{V_1(D_t)}{D_t}, \frac{V_2(D_t)}{D_t}, \ldots\right)\right] = E_\alpha[-\alpha \log(1 - V_n) f(V_1, V_2, \ldots)] \tag{17}$$

Immediately from Proposition 5, we draw the following consequences. First, summing over n gives

$$E_\alpha\left[f\left(\frac{V_1(D_t)}{D_t}, \frac{V_2(D_t)}{D_t}, \ldots\right)\right] = E_\alpha\left[\left(-\sum_n \alpha \log(1 - V_n)\right) f(V_1, V_2, \ldots)\right] \tag{18}$$

which is the analog of (15) for D_t instead of G_t. Next, an analog of (9) for D_t instead of t can be read from (17) as follows: for each $n = 1, 2, \ldots$

$$P_\alpha\left(D_t - G_t = V_n(D_t) \,\middle|\, \frac{V_m(D_t)}{D_t} = u_m, m = 1, 2, \ldots\right) = \frac{\log(1 - u_n)}{\sum_m \log(1 - u_m)} \tag{19}$$

Note the remarkable fact that, just as in (9), the conditional distribution does not depend on α.

Finally, by taking $f = 1$ in (17), we obtain the formula

$$P_\alpha(N_{D_t} = n) = E_\alpha[-\alpha \log(1 - V_n)] \tag{20}$$

As noted in [16], combined with (14) and (16) this allows the asymptotic evaluations as $n \to \infty$:

$$P_\alpha(N_{D_t} = n) \sim \alpha P_\alpha(N_t = n) \sim \frac{\alpha \Gamma(\frac{1}{\alpha} + 1)}{\Gamma(1 - \alpha)^{1/\alpha}} \frac{1}{n^{1/\alpha}} \tag{21}$$

where $a(n) \sim b(n)$ means $a(n)/b(n) \to 1$ as $n \to \infty$. See [19, 16] for integral expressions for the distributions of N_t and N_{D_t}, and some numerical values.

In (15) and (17) we have described the law of $(V_1(T)/T, V_2(T)/T, \ldots)$ for $T = G_t$ and for $T = D_t$ by a change of measure relative to the law of this random vector for a fixed time T. By similar arguments we obtain change of measure formulae for $T = G_{H_n}$ and $T = D_{H_n}$. We now give these descriptions for $n = 1$.

Proposition 6 *For each non-negative product measurable function f,*

$$E_\alpha\left[f\left(\frac{V_1(G_{H_1})}{G_{H_1}}, \frac{V_2(G_{H_1})}{G_{H_1}}, \ldots\right)\right] = E_\alpha\left[\left(\frac{S_1}{V_1^\alpha}\right) f(V_1, V_2, \ldots)\right] \tag{22}$$

$$E_\alpha\left[f\left(\frac{V_1(D_{H_1})}{D_{H_1}}, \frac{V_2(D_{H_1})}{D_{H_1}}, \ldots\right)\right] = E_\alpha\left[\left(\alpha \log\frac{V_1}{V_2}\right) f(V_1, V_2, \ldots)\right] \tag{23}$$

As checks, we recall from [16, Props. 10 and 8] that under P_α the distribution of S_1/V_1^α is standard exponential, whereas the distribution of V_2/V_1 is beta$(\alpha, 1)$. Therefore, both S_1/V_1^α and $\alpha \log(V_1/V_2)$ are random variables whose P_α expectation equals 1, as implied by (22) and (23) for $f = 1$.

3 Proofs

3.1 Admissible times

The foundation for the proof of Proposition 3 is a scaling argument which may prove useful in other contexts. The following theorem presents the conclusion of this argument in a fairly general setting.

Recall that a real or vector-valued process $(X_t, t > 0)$ is called *β-self-similar* for some $\beta \in \mathbb{R}$ if for every $c > 0$

$$(X_{ct}, t > 0) \stackrel{d}{=} (c^\beta X_t, t > 0) \tag{24}$$

See [20] for a survey of the literature of these processes. Note that (X_t) is β-self-similar iff the process (Y_t) defined by $Y_t = t^{-\beta} X_t$ is 0-self-similar, that is to say, for every $c > 0$

$$(Y_{ct}, t > 0) \stackrel{d}{=} (Y_t, t > 0) \tag{25}$$

This definition of 0-self-similarity makes sense even for Y with values in an abstract measurable space where there is no notion of scalar multiplication. Suppose now that X is viewed as a measurable map from the basic probability space to a suitable path space $(S, \mathcal{S})$, e.g. $S = C[0, \infty)$ and $\mathcal{S}$ the σ-field generated by coordinate maps, assuming X has continuous paths. Suppose (X_t) is β-self-similar. Let $(\mathbf{X}_t, t > 0)$ denote the path valued process defined by letting $\mathbf{X}_t$ be the rescaling of $\mathbf{X}$ that maps time t to time 1, that is

$$\mathbf{X}_t(s) = t^{-\beta} X_{st} \qquad (s \geq 0) \tag{26}$$

Then it is easily verified that $(\mathbf{X}_t, t > 0)$ is 0-self-similar.

It is this kind of 0-self-similar process which we have in mind for applications of the following theorem.

Theorem 7 *Let* $(\mathbf{X}_t, t > 0)$ *be a jointly measurable 0-self-similar process with values in an arbitrary measurable space* $(S, \mathcal{S})$. *Let* $\theta_s = \Theta(\mathbf{X}_s)$ *for a non-negative* $\mathcal{S}$*-measurable function* Θ *defined on* S, *let*

$$A_t = \int_0^t \theta_s ds, \qquad (t \geq 0) \tag{27}$$

$$\alpha_u = \inf\{t : A_t > u\} \qquad (u \geq 0) \tag{28}$$

Suppose that $0 < A_1 < \infty$ *a.s. Then* $0 < \alpha_u < \infty$ *a.s. for every* $u > 0$, *and for all non-negative product measurable* ψ *defined on* $S \times [0, \infty)$

$$E\left[\psi(\mathbf{X}_{\alpha_1}, 1/\alpha_1)\right] = E\left[\psi(\mathbf{X}_1, A_1)\frac{\theta_1}{A_1}\right] \tag{29}$$

Remarks. According to (29), the law of $(\mathbf{X}_{\alpha_1}, 1/\alpha_1)$ on the product space $S \times [0, \infty)$ is absolutely continuous with respect to that of $(\mathbf{X}_1, A_1)$, with Radon-Nikodym density g defined by

$$g(\mathbf{X}_1, A_1) = \frac{E\left[\theta_1 | \mathbf{X}_1, A_1\right]}{A_1}$$

It follows that for an arbitrary product measurable map Ψ whose range can be any measurable space,

$$\Psi(\mathbf{X}_{\alpha_1}, 1/\alpha_1) \stackrel{d}{=} \Psi(\mathbf{X}_1, A_1) \text{ iff } E\left[\frac{\theta_1}{A_1}\middle| \Psi(\mathbf{X}_1, A_1)\right] = 1 \tag{30}$$

For $\Psi(\mathbf{x}, a)$ such that a can be recovered as a measurable function of $\Psi(\mathbf{x}, a)$, condition (30) reduces to

$$E\left[\theta_1 | \Psi(\mathbf{X}_1, A_1)\right] = A_1 \tag{31}$$

In particular, since it follows immediately from the 0-self-similarity of the process (θ_s) that

$$1/\alpha_1 \stackrel{d}{=} A_1 \tag{32}$$

we learn from (30) that

$$E[\theta_1 | A_1] = A_1 \tag{33}$$

Taking $\mathbf{X}_t = \theta_t$ shows that the identity (33) holds for an arbitrary non-negative 0-self-similar process (θ_t) and $A_1 = \int_0^1 \theta_s ds$. See [18, 17] for further developments and applications of this identity. Formula (29) is an abstract version of a result of Yor [26] in the case that $(\mathbf{X}_t)$ is the path-valued process derived by the scaling transformation (26) starting from a Brownian motion (X_t). A consequence of (29) is the following variation of the result of [26] for Brownian motion.

Corollary 8 *Let* $(X_t, t \geq 0)$ *a* β*-self-similar process and let* $(\theta_t, t \geq 0)$ *be such that for each* $c > 0$

$$(X_{ct}, \theta_{ct}; t \geq 0) \stackrel{d}{=} (c^\beta X_t, \theta_t; t \geq 0) \tag{34}$$

Then, with A_1 *and* α_1 *defined as in* (27) *and* (28), *for all non-negative measurable functions* F *defined on the path space*

$$E\left[F\left(\frac{X_{t\alpha_1}}{\alpha_1^\beta}; t \geq 0\right)\right] = E\left[\frac{\theta_1}{A_1} F(X_t; t \geq 0)\right] \tag{35}$$

Proof of Theorem 7. The following proof of (29) is a simple adaptation of the argument in [26]. Since the bivariate process $((\mathbf{X}_t, \frac{A_t}{t}), t \geq 0)$ is also 0-self-similar, it suffices to prove (29) for ψ of the form $\psi(\mathbf{x}, a) = \phi(\mathbf{x})$ for an arbitrary non-negative $\mathcal{S}$-measurable function ϕ. For h a non-negative Borel function with $\int_0^\infty s^{-1}h(s)ds < \infty$, consider the quantity

$$Q = \int_0^\infty ds\, h(s) E\left[\frac{\theta_s}{A_s}\phi(\mathbf{X}_s)\right] \tag{36}$$

On the one hand, the assumption that $(\mathbf{X}_s)$ is 0-self-similar and the definitions of θ_s and A_s imply that $((\theta_s, A_s/s, \mathbf{X}_s), s > 0)$ is 0-self-similar. So $(\theta_s, A_s, \mathbf{X}_s) \stackrel{d}{=} (\theta_1, sA_1, \mathbf{X}_1)$ and we can compute

$$Q = \left(\int_0^\infty \frac{ds}{s} h(s)\right) E\left[\frac{\theta_1}{A_1}\phi(\mathbf{X}_1)\right] \tag{37}$$

On the other hand, using Fubini, a time change, and using scaling again to see that $(\alpha_t, \mathbf{X}_{\alpha_t}) \stackrel{d}{=} (t\alpha_1, \mathbf{X}_{\alpha_1})$, we can compute

$$\begin{aligned} Q &= E\left[\int_0^\infty \frac{dt}{t} h(\alpha_t)\phi(\mathbf{X}_{\alpha_t})\right] \\ &= E\left[\int_0^\infty \frac{dt}{t} h(t\alpha_1)\phi(\mathbf{X}_{\alpha_1})\right] \\ &= \left(\int_0^\infty \frac{ds}{s} h(s)\right) E\left[\phi(\mathbf{X}_{\alpha_1})\right] \end{aligned} \tag{38}$$

Comparison of (38) with (37) yields (29) for $\psi(\mathbf{x}, a) = \phi(\mathbf{x})$, as was to be proved. □

Proposition 9 *Suppose that Z is the closure of the random set of zeros of a β-self-similar process $(X_t, t \geq 0)$, and assume that the Lebesgue measure of Z is 0 almost surely. Let $V_1(t) \geq V_2(t) \geq \cdots$ be the ranked lengths of the component intervals of $[0, t]\backslash Z$, and put $V_n = V_n(1)$. Let $\mathbf{X}_t$ be the 0-self-similar path valued process defined as in (26) by $\mathbf{X}_t(s) = t^{-\beta}X_{st}, s \geq 0$, let $\theta_s = \Theta(\mathbf{X}_s)$ for a non-negative $\mathcal{S}$-measurable function Θ, and for $t \geq 0$ and $u \geq 0$, let $A_t = \int_0^t \theta_s ds$, assume that $0 < A_1 < \infty$ almost surely, and let $\alpha_u = \inf\{t : A_t > u\}$. Then*

$$E\left[F\left(\frac{V_n(\alpha_1)}{\alpha_1}, n \geq 1\right)\right] = E\left[F(V_n, n \geq 1)\frac{\theta_1}{A_1}\right] \tag{39}$$

for all non-negative product measurable functions F. Consequently, α_1 is admissible, meaning

$$\left(\frac{V_1(\alpha_1)}{\alpha_1}, \frac{V_2(\alpha_1)}{\alpha_1}, \ldots\right) \stackrel{d}{=} (V_1, V_2, \ldots) \tag{40}$$

if and only if

$$E\left[\frac{\theta_1}{A_1}\,\middle|\, V_1, V_2, \ldots\right] = 1. \tag{41}$$

Proof. Since for each n, and every $t > 0$, $V_n(t)/t = f_n(\mathbf{X}_t)$ for a measurable function f_n which does not depend on t, formula (39) follows immediately from the previous theorem. □

Note that in case A_1 is a measurable function of $(V_n, n \geq 1)$, the condition (41) becomes

$$E[\theta_1 | V_1, V_2, \ldots] = A_1. \tag{42}$$

Corollary 10 *Let A_t be the time spent positive by a standard Brownian motion B up to time t, so α_1 is the first instant that B has spent time 1 positive. Then α_1 is admissible for the zero set of B.*

Proof. We show that (41) holds. Clearly, it suffices to show that

$$E\left[\theta_1 \middle| A_1, V_1, V_2, \ldots\right] = A_1 \tag{43}$$

where $\theta_1 = 1(B_1 > 0)$. Let ε_n be the indicator of the event that B is positive on the interval whose length is V_n. Since the V_n are a.s. all distinct, there are a.s. no quibbles about the definition of the ε_n. By Itô's excursion theory, the ε_n are independent Bernoulli($\frac{1}{2}$) variables, independent of $(G_1, V_1, V_2, V_3, \ldots)$, and by definition

$$\theta_1 = \sum_n \varepsilon_n 1(1 - G_1 = V_n) \text{ and } A_1 = \sum_n \varepsilon_n V_n$$

so we have, by the sampling property (9),

$$E(\theta_1 | \varepsilon_1, \varepsilon_2, \ldots, V_1, V_2, \ldots) = \sum_n \varepsilon_n P(1 - G_1 = V_n | V_1, V_2, \ldots)$$

$$= \sum_n \varepsilon_n V_n = A_1$$

and (43) follows. □

Remark 11 It is clear from the above proof that the conclusion of Corollary 10 holds just as well for B a skew Brownian motion or a skew Bessel process, as discussed in Section 4.

Remark 12 As a companion to (43) we note that the sampling property (9) and [25, Exercise 3.4] imply that if $V_1, V_2, \ldots$ are the ranked interval lengths generated by the zero set of a Bessel process $(R_t, 0 \leq t \leq 1)$ of dimension $2 - 2\alpha$ started at $R_0 = 0$ then for $x > 0$

$$P(R_1 \in dx \,|\, V_1, V_2, \ldots) = x\, dx \sum_{n=1}^{\infty} \exp\left(-\frac{x^2}{2V_n}\right)$$

Corollary 13 *In the setting of Proposition 9, the random time $H_n := \inf\{t : V_n(t) \geq 1\}$ is admissible for Z iff*

$$P(1 - G_1 = V_n | V_1, V_2, \ldots) = V_n \tag{44}$$

Proof. Observe that for each n the process

$$\theta_s := 1(s - G_s = V_n(s)) \tag{45}$$

is of the form $\theta_s = \Theta(\mathbf{X}_s)$ required in Theorem 7 and Proposition 9. Moreover, as observed in [17], the corresponding A_t is just

$$V_n(t) = \int_0^t ds\, 1(s - G_s = V_n(s)) \tag{46}$$

so the corresponding α_1 equals H_n as defined in (7). □

In particular, H_n is admissible for every n iff (44) holds for every n. We then say that Z has the *sampling property*. For Z the range of a stable(α) subordinator, the sampling property of Z was established in [15] while the admissibility of H_n for all n was shown in [16]. Neither of these results seems obvious without some calculation. In [17] we give examples of various 0-self-similar sets Z, some with and some without the sampling property. It would be interesting to characterize all 0-self-similar sets Z with the sampling property, but we have no idea how to do this.

Proof of Proposition 3. Note first that if (T_n) is a sequence of admissible times, and T_n converges in probability as $n \to \infty$ to T with $T > 0$ a.s., then T is admissible. By this observation and Proposition 2, it suffices to prove Proposition 3 for

$$A_t = \sum_{k=1}^{p} c_k V_k(t)$$

In this case we have from (46)

$$\theta_t = \sum_{k=1}^{p} c_k 1(t - G_t = V_k(t))$$

so the sampling property and linearity of conditional expectations imply (42). □

The class of admissible times is preserved under certain homogeneous transformations described in the following proposition.

Proposition 14 *In the setting of Proposition 9, with Z the closure of the random set of zeros of a β-self-similar process $(X_t, t \geq 0)$, the Lebesgue measure of Z equal to 0 almost surely, and $\mathbf{X}_t$ the 0-self-similar path valued process defined by $\mathbf{X}_t(s) = t^{-\beta} X_{st}, s \geq 0$, suppose for each $1 \leq j \leq k$ that $\theta_s^{(j)} = \Theta^{(j)}(\mathbf{X}_s)$ for a non-negative $\mathcal{S}$-measurable function $\Theta^{(j)}$, and for $t \geq 0$ and $u \geq 0$ let $A_t^{(j)} = \int_0^t \theta_s^{(j)} ds$ be such that $0 < A_1^{(j)} < \infty$ almost surely, and define $\alpha_u^{(j)} = \inf\{t : A_t^{(j)} > u\}$. Suppose further for each $1 \leq j \leq k$ that $A_1^{(j)}$ is $\mathcal{V}$-measurable, where $\mathcal{V}$ is the σ-field generated by $V_1, V_2, \ldots$, and that $\alpha_1^{(j)}$ is admissible for Z. Let $f : \mathbb{R}_+^k \to \mathbb{R}_+$ be an increasing function in each variable such that*

$$f(cx_1, cx_2, \ldots, cx_k) = cf(x_1, x_2, \ldots, x_k) \tag{47}$$

and f is differentiable on $(0,\infty)^k$, and let $A_t := f(A_t^{(1)}, \ldots, A_t^{(k)})$. Then $\alpha_1 := \inf\{t : A_t > 1\}$ is admissible.

Proof. By calculus $A_t = \int_0^t \theta_s ds$ where

$$\theta_s = \sum_{i=1}^{k} f_i'(A_s^{(1)}, \ldots, A_s^{(k)}) \theta_s^{(i)}$$

Thus we can compute

$$E[\theta_1|\mathcal{V}] = \sum_{i=1}^{k} f_i'(A_1^{(1)},\ldots,A_1^{(k)})E[\theta_1^{(i)}\,|\,\mathcal{V}]$$
$$= \sum_{i=1}^{k} f_i'(A_1^{(1)},\ldots,A_1^{(k)})A_1^{(i)}$$

by (42). But, from the hypotheses on f we deduce that $\sum_{i=1}^{k} f_i'(x_1,\ldots,x_k)x_i = f(x_1,\ldots,x_k)$ so we obtain $E[\theta_1\,|\,\mathcal{V}] = A_1$, as in (42). Therefore, α_1 is admissible. □

Note that the class of functions f considered above is much larger than the class of functions of the form $f(x) = \sum_{i=1}^{k} c_i x_i$. For instance, one can take

$$f_p(x_1,\ldots,x_k) = \left(\sum_{i=1}^{k}(c_i x_i)^p\right)^{1/p}$$

for $p > 0$ and positive constants c_i. By passage to the limit, it can be deduced that the conclusion of Proposition 14 also holds for

$$f(x_1,\ldots,x_k) = \max_{1\le i\le k} x_i$$

3.2 The lengths at time D_t

Proof of Proposition 5. Let $\mathbf{V}(T) = (V_1(T), V_2(T),\ldots)$ denote the sequence of ranked lengths of component intervals of $[0,T]\backslash Z$ for Z the closed range of a stable subordinator (τ_s). By scaling, the distribution of $\mathbf{V}(D_t)/D_t$ for fixed $t > 0$ does not depend on t. So let us write simply D for D_1 and G for G_1, and compute the law of $\mathbf{V}(D)/D$. Recall that the sequence $\mathbf{V}(1)$ contains the term $1-G$ as $1-G = V_N(1)$ for a random index N. The sequence $\mathbf{V}(D)$ is derived from $\mathbf{V}(1)$ by first substituting $D-G$ for this term, then reranking. Let (S_t) be the local time inverse of (τ_s). Let $S = S_1$. So $S^{-1/\alpha} \stackrel{d}{=} \tau_1$. Consider the three point processes N_1, N_G, and N_D on $(0,\infty)$ defined as follows for $T = 1$, $T = G$ or $T = D$:

$$N_T(\cdot) = \sum_n 1(S^{-1/\alpha}V_n(T) \in \cdot)$$

Let $X := S^{-1/\alpha}(1-G)$ and $Y := S^{-1/\alpha}(D-G)$. Then

$$N_G = N_1 - \delta_X = N_D - \delta_Y$$

where $\delta_W(\cdot) = 1(W \in \cdot)$. According to Theorems 2.1 and 1.2 of [15], P_α governs N_1 as a Poisson random measure with intensity measure Λ_α on $(0,\infty)$ where Λ_α is the stable(α) Lévy measure, and given N_1 the point X is a size-biased pick from the points of N_1. That is to say

$$P_\alpha(N_G \in dn, X \in dx) = \frac{x}{\Sigma n + x} P_\alpha(N_1 \in dn)\Lambda_\alpha(dx) \tag{48}$$

where for a counting measure n on $(0,\infty)$, $\Sigma n = \int_0^\infty x n(dx)$ is the sum of locations of the points of n. Let

$$R := \frac{Y}{X} = \frac{D-G}{1-G}$$

From asymptotic renewal theory [3], or by the last exit decomposition at time G, there is the formula

$$P_\alpha(G \in dx, D \in dy) = \frac{\alpha}{\Gamma(\alpha)\Gamma(1-\alpha)} \frac{x^{\alpha-1}}{(y-x)^{\alpha+1}}\, dx\, dy \qquad (0 < x < 1 < y < \infty) \tag{49}$$

which implies that G and R are independent, with

$$P_\alpha(R \in dr) = \frac{\alpha}{r^{\alpha+1}}\, dr \qquad (r > 1) \tag{50}$$

The last exit decomposition at time G and scaling imply further that G, N_G and R are mutually independent. Since S is a measurable function of G and N_G, so is X, and we can compute for $y > x$

$$P_\alpha(Y \in dy | N_G, X = x) = P_\alpha(XR \in dy | N_G, X = x) = P_\alpha(xR \in dy)$$

$$= P_\alpha(R \in \frac{dy}{x}) = \alpha \left(\frac{x}{y}\right)^{\alpha+1} \frac{dy}{x}$$

and hence

$$\begin{aligned}
P_\alpha(N_G \in dn, Y \in dy) &= \int_0^y P_\alpha(N_G \in dn, X \in dx, Y \in dy) \\
&= \left(\int_0^y \frac{x}{\Sigma n + x} \Lambda_\alpha(dx) P_\alpha(Y \in dy | N_G, X = x)\right) P_\alpha(N_1 \in dn) \\
&= \left(\int_0^y \frac{x}{\Sigma n + x} \frac{C\alpha\, dx}{x^{\alpha+1}} \alpha \left(\frac{x}{y}\right)^{\alpha+1} \frac{1}{x}\right) P_\alpha(N_1 \in dn)\, dy \\
&= \alpha \left(\int_0^y \frac{dx}{\Sigma n + x}\right) P_\alpha(N_1 \in dn) \Lambda_\alpha(dy) \\
&= \alpha \log\left(\frac{\Sigma n + y}{\Sigma n}\right) P_\alpha(N_1 \in dn) \Lambda_\alpha(dy)
\end{aligned}$$

That is to say

$$P_\alpha(N_G \in dn, Y \in dy) = \rho(y | n + \delta_y) P_\alpha(N_1 \in dn) \Lambda_\alpha(dy) \tag{51}$$

where for a counting measure m

$$\rho(y|m) = \alpha \log\left(\frac{\Sigma m}{\Sigma m - y}\right)$$

Since $N_D = N_G + \delta_Y$ and N_1 is a Poisson measure with intensity Λ_α, the Palm formula of [15, Lemma 2.2] shows that (51) can be recast as

$$P_\alpha(N_D \in dm, Y \in dy) = \rho(y|m) P_\alpha(N_1 \in dm) \Lambda_\alpha(dy) \tag{52}$$

which implies that

$$P_\alpha(N_D \in dm) = \rho(m) P_\alpha(N_1 \in dm) \tag{53}$$

where

$$\rho(m) = \int \rho(y|m) m(dy) = \alpha \sum_{y:m\{y\}=1} \log\left(\frac{\Sigma m}{\Sigma m - y}\right).$$

Now

$$\frac{\mathbf{V}(T)}{T} = \frac{S^{-1/\alpha}\mathbf{V}(T)}{S^{-1/\alpha}T}$$

Since for $T = 1$ and $T = D$, both $S^{-1/\alpha}\mathbf{V}(T)$ and $S^{-1/\alpha}T = \sum_n S^{-1/\alpha}V_n(T)$ are measurable functions of N_T, so is $\mathbf{V}(T)/T$. Since also

$$\rho(N_T) = \alpha \sum_i \log\left(\frac{T}{T - V_i(T)}\right) = -\alpha \sum_i \log\left(1 - \frac{V_i(T)}{T}\right) \tag{54}$$

is a function of $\mathbf{V}(T)/T$, a change of variables in (53) yields (18). A similar manipulation of (52) yields (17). □

As noted in [16], formula (9) implies that for every non-negative measurable function f defined on $[0,1]$,

$$E_\alpha\left[\sum_n f(V_n)\right] = E_\alpha\left[\frac{f(1-G_1)}{(1-G_1)}\right] = \frac{1}{\Gamma(\alpha)\Gamma(1-\alpha)} \int_0^1 du f(u) u^{-\alpha-1}(1-u)^{\alpha-1} \tag{55}$$

where the last expression is obtained from the beta$(\alpha, 1-\alpha)$ density of G_1. The consequence of (18), that

$$E_\alpha\left(-\alpha \sum_n \log(1 - V_n)\right) = 1$$

therefore amounts to the formula

$$\frac{\alpha}{\Gamma(\alpha)\Gamma(1-\alpha)} \int_0^1 du(-\log(1-u)) u^{-\alpha-1}(1-u)^{\alpha-1} = 1 \tag{56}$$

This identity can be checked directly as follows. Expanding

$$-\log(1-u) = u + \frac{u^2}{2} + \frac{u^3}{3} + \cdots$$

allows the left side of (56) to be evaluated as

$$\frac{\alpha}{\Gamma(\alpha)\Gamma(1-\alpha)}\left(B(1-\alpha,\alpha) + \frac{1}{2}B(2-\alpha,\alpha) + \frac{1}{3}B(3-\alpha,\alpha) + \cdots\right)$$

where $B(a,b) = \Gamma(a)\Gamma(b)/\Gamma(a+b)$ is the beta function, so (56) reduces to

$$\alpha\left(1 + \frac{1-\alpha}{2!} + \frac{(1-\alpha)(2-\alpha)}{3!} + \cdots\right) = 1$$

which can be seen by letting $x \uparrow 1$ in the formula

$$1 - (1-x)^\alpha = \alpha x + \alpha(1-\alpha)\frac{x^2}{2!} + \alpha(1-\alpha)(2-\alpha)\frac{x^3}{3!} + \cdots \tag{57}$$

obtained from the binomial expansion of $(1-x)^\alpha$. See [14] for an interpretation in terms of a stable(α) subordinator of the discrete distribution with the generating function (57).

A number of variations of the identity (56) can be obtained as follows. Since G_1 has beta$(\alpha, 1-\alpha)$ distribution, if T is an independent exponential variable, then TG_1 has gamma(α) distribution. Therefore, for $\lambda > -1$,

$$E_\alpha\left[\frac{1}{1+\lambda G_1}\right] = \int_0^\infty dt\, e^{-t} E_\alpha(e^{-t\lambda G_1}) = E_\alpha[\exp(-\lambda T G_1)] = (1+\lambda)^{-\alpha} \tag{58}$$

Take $\lambda = (1-x)/x$ in (58) to obtain

$$E_\alpha\left[(x+(1-x)G_1)^{-1}\right] = x^{\alpha-1} \quad (0<\alpha<1, x>0). \tag{59}$$

Integration of (59) with respect to dx over $0 < x < a$ yields the formula

$$E_\alpha\left[\frac{1}{1-G_1}\log\left(1+\frac{a(1-G_1)}{G_1}\right)\right] = \frac{a^\alpha}{\alpha} \tag{60}$$

which reduces to (56) for $a = 1$. For later reference, we note also the following elementary formula. For an arbitrary non-negative Borel f:

$$E_\alpha\left[\frac{1}{1-G_1} f\left(\frac{1-G_1}{G_1}\right)\right] = \frac{1}{\Gamma(\alpha)\Gamma(1-\alpha)} \int_0^\infty \frac{dv}{v^{\alpha+1}} f(v) \tag{61}$$

3.3 The lengths at times G_{H_1} and D_{H_1}

In this section, we prove Proposition 6. We can assume that Z is the zero set of $\rho := (\rho(u), u \geq 0)$ where under P_α the process ρ is a Bessel process of dimension $2-2\alpha$ started at $\rho(0) = 0$. Let π denote the Bessel bridge of dimension $2-2\alpha$ defined by $\pi_u := \rho(uG_1)/\sqrt{G_1}, 0 \leq u \leq 1$ and let $\tilde\rho$ be the process defined by $\tilde\rho_u := \rho(uG_{H_1})/\sqrt{G_{H_1}}, 0 \leq u \leq 1$.

Proof of (22). This formula is a consequence of (15) and the following absolute continuity relationship between the laws of π and $\tilde\rho$ on $C[0,1]$: for every measurable function $F : C[0,1] \to \mathbb{R}^+$

$$E_\alpha[F(\tilde\rho)] = \gamma_\alpha E_\alpha[(V_1(\pi))^{-\alpha} F(\pi)] \tag{62}$$

where $V_1(\pi)$ denotes the longest excursion interval of the bridge π and

$$\gamma_\alpha := 1/E_\alpha[(V_1(\pi))^{-\alpha}] = E_\alpha[(1-G_1)^\alpha] = \frac{1}{\alpha\Gamma(\alpha)\Gamma(1-\alpha)} = \frac{\sin(\pi\alpha)}{\pi\alpha} \tag{63}$$

Formula (62) is a consequence of the following identity, which we obtain from Corollary 8 with the help of (46):

$$E_\alpha\left[F\left(\frac{\rho(uH_1)}{\sqrt{H_1}}; 0 \leq u \leq 1\right)\right] = E_\alpha\left[\frac{1(1-G_1 = V_1)}{1-G_1} F(\rho(u); 0 \leq u \leq 1)\right] \tag{64}$$

To obtain (62) from (64), observe that G_{H_1}/H_1 is the last zero before time 1 of $(\rho(uH_1)/\sqrt{H_1}; 0 \leq u \leq 1)$, and consequently

$$E_\alpha[F(\tilde\rho)] = E_\alpha\left[\frac{1(1-G_1 = V_1)}{1-G_1} F(\pi)\right] \tag{65}$$

Formula (62) now appears as a consequence of

$$E_\alpha\left[\frac{1(1-G_1=V_1)}{1-G_1}\,\middle|\,\pi\right]=\frac{\gamma_\alpha}{(V_1(\pi))^\alpha} \tag{66}$$

To check (66), evaluate the left side of (66) as

$$E_\alpha\left[\frac{1\{(1-G_1)/G_1>V_1(\pi)\}}{1-G_1}\,\middle|\,\pi\right]=h_\alpha(V_1(\pi))$$

where

$$h_\alpha(v):=E_\alpha\left[\frac{1}{1-G_1}1\left(\frac{1-G_1}{G_1}>v\right)\right]=(\alpha\Gamma(\alpha)\Gamma(1-\alpha)v^\alpha)^{-1},$$

the last equality being a consequence of (61). □

Proof of (23). For $t>0$ and $n=1,2,\ldots$ let $R_n(t):=V_{n+1}(t)/V_n(t)$. Since H_1 is admissible,

$$(R_1(H_1),R_2(H_1),\ldots)\stackrel{d}{=}(R_1(1),R_2(1),\ldots). \tag{67}$$

According to Proposition 8 of [16], the $R_n(1)$ are independent, and $R_n(1)$ has a beta$(n\alpha,1)$ distribution. Now

$$R_1(D_{H_1})=\frac{V_2(H_1)}{D_{H_1}-G_{H_1}}=R_1(H_1)(D_{H_1}-G_{H_1})^{-1} \tag{68}$$

and $R_m(D_{H_1})=R_m(H_1)$ for $m\geq 2$. Since $D_{H_1}-G_{H_1}$ is independent of the sequence $(V_1(H_1),V_2(H_1),\ldots)$, for a generic non-negative product measurable f, we obtain

$$E_\alpha[f(V_1(D_{H_1}),V_2(D_{H_1}),\ldots)]=E_\alpha[\xi_\alpha(R_1(H_1))\,f(V_1(H_1),V_2(H_1),\ldots)] \tag{69}$$

and hence from (67)

$$E_\alpha\left[f\left(\frac{V_1(D_{H_1})}{D_{H_1}},\frac{V_2(D_{H_1})}{D_{H_1}},\ldots\right)\right]=E_\alpha[\xi_\alpha(V_2/V_1)\,f(V_1,V_2,\ldots)] \tag{70}$$

where

$$\xi_\alpha(x):=\frac{P_\alpha(R_1(D_{H_1})\in dx)}{P_\alpha(R_1(1)\in dx)}=-\alpha\log x \tag{71}$$

The last equality follows by elementary computation from the fact that under P_α the distribution of $R_1(1)$ is beta$(\alpha,1)$ while $P_\alpha(D_{H_1}-G_{H_1}>t)=t^{-\alpha}$ for $t>1$. □

To conclude this section we note that there are analogs of the above formulae for H_n instead of H_1. For example, formula (22) is modified by replacing $S_1V_1^{-\alpha}$ by $S_1(V_n^{-\alpha}-V_{n-1}^{-\alpha})$, which is also exponentially distributed [17, Prop. 10], and formula (62) is modified by replacing $V_1^{-\alpha}$ by $V_n^{-\alpha}-V_{n-1}^{-\alpha}$.

4 Generalized arc-sine laws.

In this section, we assume that $0<\alpha<1, 0<p<1$, and let $P_{\alpha,p}$ govern a real-valued process $(B_t,t\geq 0)$ with continuous paths, such that

(i) the zero set Z of B is the range of a stable (α) subordinator, and

(ii) given $|B|$, the signs of excursions of B away from zero are chosen independently of each other to be positive with probability p and negative with probability $q:=1-p$.

For example, B could be any of the following:

- an ordinary Brownian motion ($\alpha = p = \frac{1}{2}$) [11]
- a *skew Brownian motion* ($\alpha = \frac{1}{2}, 0 < p < 1$) [21, 6, 2, 1]
- a *symmetrized Bessel process* of dimension $2 - 2\alpha$ [10]
- a *skew Bessel process of dimension* $2 - 2\alpha$ [2, 23]

For $t > 0$ let

$$A_t := \int_0^t 1(B_s > 0)\, ds \tag{72}$$

denote the time spent positive by B up to time t. See the papers cited above for background and motivation for the study of this process. For any random time T which is a measurable function of $|B|$,

$$A_T = \int_0^T 1(B_s > 0)\, ds = \sum_n \varepsilon_n(T) V_n(T) \tag{73}$$

where under $P_{\alpha,p}$ the $\varepsilon_n(T)$ are independent indicators of events with probability p, independent of the sequence of ranked lengths $(V_n(T), n = 1, 2, \ldots)$ of component intervals of $[0, T] \backslash Z$. Consequently, the $P_{\alpha,p}$ distribution of A_T/T is the same for such T that are admissible for the zero set of B, and this common distribution is the $P_{\alpha,p}$ distribution of $A := A_1$. This is Lamperti's [9] generalized arc-sine distribution on $[0, 1]$, determined by its Stieltjes transform

$$E_{\alpha,p}\left[\frac{1}{\lambda + A}\right] = \frac{p(1+\lambda)^{\alpha-1} + q\lambda^{\alpha-1}}{p(1+\lambda)^{\alpha} + q\lambda^{\alpha}} \qquad (\lambda > 0) \tag{74}$$

Let $P^{\text{br}}_{\alpha,p}$ denote the standard bridge law obtained by conditioning $P_{\alpha,p}$ on $(1 \in Z)$. If $P_{\alpha,p}$ governs B as a skew Bessel process, $P^{\text{br}}_{\alpha,p}$ governs B as a skew Bessel bridge of length 1. According to formula (4.b') of [2],

$$E^{\text{br}}_{\alpha,p}\left[\frac{1}{(1+\lambda A)^{\alpha}}\right] = \frac{1}{p(1+\lambda)^{\alpha} + q} \qquad (\lambda > 0) \tag{75}$$

Lamperti [9] inverted the Stieltjes transform (74) to obtain the corresponding density on $[0, 1]$, which is reproduced in [15] and [23]. We do not know how to invert (75) to obtain such an explicit formula in the bridge case for general α with $0 < \alpha < 1$, but it is a famous result of Lévy [11] that for the standard Brownian bridge, with $\alpha = p = 1/2$, the distribution of A is simply uniform on $[0, 1]$.

We note that the $P_{\alpha,p}$ distribution of A is uniquely determined by formula (75), since by differentiating k times we obtain for $k = 1, 2, \ldots$

$$E^{\text{br}}_{\alpha,p}\left[\frac{\alpha(\alpha+1)\cdots(\alpha+k-1)A^k}{(1+\lambda A)^{\alpha+k}}\right] = (-1)^k \frac{d^k}{d\lambda^k}\left(\frac{1}{p(1+\lambda)^{\alpha} + q}\right) \qquad (\lambda > 0) \tag{76}$$

so we recover the moments

$$E^{\text{br}}_{\alpha,p}(A^k) = \frac{(-1)^k}{\alpha(\alpha+1)\cdots(\alpha+k-1)} \frac{d^k}{d\lambda^k}\left(\frac{1}{p(1+\lambda)^{\alpha} + q}\right)\Bigg|_{\lambda=0} \tag{77}$$

In particular, from (74) and (77), for all $0 < \alpha < 1$ and $0 < p < 1$, we obtain the means

$$E^{br}_{\alpha,p}(A) = E_{\alpha,p}(A) = p \tag{78}$$

which is also obvious from (72) and $P_{\alpha,p}(B_t > 0) = P^{br}_{\alpha,p}(B_t > 0) = p$ for all $0 < t < 1$, and the variances

$$Var^{br}_{\alpha,p}(A) = \frac{(1-\alpha)pq}{1+\alpha} < (1-\alpha)pq = Var_{\alpha,p}(A) \tag{79}$$

The inequality between the variances can be understood intuitively as follows. Conditioning to return to zero at time 1 tends to make the intervals smaller and more evenly distributed in length. So there is less variability in the fraction of time spent positive. For fixed p, as α increases from $0+$ to $1-$, both variances decrease, from the variance pq of a Bernoulli(p) variable ϵ_p at $\alpha = 0+$, down to variance 0 at $\alpha = 1-$. Consequently, under either $P_{\alpha,p}$ or $P^{br}_{\alpha,p}$

$$A \stackrel{d}{\rightarrow} \begin{cases} p & \text{as } \alpha \uparrow 1 \\ \epsilon_p & \text{as } \alpha \downarrow 0 \end{cases} \tag{80}$$

where $\stackrel{d}{\rightarrow}$ denotes convergence in distribution. This behaviour can also be understood from the representation (73) and the observation that under either $P_{\alpha,p}$ or $P^{br}_{\alpha,p}$

$$V_1(1) \stackrel{d}{\rightarrow} \begin{cases} 0 & \text{as } \alpha \uparrow 1 \\ 1 & \text{as } \alpha \downarrow 0 \end{cases} \tag{81}$$

See [16] for details and further references concerning the exact distribution of $V_1(1)$ under $P_{\alpha,p}$ and $P^{br}_{\alpha,p}$.

Let $G := G_1$ be the time of the last zero of B before time 1. To conclude this section, we record the following proposition which describes the $P_{\alpha,p}$ distribution of A_G by a surprisingly simple density relative to the $P_{\alpha,p}$ distribution of $A := A_1$ discussed above. Combined with Lamperti's formula for the density of A_1, this yields an explicit formula for the density of A_G relative to Lebesgue measure.

Proposition 15 *For all* $0 < \alpha < 1$ *and* $0 < p < 1$,

$$P_{\alpha,p}(A_G \in dx) = \frac{1-x}{1-p} P_{\alpha,p}(A_1 \in dx) \qquad (0 < x < 1) \tag{82}$$

Proof. Write E for $E_{\alpha,p}$. Then for all Borel measurable $f : [0,1] \to [0,\infty)$

$$\begin{aligned} (1-p)E[f(A_G)] &= E\left[f(A_G)1_{(B_1 \le 0)}\right] \\ &= E\left[f(A_1)1_{(B_1 \le 0)}\right] \\ &= E\left[f(A_1)(1-A_1)\right] \end{aligned}$$

where the first equality is due to the independence of A_G and the event $(B_1 < 0)$, the second is obvious, and the third is deduced from the formula

$$P_{\alpha,p}(B_1 \le 0 \,|\, A_1) = 1 - A_1 \tag{83}$$

which, as noted in [15], is an easy consequence of the sampling property (9). □

As a consequence of (82), the moments of A_G can be expressed simply in terms of those of $A := A_1$ which can be read from (74). Assume now for simplicity that B is a skew Bessel process under $P_{\alpha,p}$. As noted in [2], we can write

$$A_G = GA^{\text{br}} \tag{84}$$

where G has beta$(\alpha, 1-\alpha)$ distribution, and A^{br} is the fraction of time spent positive by the skew Bessel bridge of length 1 obtained by rescaling of B on the random interval $[0, G]$. So the $P_{\alpha,p}$ distribution of A^{br} is identical to the $P^{\text{br}}_{\alpha,p}$ distribution of $A := A_1$ discussed before. In principle, (84) determines this distribution of A^{br} in terms of the distributions of G and A_G just described. This gives an alternative formula to (77) for computing moments of A^{br}, hence some tricky algebraic identities, but unfortunately does not seem to yield any more explicit description of the law of A^{br}.

References

[1] M. Barlow. Skew brownian motion and a one-dimensional differential equation. *Stochastics*, 25:1–2, 1988.

[2] M. Barlow, J. Pitman, and M. Yor. Une extension multidimensionnelle de la loi de l'arc sinus. In *Séminaire de Probabilités XXIII*, pages 294–314. Springer, 1989. Lecture Notes in Math. 1372.

[3] E. B. Dynkin. Some limit theorems for sums of independent random variables with infinite mathematical expectations. *IMS-AMS Selected Translations in Math. Stat. and Prob.*, 1:171–189, 1961.

[4] B. Fristedt and S. J. Taylor. Constructions of local time for a Markov process. *Z. Wahrsch. Verw. Gebiete*, 62:73 - 112, 1983.

[5] P. Greenwood and J. Pitman. Construction of local time and Poisson point processes from nested arrays. *Journal of the London Mathematical Society*, 22:182–192, 1980.

[6] J. M. Harrison and L. A. Shepp. On skew Brownian motion. *The Annals of Probability*, 9:309 - 313, 1981.

[7] J. F. C. Kingman. Random discrete distributions. *J. Roy. Statist. Soc. B*, 37:1–22, 1975.

[8] F.B. Knight. On the duration of the longest excursion. In E. Cinlar, K.L. Chung, and R.K. Getoor, editors, *Seminar on Stochastic Processes*, pages 117–148. Birkhäuser, 1985.

[9] J. Lamperti. An occupation time theorem for a class of stochastic processes. *Trans. Amer. Math. Soc.*, 88:380 - 387, 1958.

[10] J. Lamperti. An invariance principle in renewal theory. *Ann. Math. Stat.*, 33:685 - 696, 1962.

[11] P. Lévy. Sur certains processus stochastiques homogènes. *Compositio Math.*, 7:283–339, 1939.

[12] M. Perman. Order statistics for jumps of normalized subordinators. *Stoch. Proc. Appl.*, 46:267–281, 1993.

[13] M. Perman, J. Pitman, and M. Yor. Size-biased sampling of Poisson point processes and excursions. *Probability Theory and Related Fields*, 92:21–39, 1992.

[14] J. Pitman. Partition structures derived from Brownian motion and stable subordinators. Technical Report 346, Dept. Statistics, U.C. Berkeley, 1992. To appear in *Bernoulli*.

[15] J. Pitman and M. Yor. Arcsine laws and interval partitions derived from a stable subordinator. *Proc. London Math. Soc. (3)*, 65:326–356, 1992.

[16] J. Pitman and M. Yor. The two-parameter Poisson-Dirichlet distribution derived from a stable subordinator. Technical Report 433, Dept. Statistics, U.C. Berkeley, 1995. To appear in *The Annals of Probability*.

[17] J. Pitman and M. Yor. Random discrete distributions derived from self-similar random sets. *Electronic J. Probability*, 1:Paper 4, 1–28, 1996.

[18] J. Pitman and M. Yor. Some conditional expectations given an average of a stationary or self-similar random process. Technical Report 438, Dept. Statistics, U.C. Berkeley, 1996. In preparation.

[19] C.L. Scheffer. The rank of the present excursion. *Stoch. Proc. Appl.*, 55:101–118, 1995.

[20] M. S. Taqqu. A bibliographical guide to self-similar processes and long-range dependence. In *Dependence in Probab. and Stat.: A Survey of Recent Results; Ernst Eberlein, Murad S. Taqqu (Ed.)*, pages 137-162. Birkhäuser (Basel, Boston), 1986.

[21] J. Walsh. A diffusion with a discontinuous local time. In *Temps Locaux*, volume 52-53 of *Astérisque*, pages 37–45. Soc. Math. de France, 1978.

[22] S. Watanabe. On time inversion of one-dimensional diffusion processes. *Z. Wahrsch. Verw. Gebiete*, 31:115–124, 1975.

[23] S. Watanabe. Generalized arc-sine laws for one-dimensional diffusion processes and random walks. In *Proceedings of Symposia in Pure Mathematics*, volume 57, pages 157-172. A.M.S., 1995.

[24] J.G. Wendel. Zero-free intervals of semi-stable Markov processes. *Math. Scand.*, 14:21 - 34, 1964.

[25] M. Yor. *Some Aspects of Brownian Motion.* Lectures in Math., ETH Zürich. Birkhaüser, 1992. Part I: Some Special Functionals.

[26] M. Yor. Random Brownian scaling and some absolute continuity relationships. In E. Bolthausen, M. Dozzi, and F. Russo, editors, *Seminar on Stochastic Analysis, Random Fields and Applications. Centro Stefano Franscini, Ascona, 1993*, pages 243–252. Birkhäuser, 1995.

Some remarks about the joint law of Brownian motion and its supremum

Marc Yor

Laboratoire de Probabilités - Université Paris VI - 4, Place Jussieu - Tour 56 3ème Etage - 75252 PARIS CEDEX 05

Introduction.

Let $(B_t, t \geq 0)$ be a standard 1-dimensional Brownian motion starting from 0, and denote by $S_t = \sup_{s \leq t} B_s$, $t \geq 0$, its one-sided supremum.

The aim of this Note is to give a simple proof, and equivalent formulations of a striking remark due to Seshadri [7] (see also Lépingle [5]).
No novelty claim is made, but Seshadri's remark probably deserves to be more widely known (see, e.g., Rogers-Satchell [6] for some consequences) ; moreover, the arguments developed below are very different from those in [7], which hinge on some "foliation" property of certain exponential families.

Theorem 1 *(Seshadri) : Let* $t > 0$ *be fixed.*

Then, the two variables $S_t(S_t - B_t)$ *and* B_t *are independent, and, moreover :*

$$S_t(S_t - B_t) \overset{(law)}{=} \frac{t}{2}\, \mathbf{e}, \tag{1}$$

where $\mathbf{e}$ *is a standard exponential variable (i.e. :* $P(\mathbf{e} \in dt) = dt\, e^{-t}$).

Obviously, this result may be immediately derived from the well-known formula for the joint law of (S_t, B_t), which we present as follows :

$$P(S_t \in dx\ ;\ S_t - B_t \in dy) = \left(\frac{2}{\pi t^3}\right)^{1/2} (x+y) \exp\left(-\frac{(x+y)^2}{2t}\right) dx\, dy. \tag{2}$$

However, we find it more interesting to derive the Theorem as a consequence of some elementary considerations about the supremum of a Brownian bridge ; this is done in Section 1.
In Section 2, we show how, using some algebraic relations between beta and gamma variables, Seshadri's remark may be deduced from the uniform distribution on $[0, R_t \equiv 2S_t - B_t]$ of either S_t or $S_t - B_t$. Finally, in Section 3,

we show how Denisov's path decomposition [1] of $(B_u, u \le 1)$ before and after the unique time θ^* (<1) at which $B_{\theta^*} = \sup_{u\le 1} B_u$ also allows to recover *(1)*.

1. The distribution of the suprema of Brownian bridges.

To start with, we give an easy, although helpful, criterion of independence between a Brownian functional F and B_1.

Proposition : *Let* $F : C[0,1] \longrightarrow \mathbb{R}$ *be a continuous functional on the canonical space* $C[0,1]$, *endowed with the topology of uniform convergence on* $[0,1]$. *Then, the following properties are equivalent :*

i) $F(B_u, u \le 1)$ *and* B_1 *are independent ;*

ii) *The law of* $F(B_u + cu \,;\, u \le 1)$ *does not depend on* c, *as* c *varies in* $\mathbb{R}$ *;*

iii) *The law of* $F(b_u + xu \,;\, u \le 1)$ *does not depend on* x, *as* x *varies in* $\mathbb{R}$, *and* $(b_u, u \le 1)$ *denotes the standard brownian bridge.*

Proof : The equivalence between i) and ii) follows easily from the Cameron-Martin relationship between the laws of $(B_u, u \le 1)$ and $(B_u + cu \,;\, u \le 1)$.

The equivalence between i) and iii) follows from the well-known representation : $B_u = b_u + uB_1$, $u \le 1$, where $(b_u, u \le 1)$ is a Brownian bridge independent from B_1. □

In order to prove the Theorem, we need only show, using the equivalence between i) and iii) in the Proposition, that :

$$(3)_x \qquad S_x(S_x - x) \overset{(law)}{=} \frac{1}{2}\, e, \quad \text{where : } S_x = \sup_{u\le 1}(b_u + xu).$$

[For $x = 0$, $(3)_o$ is the well-known fact that $(S_o)^2 \overset{(law)}{=} \frac{1}{2}\, e$;

note also that $(b_u + xu, u \le 1)$ is the brownian bridge $0 \longrightarrow x$ on the time-interval $[0,1]$].

It is immediate that $(3)_x$ is equivalent to :

$$(4)_x \qquad S_x \overset{(law)}{=} \frac{x}{2} + \left(\frac{x^2}{4} + \frac{e}{2}\right)^{1/2}.$$

Using $(b_u, u < 1) \overset{(law)}{=} \left((1-u) B_{u/_{1-u}}, u < 1\right)$, we obtain :

$$(5)_x \qquad S_x \overset{(law)}{=} \sup_{t\geq 0}\left(\frac{B_t + tx}{1+t}\right) \overset{(law)}{=} \sup_{u\geq 0}\left(\frac{B_u + x}{1+u}\right)$$

where, for the last equality in law, we have used the fact that $(uB_{1/_u}, u > 0)$ is also a Brownian motion.

Thus, for any $a > x$, we have :

$$P(S_x < a) = P\left(\sup_{u\geq 0} \frac{B_u + x}{1+u} < a\right) = P(\forall u \geq 0,\ B_u + x < a(1+u))$$

$$(6) \qquad = P\left(\sup_{u\geq 0} (B_u - au) < a - x\right).$$

We now use the well-known

Lemma 1 : *If* $(M_t, t \geq 0)$ *is a continuous,* $\mathbb{R}_+$ *valued martingale such that* $M_t \xrightarrow[t\to\infty]{} 0$ *, and* $M_0 = 1$*, then :* $\sup_{t\geq 0} M_t \overset{(law)}{=} 1/_U$ *, where* U *is uniform on* $[0,1]$.

as well as the following consequence, which goes back to Doob.

Corollary : *For* $a > 0$, $\sup_{u\geq 0}(B_u - au) \overset{(law)}{=} \frac{1}{2a}\, \mathbf{e}$.

Proof : Apply the Lemma to : $M_u = \exp(2a(B_u - au))$. □

We then go back to *(6)* to end the proof of $(4)_x$ by writing :

$$P(S_x < a) = P\left(\frac{1}{2a}\, \mathbf{e} < a - x\right) = P\left(\frac{1}{2}\, \mathbf{e} < (a - \frac{x}{2})^2 - \frac{x^2}{4}\right)$$

The proof of $(4)_x$ now follows. □

We now make a few comments on some of the assertions found above :

a) in the statement of the Proposition, the hypothesis that F is continuous on $C([0,1])$ serves to ensure that the law of $F(b_u + xu, u \leq 1)$ does not depend on x, for every $x \in \mathbb{R}$.

b) A sufficient condition for iii) to be satisfied is, of course, that :

$$F(b_u + xu, u \leq 1) = G(b_u, u \leq 1)\ ,$$

for some functional G independent of x. This is satisfied if F, as defined on the canonical space $C[0,1]$, where $X_u(\omega) \equiv \omega(u)$, $u \leq 1$, is measurable with respect to $\mathcal{F}' = \sigma\{X_u - uX_1 ; u \leq 1\}$.

But, Seshadri's remark shows that this condition is only sufficient, and not necessary to ensure that $F(B_u, u \leq 1)$ and B_1 be independent.

Furthermore, from Theorem 1, one can construct many other r.v's which are independent from B_1, although they are not measurable with respect to $(b(u), u \leq 1)$. The following is a finite dimensional example :

take $0 = t_0 < t_1 < \dots < t_{k+1} = t$; then, the vector

$(S_{(t_j,t_{j+1})} - B_{t_j})(S_{(t_j,t_{j+1})} - B_{t_{j+1}})$; $j = 0,\dots,k$, is independent from B_1.

(We use the notation $S_{(u,v)} = \sup_{u \leq s \leq v} B_s$.)

This assertion follows from Theorem 1, used together with the independence of the increments of B.

c) Different applications of the Lemma are given in [4], where the following consequences are shown :

for $a > 0$,
$$\int_0^\infty ds \exp\left(B_s - \frac{as}{2}\right) \overset{(law)}{=} 2/Z_a ,$$

where Z_a denotes a gamma variable with parameter a, i.e :

$$P(Z_a \in dt) = \frac{t^{a-1} e^{-t} dt}{\Gamma(a)} .$$

2. <u>Going from</u> $(2S_t - B_t)$ <u>to</u> $S_t(S_t - B_t)$.

It is easily shown, using formula *(2)* for instance, that the joint law of $(S_t, S_t - B_t)$ is a consequence of the following subproducts of Pitman's celebrated theorem : $R_t \overset{def}{=} 2S_t - B_t \equiv S_t + (S_t - B_t)$, $t \geq 0$, is a 3-dimensional Bessel process, and, for every t, both S_t and $(S_t - B_t)$ are uniformly distributed on $[0, R_t]$. (More generally, this holds whenever t is replaced by any stopping time T w.r. to the natural filtration of R).

Hence, we can write *(2)* in the random variables "algebraic" form :

(2') $$(S_t, S_t - B_t) \overset{(law)}{=} R_t(U, 1-U),$$

where U is uniform on $[0,1]$, and independent from $R_t \overset{(law)}{=} \sqrt{t}\,|N^{(3)}|$, with $N^{(3)}$ a 3-dimensional Gaussian variable, the 3 components of which are independent $N(0,1)$ variables.

We are now in a position to give another proof of Theorem 1 as well as other remarks of the same ilk

Theorem 2 : *(We keep the previous notation). Let* $t > 0$.

Define the 3 "remainders" ρ_t , ρ'_t , *and* ρ''_t *as follows :*

$$R_t = (2S_t - B_t)^2 = S_t^2 + \rho'_t = (S_t - B_t)^2 + \rho''_t = B_t^2 + \rho_t.$$

Obviously, one has :

$$\rho'_t = (3S_t - B_t)(S_t - B_t) \qquad ; \rho''_t = (3S_t - 2B_t)S_t \qquad ; \rho_t = 4S_t(S_t - B_t).$$

Then, the following identities hold :

$$(S_t^2 , \rho'_t) \overset{(law)}{=} ((S_t - B_t)^2 , \rho''_t) \overset{(law)}{=} (B_t^2, \rho_t) \overset{(law)}{=} t(N^2, (N')^2 + (N'')^2)$$

where N, N' *and* N'' *are 3 independent* $N(0,1)$ *variables.*

Concerning the third pair (B_t^2, ρ_t) , *more precisely, the r.v's* B_t *and* $S_t(S_t - B_t)$ *are independent.*

Proof : i) We shall only prove the last assertion, since the two first ones, which amount to :

$$(S_t^2, \rho'_t) \overset{(law)}{=} ((S_t - B_t)^2, \rho''_t) \overset{(law)}{=} t(N^2, (N')^2 + (N'')^2)$$

may be obtained by using the same arguments.

ii) Our proof will consist in using the identity in law :

(7) $$(Z_a ; Z_b) \overset{(law)}{=} Z_{a+b}(Z_{a,b} ; 1 - Z_{a,b})$$

where Z_a and Z_b are two independent gamma variables, with respective parameters a and b, and $Z_{a,b}$ is a beta variable with parameters (a,b). We shall use (7) for $a = 1$, and $b = 1/2$, in the following form :

if U is uniform on $[0,1]$, then $V = 1-2U$ is uniform on $[-1,1]$, and moreover :

$$4U(1-U) \equiv 1-V^2 \overset{(law)}{=} Z_{1,\frac{1}{2}} .$$

Consequently, from *(4')*, we deduce :

$$\begin{aligned}(S_t(S_t-B_t),B_t) &\equiv (R_t^2\, U(1-U)\ ,\ R_tU - R_t(1-U))\\ &\equiv \left((\tfrac{R_t^2}{4})\ 4U(1-U),\ R_t(2U-1)\right)\\ &\equiv \left((\tfrac{R_t^2}{4})\ (1-V^2),\ -R_tV\right).\end{aligned}$$

To finish the proof, we take $t = 1$, and we obtain :

$$\text{(8)}\qquad \left(\frac{R_1^2}{2}\,(1-V^2)\ ,\ \frac{R_1^2}{2}\,V^2\right) \overset{(law)}{=} (Z_{3/2}Z_{1,1/2}\ ,\ Z_{3/2}\,(1-Z_{1,1/2}))$$

where on the r.h.s, the beta and gamma variables are assumed to be independent.

Finally, reading *(7)* from right to left, the joint law found in *(8)* is that of $(Z_1,Z_{1/2})$, which ends the proof. □

3. Karatzas-Shreve trivariate identity and Denisov's decomposition.

3.1. Using Lévy's equivalence theorem :

$$(S_t,S_t-B_t\ ;\ t \ge 0) \overset{(law)}{=} (L_t,|B_t|\ ;\ t \ge 0),$$

where $(L_t,t \ge 0)$ denotes the local time of $(B_t,t \ge 0)$ at 0, one may immediately translate Theorem 1 as follows :

fix $t > 0$; then, L_tB_t is a bilateral exponential variable, which is independent of $L_t - |B_t|$.

3.2. Another relation between the joint laws of (B_1,L_1) and (B_1,S_1) was noticed by Karatzas and Shreve ([3], p. 425, Remark 3.12) :

$$\text{(9)}\qquad (B_1^+ + \tfrac{1}{2}L_1\ ,\ B_1^- + \tfrac{1}{2}L_1\ ,\ A_o^+) \overset{(law)}{=} (S_1,S_1-B_1,\theta_o^+).$$

where $A_o^+ = \int_0^1 ds\, 1_{(B_s > 0)}$ and θ_o^+ is the unique time $t < 1$ at which B_t equals S_1.

This trivariate identity is shown in [2] to be a particular consequence of Bertoin's rearrangement of positive and negative excursions for Brownian motion (with or without drift).
Karatzas and Shreve [4] also explained *(9)* via a Sparre-Andersen type transformation.
We now remark that, using *(9)*, Theorem 1 may be translated as follows :

$$(B_1^+ + \frac{1}{2} L_1)(B_1^- + \frac{1}{2} L_1) \text{ is independent of } B_1 ,$$

or, equivalently :

$$\textit{(10)} \qquad \frac{1}{2} L_1 (|B_1| + \frac{1}{2} L_1) \text{ is independent of } B_1.$$

Now, using again Lévy's equivalence theorem recalled in 3.1 above, *(10)* is equivalent to :

$$\textit{(11)} \qquad S_1((S_1 - B_1) + \frac{1}{2} S_1) \text{ is independent of } (S_1 - B_1) ,$$

which is precisely the result in Theorem 2 concerning the "second remainder".

3.3. Finally, we also remark that Denisov's path decomposition [1] of $(B_u, u \le 1)$ before and after time θ_o^+ also yields at least a part of Theorem 1, in particular the identity in law *(1)*.

Indeed, from [1], one deduces :

$$(S_1, S_1 - B_1, \theta_o^+) \overset{(law)}{=} (\sqrt{1-A}\, m_1, \sqrt{A}\, m'_1, A)$$

where A, m_1 and m'_1 are independent, A is arc sine distributed, and

$$m_1 \overset{(law)}{=} m'_1 \overset{(law)}{=} \sqrt{2\mathbf{e}}.$$

Hence, $S_1(S_1 - B_1) \overset{(law)}{=} (A(1-A)4\, \mathbf{e}\mathbf{e}')^{1/2}$, where on the r.h.s., A, $\mathbf{e}$ and $\mathbf{e}'$ are independent.

Since $A \overset{(law)}{=} \cos^2(\theta)$, with θ uniform on $[0, 2\pi[$, it follows that :

$A(1-A) \overset{(law)}{=} \frac{1}{4} A$, hence :

(12) $$S_1(S_1-B_1) \overset{(law)}{=} (A\, ee')^{1/2}.$$

Next, we shall use

Lemma 2 : *For any* $r > 0$, *the following identity in law holds :*

(13) $$\frac{1}{2} Z_{2r} \overset{(law)}{=} (Z_{r,1/2}\; Z_{r+1/2}\; Z'_{r+1/2})^{1/2}$$

and, in particular :

(14) $$\frac{1}{2} e \overset{(law)}{=} (A\, ee')^{1/2},$$

where on the r.h. sides, the three r.v's are independent.

Proof : From the duplication formula for the gamma function, one deduces :

$$Z_{2r}^2 \overset{(law)}{=} 4\, Z_{r+1/2}\; Z_r$$

(see [9], p. 112, Lemma 8.1.).

Then, *(13)* follows as a consequence of *(7)*. *Finally, (14)* follows from *(13)*, for $r = 1/2$. □

Now, from *(12)* and *(14)*, we recover the identity in law *(1)*.

References

[1] **I.V. Denisov** : A random walk and a Wiener process near a maximum. *Teo. Veroyat i. Prim. 28, p. 821-824.*

[2] **P. Embrechts, L.C.G. Rogers, M. Yor** : A proof of Dassios' representation of the α-quantile of Brownian motion with drift. *Ann. App. Prob.* 5, *n° 3, p. 757-767, (1995).*

[3] I. Karatzas, M. Shreve : Brownian Motion and Stochastic Calculus.
Springer, Berlin (1987).

[4] I. Karatzas, M. Shreve : A decomposition of the Brownian path.
Stat. Prob. Lett 5, p. 87-94 (1987).

[5] D. Lépingle : Un schéma d'Euler pour équations différentielles stochastiques réfléchies.
C.R.A.S. Paris, 316, p. 601-605, 1993.

[6] L.C.G. Rogers, S.E. Satchell : Estimating variance from high, low and closing prices.
The Annals of App. Prob., vol. 1, n° 4, p. 504-512, 1991.

[7] V. Seshadri : Exponential models, Brownian motion and independence.
Can. J. of Stat., 16, p. 209-221, 1988.

[8] M. Yor : Sur certaines fonctionnelles exponentielles du mouvement brownien réel.
J. App. Prob, 29,; p. 202-208 (1992).

[9] M. Yor : Some Aspects of Brownian motion, Part I : Some special functionals.
Lect. in Maths. E.T.H. Zurich, Birkhaüser (1992).

A characterization of Markov solutions for stochastic differential equations with jumps

Anne Estrade

Introduction

It is well known that solutions of stochastic differential equations such as

$$X_0 = x \; ; \; dX_t = f(X_{t-})\, dZ_t \, , \tag{1}$$

where Z is a Lévy process, are Markov processes. A converse result has been obtained by Jacod and Protter [6] as follows : consider the stochastic differential equations $(1)_x$ driven by the same semimartingale Z with initial conditions x and never-vanishing coefficient f. It is proved that, if the solutions X^x of $(1)_x$ are time homogeneous Markov processes with the same transition semigroup for all x, then Z is a Lévy process.

The present work is in the spirit of Jacod and Protter's converse problem. We obtain a converse result for stochastic differential equations with jumps between manifolds. More precisely we will look at the equations studied by Cohen [4] for which it is already known that solutions are Markov processes provided the driving semimartingale is a Lévy process.

The main interest of this paper is in the consequences of this converse result. In fact we are able to establish a characterization of diffusions with jumps : usually, diffusions are constructed as Markov solutions of stochastic differential equations. What we prove here is that the only time homogeneous Markov processes obtained as solutions of stochastic differential equations are those arising from equations driven by Lévy processes.

The method is an extension of [6]. The principle consists in "inverting" the stochastic differential equation and writing the driving process as an additive functional of the solution; the Markov property of the solution then yields the conclusion. To invert the stochastic differential equation, some inverting assumptions are required, similar to the "never-vanishing coefficient f" assumption in [6].

The paper is divided into two main sections. In section 1, we establish the method to prove that the driving process is Lévy. In section 2, we characterize the diffusions with jumps, first in a manifold and then, as a special case, in $\mathbf{R}^d$.

In the following $(\Omega, \mathcal{F}, \mathcal{F}_t, \mathbf{P})$ will be a filtered probability space with $(\mathcal{F}_t)_{t\geq 0}$ a right continuous filtration containing all $\mathbf{P}$-zero measure sets of $\mathcal{F}$.

1 A criterion to be a Lévy process

Let M be a finite dimensional manifold and $(X^x)_{x\in M}$ a collection of $(\Omega, \mathcal{F}, \mathcal{F}_t, \mathbf{P})$-adapted càdlàg semimartingales with values in M such that $X_0^x = x$ for all x in M.

Let $(\hat{\Omega}, \hat{\mathcal{F}}, \hat{\mathcal{F}}_t)$ be the canonical space of càdlàg M-valued functions, equipped with the canonical process $\hat{X}$ ($\hat{X}_t(\omega) = \omega(t)$, for $t \geq 0$ and ω in $\hat{\Omega}$) and the natural filtration $(\hat{\mathcal{F}}_t)_{t\geq 0}$ of $\hat{X}$. We will also denote by $(\theta_t)_{t\geq 0}$ the semigroup of translations on $\hat{\Omega}$ ($\theta_t(\omega)(.) = \omega(t+.)$, for $t \geq 0$ and ω in $\hat{\Omega}$) and by P^x the probability measure on $(\hat{\Omega}, \hat{\mathcal{F}})$ which is the law of X^x for all x in M.

Finally let Z be an $\mathbf{R}^d$-valued càdlàg semimartingale adapted to $(\Omega, \mathcal{F}, \mathcal{F}_t, \mathbf{P})$ with $Z_0 = 0$. We recall the usual definition of a Lévy process.

Definition 1 *A process Z on $(\Omega, \mathcal{F}, \mathcal{F}_t, \mathbf{P})$ is called a Lévy process if it is a càdlàg adapted process such that $\mathbf{P}(Z_0 = 0) = 1$, and for all $s, t \geq 0$, the variable $Z_{t+s} - Z_t$ is independent from the $(Z_u\,;\, 0 \leq u \leq t)$ and has the same distribution as Z_s.*

We are now able to give the main result of this section.

Proposition 1 *Assume that there exists an $(\hat{\Omega}, \hat{\mathcal{F}}, \hat{\mathcal{F}}_t)$-adapted process $(A_t)_{t\geq 0}$ with values in $\mathbf{R}^d$ such that*
(i) $\forall x \in M$, $P^x(A_0 = 0$ and $A_{s+t} = A_t + A_s \circ \theta_t\,, \forall s, t \geq 0) = 1$;
(ii) $\forall x \in M$, $\mathbf{P}(Z_t = A_t(X^x)\,, \forall t \geq 0) = 1$.
If the X^x are time homogeneous Markov processes with transition semigroup independent of x, then Z is a Lévy process.
This proposition is very similar to the result in [6]. The generalization consits in replacing the explicit formula giving Z in terms of X by a condition assuring that Z is an additive functional of X. It is also close in spirit to theorem 6.27 of [2] where the local characteristics of an additive semimartingale based on a Markov process are described.

Proof of proposition 1 : Take a bounded Borel function f on $\mathbf{R}^d$ and compute $E^x(f(A_{t+s} - A_t)/\hat{\mathcal{F}}_t)$ for $s, t \geq 0$ and some x in M. Using the additive property (i) of A and the Markov property of $\hat{X}$ on $(\hat{\Omega}, \hat{\mathcal{F}}_t, P^x)$, we get

$$E^x(f(A_{t+s} - A_t)/\hat{\mathcal{F}}_t) = E^x(f(A_s) \circ \theta_t)/\hat{\mathcal{F}}_t) = E^{\hat{X}_t}(f(A_s)).$$

By (ii), the $\mathbf{P}$-distribution of Z equals the P^x-distribution of A, for all x in M. Then we get

$$E^x(f(A_{t+s} - A_t)/\hat{\mathcal{F}}_t) = \mathbf{E}(f(Z_s)).$$

This proves that under P^x, $A_{t+s} - A_t$ is independent from $\hat{\mathcal{F}}_t$ and hence from $(A_u\,;\, 0 \leq u \leq t)$, and has the same distribution as A_s. Finally, use (ii) again and the proposition follows. □

2 Stochastic differential equations in manifolds

We will be concerned with stochastic differential equations driven by d-dimensional càdlàg semimartingales, whose solutions live in a d-dimensional manifold ($\mathbf{R}^d$ included !). We will use the formalism introduced by Cohen and studied with respect to Markov property in [4]. Such equations can also be studied with the formalism of Kurtz, Pardoux and Protter in [5] but with restricted possibilities for the jumps of the solution (at a jump time s, in [5] the X_s term is given as the end point of an ordinary differential equation starting at X_{s-} with a coefficient linearly depending on ΔZ_s, whereas in [4], X_s is given by any function of X_{s-} and ΔZ_s).

2.1 Definitions and properties

Let us first recall some of Cohen's results. In the following, M will be a smooth manifold of dimension d.

Definition 2 *A map $\psi : M \times \mathbf{R}^d \longrightarrow M$ is called a jump coefficient if*
(i) $\forall x \in M$, $\psi(x,0) = x$;
(ii) ψ is C^3 in a neighborhood of $M \times \{0\}$ in $M \times \mathbf{R}^d$.

Suppose we are given a d-dimensional càdlàg semimartingale Z, a jump coefficient ψ according to the previous definition and a fixed point $x \in M$. In [4], a meaning is given to the following stochastic differential equation

$$X_0 = x \; ; \; \overset{\Delta}{d}X = \psi(X, \overset{\Delta}{d}Z) \tag{2}$$

by the prescription that the process X is a solution of (2) if X is an M-valued semimartingale such that, for any embedding $(x^\alpha)_{1\leq\alpha\leq m}$ of M in $\mathbf{R}^m$, one has

$$\begin{aligned} \forall \alpha = 1, ..., m \qquad X_t^\alpha &= x^\alpha + \int_0^t \frac{\partial \psi^\alpha}{\partial z^i}(X_{s-},0)\, dZ_s^i \\ &+ \frac{1}{2}\int_0^t \frac{\partial^2 \psi^\alpha}{\partial z^i \partial z^j}(X_s,0)\, d<Z^{ic}, Z^{jc}>_s \\ &+ \sum_{s\leq t}\left(\psi^\alpha(X_{s-}, \Delta Z_s) - X_{s-}^\alpha - \frac{\partial \psi^\alpha}{\partial z^i}(X_{s-},0)\,\Delta Z_s^i\right) \end{aligned} \tag{3}$$

In the following, the summation convention on repeated indices will be in force; sums on i and j will run from 1 to d and sums on α and β from 1 to m. Also Z^c will denote the continuous martingale part of any real semimartingale Z.

It is established in [4] that the equation (2) admits a unique, possibly exploding solution X^x. Moreover, if Z is a Lévy process then X^x is an homogeneous Markov process with transition semigroup independent of x.

2.2 Converse result

To obtain a converse to this result, we need some inverting assumptions. The first one deals with the jump coefficient ψ of the stochastic differential equation (2).

Definition 3 *A jump coefficient ψ is said to be invertible if for all x in M, the differential at 0 of $\psi(x,.) : z \in \mathbf{R}^d \longrightarrow \psi(x,z) \in M$, which we denote by $d_z\psi(x,0)$, is an isomorphism from $\mathbf{R}^d$ onto T_xM.*

As promised, we now give a characterization of jump diffusions in manifolds.

Theorem 2 *Let Z be a d-dimensional semimartingale, let ψ be an invertible jump coefficient and, for all x in M, let X^x be the unique solution of the equation*

$$X_0 = x \; ; \; \overset{\triangle}{d}X = \psi(X, \overset{\triangle}{d}Z).$$

Suppose that

$$\forall x \in M \; , \; \mathbf{P}((X^x_{t-}, X^x_t) \in \mathcal{V}_\psi, \forall t > 0) = 1 \tag{4}$$

where

$$\mathcal{V}_\psi = \{(x,y) \in M \times M \; ; \; \text{there exists a unique } z \text{ in } \mathbf{R}^d \text{ such that } y = \psi(x,z)\}.$$

Then, the X^x are time homogeneous Markov processes with the same transition semigroup for all x in M if and only if Z is a Lévy process.

Proof : If Z is a Lévy process we already know by prop.1 of [4] that the X^x are time homogeneous Markov processes. We will prove the converse result.

The procedure is to write Z as an additive functional of X^x in order to show that the hypothesis of proposition 1 is valid and then the result follows immediatly.

For $x \in M$, by definition 3, $d_z\psi(x,0)$ is an isomorphism from $\mathbf{R}^d$ onto T_xM; denote by $\Phi(x)$ the inverse isomorphism and for $(x,y) \in \mathcal{V}_\psi$ denote by $\Gamma(x,y)$ the unique z in $\mathbf{R}^d$ such that $y = \psi(x,z)$.

Recall the notations introduced in the first part concerning the canonical space $\hat{\Omega}$ of all càdlàg M-valued functions. We choose as a candidate for our additive process on $(\hat{\Omega}, \hat{\mathcal{F}}, \hat{\mathcal{F}}_t, P^x)$ the following :

$$\begin{aligned} A_t &= \int_0^t \Phi_\alpha(\hat{X}_{s-})\, d\hat{X}^\alpha_s \\ &- \frac{1}{2}\int_0^t \Phi(\hat{X}_s) \circ d^2_{zz}\psi(\hat{X}_s, 0) \circ (\Phi_\alpha(\hat{X}_s) \otimes \Phi_\beta(\hat{X}_s))\, d < \hat{X}^{\alpha c}, X^{\beta c} >_s \\ &+ \sum_{s \le t} \left(\Gamma(\hat{X}_{s-}, \hat{X}_s) - \Phi_\alpha(\hat{X}_{s-})\, \Delta \hat{X}^\alpha_s\right) \end{aligned} \tag{5}$$

for any embedding $(x^\alpha)_{1 \le \alpha \le m}$ of M into $\mathbf{R}^m$.

Following [2] th.3.12, there exists a version of A such that (5) is valid for every probability P^x, $x \in M$.

Take any x in M. An easy computation based on the stochastic differential equation (2) solved by X^x yields

$$\begin{aligned} A_t(X^x) &= \int_0^t dZ_s + \sum_{s \le t} (\Gamma(X^x_{s-}, X^x_s) - \Delta Z_s) \\ &= Z_t \end{aligned}$$

since $\forall s \geq 0, \Delta Z_s = \Gamma(X^x_{s-}, X^x_s)$. This proves that the process A satisfies condition (ii) of prop.1.

On the other hand, A is clearly additive and so also satisfies condition (i) of prop.1.

□

Before we study the geometrical aspect of the inverting assumptions, let us look at the special case where the manifold M is the whole of $\mathbf{R}^d$.

2.3 The vectorial case

We now take $M = \mathbf{R}^d$ and choose $f \in \mathcal{C}^3(\mathbf{R}^d, \mathbf{R}^{d\times d})$ such that $f(x) \in GL(d)$, $\forall x \in \mathbf{R}^d$. We define a jump coefficient ψ as follows :

$$\forall x, z \in \mathbf{R}^d \; ; \; \psi(x, z) = \psi(x, z, 1)$$

where

$$(\psi(x, z, u))_{0\leq u\leq 1} = (y(u))_{0\leq u\leq 1}$$

is the (possibly exploding) unique solution of the ordinary differential equation

$$y(0) = x \; ; \; \frac{dy}{du}(u) = f(y(u)).z \tag{6}$$

Since $d_z\psi(x, 0) = f(x)$, ψ is an invertible jump coefficient and the stochastic differential equation (2) becomes :

$$\begin{aligned} \forall \alpha = 1, ..., d \qquad X^\alpha_t &= x + \int_0^t f^\alpha_i(X_{s-})\, dZ^i_s \\ &+ \frac{1}{2}\int_0^t \frac{\partial f^\alpha_i}{\partial x^\beta}(X_s) f^\beta_j(X_s)\, d < Z^{ic}, Z^{jc} >_s \\ &+ \sum_{s\leq t} \left(\psi^\alpha(X_{s-}, \Delta Z_s) - X^\alpha_{s-} - f^\alpha_i(X_{s-})\, \Delta Z^i_s\right) . \end{aligned} \tag{7}$$

It is of the type introduced by Kurtz, Pardoux and Protter in [5]. To be able to apply theorem 2, we must verify the condition (4). A sufficient condition for this is :

$$\exists F \in \mathcal{C}^1(\mathbf{R}^d, \mathbf{R}^d) \text{ such that } \forall x \in \mathbf{R}^d \,,\; dF(x) = (f(x))^{-1}. \tag{8}$$

In fact, if (8) holds, at every jump time s, there will be exactly one z in $\mathbf{R}^d$ such that $X^x_s = \psi(X^x_{s-}, z)$, which is given by $z = \Delta Z_s = F(X^x_s) - F(X^x_{s-})$.

Remarks :

a) In the one-dimensional case, condition (8) reduces to "$\forall x \in \mathbf{R}^d$, $f(x) \neq 0$" since existence of a primitive F of $1/f$ is then assured.

b) One can use the definition of a closed 1-form (see [1] p.207) to give the following equivalent form of condition (8), where $g(x)$ denotes the inverse matrix of $f(x)$:

$$\forall i, \alpha, \beta \in \{1, ..., d\}, \forall x \in \mathbf{R}^d \,,\; \frac{\partial g^i_\alpha(x)}{\partial x^\beta} = \frac{\partial g^i_\beta(x)}{\partial x^\alpha}.$$

2.4 Comments on the inverting assumptions

As a conclusion let us comment about the inverting assumptions we have required. First of all, we give an example.

An example of invertible jump coefficient. Suppose there exist $(L_1, ..., L_d)$ vector fields on M such that for all x in M, $(L_i(x))_{1\leq i\leq d}$ is a basis of T_xM. Define the map ψ by

$$(x,z) \in M \times \mathbf{R}^d \mapsto \psi(x,z) = Exp_x(\sum_i z^i L_i(x))$$

where Exp_x denotes the exponential mapping at x. Then ψ is clearly a jump coefficient as defined in definition 2. Moreover, for all x in M, the differential of $\psi(x,.)$ at 0 is given by

$$h \in \mathbf{R}^d \mapsto d_z\psi(x,0).h = \sum_i h^i L_i(x)$$

and therefore is an isomorphism. According to definition 3, ψ is an invertible jump coefficient. Next we look at condition (4) of theorem 2.

Condition (4) is a "usual" condition. Let us recall a paper with M.Pontier [7] where the horizontal lift of a càdlàg manifold valued semimartingale is defined. This lift exists provided the semimartingale X satisfies some hypothesis (H) very similar to condition (4).
(H) : *with probability one, there exists one and only one geodesic curve between* X_{t-} *and* X_t *for all* $t \geq 0$.
Note that this hypothesis is stronger than that given in [7], but it is actually the right one. We have just become aware of this error in [7]. However all the results theirein are valid under the correct hypothesis (H).

Let us also mention that in [3] another horizontal lift is defined for all càdlàg manifold valued semimartingale without any hypothesis (H). However, as we will see, we cannot deal here without condition (4).

Condition (4) is essential for th.2. We give an example where (4) is not fulfilled and theorem 2 does not apply.

Let M be the unit circle : $M = \{e^{i\theta} \; ; \; \theta \in [0, 2\pi[\}$ and take for jump coefficient $\psi : (x,z) \in M \times \mathbf{R} \mapsto \psi(x,z) = xe^{i\pi z}$.

Let $(T_n)_{n\geq 1}$ be a sequence of random exponential times and, independently, $(Y_n)_{n\geq 1}$ be a sequence of independent random variables with the same Poisson distribution. We define the processes Z and $\tilde{Z}$ by

$$\begin{aligned} Z_t &= \sum_{n\geq 1} \mathbf{1}_{T_n\leq t} Y_n \\ \tilde{Z}_t &= \sum_{n\geq 1} \mathbf{1}_{T_n\leq t}(Y_n + 2n\mathbf{1}_{Y_{n+1}=0}). \end{aligned}$$

Process Z is a Lévy process (actually a compound Poisson process) whereas $\tilde{Z}$ is not a Lévy process (the variables $(Y_n + 2n\mathbf{1}_{Y_{n+1}=0})_n$ are neither independent nor are they identically distributed).

Now let X^x be the solution of

$$X_0 = x \; ; \; \overset{\triangle}{d}X = \psi(X, \overset{\triangle}{d}\tilde{Z}). \tag{9}$$

By the definition of $\tilde{Z}$ and ψ, it is clear that $X_t^x = x\, e^{i\pi\tilde{Z}_t}$. But, since $\pi\tilde{Z}_t - \pi Z_t \in 2\pi\mathbb{N}$, one also has $X_t^x = x\, e^{i\pi Z_t}$. This proves that X^x is the solution of (9) where $\tilde{Z}$ has been replaced by Z, and since Z is Lévy, X^x is a Markov process with transition semigroup independent of x. Hence we obtain a Markov solution of a stochastic differential equation driven by a non-Lévy process.

Of course, the assumption (4) of th.2 is not valid since $\mathcal{V}_\psi$ is empty !

References

[1] H. Cartan. Cours de calcul différentiel. Hermann, Paris, 1977.

[2] E. Çinlar , J. Jacod , P. Protter , M.J. Sharpe. Semimartingales and Markov processes. *Zeitschrift für Wahrsch.*, 54 : 161-220, 1980.

[3] S. Cohen. Géométrie différentielle stochastique avec sauts 2 : discrétisation et applications des eds avec sauts. *Stochastics and Stochastic Reports*, vol.56 : 205-225, 1996.

[4] S. Cohen. Some Markov properties of stochastic differential equations with jumps. *Séminaire de Proba. XXIX*, LNM 1613 : 181-193, 1995.

[5] T.G. Kurtz, E. Pardoux, P. Protter. Stratonovich stochastic differential equations driven by general semimartingales. *Annales de l'Institut Henri Poincaré*, 31(2) : 351–378, 1995.

[6] J. Jacod , P. Protter. Une remarque sur les équations différentielles stochastiques à solutions markoviennes. *Séminaire de Proba. XXV*, LNM 1485 : 138-139, 1991.

[7] M. Pontier , A. Estrade. Relèvement horizontal d'une semi-martingale càdlàg. *Séminaire de Proba. XXVI*, LNM 1526 : 127-145, 1992.

[8] P. Protter. Stochastic integration and differential equations : a new approach. Springer-Verlag, 1990.

Anne Estrade
URA MAPMO - DÉPARTEMENT DE MATHÉMATIQUES
Université d'Orléans - BP 6759
45067 ORLEANS CEDEX 2

DIFFEOMORPHISMS OF THE CIRCLE AND THE BASED STOCHASTIC LOOP SPACE

R. Léandre

Let $L_x(M)$ be the based loop space of a compact manifold M, that is, the space of all continuous applications γ from the circle $S^1 = \mathbf{R}/\mathbf{Z}$ to the manifold M such that $\gamma_0 = x$.

Let α be a diffeomorphism of S^1 fixing the point $0 \in S^1$. We get an application $\tilde{\alpha} : L_x(M) \to L_x(M)$ by putting:

$$\tilde{\alpha}(\gamma) = \gamma \circ \alpha \, .$$

In other words, the space of diffeomorphisms of the circle acts on the based loop space by reparametrization of the loop.

If we put on M a Riemannian structure, we can consider the law P_x of the Brownian bridge; it is a measure on $L_x(M)$. $\tilde{\alpha}$ does not preserve P_x; more precisely, P_x and its image by $\tilde{\alpha}$ are in general singular with respect to each other. So the group of diffeomorphisms of the circle does not act on the Brownian bridge.

The purpose of this paper is to make diffeomorphisms act on a space of stochastic loops by using the theory of quasi-invariant measures on the group of diffeomorphisms of the circle due to [Sh], [K] and [M.M].

Construction of the probabilistic model.

Let $(B_s)_{0 \leqslant s \leqslant 1}$ be a Brownian motion. Using B, we can define a (random) diffeomorphism of $[0,1]$ by

$$\phi(B)_t = \frac{\int_0^t \exp B_s \, ds}{\int_0^1 \exp B_s \, ds} \, . \tag{1}$$

The law of $\phi(B)$ is a measure $dP(\phi)$ on the set of increasing diffeomorphisms of $[0,1]$. (See [K], [M.M].)

Consider the Brownian bridge measure dP_x on $L_x(M)$. We introduce another measure on $L_x(M)$ which takes care of all possible orientation-preserving reparametrizations of the loops. Denoting by $F(\gamma_{s_1}, \ldots, \gamma_{s_r})$ a bounded cylindrical function on $L_x(M)$, we put

$$E_{tot}[F] = E_{P(\phi)}\left[E_{P_x}[F(\gamma_{\phi^{-1}_{s_1}}, \ldots, \gamma_{\phi^{-1}_{s_r}})]\right] \, . \tag{2}$$

In other words, considering the Brownian bridge γ and the family of all possible time-changes ϕ, we work on all time-changed Brownian bridges $\gamma_{\phi_s^{-1}}$, and we average in γ and in ϕ. From a loop $\gamma_{\phi_s^{-1}}$, the associated clock ϕ_s^{-1} can be recovered via the (Riemannian) quadratic variation $|\phi_s'^{-1}|Id$ of the path.

The action of $\tilde{\alpha}$ can be seen as the transformation $\phi \mapsto \alpha^{-1} \circ \phi$ at the level of the time-change.

In the sequel, we will denote by $\mathrm{Diff}_0^{\infty,+}(S^1)$ the group of all diffeomorphisms of the circle preserving the orientation and the origin $0 \in S^1$.

Infinitesimal action of $\mathrm{Diff}_0^{\infty,+}(S^1)$.

The bijection given by (1) can be seen at another level:

$$B_t = \mathrm{Log}\,\phi_t' - \mathrm{Log}\,\phi_0' \,;$$

considering the diffeomorphism $\alpha^{-1}\phi_t = \phi_{\alpha,t}$, so that

$$\phi_{\alpha,t}' = (\alpha^{-1})'(\phi_t)\phi_t' \,,$$

we get

$$(3) \qquad \begin{aligned} \mathrm{Log}\,\phi_{\alpha,t}' - \mathrm{Log}\,\phi_{\alpha,0}' &= \mathrm{Log}\,\phi_t' - \mathrm{Log}\,\phi_0' + \mathrm{Log}\,(\alpha^{-1})'(\phi_t) - \mathrm{Log}\,(\alpha^{-1})'(0) \\ &= B_t + u_{\alpha,t} \,. \end{aligned}$$

Let $(k_t)_{t\in S^1}$ be a smooth vector field on S^1 such that $k_0 = 0$. Putting

$$\alpha_{\lambda,t} = [\exp \lambda k]_t \,,$$

formula (3) can be written for $\alpha_{\lambda,t}$. Differentiating it with respect to λ at $\lambda = 0$, we get infinitesimally a vector field (see [N]) on the Brownian motion B, of the form $j(\phi_t) - j(0)$ for some smooth function j on $[0,1]$. This vector field is anticipative and has finite energy; denote it by $H(k(\phi))$ and consider the derivative D_H associated with H:

$$(4) \qquad D_H\phi_t = \frac{\int_0^t (\exp B_s)\, H_s\, ds}{\int_0^1 \exp B_s\, ds} - \frac{\int_0^t \exp B_s\, ds\ \int_0^1 (\exp B_s)\, H_s\, ds}{\int_0^1 \exp B_s\, ds\ \int_0^1 \exp B_s\, ds} \,.$$

Its kernels are bounded; therefore $H(k(\phi))$ is bounded in all Sobolev spaces of first order (see [N]). Hence, we get the following integration by parts formula (see [N]):

$$(5) \qquad E_B[<dF, H(k(\phi))>] = E_B[F \operatorname{div} H(k(\phi))] \,,$$

where $\operatorname{div} H(k(\phi))$ is bounded in all L^p-spaces.

Consider now all functionals on $L_x(M)$ that are linear combinations of products of functionals of the type $\tilde{F} = \int_0^1 F(\gamma_s) h_s\, ds$, where h is deterministic and smooth on S^1 and where F is bounded. These functionals form a dense linear subspace K of L^2_{tot}.

Let us compute $\tilde{F}(\tilde{\alpha}(\gamma))$:

$$\tilde{F}(\tilde{\alpha}(\gamma)) = \int_0^1 F(\gamma_{\alpha_s}) h_s\, ds = \int_0^1 F(\gamma_s) h_{\alpha_s^{-1}}\, d\alpha_s^{-1} \,.$$

In particular, if $\alpha_s = [\exp \lambda k]_s$, $\tilde{F}(\tilde{\alpha}(\gamma))$ is differentiable in λ; the differential can be seen as the action of a formal vector field $k.\gamma'$ defined on $L_x(M)$.

This formal vector field can be considered as an operator H_k from the vector space K to L^2_{tot}. By (5) and the Fubini theorem in (2), we get for $F \in K$:

$$E_{tot}[<dF, H_k>] \leqslant C \, \|F\|_{L^2_{tot}} .$$

We deduce, since K is dense in L^2_{tot}:

THEOREM: The formal vector field $H_k = k_. \gamma'_.$ can be seen as a densely defined, closable operator on L^2_{tot}.

REMARK: In [L], we have defined the "formal vector field" $c'_.$ over γ, by using the rotational invariance of the B.H.K. measure on the free loop space. It is the generator of a periodic group of isometries, hence it is antisymmetric, therefore closable. But here we do not have a Haar measure on the space of diffeomorphisms of the circle, so we cannot expect to obtain the "vector field" $k_. \gamma'_.$ as some antisymmetric operator on $L_x(M)$ for a suitable measure. We only have a quasi-invariant measure on the space of diffeomorphisms of the circle; and this allows to define the adjoint to the formal vector field $k_. \gamma'_.$ by integration by parts. This leads to the next chapter.

Global action of $\mathrm{Diff}_0^{\infty,+}(S^1)$.

Let us come back to the transformation (3) and suppose that α is not too far from identity in the C^3 sense. From (4), the finite-variation norm of Du_α is smaller than $c < 1$. Therefore $Id + Du_\alpha$ is a bijection from the space of curves with an L^1-derivative to itself. And if $(Id + Du_\alpha)H$ has finite energy, H has finite energy too. This shows that $Id + Du_\alpha$ is a bijection from the Cameron-Martin space to itself. On the other hand, the transformation acts bijectively on Brownian paths, for it is already bijective at the level of the diffeomorphisms of the circle. Therefore the law of $(Id + u_\alpha)(B_.)$ is equivalent to the law of B (see [N]).

To get rid of the restriction that α is close to identity, it suffices to notice that the space of C^3-diffeomorphisms is arcwise-connected and to express every diffeomorphism α as a product of finitely many diffeomorphisms close to identity. Thus the law of $\alpha^{-1} \circ \phi$ is equivalent to the law of ϕ.

Therefore the mapping $\tilde{\alpha}$ is an absolutely continuous transformation of the loop space endowed with the probability P_{tot}. Denote by J_α the quasi-invariance density.

THEOREM: The mapping

$$\psi_\alpha : F \mapsto J_\alpha^{1/2} \, F(\tilde{\alpha}(\gamma))$$

is a unitary transformation of L^2_{tot}. And $\alpha \mapsto \psi_\alpha$ is a unitary representation of $\mathrm{Diff}_0^{\infty,+}(S^1)$ over L^2_{tot}.

A proof can be found in [K] and other proofs of the quasi-invariance formula in [K] or in [M.M].

Chen forms.

For almost all $\gamma \in L_x(M)$, the parallel transport τ_s along γ is defined (using the Levi-Civita connection) and compatible with the time-changing.

Let $\Omega_{.}(M)$ be the space of all forms on M with degree larger than 1; elements of $\Omega_{.}(M)^n$ will be denoted $\tilde{\omega} = (\omega_1, ..., \omega_n)$. On $\Omega_{.}(M)^n$ it is possible to define the stochastic Chen form

$$\sigma(\tilde{\omega}) = \int_{0<s_1<...<s_n<1} \omega_1(d\gamma_{s_1}, .) \wedge ... \wedge \omega_n(d\gamma_{s_n}, .)$$

(see [J.L]). The tangent space to $L_x(M)$ at γ, called $T_\gamma L_x(M)$, is the space of all vector fields along γ having the form $\tau_s H_s$, where H has finite energy and verifies $H_0 = H_1 = 0$ (see [B], [J.L]). $T_\gamma L_x(M)$ is endowed with the norm $\int_0^1 \|H_s'\|^2 ds$. Every $\alpha \in \mathrm{Diff}_0^{\infty,+}(S^1)$ acts as an operator from $T_\gamma L_x(M)$ to $T_{\tilde{\alpha}(\gamma)} L_x(M)$ by time-changing. This operator is continuous, but is not an isometry. To make it an isometry, we change the Hilbert structure by

$$\|H\|_\gamma^2 = \int_0^1 \|H_s'\|^2 \left(\tfrac{d}{ds} <\gamma,\gamma>_s\right)^{-1} ds \, .$$

This formula makes sense because the quadratic variation $<\gamma,\gamma>_s Id$ of of γ has finite energy; it yelds the isometry property because the quadratic variation of γ behaves nicely under time-changing.

Moreover, noticing that stochastic integration too is compatible with time-changing, we get the following stochastic analogue to a remark of [G.J.P]:

THEOREM: $\sigma(\tilde{\omega})$ is invariant under the action of $\mathrm{Diff}_0^{\infty,+}(S^1)$.

REMARK: Both these results, isometry and invariance, extend to the case of the free loop space. This can be done by using the measure introduced by [M.M] on the full space of diffeomorphisms of the circle.

Bibliography.

[B] Bismut J.M.: Large deviations and the Malliavin calculus. Progress in Math. 45, Birkhäuser (1984).

[G.J.P] Getzler E., Jones J.D.S., Petrack S.: Differential forms on loop spaces and the cyclic bar complex. Topology 30, 339–373 (1971).

[J.L] Jones J.D.S., Léandre R.: L^p Chen forms on loop spaces. In "Stochastic Analysis", M. T. Barlow and N. H. Bingham editors, 104–162, Cambridge University Press (1991).

[K] Kosyak A.V.: Irreducible regular gaussian representations of the group of the interval and circle diffeomorphism. J. Funct. Anal. 125, 493–547 (1994).

[L] Léandre R.: Invariant Sobolev Calculus on the free loop space. To appear in Acta Applicandae Mathematicae.

[M.M] Malliavin M.P., Malliavin P.: An infinitesimally quasi-invariant measure on the group of diffeomorphisms of the circle. In "Integration on loop groups", Publication de l'Université Paris VI (1990).

[N] Nualart D.: The Malliavin calculus and related topics. Springer (1995).

[Sh] Shavgulidze E.T.: Distributions on infinite-dimensional spaces and second quantization in string theories II. In "Vth International Vilnius Conference in Probability and Mathematical Statistics", 359-360 (1989).

Département de Mathématiques. Institut Élie Cartan.
Faculté des Sciences. Université de Nancy I.
54 000 Vandœuvre lès Nancy. France

Corrections de l'article de F. Coquet et J. Mémin "Vitesse de convergence en loi pour des solutions d'équations différentielles stochastiques vers une diffusion" (Séminaire de Probabilités XXVIII)

1) Enoncé du lemme 3 p. 285 : il faut remplacer l'inégalité (19) par :

$$\overline{P}(\|\overline{X}-\overline{Y}_\tau\|_T>\epsilon)\le K/\epsilon^2(\alpha+\overline{P}(\|\tau-Id\|_T>\alpha))+\overline{P}([M^\beta]_T>4T).$$

2) Ensuite p. 286 (21) doit être remplacé par :

$$\begin{aligned}\overline{P}\Big(\|\overline{X}-\overline{Y}_{\tau_.}\|_T>\epsilon\Big)&\le 4/\epsilon^2\Big(\overline{E}\Big(\|\overline{X}-\overline{X}^n\|_T^2\Big)+\overline{E}\Big[\|\overline{Y}-\overline{Y}^n\|^2_{\tau_T\wedge(T+\alpha)}\Big]\Big)\\&\quad+\overline{P}(\tau_T>T+\alpha)+\overline{P}(S>T)\end{aligned}\qquad(21)$$

avec $S=inf\{t/[W_\tau]_t>4T\}$. Les calculs estimant $\overline{E}(\|\overline{X}-\overline{X}^n\|_T^2)$ sont entâchés d'une erreur à la troisième ligne que nous a signalée L. Vostrikova. La démonstration suivante remplace la précédente jusqu'à l'inégalité (23).

$$\begin{aligned}\overline{E}\Big(\|\overline{X}-\overline{X}^n\|^2_{T\wedge S}\Big)&=\overline{E}\Big(\Big\|\int_0^{\cdot}(\sigma(\overline{X}_{s-})-\sigma^n(\overline{X}^n_{s-}))dW_{\tau_s}\Big\|^2_{T\wedge S}\Big)\\&\le 4\overline{E}\Big(\int_0^{T\wedge S}(\sigma(\overline{X}_{s-})-\sigma^n(\overline{X}^n_{s-}))^2d[W_\tau]_s\Big)\\&\le 8\overline{E}\Big(\int_0^{T\wedge S}(\sigma(\overline{X}_{s-})-\sigma(\overline{X}^n_{s-}))^2d[W_\tau]_s\Big)\\&\quad+8\overline{E}\Big(\int_0^{T\wedge S}(\sigma(\overline{X}^n_{s-})-\sigma^n(\overline{X}^n_{s-}))^2d[W_\tau]_s\Big)\end{aligned}\qquad(22)$$

Calculons le dernier terme du membre de droite de (22) :

$$\begin{aligned}&\overline{E}\Big(\int_0^{T\wedge S}(\sigma(\overline{X}^n_{s-})-\sigma^n(\overline{X}^n_{s-}))^2d[W_\tau]_s\Big)\\&=\sum_{k=0}^{n-1}\overline{E}\Big(\int_{kT/n\wedge S}^{(k+1)T/n\wedge S}(\sigma(\overline{X}^n_{s-})-\sigma(\overline{X}^n_{kT/n}))^2d[W_\tau]_s\Big)\\&\le L^2\sum_{k=0}^{n-1}\overline{E}\Big(\int_{kT/n\wedge S}^{(k+1)T/n\wedge S}(\overline{X}^n_{s-}-\overline{X}^n_{kT/n})^2\wedge 4N^2/L^2d[W_\tau]_s\Big)\\&\le L^2N^2\sum_{k=0}^{n-1}\overline{E}\Big(\int_{kT/n\wedge S}^{(k+1)T/n\wedge S}(W_{\tau_{s-}}-W\tau_{kT/n})^2\wedge 4/L^2d[W_\tau]_s\Big)\end{aligned}$$

Notant $B(\alpha) = \{\omega : \sup_{t\leq T} |\tau_t(\omega) - t| > \alpha\}$ et $I_k = [kT/n - \alpha, (k+1)T/n + \alpha]$, la ligne précédente devient :

$$\begin{aligned}
&\leq O(\overline{P}(B(\alpha))) \\
&+ 2L^2N^2 \sum_{k=0}^{n-1} \overline{E}\Big(\sup_{t\in I_k}(W_t - W_{kT/n})^2 1_{B^c(\alpha)}([W_\tau]_{(k+1)T/n\wedge S} - [W_\tau]_{kT/n\wedge S})\Big) \\
&\leq O(\overline{P}(B(\alpha))) \\
&+ 2L^2N^2 \sum_{k=0}^{n-1} \overline{E}(\sup_{t\in I_k}(W_t - W_{kT/n})^4)^{1/2}\Big(\overline{E}(([W_\tau]_{(k+1)T/n\wedge S} - [W_\tau]_{kT/n\wedge S})^2 1_{B^c(\alpha)})\Big)^{1/2} \\
&\leq O(\overline{P}(B(\alpha))) + 2KL^2N^2 \sum_{k=0}^{n-1}(2\alpha + T/n)^2
\end{aligned}$$

(On a écrit que $\overline{E}(\sup_{t\in I_k}(W_t - W_{kT/n})^4) \leq K^2(2\alpha + T/n)^2$ et que $\overline{E}\Big(([W_\tau]_{(k+1)T/n\wedge S} - [W_\tau]_{kT/n})^2 1_{B^c(\alpha)}\Big) \leq \overline{E}\Big(([W]_{(k+1)T/n+\alpha} - [W]_{kT/n-\alpha})^2\Big)$).
En prenant $n = [1/\alpha]$, l'inégalité précédente est alors majorée par :

$$O(\overline{P}(B(\alpha))) + 36KL^2N^2(1+T)\alpha$$

On a donc en définitive

$$\overline{E}\Big(\|\overline{X} - \overline{X}^n\|^2_{T\wedge S}\Big) \leq O(\overline{P}(\|\tau - Id\|_T > \alpha)) + O(\alpha) + 8L^2\int_0^T \overline{E}\Big(\|\overline{X} - \overline{X}^n\|^2_s\Big)ds,$$

d'où, en utilisant le lemme 1 avec $U = \|\overline{X} - \overline{X}^n\|^2$ et $A = [W_\tau]$, on obtient :

$$\overline{E}\Big(\|\overline{X} - \overline{X}^n\|^2_{T\wedge S}\Big) \leq O(\overline{P}(\|\tau - Id\|_T > \alpha) + \alpha). \tag{23}$$

La modification de l'inégalité (19) ne change rien à l'estimation finale, le terme $\overline{P}(S > T) = \overline{P}([M^\beta]_T > 4T)$ étant déjà comptabilisé dans le lemme 2.

Errata : Projection d'une diffusion réelle sur sa filtration lente.

Catherine RAINER
Laboratoire de Probabilités, Université Paris VI, tour 56, 3ème étage,
4, place Jussieu, Paris 75252 Cedex 05.

Une erreur gênante nous a échappé dans la formule de la 14ème ligne de l'introduction, et nous profitons de la rectification pour ajouter à sa suite une remarque utile. A la fin du paragraphe d'introduction, on doit donc lire :

...Dans le cas général, on montrera que

$$^{o}X_t = \frac{1}{2}\Big(\frac{1_{\{X_t>0\}}}{N_+(]\mathbf{t}-\mathbf{G_t},+\infty])} - \frac{1_{\{X_t<0\}}}{N_-(]\mathbf{t}-\mathbf{G_t},+\infty])}\Big),$$

où N_+ resp. (N_-) est la mesure de Lévy des excursions positives (resp. négatives) de (X_t) en dehors de zéro.

Remarque : Lorsque (X_t) est symétrique par rapport à zéro, les deux mesures de Lévy sont égales à une demi fois la mesure de Lévy des zéros de (X_t), $\overline{N}$, et on a la formule encore plus simple suivante :

$$^{o}X_t = \frac{\mathbf{sgn}X_t}{\overline{N}(]t-G_t,+\infty])}.$$

Vol. 1559: V. G. Sprindžuk, Classical Diophantine Equations. XII, 228 pages. 1993.

Vol. 1560: T. Bartsch, Topological Methods for Variational Problems with Symmetries. X, 152 pages. 1993.

Vol. 1561: I. S. Molchanov, Limit Theorems for Unions of Random Closed Sets. X, 157 pages. 1993.

Vol. 1562: G. Harder, Eisensteinkohomologie und die Konstruktion gemischter Motive. XX, 184 pages. 1993.

Vol. 1563: E. Fabes, M. Fukushima, L. Gross, C. Kenig, M. Röckner, D. W. Stroock, Dirichlet Forms. Varenna, 1992. Editors: G. Dell'Antonio, U. Mosco. VII, 245 pages. 1993.

Vol. 1564: J. Jorgenson, S. Lang, Basic Analysis of Regularized Series and Products. IX, 122 pages. 1993.

Vol. 1565: L. Boutet de Monvel, C. De Concini, C. Procesi, P. Schapira, M. Vergne. D-modules, Representation Theory, and Quantum Groups. Venezia, 1992. Editors: G. Zampieri, A. D'Agnolo. VII, 217 pages. 1993.

Vol. 1566: B. Edixhoven, J.-H. Evertse (Eds.), Diophantine Approximation and Abelian Varieties. XIII, 127 pages. 1993.

Vol. 1567: R. L. Dobrushin, S. Kusuoka, Statistical Mechanics and Fractals. VII, 98 pages. 1993.

Vol. 1568: F. Weisz, Martingale Hardy Spaces and their Application in Fourier Analysis. VIII, 217 pages. 1994.

Vol. 1569: V. Totik, Weighted Approximation with Varying Weight. VI, 117 pages. 1994.

Vol. 1570: R. deLaubenfels, Existence Families, Functional Calculi and Evolution Equations. XV, 234 pages. 1994.

Vol. 1571: S. Yu. Pilyugin, The Space of Dynamical Systems with the C^0-Topology. X, 188 pages. 1994.

Vol. 1572: L. Göttsche, Hilbert Schemes of Zero-Dimensional Subschemes of Smooth Varieties. IX, 196 pages. 1994.

Vol. 1573: V. P. Havin, N. K. Nikolski (Eds.), Linear and Complex Analysis – Problem Book 3 – Part I. XXII, 489 pages. 1994.

Vol. 1574: V. P. Havin, N. K. Nikolski (Eds.), Linear and Complex Analysis – Problem Book 3 – Part II. XXII, 507 pages. 1994.

Vol. 1575: M. Mitrea, Clifford Wavelets, Singular Integrals, and Hardy Spaces. XI, 116 pages. 1994.

Vol. 1576: K. Kitahara, Spaces of Approximating Functions with Haar-Like Conditions. X, 110 pages. 1994.

Vol. 1577: N. Obata, White Noise Calculus and Fock Space. X, 183 pages. 1994.

Vol. 1578: J. Bernstein, V. Lunts, Equivariant Sheaves and Functors. V, 139 pages. 1994.

Vol. 1579: N. Kazamaki, Continuous Exponential Martingales and *BMO*. VII, 91 pages. 1994.

Vol. 1580: M. Milman, Extrapolation and Optimal Decompositions with Applications to Analysis. XI, 161 pages. 1994.

Vol. 1581: D. Bakry, R. D. Gill, S. A. Molchanov, Lectures on Probability Theory. Editor: P. Bernard. VIII, 420 pages. 1994.

Vol. 1582: W. Balser, From Divergent Power Series to Analytic Functions. X, 108 pages. 1994.

Vol. 1583: J. Azéma, P. A. Meyer, M. Yor (Eds.), Séminaire de Probabilités XXVIII. VI, 334 pages. 1994.

Vol. 1584: M. Brokate, N. Kenmochi, I. Müller, J. F. Rodriguez, C. Verdi, Phase Transitions and Hysteresis. Montecatini Terme, 1993. Editor: A. Visintin. VII. 291 pages. 1994.

Vol. 1585: G. Frey (Ed.), On Artin's Conjecture for Odd 2-dimensional Representations. VIII, 148 pages. 1994.

Vol. 1586: R. Nillsen, Difference Spaces and Invariant Linear Forms. XII, 186 pages. 1994.

Vol. 1587: N. Xi, Representations of Affine Hecke Algebras. VIII, 137 pages. 1994.

Vol. 1588: C. Scheiderer, Real and Étale Cohomology. XXIV, 273 pages. 1994.

Vol. 1589: J. Bellissard, M. Degli Esposti, G. Forni, S. Graffi, S. Isola, J. N. Mather, Transition to Chaos in Classical and Quantum Mechanics. Montecatini Terme, 1991. Editor: S. Graffi. VII, 192 pages. 1994.

Vol. 1590: P. M. Soardi, Potential Theory on Infinite Networks. VIII, 187 pages. 1994.

Vol. 1591: M. Abate, G. Patrizio, Finsler Metrics – A Global Approach. IX, 180 pages. 1994.

Vol. 1592: K. W. Breitung, Asymptotic Approximations for Probability Integrals. IX, 146 pages. 1994.

Vol. 1593: J. Jorgenson & S. Lang, D. Goldfeld, Explicit Formulas for Regularized Products and Series. VIII, 154 pages. 1994.

Vol. 1594: M. Green, J. Murre, C. Voisin, Algebraic Cycles and Hodge Theory. Torino, 1993. Editors: A. Albano, F. Bardelli. VII, 275 pages. 1994.

Vol. 1595: R.D.M. Accola, Topics in the Theory of Riemann Surfaces. IX, 105 pages. 1994.

Vol. 1596: L. Heindorf, L. B. Shapiro, Nearly Projective Boolean Algebras. X, 202 pages. 1994.

Vol. 1597: B. Herzog, Kodaira-Spencer Maps in Local Algebra. XVII, 176 pages. 1994.

Vol. 1598: J. Berndt, F. Tricerri, L. Vanhecke, Generalized Heisenberg Groups and Damek-Ricci Harmonic Spaces. VIII, 125 pages. 1995.

Vol. 1599: K. Johannson, Topology and Combinatorics of 3-Manifolds. XVIII, 446 pages. 1995.

Vol. 1600: W. Narkiewicz, Polynomial Mappings. VII, 130 pages. 1995.

Vol. 1601: A. Pott, Finite Geometry and Character Theory. VII, 181 pages. 1995.

Vol. 1602: J. Winkelmann, The Classification of Three-dimensional Homogeneous Complex Manifolds. XI, 230 pages. 1995.

Vol. 1603: V. Ene, Real Functions – Current Topics. XIII, 310 pages. 1995.

Vol. 1604: A. Huber, Mixed Motives and their Realization in Derived Categories. XV, 207 pages. 1995.

Vol. 1605: L. B. Wahlbin, Superconvergence in Galerkin Finite Element Methods. XI, 166 pages. 1995.

Vol. 1606: P.-D. Liu, M. Qian, Smooth Ergodic Theory of Random Dynamical Systems. XI, 221 pages. 1995.

Vol. 1607: G. Schwarz, Hodge Decomposition – A Method for Solving Boundary Value Problems. VII, 155 pages. 1995.

Vol. 1608: P. Biane, R. Durrett, Lectures on Probability Theory. Editor: P. Bernard. VII, 210 pages. 1995.

Vol. 1609: L. Arnold, C. Jones, K. Mischaikow, G. Raugel, Dynamical Systems. Montecatini Terme, 1994. Editor: R. Johnson. VIII, 329 pages. 1995.

Vol. 1610: A. S. Üstünel, An Introduction to Analysis on Wiener Space. X, 95 pages. 1995.

Vol. 1611: N. Knarr, Translation Planes. VI, 112 pages. 1995.

Vol. 1612: W. Kühnel, Tight Polyhedral Submanifolds and Tight Triangulations. VII, 122 pages. 1995.

Vol. 1613: J. Azéma, M. Emery, P. A. Meyer, M. Yor (Eds.), Séminaire de Probabilités XXIX. VI, 326 pages. 1995.

Vol. 1614: A. Koshelev, Regularity Problem for Quasilinear Elliptic and Parabolic Systems. XXI, 255 pages. 1995.

Vol. 1615: D. B. Massey, Lê Cycles and Hypersurface Singularities. XI, 131 pages. 1995.

Vol. 1616: I. Moerdijk, Classifying Spaces and Classifying Topoi. VII, 94 pages. 1995.

Vol. 1617: V. Yurinsky, Sums and Gaussian Vectors. XI, 305 pages. 1995.

Vol. 1618: G. Pisier, Similarity Problems and Completely Bounded Maps. VII, 156 pages. 1996.

Vol. 1619: E. Landvogt, A Compactification of the Bruhat-Tits Building. VII, 152 pages. 1996.

Vol. 1620: R. Donagi, B. Dubrovin, E. Frenkel, E. Previato, Integrable Systems and Quantum Groups. Montecatini Terme, 1993. Editors:M. Francaviglia, S. Greco. VIII, 488 pages. 1996.

Vol. 1621: H. Bass, M. V. Otero-Espinar, D. N. Rockmore, C. P. L. Tresser, Cyclic Renormalization and Auto-morphism Groups of Rooted Trees. XXI, 136 pages. 1996.

Vol. 1622: E. D. Farjoun, Cellular Spaces, Null Spaces and Homotopy Localization. XIV, 199 pages. 1996.

Vol. 1623: H.P. Yap, Total Colourings of Graphs. VIII, 131 pages. 1996.

Vol. 1624: V. Brînzănescu, Holomorphic Vector Bundles over Compact Complex Surfaces. X, 170 pages. 1996.

Vol.1625: S. Lang, Topics in Cohomology of Groups. VII, 226 pages. 1996.

Vol. 1626: J. Azéma, M. Emery, M. Yor (Eds.), Séminaire de Probabilités XXX. VIII, 382 pages. 1996.

Vol. 1627: C. Graham, Th. G. Kurtz, S. Méléard, Ph. E. Protter, M. Pulvirenti, D. Talay, Probabilistic Models for Nonlinear Partial Differential Equations. Montecatini Terme, 1995. Editors: D. Talay, L. Tubaro. X, 301 pages. 1996.

Vol. 1628: P.-H. Zieschang, An Algebraic Approach to Association Schemes. XII, 189 pages. 1996.

Vol. 1629: J. D. Moore, Lectures on Seiberg-Witten Invariants. VII, 105 pages. 1996.

Vol. 1630: D. Neuenschwander, Probabilities on the Heisenberg Group: Limit Theorems and Brownian Motion. VIII, 139 pages. 1996.

Vol. 1631: K. Nishioka, Mahler Functions and Transcendence. VIII, 185 pages. 1996.

Vol. 1632: A. Kushkuley, Z. Balanov, Geometric Methods in Degree Theory for Equivariant Maps. VII, 136 pages. 1996.

Vol.1633: H. Aikawa, M. Essén, Potential Theory – Selected Topics. IX, 200 pages. 1996.

Vol. 1634: J. Xu, Flat Covers of Modules. IX, 161 pages. 1996.

Vol. 1635: E. Hebey, Sobolev Spaces on Riemannian Manifolds. X, 116 pages. 1996.

Vol. 1636: M. A. Marshall, Spaces of Orderings and Abstract Real Spectra. VI, 190 pages. 1996.

Vol. 1637: B. Hunt, The Geometry of some special Arithmetic Quotients. XIII, 332 pages. 1996.

Vol. 1638: P. Vanhaecke, Integrable Systems in the realm of Algebraic Geometry. VIII, 218 pages. 1996.

Vol. 1639: K. Dekimpe, Almost-Bieberbach Groups: Affine and Polynomial Structures. X, 259 pages. 1996.

Vol. 1640: G. Boillat, C. M. Dafermos, P. D. Lax, T. P. Liu, Recent Mathematical Methods in Nonlinear Wave Propagation. Montecatini Terme, 1994. Editor: T. Ruggeri. VII, 142 pages. 1996.

Vol. 1641: P. Abramenko, Twin Buildings and Applications to S-Arithmetic Groups. IX, 123 pages. 1996.

Vol. 1642: M. Puschnigg, Asymptotic Cyclic Cohomology. XXII, 138 pages. 1996.

Vol. 1643: J. Richter-Gebert, Realization Spaces of Polytopes. XI, 187 pages. 1996.

Vol. 1644: A. Adler, S. Ramanan, Moduli of Abelian Varieties. VI, 196 pages. 1996.

Vol. 1645: H. W. Broer, G. B. Huitema, M. B. Sevryuk, Quasi-Periodic Motions in Families of Dynamical Systems. XI, 195 pages. 1996.

Vol. 1646: J.-P. Demailly, T. Peternell, G. Tian, A. N. Tyurin, Transcendental Methods in Algebraic Geometry. Cetraro, 1994. Editors: F. Catanese, C. Ciliberto. VII, 257 pages. 1996.

Vol. 1647: D. Dias, P. Le Barz, Configuration Spaces over Hilbert Schemes and Applications. VII. 143 pages. 1996.

Vol. 1648: R. Dobrushin, P. Groeneboom, M. Ledoux, Lectures on Probability Theory and Statistics. Editor: P. Bernard. VIII, 300 pages. 1996.

Vol. 1649: S. Kumar, G. Laumon, U. Stuhler, Vector Bundles on Curves – New Directions. Cetraro, 1995. Editor: M. S. Narasimhan. VII, 193 pages. 1997.

Vol. 1650: J. Wildeshaus, Realizations of Polylogarithms. XI, 343 pages. 1997.

Vol. 1651: M. Drmota, R. F. Tichy, Sequences, Discrepancies and Applications. XIII, 503 pages. 1997.

Vol. 1652: S. Todorcevic, Topics in Topology. VIII, 153 pages. 1997.

Vol. 1653: R. Benedetti, C. Petronio, Branched Standard Spines of 3-manifolds. VIII, 132 pages. 1997.

Vol. 1654: R. W. Ghrist, P. J. Holmes, M. C. Sullivan, Knots and Links in Three-Dimensional Flows. X, 208 pages. 1997.

Vol. 1655: J. Azéma, M. Emery, M. Yor (Eds.), Séminaire de Probabilités XXXI. VIII, 329 pages. 1997.

General Remarks

Lecture Notes are printed by photo-offset from the master-copy delivered in camera-ready form by the authors. For this purpose Springer-Verlag provides technical instructions for the preparation of manuscripts.

Careful preparation of manuscripts will help keep production time short and ensure a satisfactory appearance of the finished book. The actual production of a Lecture Notes volume normally takes approximately 8 weeks.

Authors receive 50 free copies of their book. No royalty is paid on Lecture Notes volumes.

Authors are entitled to purchase further copies of their book and other Springer mathematics books for their personal use, at a discount of 33,3 % directly from Springer-Verlag.

Commitment to publish is made by letter of intent rather than by signing a formal contract. Springer-Verlag secures the copyright for each volume.

Addresses:

Professor A. Dold
Mathematisches Institut
Universität Heidelberg
Im Neuenheimer Feld 288
D-69120 Heidelberg
Federal Republic of Germany

Professor F. Takens
Mathematisch Instituut
Rijksuniversiteit Groningen
Postbus 800
NL-9700 AV Groningen
The Netherlands

Springer-Verlag, Mathematics Editorial
Tiergartenstr. 17
D-69121 Heidelberg
Federal Republic of Germany
Tel.: *49 (6221) 487-410